FRACTIONAL CALCULUS
Models and Numerical Methods

Second Edition

Series on Complexity, Nonlinearity and Chaos

ISSN 2010-0019

Series Editor: Albert C.J. Luo
(Southern Illinois University Edwardsville, USA)

Aims and Scope

The books in this series will focus on the recent developments, findings and progress on fundamental theories and principles, analytical and symbolic approaches, computational techniques in nonlinear physical science and nonlinear mathematics.

Topics of interest in Complexity, Nonlinearity and Chaos include but not limited to:

- New findings and discoveries in nonlinear physics and mathematics,
- Complexity and mathematical structures in nonlinear physics,
- Nonlinear phenomena and observations in nature,
- Computational methods and simulations in complex systems,
- New theories, and principles and mathematical methods,
- Stability, bifurcation, chaos and fractals in nonlinear physical science.

Vol. 1 Ray and Wave Chaos in Ocean Acoustics: Chaos in Waveguides
D. Makarov, S. Prants, A. Virovlyansky & G. Zaslavsky

Vol. 2 Applications of Lie Group Analysis in Geophysical Fluid Dynamics
N. H. Ibragimov & R. N. Ibragimov

Vol. 3 Fractional Calculus: Models and Numerical Methods
D. Baleanu, K. Diethelm & J. J. Trujillo

Vol. 4 Asymptotic Integration and Stability: For Ordinary, Functional and Discrete Differential Equations of Fractional Order
D. Baleanu & O. G. Mustafa

Vol. 5 Fractional Calculus: Models and Numerical Methods (Second Edition)
D. Baleanu, K. Diethelm, E. Scalas & J. J. Trujillo

Series on Complexity, Nonlinearity and Chaos – Vol. 5

FRACTIONAL CALCULUS

Models and Numerical Methods

Second Edition

Dumitru Baleanu
Çankaya University, Turkey
& Institute of Space Sciences, Romania

Kai Diethelm
Technische Universität Braunschweig, Germany
& GNS mbH, Germany

Enrico Scalas
University of Sussex, UK
& Basque Center for Applied Mathematics, Spain

Juan J. Trujillo
University of La Laguna, Spain

World Scientific

NEW JERSEY • LONDON • SINGAPORE • BEIJING • SHANGHAI • HONG KONG • TAIPEI • CHENNAI • TOKYO

Published by

World Scientific Publishing Co. Pte. Ltd.

5 Toh Tuck Link, Singapore 596224

USA office: 27 Warren Street, Suite 401-402, Hackensack, NJ 07601

UK office: 57 Shelton Street, Covent Garden, London WC2H 9HE

Library of Congress Cataloging-in-Publication Data
Names: Baleanu, D. (Dumitru)
Title: Fractional calculus : models and numerical methods / by Dumitru Baleanu (Çankaya University, Turkey & Institute of Space Science, Romania) [and three others].
Description: 2nd edition. | New Jersey : World Scientific, 2016. | Series: Series on complexity, nonlinearity and chaos ; volume 5 | Includes bibliographical references and index.
Identifiers: LCCN 2016030115 | ISBN 9789813140035 (hardcover : alk. paper)
Subjects: LCSH: Fractional calculus.
Classification: LCC QA314 .F73 2016 | DDC 515/.83--dc23
LC record available at https://lccn.loc.gov/2016030115

British Library Cataloguing-in-Publication Data
A catalogue record for this book is available from the British Library.

Printed in Singapore by B & Jo Enterprise Pte Ltd

To our families for their support and encouragement

Preface

Fractional Calculus deals with the study of so-called fractional order integral and derivative operators over real or complex domains and their applications. It has its roots in 1695, in a letter from de l'Hospital to Leibniz. Questions such as "What is understood by Fractional Derivative?" or "What does the derivative of order $1/3$ or $\sqrt{2}$ of a function mean?" motivated many brilliant scientists to focus their attention on this topic during the 18th and 19th centuries. For instance, we can mention Euler (1738, [228]), Laplace (1812, [356]), Fourier (1822, [244]), Abel (1823, [3]), Liouville (1832–1855, [375, 376]), Grünwald (1867, [275]), Letnikov (1868–1872, [364–366]), Riemann (1876, [518]), Laurent (1884, [360]), or Heaviside (1893–1912, [291, 292]).

It is well known that Abel implicitly applied fractional calculus in 1823 in connection with the tautochrone problem, which was modeled through a certain integral equation with a weak singularity of the type that appears in the so-called Riemann-Liouville fractional integral [3]. Therefore he can be considered the first scholar who investigated an interesting physical problem using techniques from what we today call fractional calculus. Later, Liouville tried to apply his definitions of fractional derivatives to different problems [375]. On the other hand, in 1882 Heaviside introduced a so-called operational calculus which reconciliated the fractional calculus with the explicit solution of some diffusion problems. Particularly, his techniques were applied to the theory of the transmission of electrical currents in cables [291]. For more historic facts about the development of the fractional calculus during these two centuries, the monographs by Oldham and Spanier

[465], by Ross [531], by Miller and Ross [430] and Samko *et al.* [542] can be consulted.

During the 20th century, up to 1985, we can list some of the pioneer researchers in this topic, such as Weyl (1917, [625]), Hardy (1917–1928, [282, 283]), Littlewood (1925–1928, [284, 285]), Levy (1937, [367]), Zygmud (1934–1945, [652, 653]), M. Riesz (1936–1949, [519, 520]), Doetsch (1937, [204]), Erdélyi (1939–1965, [224, 225]), Kober (1940, [346]), Widder (1941, [629]), Rabotnov (1948–1980, [487, 513]), Feller (1943–1971, [231, 232]), Maraval (1956–1971, [404, 405]), Sneddon (1957–1979, [570–572]), Gorenflo (since 1965, [267]), Caputo (since 1966, [140]), Dzherbashyan (1970, [214]), Samko (since 1967, [540]), Srivastava (since 1968, [574, 575]), Oldham (1969, [464]), Osler (1970, [468]), Caputo and Mainardi (since 1971, [145]), Love (since 1971, [380]), Oldham and Spanier (1974, [465]), Mathai and Saxena (since 1978, [411]), Ross (since 1974, [530]), McBride (since 1979, [413]), Nigmatullin (since 1979, [456]), Oustaloup (since 1981, [469, 470]), Bagley and Torvik (since 1983, [605, 605]), among others. As we will argue below, around 1985, new and fertile applications of fractional differential equations emerged and this field became part of applied sciences and engineering.

A fractional derivative is just an operator which generalizes the ordinary derivative, such that if the fractional derivative is represented by the operator symbol D^α then, when $\alpha = 1$, it coincides with the ordinary differential operator D. As a matter of fact, there are many different ways to set up a fractional derivative, and, nowadays, it is usual to see many different definitions. Here we must remark that, when we speak of fractional calculus, or fractional operators, we are not speaking of fractional powers of operators, except when we are working in very special functional spaces, such as the Lizorkin spaces.

Fractional differential equations, that is, those involving real or complex order derivatives, have assumed an important role in modeling the anomalous dynamics of many processes related to complex systems in the most diverse areas of science and engineering. However the interest in the specific topic of fractional calculus surged only at the end of the last century.

The theoretical interest in fractional differential equations as a mathematical challenge can be traced back to 1918, when O'Shaughnessy [467] gave an explicit solution to the differential equation $y^{(\alpha} = y/x$, after he himself had suggested the problem. In 1919, Post [495] proposed a completely different solution. Note that, at that time, this problem was not rigorously defined, since there was no mention of what fractional derivative was being used in the proposed differential equation. This explains why both authors found such different solutions, and why neither of them was wrong.

As one would expect, since a fractional derivative is a generalization of the ordinary derivative, it is going to lose many of its basic properties; for example, it loses its clear geometric or physical interpretation, the index law is only valid when working in specific functional spaces, the derivative of the product of two functions is difficult to compute, and the chain rule cannot be straightforwardly applied. It is natural to ask, then, what properties of fractional derivatives make them so suitable for modeling certain complex systems. We think the answer lies in the property exhibited by such systems of "non-local dynamics", that is, the processes' dynamics have a certain degree of memory and fractional operators are non-local, while the ordinary derivative is a local operator.

In 1974, after a joint research activity, Oldham and Spanier published the first monograph devoted to fractional operators and their applications in problems of mass and heat transfer [465]. In 1974, the First Conference on Fractional Calculus and its Applications took place at the University of New Haven, organized by B. Ross who edited the corresponding proceedings [530]. We can think of this year as the beginning of a new age for fractional calculus.

The stochastic interpretation for the fundamental solution of the ordinary diffusion equation in terms of Brownian motion has been known since the early years of 20th century. Physicists often mention Einstein as the pioneer in this field [217]. Indeed, Einstein's paper on Brownian motion had a large success and motivated further experimental work on the atomic and molecular hypothesis. However, five years before Einstein, L. Bachelier published his thesis on price fluctuations at the Paris stock exchange [47]. In this thesis, the connection was already made clear between Brownian motion and the diffusion equation. These results considered the position of a diffusing object as the sum of independent and identically distributed random variables leading to a Gaussian distribution in the asymptotic limit by virtue of the central limit theorem which was refined in the first half of the 20th century as well [233].

In 1949, Gnedenko and Kolmogorov [259] introduced a generalization of the classical central limit theorem for sums of random variables with infinite second moment converging to α-stable random variables. Almost simultaneously, Lévy and Feller also wrote seminal contributions leading to some controversy on priority [233]. In 1965, Montroll and Weiss [440] introduced a process in physics, later called continuous time random walk (CTRW) by Scher [439, 554, 555]. This process turned out to be very useful for the theoretical description of anomalous diffusion phenomena associated to certain materials [92]. CTRWs (also known as compound renewal processes in the mathematical community) are a generalization of the above mentioned method for normal diffusion processes. Therefore, they became the tool of choice for many applied scientists in order to characterize and describe anomalous diffusive processes from the mid-20th century until today.

The use of Laplace and Fourier integral transforms helps us in proving that, for a sub-diffusive process, the density function $u(x,t)$ of finding the diffusing particle in x at time t is the fundamental solution of the following time-fractional diffusion equation:

$$\Delta_x^2 u = k D_t^\beta u. \tag{1}$$

Such a connection lets us consider the CTRW models of the subdiffusive process as fractional differential models. Among the papers dealing with this fact there are Balakrishnan (1985, [52]), Wyss (1986, [635]), Schneider and Wyss (1989, [636]), Fujita (1990, [246–248]), Shlesinger, Zalavsky, and Klafter (1993, [565]), Metzler, Glöckle, and Nonnenmacher (1994, [424]), Zaslavsky (1994, [642]), Engheta (1996, [220]), Klafter, Shlesinger, and Zumofen (1996, [342]), Metzler and Nonnenmacher (1997, [427]), Metzler, Barkai, and Klafter (1999, [423]), Hilfer (2000, [304]), Anton (1995, [306]), and many more. For a comprehensive review, we recommend the excellent papers by Metzler and Klafter (2000–2004, [425, 426]). In these papers, the reader finds an accurate description of fractional diffusion models as well as a clear explanation of the role played by other linear or non-linear diffusive fractional models, such as the fractional Fokker-Planck equation.

The relationship between CTRWs and fractional diffusion will be dealt with in Chapters 6 and 7, as well as in several sections of Chapter 5. We must also point out that this is perhaps the first monograph presenting modern numerical methods used to solve fractional differential equations (see Chapters 2 and 3).

As a result of many investigations in different areas of applied sciences and engineering and as a consequence of the relationship between CTRWs and diffusion-type pseudo-differential equations, new fractional differential models were used in a great number of different applied fields. We can mention material science, physics, astrophysics, optics, signal processing and control theory, chemistry, transport phenomena, geology, bioengineering and medicine, finance, wave and diffusion phenomena, dissemination of atmospheric pollutants, flux of contaminants transported by subterranean waters through different strata, chaos, and so on. Also, the reader can find many more references, *e. g.*, in the monographs, [221, 304, 412, 474, 487, 595, 510, 117, 335, 392, 537, 560] and [68, 139, 154, 186, 398, 438, 549, 598, 588, 341, 100, 399]

The idea that physical phenomena, such as anomalous diffusive or wave processes, can be described with fractional differential models raises, at least, the following three fundamental questions:

- Are mathematical models with fractional space and/or time derivatives consistent with the fundamental laws and well known symmetries of the nature?
- How can the fractional order of differentiation be observed experimentally or how does a fractional derivative emerge from microscopic models?
- Once a fractional calculus model is available, how can a fractional order equation be solved (exactly or approximately)?

Of course, here, we must mention the very important contributions in nonlinear non-fractional differential models which were more studied by mathematicians than used by applied researchers, at least to describe the dynamics of processes within anomalous media, but this is beyond the scope of this book. However, we must remark the fractional differential models are a complementary tool to classical methods. The reader can consult the paper [114], where it is shown that strongly non-differentiable functions can be solutions of elementary fractional equations.

During the last 25 years there has been a spectacular increase in the use of fractional differential models to simulate the dynamics of many different anomalous processes, especially those involving ultra-slow diffusion. The following table is only based on the Scopus database, but it reflects this state of affairs clearly:

Table 1 Evolution in the number of publications on fractional differential equations and their applications

Words in title or abstract	**1960–1980**	**1981–1990**	**1991–2000**	**2001–2010**
Fractional Brownian Motion	2	38	532	1295
Anomalous Diffusion	185	261	626	1205
Anomalous Relaxation	21	23	70	61
Superdiffusion or Subdiffusion	0	22	121	521
Anomalous Dynamics Anomalous Processes Fractional Models Fractional Relaxation Fractional Kinetics Fractional Dynamics	11	24	128	443
Fractional Differential Equation Fractional Fokker-Planck Equation Fractional Diffusion Equation	1	1	74	943

On the other hand, and from a mathematical point of view, during the last five years we have been able to find many interesting publications connected with applications of classical fixed point theorems on abstract spaces to study the existence and uniqueness of solutions of many kinds of initial value problems and boundary value problems for fractional operators. See, *e. g.*, [29, 354, 13, 455, 294, 12, 9, 592, 80, 192, 196].

For these reasons, we expect that fractional differential models will play an important role in the near future in the description of the dynamics of many complex systems. From our point of view, despite the attention given to it until the moment by many authors, only a few steps have been taken toward what may be called a clear and coherent theory of fractional differential equations that supports the widespread use of this tool in the applied sciences in a manner analogous to the classical case. Therefore we can find here a great number of both theoretical and applied open problems. For example, we think that three important kinds of such challenges, among others, are the following:

- In spite of the fact that there were several attempts to formulate a deterministic approach of fractional differential models in many different areas of science and engineering, in general there have not been many rigorous justifications of such models. A deterministic approach to fractional differential models including a clear justification, as in the classical case, is an important open problem. Such an objective comes motivated mainly by the need to take into account the macroscopic behavior of anomalous processes not connected to stochastic theories. An example could be when studying the dynamics of ultrasonic waves through very irregular media, and there are many other potential examples. The first steps in this direction may have been done recently; see for example, [30, 526, 639, 638, 388, 525].
  In addition, we recall the paper [459] where memory, expressed in terms of fractional operators, emerges from the initial model that does not "hint" to the presence of memory. Also, as an example, one can point to the literature on dielectric relaxation based on fractional kinetics [457] where differential equations containing non-integer operators describing the kinetic phenomena emerge from the self-similar structure of the medium considered. This approach recently received also its experimental confirmation [458].
- The introduction of a suitable fractional Laplacian for Dirichlet and Neumann problems associated to isotropic and anisotropic media. We must remark that, in the literature, at least three different approaches were used to solve such a problem in the isotropic case, namely the application of the well-known fractional power of operators, the hyper-singular inverse of one of the Riesz fractional integral operators of potential, and the characterization by means of the corresponding Fourier transform. The first two cases

do not allow to work in wide functional spaces, whereas in the third one, the possibility exists not to determine with enough rigor the fractional Laplacian in the spatial field. We refer to [407, 541, 135, 134, 146, 136] for further details.

- The development of suitable and well-founded numerical methods to solve fractional ordinary and partial differential equations, so that applied researchers can refine their results, as in the classical case.
  We will devote two chapters of this book to this problem, where the reader can find a number of relevant references. In any case, this is still an important open field whose development will allow quicker advances in applied fields.

Until now, our attention was focused on fractional differential equations and their applications. However, fractional operators had been mainly used for other objectives in the past. We will illustrate this fact with the following two examples connected with potential theory: the n-dimensional fractional operators introduced by Riesz (1932–1945), as a generalization of the Riemann-Liouville integral operators, were used to write the solution of certain ordinary partial differential equations explicitly [519]; Erdélyi and Sneddon (1960–1966) used the so-called Erdélyi-Kober fractional integral operators to solve explicitly certain dual integral equations (see [225, 571]).

Following such ideas, many other authors used fractional operators to generalize certain classical theories or to simplify classical problems. So, many special functions were expressed in terms of elementary functions by using fractional operators by Kiryakova [340]; the singularities of certain known ordinary differential equations were avoided in the framework of fractional calculus (see [528, 524]), or Riewe, Agrawal, Klimek, Baleanu *et al.* have initiated a fractional generalization of variational theory (see [521, 343, 15, 55, 63, 506, 34, 43, 322, 73, 587]), etc. Remarkable are the results obtained in control theory through the fractional generalization of the well-known PID controllers (see, for example, [474, 594, 595, 490, 473, 537, 438, 139]).

Therefore, we can use the label "fractional calculus models" when we refer to a generalization of a classical theory in the framework of fractional calculus. In this sense, in Chapter 4 and in a part of Chapter 5, we develop fractional generalizations of important classical theories. Below, we are going to explain such issues with more detail.

This book consists of a total of seven chapters, one appendix and an extensive bibliography.

Chapter 1 contains preliminary material that can be skipped by informed readers. This chapter is here to help the reader in reducing the use of external sources.

As we have seen, fractional-order models are a generalization of classical integer-order models. However, it turns out that these models are also in need of more general techniques in order to provide analytical solutions in closed form and/or qualitative studies of the solutions. As in the classical case, such techniques are not enough in many practical relevant cases. Thus there is a substantial demand for efficient numerical techniques to handle fractional derivatives and integrals and equations involving such operators. Many algorithms were proposed for this purpose in the last few decades, but they tend to be scattered across a large number of different publications and, moreover, an appropriate and rigorous convergence analysis is often not available. Thus, a user who needs a numerical scheme for a particular problem often has difficulties in finding a suitable method. As a partial remedy to this state of affairs, in Chapters 2 and 3 we collect the most important numerical methods for practically relevant tasks. We have focused our attention on those algorithms whose behavior is well understood and that have proven to be reliable and efficient.

The fourth chapter is devoted to generalize the classical theory of Stirling numbers of first $s(n,k)$ and second kind $S(n,k)$ in the framework of the fractional calculus, basically using fractional differential and integral operators. Such special numbers play a very important role in connection with many applications, in particular in computing finite difference schemes and in numerical approximation methods. Such generalizations have been an open problem whose solution was approached by Butzer and collaborators, [120–124, 288], during the last years of the 20th century. We have worked out the mentioned generalization with respect to both parameters, n and k, so that they can be real or complex numbers, but keeping almost any known property corresponding to the classical numbers. Moreover, in this chapter, we introduce a number of important applications; for instance we connect the generalized Stirling functions with the corresponding infinity differences and with the fractional Hadamard derivative or with the fractional Liouville operators. On the other hand, we must remark that our treatment of this issue is not the ultimate one, even if we believe that our theory can open new and interesting perspectives to apply such results to approach to the calculus of infinite difference or fractional difference equations. The latter could be very important in the context of modeling the dynamics of anomalous processes.

Classical calculus of variations as a branch of mathematics is recognized for its fundamental contributions in various areas of physics and engineering. The history of variational calculus started already with problems well-known to Greek philosophers as well as scientists and contains illustrative contributions to the evolution of the science and engineering. During the last decade, when fractional calculus started to be applied intensively to various problems related to real world applications, it was pointed out that it should be applied also to variational problems. As a result of this fusion the theory of fractional variational principles was created. This new theory consists of two parts, the first one is related to the mathematical generalization of the classical theory of calculus of variations and the second one involves the applications.

The fractional Euler-Lagrange equations recently studied are a new set of differential equations involving both the left and the right fractional derivatives.

As a result of interaction among fractional calculus, delay theory and time scales calculus, we observed that the new theory started to be generalized according to new results obtained in these fields. Also the applications of this new theory in so called fractional differential geometry started to be reported as well as with some promising generalizations of the classical formalisms in physics and in control theory.

An important feature of fractional variational principles is that they contain classical ones as a particular case when fractional operators converge to ordinary differential operators. Besides, fractional optimal control is largely developed and fractional numerical methods started to be applied to solve the fractional Euler-Lagrange equations. At this stage, we are confident that fractional variational principles will lead to new discoveries in several fields.

In Chapter 5, we introduce the reader of this book to the extension of variational calculus within the framework of fractional calculus, presenting a theory of fractional variational principles. Moreover, in the first part of the mentioned chapter, we consider the study of solutions for the corresponding fractional Euler-Lagrange equations, a new set of fractional differential equations involving both the left and right fractional derivatives.

In the second part of this chapter, we study the discrete and continuous case of the called fractional Hamiltonian dynamics, which generalizes the classical dynamics of Hamiltonian systems.

As already discussed, there is a deep connection between the fractional diffusion equation and the stochastic models for anomalous diffusion called CTRWs (continuous-time random walks). These processes, as discussed by physicists, are an instance of semi-Markov processes. This mild generalization already leads to an infinite memory in time. Considered non-physical by several authors, spatial non-locality is connected to the power-law behavior of the distribution of jumps. All these phenomena are described by means of a suitable stochastic process, the fractional compound Poisson process with symmetric α-stable jumps which makes quite simple the proof of the generalized central limit theorem. This is the subject we study in Chapter 6.

In Chapter 7, we present an overview of the application of CTRWs to finance. In particular we give a brief presentation on the application of CTRWs, and implicitly fractional models, to option pricing and we point the reader to other applications such as insurance risk evaluation and economic growth models.

When tick-by-tick prices are considered, not only price jumps, but also inter-trade durations seem to vary at random. Therefore, as a first approximation, it is possible to describe durations as independent and identically distributed random variables. In this framework, position is replaced by log-price and jumps in position by tick-by-tick log-returns. The interesting case comes when inter-trade durations do not follow a exponential distribution.

The material covered in the seven chapters is complemented by an appendix where we explicitly provide the implementation of the algorithms described in the previous chapters, in several common programming languages.

Finally we include an extensive bibliography which, however, is far from being exhaustive.

During the time we have dedicated to write this monograph in this present form, the authors have gratefully received invaluable suggestions and comments from researches at many different academic institutions and research centers around the world. Special mention ought to be made of the help and assistance so generously and meticulously provided by colleagues Thabet Abdeljawad, Mohamed Herzallah, Fahd Jarad, Sami I. Muslih, Eqab M. Rabei, Margarita Rivero, Luis Rodríguez-Germá, and Luis Vázquez.

Here, we would like to remember the great inspiration and support we received from Professors Om P. Agrawal at SIU Carbondale, Paul Butzer at RWTH Aachen University, Rudolf Gorenflo at Freie Universität Berlin, Anatoly A. Kilbas at State University of Belarus, Raoul R. Nigmatullin at University of Kazan, Hari M. Srivastava at the University of Victoria, J. A. Tenreiro Machado at ISEP Porto, George Zaslavsky at Courant Institute, Neville J. Ford at the University of Chester and Alan D. Freed at Saginaw Valley State University.

Finally, the authors would like to thankfully acknowledge the financial grants and support for this book project, which were awarded by the MICINN of Spain (Projects No. MTM2007/60246 and MTM2010/16499), the Belarusian Foundation for Funding Scientific Research (Project No. F10MC-24), the Research Promotion Plan 2010 of Universitat Jaume I and the Basque Center for Applied Mathematics in Spain, Çankaya University of Turkey, and Technische Universität Braunschweig, Germany, among others.

August 2011

Dumitru Baleanu
Çankaya University and
Institute of Space Sciences, Magurele-Bucharest

Kai Diethelm
Gesellschaft für Numerische Simulation mbH, Braunschweig, and
Technische Universität Braunschweig

Enrico Scalas
East Piedmont University and
BCAM (Basque Center for Applied Mathematics)

Juan J. Trujillo
Universidad de La Laguna

Preface to the Second Edition

The first edition of our book was well received by the scientific community. However, the rapid progress in the area of fractional calculus and its applications has made an update necessary. We are therefore very happy that World Scientific has agreed to publish a second edition.

In comparison to the first edition, we have extended the description of the numerical methods, in particular in Chapter 2; some recently developed techniques like spectral methods are presented now. Moreover, we have added a new chapter, Chapter 8, that describes applications of fractional calculus based models in the life sciences.

Throughout the text, we have corrected a number of misprints and updated the references in order to cover relevant recent developments.

December 2015

Dumitru Baleanu
Çankaya University and
Institute of Space Sciences, Magurele-Bucharest

Kai Diethelm
Gesellschaft für Numerische Simulation mbH, Braunschweig, and
Technische Universität Braunschweig

Enrico Scalas
University of Sussex and
BCAM (Basque Center for Applied Mathematics)

Juan J. Trujillo
Universidad de La Laguna

Contents

Chapter 1

Preliminaries

This chapter is preliminary in character and contents. We give here the definitions and some properties of several fractional integrals and fractional derivatives of different types. Also we give the definition and properties of some special functions which will be used in this book.

More detailed information about the content of this chapter may be found, for example, in the works of Erdélyi *et al.* [226], Copson [161], Riesz [520], Doetsch [204], Sneddon [570], Zemanian [646], McBride [413], Samko *et al.* [542], Kiryakova [340], Podlubny [487], Butzer *et al.* [125–127], Kilbas *et al.* [335], Caponetto *et al.* [139], Diethelm [186], Mainardi [398], Monje *et al.* [438], Duarte Ortigueira [209] and Tarasov [588].

In general, the results we present in this chapter will be considered for "suitable functions". Precise details can be found, *e. g.*, in the above mentioned references.

First of all, let $\Omega = [a, b]$ $(-\infty \leq a < b \leq \infty)$ be a finite or infinite interval of the real axis $\mathbb{R}$. We denote by $L_p(a, b)$ $(1 \leq p \leq \infty)$ the set of those Lebesgue complex-valued measurable functions f on Ω for which $\|f\|_p < \infty$, where

$$\|f\|_p = \left(\int_a^b |f(t)|^p dt \right)^{1/p} \quad (1 \leq p < \infty) \tag{1.0.1}$$

and

$$\|f\|_\infty = \text{esssup}_{a \leq x \leq b} |f(x)|. \tag{1.0.2}$$

Here $\text{esssup}|f(x)|$ is the essential maximum of the function $|f(x)|$.

1.1 Fourier and Laplace Transforms

In this section we present definitions and some properties of one- and multi-dimensional Fourier and Laplace transforms.

We begin with the one-dimensional case. The Fourier transform of a function $\varphi(x)$, of a real variable, is defined by

$$(\mathcal{F}\varphi)(\kappa) = \mathcal{F}[\varphi(x)](\kappa) = \hat{\varphi}(\kappa) = \int_{-\infty}^{\infty} e^{i\kappa x}\varphi(x)dx, \tag{1.1.1}$$

with $x, \kappa \in \mathbb{R}$. The inverse Fourier transform is given by the formula

$$(\mathcal{F}^{-1}g)(x) = \mathcal{F}^{-1}[g(\kappa)](x) = \frac{1}{2\pi}\hat{g}(-x) = \frac{1}{2\pi}\int_{-\infty}^{\infty} e^{-i\kappa x}g(\kappa)d\kappa. \tag{1.1.2}$$

Each of the transforms (1.1.1) and (1.1.2) is inverse to the other one,

$$\mathcal{F}^{-1}\mathcal{F}\varphi = \varphi, \quad \mathcal{F}\mathcal{F}^{-1}g = g, \tag{1.1.3}$$

also the following simple relation is valid

$$(\mathcal{F}\mathcal{F}\varphi)(x) = \varphi(-x). \tag{1.1.4}$$

The rate of decrease of $(\mathcal{F}\varphi)(x)$ at infinity is connected with the smoothness of the function $\varphi(x)$.

Other well known properties of the Fourier transform are

$$\mathcal{F}[D^n\varphi(x)](\kappa) = (-i\kappa)^n(\mathcal{F}\varphi)(\kappa) \quad (n \in \mathbb{N}) \tag{1.1.5}$$

and

$$D^n(\mathcal{F}\varphi)(\kappa) = (i\kappa)^n\mathcal{F}[\varphi(x)](\kappa) \quad (n \in \mathbb{N}) \tag{1.1.6}$$

where D^n denotes the classical differential operator of order n.

The Fourier convolution operator of two functions h and φ is defined by the integral

$$h * \varphi = (h * \varphi)(x) = \int_{-\infty}^{\infty} h(x-t)\varphi(t)dt \quad (x \in \mathbb{R}). \tag{1.1.7}$$

It has the commutative property

$$h * \varphi = \varphi * h. \tag{1.1.8}$$

and is connected to the Fourier transform operator by

$$\left(\mathcal{F}(h * \varphi)\right)(\kappa) = (\mathcal{F}h)(\kappa) \cdot (\mathcal{F}\varphi)(\kappa). \tag{1.1.9}$$

The n-dimensional Fourier transform of a function $\varphi(\mathbf{x})$ of $\mathbf{x} \in \mathbb{R}^n$ is defined by

$$(\mathcal{F}\varphi)(\kappa) = \mathcal{F}[\varphi(\mathbf{x})](\kappa) = \hat{\varphi}(\kappa) = \int_{\mathbb{R}^n} e^{i\kappa\cdot\mathbf{x}}\varphi(\mathbf{x})d\mathbf{x}, \tag{1.1.10}$$

with $\mathbf{k} \in \mathbb{R}^n$, while the corresponding inverse Fourier transform is given by the formula

$$(\mathcal{F}^{-1}g)(\mathbf{x}) = \mathcal{F}^{-1}[g(\kappa)](\mathbf{x}) = \frac{1}{(2\pi)^n}\hat{g}(-\mathbf{x})$$
$$= \frac{1}{(2\pi)^n}\int_{\mathbb{R}^n} e^{-i\mathbf{x}\cdot\kappa}g(\kappa)d\kappa. \tag{1.1.11}$$

The integrals in (1.1.10) and (1.1.11) have the same properties as those of the one-dimensional ones in (1.1.1) and (1.1.2). They converge absolutely, *e. g.*, for functions $\varphi, g \in L_1(\mathbb{R}^n)$ and in the norm of the space $L_2(\mathbb{R}^n)$ for $\varphi, g \in L_2(\mathbb{R}^n)$.

If Δ is the n-dimensional Laplace operator

$$\Delta = \frac{\partial^2}{\partial x_1^2} + \cdots + \frac{\partial^2}{\partial x_n^2}. \tag{1.1.12}$$

then

$$(\mathcal{F}[\Delta\varphi])(\kappa) = -|\kappa|^2(\mathcal{F}\varphi)(\kappa) \quad (\kappa \in \mathbb{R}^n). \tag{1.1.13}$$

Analogous to (1.1.7), the Fourier convolution operator of two functions h and φ is defined by

$$h * \varphi = (h * \varphi)(\mathbf{x}) = \int_{\mathbb{R}^n} h(\mathbf{x}-\mathbf{t})\varphi(\mathbf{t})d\mathbf{t} \quad (\mathbf{x} \in \mathbb{R}^n), \tag{1.1.14}$$

whose Fourier transform is given by the formula (1.1.9), but with $\kappa \in \mathbb{R}^n$.

The Laplace transform of a function $\varphi(t)$ of a variable $t \in \mathbb{R}^+ = (0, \infty)$ is defined by

$$(\mathfrak{L}\varphi)(s) = \mathfrak{L}[\varphi(t)](s) = \tilde{\varphi}(s) = \int_0^\infty e^{-st}\varphi(t)dt \quad (s \in \mathbb{C}), \tag{1.1.15}$$

if the integral converges. Here $\mathbb{C}$ is the complex plane.

If the integral (1.1.15) is convergent at the point $s_0 \in \mathbb{C}$, then it converges absolutely for $s \in \mathbb{C}$ such that $\Re e(s) > \Re e(s_0)$. The infimum σ_φ of values s for which the Laplace integral (1.1.15) converges is called the abscissa of convergence.

The inverse Laplace transform is given for $x \in \mathbb{R}^+$ by the formula

$$(\mathfrak{L}^{-1}g)(x) = \mathfrak{L}^{-1}[g(s)](x) = \frac{1}{2\pi i}\int_{\gamma-i\infty}^{\gamma+i\infty} e^{sx}g(s)ds. \tag{1.1.16}$$

with $\gamma = \Re e(s) > \sigma_\varphi$.The direct and inverse Laplace transforms are inverse to each other for "sufficiently good" functions φ and g, that is

$$\mathfrak{L}^{-1}\mathfrak{L}\varphi = \varphi \quad \text{and} \quad \mathfrak{L}\mathfrak{L}^{-1}g = g. \tag{1.1.17}$$

Some simple properties of the Laplace transform analogous to those given for the Fourier transform are the following

$$\mathfrak{L}[D^n\varphi(t)](s) = s^n(\mathfrak{L}\varphi)(s) - \sum_{j=0}^{n-1} s^{n-j-1}(D^j\varphi)(0) \quad (n \in \mathbb{N}). \tag{1.1.18}$$

and

$$D^n(\mathfrak{L}\varphi)(s) = (-1)^n\mathfrak{L}[t^n\varphi(t)](s) \quad (n \in \mathbb{N}). \tag{1.1.19}$$

The Laplace convolution operator of two functions $h(t)$ and $\varphi(t)$, given on $\mathbb{R}^+$, is defined for $x \in \mathbb{R}^+$ by the integral

$$h * \varphi = (h * \varphi)(x) = \int_0^x h(x-t)\varphi(t)dt, \tag{1.1.20}$$

which has the commutative property

$$h * \varphi = \varphi * h. \tag{1.1.21}$$

and

$$(\mathfrak{L}(h * \varphi)(s) = (\mathfrak{L}h)(s) \cdot (\mathfrak{L}\varphi)(s). \tag{1.1.22}$$

The n-dimensional Laplace transform of a function $\varphi(\mathbf{t})$ of $\mathbf{t} \subset \mathbb{R}^n_+$ is a simple generalization of the one-dimensional case, as with the Fourier transform.

1.2 Special Functions and Their Properties

In this section we present the definitions and some properties of special known functions as the Euler Gamma function, Mittag-Leffler functions, etc. To get a more extensive study about this topic consult, *e. g.*, the above mentioned books.

1.2.1 *The Gamma function and related special functions*

The Euler Gamma function $\Gamma(z)$ is defined by the so-called Euler integral of the second kind

$$\Gamma(z) = \int_0^\infty t^{z-1} e^{-t} dt \quad (\Re e(z) > 0), \tag{1.2.1}$$

where $t^{z-1} = e^{(z-1)\log(t)}$. This integral is convergent for all complex $z \in \mathbb{C}$ with $\Re e(z) > 0$. For this function we have the reduction formula

$$\Gamma(z+1) = z\Gamma(z) \qquad (\Re e(z) > 0); \tag{1.2.2}$$

using this relation, the Euler Gamma function is extended to the half-plane $\Re e(z) \leq 0$ ($\Re e(z) > -n;\ \ n \in \mathbb{N};\ \ z \notin \mathbb{Z}_0^- = \{0, -1, -2, \ldots\}$) by

$$\Gamma(z) = \frac{\Gamma(z+n)}{(z)_n}. \tag{1.2.3}$$

Here $(z)_n$ is the Pochhammer symbol, defined for complex $z \in \mathbb{C}$ and non-negative integer, with $n \in \mathbb{N}$, by

$$(z)_0 = 1 \quad \text{and} \quad (z)_n = z(z+1)\cdots(z+n-1). \tag{1.2.4}$$

Equations (1.2.2) and (1.2.4) yield

$$\Gamma(n+1) = (1)_n = n! \quad (n \in \mathbb{N}_0) \tag{1.2.5}$$

with (as usual) $0! = 1$.

We also indicate some other properties of the Gamma function such as

$$\Gamma(z)\Gamma(1-z) = \frac{\pi}{\sin(\pi z)} \quad (z \notin \mathbb{Z}_0;\ 0 < \Re e(z) < 1), \tag{1.2.6}$$

$$\Gamma\left(\frac{1}{2}\right) = \sqrt{\pi}, \tag{1.2.7}$$

the Legendre duplication formula

$$\Gamma(2z) = \frac{2^{2z-1}}{\sqrt{\pi}}\Gamma(z)\Gamma\left(z+\frac{1}{2}\right) \quad (z \in \mathbb{C}), \tag{1.2.8}$$

and Stirling's asymptotic formula, for $|\arg(z)| < \pi;\ |z| \to \infty$

$$\Gamma(z) = (2\pi)^{1/2} z^{z-1/2} e^{-z}\left[1 + O\left(\frac{1}{z}\right)\right]. \tag{1.2.9}$$

In particular, Eq. (1.2.9) implies the well known results

$$n! = (2\pi n)^{1/2}\left(\frac{n}{e}\right)^n \left[1 + O\left(\frac{1}{n}\right)\right] \quad (n \in \mathbb{N},\ \ n \to \infty), \tag{1.2.10}$$

$$|\Gamma(x+iy)| = (2\pi)^{1/2}|x|^{x-1/2}e^{-x-\pi[1-\text{sign(x)y}]/2}\left[1+O\left(\frac{1}{x}\right)\right], \quad (1.2.11)$$

when $x \to \infty$, and

$$|\Gamma(x+iy)| = (2\pi)^{1/2}|y|^{x-1/2}e^{-x-\pi|y|/2}\left[1+O\left(\frac{1}{y}\right)\right], \quad (1.2.12)$$

when $y \to \infty$.

The quotient expansion of two Gamma functions at infinity is given by

$$\frac{\Gamma(z+a)}{\Gamma(z+b)} = z^{a-b}\left[1+O\left(\frac{1}{z}\right)\right] \quad (|\arg(z+a)| < \pi;\ |z| \to \infty). \quad (1.2.13)$$

The digamma function is defined as the logarithmic derivative of the Gamma-function,

$$\psi(z) = \frac{d}{dz}\log\Gamma(z) = \frac{\Gamma'(z)}{\Gamma(z)} \quad (z \in \mathbb{C}). \quad (1.2.14)$$

This function has the property

$$\psi(z+m) = \psi(z) + \sum_{k=0}^{m-1}\frac{1}{z+k} \quad (z \in \mathbb{C};\ m \in \mathbb{N}), \quad (1.2.15)$$

which, for $m = 1$, yields

$$\psi(z+1) = \psi(z) + \frac{1}{z} \quad (z \in \mathbb{C}). \quad (1.2.16)$$

Another function, related with the digamma function, is the m-th polygamma function $\psi^m(z)$, which is given by

$$\psi^m(z) = \left(\frac{d}{dz}\right)^m \psi(z) \quad (z \in \mathbb{C}\backslash\mathbb{Z}_0) \quad (1.2.17)$$

The Beta function is defined by the Euler integral of the first kind

$$B(z,w) = \int_0^1 t^{z-1}(1-t)^{w-1}dt \quad (\Re e(z) > 0;\ \Re e(w) > 0), \quad (1.2.18)$$

This function is connected to the Gamma function by the relation

$$B(z,w) = \frac{\Gamma(z)\Gamma(w)}{\Gamma(z+w)} \quad (z, w \notin \mathbb{Z}_0^- = \{0, -1, -2, ...\}). \quad (1.2.19)$$

The binomial coefficients are defined for $\alpha \in \mathbb{C}$ and $n \in \mathbb{N}$ by

$$\binom{\alpha}{0} = 1, \quad \binom{\alpha}{n} = \frac{\alpha(\alpha-1)\cdots(\alpha-n+1)}{n!} = \frac{(-1)^n(-\alpha)_n}{n!}. \quad (1.2.20)$$

In particular, when $\alpha = m, n \in \mathbb{N}_0 = \{0, 1, \cdots\}$, with $m \geq n$, we have

$$\binom{m}{n} = \frac{m!}{n!(m-n)!} \tag{1.2.21}$$

and

$$\binom{m}{n} = 0 \quad (m, n \in \mathbb{N}_0; \quad 0 \leq m < n) \tag{1.2.22}$$

If $\alpha \notin \mathbb{Z}^- = \{-1, -2, -3, \cdots\}$, the formula (1.2.20) is represented via the Gamma function by

$$\binom{\alpha}{n} = \frac{\Gamma(\alpha+1)}{n!\Gamma(\alpha-n+1)} \quad (\alpha \in \mathbb{C}; \quad \alpha \notin \mathbb{Z}^-; \quad n \in \mathbb{N}_0). \tag{1.2.23}$$

Such a relation can be extended from $n \in \mathbb{N}_0$ to arbitrary complex $\beta \in \mathbb{C}$ by

$$\binom{\alpha}{\beta} = \frac{\Gamma(\alpha+1)}{\Gamma(\alpha-\beta+1)\Gamma(\beta+1)} \quad (\alpha, \beta \in \mathbb{C}; \quad \alpha \notin \mathbb{Z}^-). \tag{1.2.24}$$

For more information on Gamma and Beta functions we refer to the standard works [4, 41, 454].

1.2.2 *Hypergeometric functions*

The Gauss hypergeometric function ${}_2F_1(a, b; c; z)$ is defined in the unit disk as the sum of the hypergeometric series

$${}_2F_1(a, b; c; z) = \sum_{k=0}^{\infty} \frac{(a)_k (b)_k}{(c)_k} \frac{z^k}{k!}, \tag{1.2.25}$$

where $|z| < 1$; $a, b \in \mathbb{C}$; $c \in \mathbb{C}\backslash\mathbb{Z}_0^-$, and $(a)_k$ is the Pochhammer symbol (1.2.4). Alternatively, the function can be given by the Euler integral representation

$${}_2F_1(a, b; c; z) = \frac{\Gamma(c)}{\Gamma(b)\Gamma(c-b)} \int_0^1 t^{b-1}(1-t)^{c-b-1}(1-zt)^{-a} dt, \tag{1.2.26}$$

when $0 < \mathfrak{Re}(b) < \mathfrak{Re}(c)$ and $|\arg(1-z)| < \pi$.

The confluent hypergeometric function is defined by

$$ {}_1F_1(a; c; z) = \sum_{k=0}^{\infty} \frac{(a)_k}{(c)_k} \frac{z^k}{k!}, \tag{1.2.27} $$

where $z, a \in \mathbb{C}$, $c \in \mathbb{C} \backslash \mathbb{Z}^-$ and $c \neq 0$; but, in contrast to the hypergeometric function in (1.2.25), this series is convergent for any $z \in \mathbb{C}$. It has the integral representation

$$ {}_1F_1(a; c; z) = \frac{\Gamma(c)}{\Gamma(a)\Gamma(c-a)} \int_0^1 t^{a-1}(1-t)^{c-a-1} e^{zt} dt \tag{1.2.28} $$

for $0 < \mathfrak{Re}(a) < \mathfrak{Re}(c)$.

The Gauss hypergeometric series (1.2.25) and (1.2.27) are extended to the generalized hypergeometric series defined by

$$ {}_pF_q(a_1, \ldots,\ a_p; b_1, \ldots,\ b_q; z) = \sum_{k=0}^{\infty} \frac{(a_1)_k \cdots (a_p)_k}{(b_1)_k \cdots (b_q)_k} \frac{z^k}{k!}, \tag{1.2.29} $$

where $a_l, b_j \in \mathbb{C}$, $b_j \neq 0, -1, -2, \ldots$ $(l = 1, \ldots, p;\ j = 1, \ldots, q)$. This series is absolutely convergent for all values of $z \in \mathbb{C}$ if $p \leq q$.

1.2.3 *Mittag-Leffler functions*

In this section we present the definitions and some properties of the Mittag-Leffler functions.

The Mittag-Leffler function of one parameter $E_\alpha(z)$ is defined by

$$ E_\alpha(z) = \sum_{k=0}^{\infty} \frac{z^k}{\Gamma(\alpha k + 1)} \quad (z \in \mathbb{C};\ \mathfrak{Re}(\alpha) > 0). \tag{1.2.30} $$

In particular, for $\alpha = 1, 2$ we have

$$ E_1(z) = e^z \quad \text{and} \quad E_2(z) = \cosh(\sqrt{z}). \tag{1.2.31} $$

The Mittag-Leffler function of two parameters $E_{\alpha,\beta}(z)$, generalizing the one in (1.2.30), is defined by

$$ E_{\alpha,\beta}(z) = \sum_{k=0}^{\infty} \frac{z^k}{\Gamma(\alpha k + \beta)} \quad (z,\ \beta \in \mathbb{C};\ \mathfrak{Re}(\alpha) > 0). \tag{1.2.32} $$

In particular when $\beta = 1$,

$$ E_{\alpha,1}(z) = E_\alpha(z) \quad (z \in \mathbb{C};\ \mathfrak{Re}(\alpha) > 0) \tag{1.2.33} $$

and

$$E_{1,2}(z) = \frac{e^z - 1}{z}, \quad \text{and} \quad E_{2,2}(z) = \frac{\sinh(\sqrt{z})}{\sqrt{z}}. \tag{1.2.34}$$

The function $E_{\alpha,\beta}(z)$ has the integral representation

$$E_{\alpha,\beta}(z) = \frac{1}{2\pi} \int_{\mathcal{C}} \frac{t^{\alpha-\beta} e^t}{t^\alpha - z} dt, \tag{1.2.35}$$

Here the path of integration $\mathcal{C}$ is a loop which starts and ends at $-\infty$ and encircles the circular disk $|t| \leq |z|^{1/\alpha}$ in the positive sense $|\arg(t)| \leq \pi$ on $\mathcal{C}$.

The following is a Mittag-Leffler function which generalizes the Mittag-Leffler functions (1.2.30) and (1.2.32) defined, for $z, \alpha, \beta, \rho \in \mathbb{C}$ and $\Re e(\alpha) > 0$, by

$$E^{\rho}_{\alpha,\beta}(z) = \sum_{k=0}^{\infty} \frac{(\rho)_k}{\Gamma(\alpha k + \beta)} \frac{z^k}{k!}, \tag{1.2.36}$$

where $(\rho)_k$ is the Pochhammer symbol (1.2.4).

In particular, when $\rho = 1$, it coincides with the Mittag-Leffler function (1.2.32), that is,

$$E^{1}_{\alpha,\beta}(z) = E_{\alpha,\beta}(z) \quad (z \in \mathbb{C}). \tag{1.2.37}$$

When $\alpha = 1$, $E^{\rho}_{1,\beta}(z)$ coincides with the confluent hypergeometric function (1.2.27), apart from a constant factor $[\Gamma(\beta)]^{-1}$, *i. e.*,

$$E^{\rho}_{1,\beta}(z) = \frac{1}{\Gamma(\beta)} {}_1F_1(\rho; \beta; z). \tag{1.2.38}$$

A comprehensive and up-to-date set of additional information regarding Mittag-Leffler functions may be found in the recent book [269].

1.3 Fractional Operators

In this section we give the definitions and some properties of fractional integrals and fractional derivatives of different kinds, such as Riemann-Liouville, Caputo, Liouville, Hadamard, Marchaud and Grünwald-Letnikov.

1.3.1 *Riemann-Liouville fractional integrals and fractional derivatives*

In this subsection we give the definitions of the Riemann-Liouville fractional integrals and fractional derivatives on a finite real interval and some of their properties.

Let $\Omega = [a, b]$ $(-\infty < a < b < \infty)$ be a finite interval on the real axis $\mathbb{R}$. The Riemann-Liouville fractional integrals ${}^{\mathrm{RL}}I_{a+}^{\alpha} f$ and ${}^{\mathrm{RL}}I_{b-}^{\alpha} f$ of order $\alpha \in \mathbb{C}$ $(\Re e(\alpha) > 0)$ are defined by

$$\left({}^{\mathrm{RL}}I_{a+}^{\alpha} f\right)(x) = \frac{1}{\Gamma(\alpha)} \int_a^x \frac{f(t)dt}{(x-t)^{1-\alpha}} \quad (x > a;\ \Re e(\alpha) > 0) \tag{1.3.1}$$

and

$$\left({}^{\mathrm{RL}}I_{b-}^{\alpha} f\right)(x) = \frac{1}{\Gamma(\alpha)} \int_x^b \frac{f(t)dt}{(t-x)^{1-\alpha}} \quad (x < b;\ \Re e(\alpha) > 0), \tag{1.3.2}$$

respectively. These integrals are called the left-sided and the right-sided fractional integrals.

The Riemann-Liouville fractional derivatives ${}^{\mathrm{RL}}D_{a+}^{\alpha} y$ and ${}^{\mathrm{RL}}D_{b-}^{\alpha} y$ of order $\alpha \in \mathbb{C}$ $(\Re e(\alpha) \geq 0)$ are defined by

$$\left({}^{\mathrm{RL}}D_{a+}^{\alpha} y\right)(x) = \left(\frac{d}{dx}\right)^n \left({}^{\mathrm{RL}}I_{a+}^{n-\alpha} y\right)(x) \quad (x > a) \tag{1.3.3}$$

and

$$\left({}^{\mathrm{RL}}D_{b-}^{\alpha} y\right)(x) = \left(-\frac{d}{dx}\right)^n \left({}^{\mathrm{RL}}I_{b-}^{n-\alpha} y\right)(x) \quad (x < b), \tag{1.3.4}$$

respectively, with $n = -[-\Re e(\alpha)]$, where $[\bullet]$ means the integral part of the argument, that is

$$n = \begin{cases} [\Re e(\alpha)] + 1 & \text{for } \alpha \notin \mathbb{N}_0, \\ \alpha & \text{for } \alpha \in \mathbb{N}_0. \end{cases} \tag{1.3.5}$$

In particular, when $\alpha = n \in \mathbb{N}_0$, then

$$({}^{\mathrm{RL}}D_{a+}^{0} y)(x) = ({}^{\mathrm{RL}}D_{b-}^{0} y)(x) = y(x), \tag{1.3.6}$$

$$({}^{\mathrm{RL}}D_{a+}^{n} y)(x) = y^{(n)}(x), \quad ({}^{\mathrm{RL}}D_{b-}^{n} y)(x) = (-1)^n y^{(n)}(x), \tag{1.3.7}$$

where $y^{(n)}(x)$ is the classical derivative of $y(x)$ of order n.

The particular cases when $a = 0$ in the left-sided fractional integral and derivative of Riemann-Liouville are often used in the literature, because in this case such fractional operators have a straightforward Laplace transform. For the sake of simplicity, this special case will be written in this book with any of the following nomenclature

$$ {}^{\mathrm{RL}}I^{\alpha}_{0+}f = {}^{\mathrm{RL}}I^{\alpha}f = J^{\alpha}f \tag{1.3.8} $$

and

$$ {}^{\mathrm{RL}}D^{\alpha}_{0+}f = {}^{\mathrm{RL}}D^{\alpha}f \tag{1.3.9} $$

If $\alpha, \beta \in \mathbb{C}$, $\Re e(\alpha) \geq 0$ and $\Re e(\beta) > 0$, the following properties can be directly verified:

$$ \left({}^{\mathrm{RL}}I^{\alpha}_{a+}(t-a)^{\beta}\right)(x) = \frac{\Gamma(\beta+1)}{\Gamma(\beta+\alpha+1)}(x-a)^{\beta+\alpha}, \tag{1.3.10} $$

$$ \left({}^{\mathrm{RL}}D^{\alpha}_{a+}(t-a)^{\beta}\right)(x) = \frac{\Gamma(\beta+1)}{\Gamma(\beta-\alpha+1)}(x-a)^{\beta-\alpha}, \tag{1.3.11} $$

$$ \left({}^{\mathrm{RL}}I^{\alpha}_{b-}(b-t)^{\beta}\right)(x) = \frac{\Gamma(\beta+1)}{\Gamma(\beta+\alpha+1)}(b-x)^{\beta+\alpha}, \tag{1.3.12} $$

$$ \left({}^{\mathrm{RL}}D^{\alpha}_{b-}(b-t)^{\beta}\right)(x) = \frac{\Gamma(\beta+1)}{\Gamma(\beta-\alpha+1)}(b-x)^{\beta-\alpha}. \tag{1.3.13} $$

For $0 < \Re e(\alpha) < 1$ this reduces to

$$ \left({}^{\mathrm{RL}}D^{\alpha}_{a+}1\right)(x) = \frac{(x-a)^{-\alpha}}{\Gamma(1-\alpha)}, \quad \left({}^{\mathrm{RL}}D^{\alpha}_{b-}1\right)(x) = \frac{(b-x)^{-\alpha}}{\Gamma(1-\alpha)}, \tag{1.3.14} $$

and for $j = 1, 2, \ldots, n = -[-\Re e(\alpha)]$, we obtain

$$ \left({}^{\mathrm{RL}}D^{\alpha}_{a+}(t-a)^{\alpha-j}\right)(x) = 0, \quad \left({}^{\mathrm{RL}}D^{\alpha}_{b-}(b-t)^{\alpha-j}\right)(x) = 0. \tag{1.3.15} $$

From (1.3.15) we derive that the equality $(D^{\alpha}_{a+}y)(x) = 0$ is valid if, and only if,

$$ y(x) = \sum_{j=1}^{n} c_j (x-a)^{\alpha-j}, $$

where $n = [\Re e(\alpha)] + 1$ and $c_j \in \mathbb{R}$ $(j = 1, \ldots, n)$ are arbitrary constants.

In particular, when $0 < \Re e(\alpha) \leq 1$, the relation $({}^{\mathrm{RL}}D^{\alpha}_{a+}y)(x) = 0$ holds if, and only if, $y(x) = c(x-a)^{\alpha-1}$ with any $c \in \mathbb{R}$.

Likewise, the equality $({}^{\mathrm{RL}}D_{b-}^{\alpha}y)(x) = 0$ is valid if, and only if,

$$y(x) = \sum_{j=1}^{n} d_j (b-x)^{\alpha-j},$$

where $d_j \in \mathbb{R}$ $(j = 1, \ldots, n)$ are arbitrary constants. In particular, when $0 < \Re e(\alpha) \leq 1$, the relation $({}^{\mathrm{RL}}D_{b-}^{\alpha}y)(x) = 0$ holds if, and only if, $y(x) = d(b-x)^{\alpha-1}$ with any $d \in \mathbb{R}$.

The next results give us an alternative representation of the fractional derivatives ${}^{\mathrm{RL}}D_{a+}^{\alpha}$ and ${}^{\mathrm{RL}}D_{b-}^{\alpha}$, $\Re e(\alpha) \geq 0$, $n = [\Re e(\alpha)] + 1$, for suitable functions $y(x)$

$$({}^{\mathrm{RL}}D_{a+}^{\alpha}y)(x) = \sum_{k=0}^{n-1} \frac{y^{(k)}(a)}{\Gamma(1+k-\alpha)}(x-a)^{k-\alpha} + \frac{1}{\Gamma(n-\alpha)} \int_a^x \frac{y^{(n)}(t)dt}{(x-t)^{\alpha-n+1}} \tag{1.3.16}$$

and

$$({}^{\mathrm{RL}}D_{b-}^{\alpha}y)(x) = \sum_{k=0}^{n-1} \frac{(-1)^k y^{(k)}(b)}{\Gamma(1+k-\alpha)}(b-x)^{k-\alpha} + \frac{(-1)^n}{\Gamma(n-\alpha)} \int_x^b \frac{y^{(n)}(t)dt}{(t-x)^{\alpha-n+1}}. \tag{1.3.17}$$

The semigroup property of the fractional integration operators ${}^{\mathrm{RL}}I_{a+}^{\alpha}$ and ${}^{\mathrm{RL}}I_{b-}^{\alpha}$ establishes that, if $\Re e(\alpha) > 0$ and $\Re e(\beta) > 0$, then the equations

$$({}^{\mathrm{RL}}I_{a+}^{\alpha}\, {}^{\mathrm{RL}}I_{a+}^{\beta} f)(x) = ({}^{\mathrm{RL}}I_{a+}^{\alpha+\beta} f)(x) \text{ and } ({}^{\mathrm{RL}}I_{b-}^{\alpha}\, {}^{\mathrm{RL}}I_{b-}^{\beta} f)(x) = ({}^{\mathrm{RL}}I_{b-}^{\alpha+\beta} f)(x) \tag{1.3.18}$$

are satisfied at almost every point $x \in [a, b]$ for $f(x) \in L_p(a, b)$ $(1 \leq p \leq \infty)$.

If $\alpha + \beta > 1$, then the relations in (1.3.18) hold at any point of $[a, b]$.

Similarly, we have the following index rule

$$({}^{\mathrm{RL}}D_{a+}^{\alpha}\, {}^{\mathrm{RL}}D_{a+}^{\beta} f)(x) = ({}^{\mathrm{RL}}D_{a+}^{\alpha+\beta} f)(x) - \sum_{j=1}^{m} ({}^{\mathrm{RL}}D_{a+}^{\beta-j} f)(a+) \frac{(x-a)^{-j-\alpha}}{\Gamma(1-j-\alpha)}, \tag{1.3.19}$$

if $\alpha, \beta > 0$ such that $n - 1 < \alpha \leq n$, $m - 1 < \beta \leq m$ $(n, m \in \mathbb{N})$ and $\alpha + \beta < n$.

For $f(x) \in L_p(a, b)$ $(1 \leq p \leq \infty)$, the composition relations

$$({}^{\mathrm{RL}}D_{a+}^{\beta}\,{}^{\mathrm{RL}}I_{a+}^{\alpha}f)(x) = I_{a+}^{\alpha-\beta}f(x) \text{ and } ({}^{\mathrm{RL}}D_{b-}^{\beta}\,{}^{\mathrm{RL}}I_{b-}^{\alpha}f)(x) = {}^{\mathrm{RL}}I_{b-}^{\alpha-\beta}f(x) \tag{1.3.20}$$

between fractional differentiation and fractional integration operators hold almost everywhere on $[a,b]$ if $\mathfrak{Re}(\alpha) > \mathfrak{Re}(\beta) > 0$. In particular, when $\beta = k \in \mathbb{N}$ and $\mathfrak{Re}(\alpha) > k$, then

$$({}^{\mathrm{RL}}D^{k\mathrm{RL}}I_{a+}^{\alpha}f)(x) = {}^{\mathrm{RL}}I_{a+}^{\alpha-k}f(x) \text{ and } ({}^{\mathrm{RL}}D^{k\mathrm{RL}}I_{b-}^{\alpha}f)(x) = (-1)^{k\mathrm{RL}}I_{b-}^{\alpha-k}f(x). \tag{1.3.21}$$

So, the fractional differentiation is an operation inverse to the fractional integration from the left, *i. e.*, if $\mathfrak{Re}(\alpha) > 0$, then the equalities

$$({}^{\mathrm{RL}}D_{a+}^{\alpha}\,{}^{\mathrm{RL}}I_{a+}^{\alpha}f)(x) = f(x) \text{ and } ({}^{\mathrm{RL}}D_{b-}^{\alpha}\,{}^{\mathrm{RL}}I_{b-}^{\alpha}f)(x) = f(x) \tag{1.3.22}$$

hold almost everywhere on $[a,b]$.

On the other hand, if $\mathfrak{Re}(\alpha) > 0$, $n = [\mathfrak{Re}(\alpha)] + 1$ and $f_{n-\alpha}(x) = ({}^{\mathrm{RL}}I_{a+}^{n-\alpha}f)(x)$, the relation

$$({}^{\mathrm{RL}}I_{a+}^{\alpha}\,{}^{\mathrm{RL}}D_{a+}^{\alpha}f)(x) = f(x) - \sum_{j=1}^{n}\frac{f_{n-\alpha}^{(n-j)}(a)}{\Gamma(\alpha-j+1)}(x-a)^{\alpha-j} \tag{1.3.23}$$

holds almost everywhere on $[a,b]$. Also, if $g_{n-\alpha}(x) = ({}^{\mathrm{RL}}I_{b-}^{n-\alpha}g)(x)$, then the formula

$$({}^{\mathrm{RL}}I_{b-}^{\alpha}\,{}^{\mathrm{RL}}D_{b-}^{\alpha}g)(x) = g(x) - \sum_{j=1}^{n}\frac{(-1)^{n-j}g_{n-\alpha}^{(n-j)}(a)}{\Gamma(\alpha-j+1)}(b-x)^{\alpha-j} \tag{1.3.24}$$

holds almost everywhere on $[a,b]$.

Let $\mathfrak{Re}(\alpha) \geq 0$, $m \in \mathbb{N}$ and $D = d/dx$, then if the fractional derivatives $(D_{a+}^{\alpha}y)(x)$ and $({}^{\mathrm{RL}}D_{a+}^{\alpha+m}y)(x)$ exist, we have

$$({}^{\mathrm{RL}}D^{m}\,{}^{\mathrm{RL}}D_{a+}^{\alpha}y)(x) = ({}^{\mathrm{RL}}D_{a+}^{\alpha+m}y)(x), \tag{1.3.25}$$

and, if the fractional derivatives $({}^{\mathrm{RL}}D_{b-}^{\alpha}y)(x)$ and $({}^{\mathrm{RL}}D_{b-}^{\alpha+m}y)(x)$ exist, then

$$({}^{\mathrm{RL}}D^{m}\,{}^{\mathrm{RL}}D_{b-}^{\alpha}y)(x) = (-1)^{m}({}^{\mathrm{RL}}D_{b-}^{\alpha+m}y)(x). \tag{1.3.26}$$

In connection with the Laplace transform, if $\mathfrak{Re}(\alpha) > 0$ and $n = [\mathfrak{Re}(\alpha)] + 1$, we have

$$(\mathfrak{L}^{\mathrm{RL}}D_{0+}^{\alpha}y)(s) = s^{\alpha}(\mathfrak{L}y)(s) - \sum_{k=0}^{n-1}s^{n-k-1}D^{k}({}^{\mathrm{RL}}I_{0+}^{n-\alpha}y)(0+) \tag{1.3.27}$$

for $(\mathfrak{Re}(s) > q_0)$.

The rules for fractional integration by parts read as follows.

(a) If $\varphi(x) \in L_p(a,b)$ and $\psi(x) \in L_q(a,b)$, then

$$\int_a^b \varphi(x) \left({}^{\mathrm{RL}}I_{a+}^{\alpha}\psi\right)(x)dx = \int_a^b \psi(x) \left({}^{\mathrm{RL}}I_{b-}^{\alpha}\varphi\right)(x)dx. \tag{1.3.28}$$

(b) If $f(x) = \left({}^{\mathrm{RL}}I_{b-}^{\alpha}h_1\right)(x)$ with some $h_1(x) \in L_p(a,b)$ and $g(x) = \left({}^{\mathrm{RL}}I_{a+}^{\alpha}h_2\right)(x)$ with some $h_2(x) \in L_q(a,b)$, then

$$\int_a^b f(x) \left({}^{\mathrm{RL}}D_{a+}^{\alpha}g\right)(x)dx = \int_a^b g(x) \left({}^{\mathrm{RL}}D_{b-}^{\alpha}f\right)(x)dx. \tag{1.3.29}$$

Here we assume $\alpha > 0$, $p \geq 1$, $q \geq 1$, and $(1/p) + (1/q) \leq 1 + \alpha$ ($p \neq 1$ and $q \neq 1$ in the case when $(1/p) + (1/q) = 1 + \alpha$).

The generalized fractional Leibniz formula for the Riemann-Liouille derivative, applied to suitable functions on $[a,b]$, reads

$$\left[{}^{\mathrm{RL}}D_{a+}^{\alpha}(fg)\right](x) = \sum_{j=0}^{\infty} \binom{\alpha}{j} ({}^{\mathrm{RL}}D_{a+}^{\alpha-j}f)(x)(D^j g)(x), \tag{1.3.30}$$

where $\alpha > 0$. Below, we present three particular cases to illustrate this property.

(a) Let $0 < \alpha < 1$, $f(x) = x$ and $g(x)$ a suitable function. Then

$$[{}^{\mathrm{RL}}D_{0+}^{\alpha}(fg)](x) = x({}^{\mathrm{RL}}D_{0+}^{\alpha}g)(x) + ({}^{\mathrm{RL}}I_{0+}^{1-\alpha}g)(x) \tag{1.3.31}$$

(b) Let $0 < \alpha < 1$, $f(x) = x^{\alpha-1}$ and $g(x)$ a suitable function. Then

$$\left[{}^{\mathrm{RL}}D_{0+}^{\alpha}(fg)\right](x) = \sum_{j=1}^{\infty} \binom{\alpha}{j} \frac{\Gamma(\alpha)}{\Gamma(j)} x^{j-1} g^{(j)}(x) \tag{1.3.32}$$

(c) Let $p \in \mathbb{N}$, $\alpha > 0$, and $f(x)$ a suitable function. Then

$$({}^{\mathrm{RL}}D_{0+}^{\alpha}t^p f)(x) = \sum_{j=0}^{p} \binom{\alpha}{j} (D^j x^p)({}^{\mathrm{RL}}D_{0+}^{\alpha-j}f)(x) \tag{1.3.33}$$

The computation of a fractional Riemann-Liouville derivative of the composition of two suitable functions can be very complicated. The corresponding formula

$$\left[{}^{\mathrm{RL}}D_{a+}^{\alpha}(f(g))\right](x) = \frac{(x-a)^{-\alpha}}{\Gamma(1-\alpha)} f(g(x)) \tag{1.3.34}$$

$$+\sum_{j=1}^{\infty}\binom{\alpha}{j}\frac{j!(x-a)^{j-\alpha}}{\Gamma(j+1-\alpha)}\sum_{i=1}^{j}[D^i f(g)](x)\sum\prod_{r=1}^{j}\frac{1}{a_r!}\left(\frac{(D^r g)(x)}{r!}\right)^{a_r},$$

where $\sum_{r=1}^{j} r a_r = j$ and $\sum_r^j a_r = i$, exhibits the complicated structure very clearly.

The above relation is a consequence of the application of (1.3.30) and the well known Faá di Bruno formula (see, *e. g.*, formula (24.1.2) in [4]) for a natural n, *viz.*

$$(D^n f(g))(x) = \sum_{m=0}^{n}(D^m f)(g(x)) \tag{1.3.35}$$

$$\times\sum(n; a_1, a_2, \ldots, a_n)\{(Dg)(x)\}^{a_1}\{(D^2 g)(x)\}^{a_2}\cdots\{(D^n g)(x)\}^{a_n}$$

summed over $a_1 + 2a_2 + \ldots + na_n$ and $a_1 + a_2 + \ldots + a_n = m$, see also Eqs. (5.2.26) and (5.2.29).

1.3.2 *Caputo fractional derivatives*

The Caputo fractional derivatives are closely related to the Riemann-Liouville derivatives. Let $[a, b]$ be a finite interval of the real line $\mathbb{R}$. For $\alpha \in \mathbb{C}$ ($\mathfrak{Re}(\alpha) \geq 0$) the Caputo fractional derivatives are defined by

$$({}^{\mathrm{C}}D_{a+}^{\alpha}y)(x) = \frac{1}{\Gamma(n-\alpha)}\int_a^x \frac{y^{(n)}(t)dt}{(x-t)^{\alpha-n+1}} = ({}^{\mathrm{RL}}I_{a+}^{n-\alpha}D^n y)(x) \tag{1.3.36}$$

and

$$({}^{\mathrm{C}}D_{b-}^{\alpha}y)(x) = \frac{(-1)^n}{\Gamma(n-\alpha)}\int_x^b \frac{y^{(n)}(t)dt}{(t-x)^{\alpha-n+1}} = (-1)^n({}^{\mathrm{RL}}I_{b-}^{n-\alpha}D^n y)(x), \tag{1.3.37}$$

where $D = d/dx$ and $n = -[-\mathfrak{Re}(\alpha)]$, *i. e.*, $n = [\mathfrak{Re}(\alpha)]+1$ for $\alpha \notin \mathbb{N}_0$, and $n = \alpha$ for $\alpha \in \mathbb{N}_0$. These derivatives are called left-sided and right-sided Caputo fractional derivatives of order α.

In particular, when $0 < \Re e(\alpha) < 1$ then

$$({}^{\mathrm{C}}D_{a+}^{\alpha}y)(x) = \frac{1}{\Gamma(1-\alpha)} \int_a^x \frac{y'(t)dt}{(x-t)^{\alpha}} = ({}^{\mathrm{RL}}I_{a+}^{1-\alpha}Dy)(x) \tag{1.3.38}$$

and

$$({}^{\mathrm{C}}D_{b-}^{\alpha}y)(x) = -\frac{1}{\Gamma(1-\alpha)} \int_x^b \frac{y'(t)dt}{(t-x)^{\alpha}} = -({}^{\mathrm{RL}}I_{b-}^{1-\alpha}Dy)(x). \tag{1.3.39}$$

The connections between the Caputo and the Riemann-Liouville derivatives are given by the relations

$$({}^{\mathrm{C}}D_{a+}^{\alpha}y)(x) = \left({}^{\mathrm{RL}}D_{a+}^{\alpha}\left[y(t) - \sum_{k=0}^{n-1} \frac{y^{(k)}(a)}{k!}(t-a)^k\right]\right)(x) \tag{1.3.40}$$

and

$$({}^{\mathrm{C}}D_{b-}^{\alpha}y)(x) = \left({}^{\mathrm{RL}}D_{b-}^{\alpha}\left[y(t) - \sum_{k=0}^{n-1} \frac{y^{(k)}(b)}{k!}(b-t)^k\right]\right)(x), \tag{1.3.41}$$

respectively.

In particular, when $0 < \Re e(\alpha) < 1$, the relations (1.3.40) and (1.3.41) take the following forms

$$({}^{\mathrm{C}}D_{a+}^{\alpha}y)(x) = \left({}^{\mathrm{RL}}D_{a+}^{\alpha}[y(t) - y(a)]\right)(x), \tag{1.3.42}$$

$$({}^{\mathrm{C}}D_{b-}^{\alpha}y)(x) = \left({}^{\mathrm{RL}}D_{b-}^{\alpha}[y(t) - y(b)]\right)(x). \tag{1.3.43}$$

If $\alpha = n \in \mathbb{N}_0$ and the usual derivative $y^{(n)}(x)$ of order n exists, then $({}^{\mathrm{C}}D_{a+}^{n}y)(x)$ coincides with $y^{(n)}(x)$, while $({}^{\mathrm{C}}D_{b-}^{n}y)(x)$ coincides with $y^{(n)}(x)$ up to the constant factor $(-1)^n$, *i. e.*,

$$({}^{\mathrm{C}}D_{a+}^{n}y)(x) = y^{(n)}(x) \text{ and } ({}^{\mathrm{C}}D_{b-}^{n}y)(x) = (-1)^n y^{(n)}(x) \quad (n \in \mathbb{N}). \tag{1.3.44}$$

The Caputo derivatives $({}^{\mathrm{C}}D_{a+}^{\alpha}y)(x)$ and $({}^{\mathrm{C}}D_{b-}^{\alpha}y)(x)$ have properties similar to those given in Eqs. (1.3.11) and (1.3.13) for the Riemann-Liouville fractional derivatives. If $\Re e(\alpha) > 0$, $n = -[-\Re e(\alpha)]$ is given by (1.3.5) and $\Re e(\beta) > n-1$, then

$$\left({}^{\mathrm{C}}D_{a+}^{\alpha}(t-a)^{\beta}\right)(x) = \frac{\Gamma(\beta+1)}{\Gamma(\beta-\alpha+1)}(x-a)^{\beta-\alpha} \tag{1.3.45}$$

and

$$\left({}^{\mathrm{C}}D_{b-}^{\alpha}(b-t)^{\beta}\right)(x) = \frac{\Gamma(\beta+1)}{\Gamma(\beta-\alpha+1)}(b-x)^{\beta-\alpha}. \tag{1.3.46}$$

However, for $k = 0, 1, \ldots, n-1$, we have

$$({}^{\mathrm{C}}D_{a+}^{\alpha}(t-a)^k)(x) = 0 \text{ and } ({}^{\mathrm{C}}D_{b-}^{\alpha}(t-a)^k)(x) = 0. \tag{1.3.47}$$

In particular,

$$({}^{\mathrm{C}}D_{a+}^{\alpha}1)(x) = 0 \text{ and } ({}^{\mathrm{C}}D_{b-}^{\alpha}1)(x) = 0. \tag{1.3.48}$$

On the other hand, if $\mathfrak{Re}(\alpha) > 0$ and $\lambda > 0$, then

$$\left({}^{\mathrm{C}}D_{a+}^{\alpha}e^{\lambda t}\right)(x) \neq \lambda^{\alpha}e^{\lambda x} \tag{1.3.49}$$

for any $a \in \mathbb{R}$.

Let $\mathfrak{Re}(\alpha) > 0$ and let $y(x)$ a suitable function, for example $y(x) \in C[a,b]$. Then If $\mathfrak{Re}(\alpha) \notin \mathbb{N}$ or $\alpha \in \mathbb{N}$, the Caputo fractional differentiation operators ${}^{\mathrm{C}}D_{a+}^{\alpha}$ and ${}^{\mathrm{C}}D_{b-}^{\alpha}$ provide operations inverse to the Riemann-Liouville fractional integration operators I_{a+}^{α} and I_{b-}^{α} from the left, that is

$$({}^{\mathrm{C}}D_{a+}^{\alpha}\,{}^{\mathrm{RL}}I_{a+}^{\alpha}y)(x) = y(x) \text{ and } ({}^{\mathrm{C}}D_{b-}^{\alpha}\,{}^{\mathrm{RL}}I_{b-}^{\alpha}y)(x) = y(x). \tag{1.3.50}$$

However, when $\mathfrak{Re}(\alpha) \in \mathbb{N}$ and $\mathfrak{Im}(\alpha) \neq 0$, we have

$$({}^{\mathrm{C}}D_{a+}^{\alpha}\,{}^{\mathrm{RL}}I_{a+}^{\alpha}y)(x) = y(x) - \frac{({}^{\mathrm{RL}}I_{a+}^{\alpha+1-n}y)(a+)}{\Gamma(n-\alpha)}(x-a)^{n-\alpha} \tag{1.3.51}$$

and

$$({}^{\mathrm{C}}D_{b-}^{\alpha}\,{}^{\mathrm{RL}}I_{a+}^{\alpha}y)(x) = y(x) - \frac{({}^{\mathrm{RL}}I_{b-}^{\alpha+1-n}y)(b-)}{\Gamma(n-\alpha)}(b-x)^{n-\alpha}. \tag{1.3.52}$$

On the other hand, if $\mathfrak{Re}(\alpha) > 0$ and $n = -[-\mathfrak{Re}(\alpha)]$ is given by (1.3.5), then under sufficiently good conditions for $y(x)$

$$({}^{\mathrm{RL}}I_{a+}^{\alpha}\;{}^{\mathrm{C}}D_{a+}^{\alpha}y)(x) = y(x) - \sum_{k=0}^{n-1}\frac{y^{(k)}(a)}{k!}(x-a)^k \tag{1.3.53}$$

and

$$({}^{\mathrm{RL}}I_{b-}^{\alpha}\;{}^{\mathrm{C}}D_{b-}^{\alpha}y)(x) = y(x) - \sum_{k=0}^{n-1}\frac{(-1)^k y^{(k)}(b)}{k!}(b-x)^k. \tag{1.3.54}$$

In particular, if $0 < \mathfrak{Re}(\alpha) \leq 1$, then

$$({}^{\mathrm{RL}}I_{a+}^{\alpha}\;{}^{\mathrm{C}}D_{a+}^{\alpha}y)(x) = y(x) - y(a) \text{ and } ({}^{\mathrm{RL}}I_{b-}^{\alpha}\;{}^{\mathrm{C}}D_{b-}^{\alpha}y)(x) = y(x) - y(b). \tag{1.3.55}$$

Under suitable conditions, the Laplace transform of the Caputo fractional derivative ${}^{\mathsf{C}}D_{0+}^{\alpha}y$ is given by

$$(\mathfrak{L}\,{}^{\mathsf{C}}D_{0+}^{\alpha}y)(s) = s^{\alpha}(\mathfrak{L}y)(s) - \sum_{k=0}^{n-1} s^{\alpha-k-1}(D^k y)(0). \tag{1.3.56}$$

In particular, if $0 < \alpha \leq 1$, then

$$(\mathfrak{L}\,{}^{\mathsf{C}}D_{0+}^{\alpha}y)(s) = s^{\alpha}(\mathfrak{L}y)(s) - s^{\alpha-1}y(0). \tag{1.3.57}$$

We have defined the Caputo derivatives on a finite interval $[a, b]$. Formulas (1.3.36) and (1.3.37) can be used for the definition of the Caputo fractional derivatives on the whole axis $\mathbb{R}$. Thus the corresponding Caputo fractional derivative of order $\alpha \in \mathbb{C}$ (with $\Re e(\alpha) > 0$ and $\alpha \notin \mathbb{N}$) can be defined as follows

$$({}^{\mathsf{C}}D_{+}^{\alpha}y)(x) = \frac{1}{\Gamma(n-\alpha)} \int_{-\infty}^{x} \frac{y^{(n)}(t)dt}{(x-t)^{\alpha+1-n}} \tag{1.3.58}$$

and

$$({}^{\mathsf{C}}D_{-}^{\alpha}y)(x) = \frac{(-1)^n}{\Gamma(n-\alpha)} \int_{x}^{\infty} \frac{y^{(n)}(t)dt}{(t-x)^{\alpha+1-n}}, \tag{1.3.59}$$

with $x \in \mathbb{R}$.

The Caputo derivatives $({}^{\mathsf{C}}D_{+}^{\alpha}y)(x)$ and $({}^{\mathsf{C}}D_{-}^{\alpha}y)(x)$ have properties similar to those that we will describe below for the operators known as Liouville derivatives. In particular, we mention the identities

$$\left({}^{\mathsf{C}}D_{+}^{\alpha}e^{\lambda t}\right)(x) = \lambda^{\alpha}e^{\lambda x} \text{ and } \left({}^{\mathsf{C}}D_{-}^{\alpha}e^{-\lambda t}\right)(x) = \lambda^{\alpha}e^{-\lambda x}. \tag{1.3.60}$$

1.3.3 *Liouville fractional integrals and fractional derivatives. Marchaud derivatives*

First of all, we present the definitions and some properties of the Liouville fractional integrals and fractional derivatives on the whole axis $\mathbb{R}$. More detailed information may be found in the bibliography.

The Liouville fractional integrals on $\mathbb{R}$ have the form

$$({}^{\mathsf{L}}I_{+}^{\alpha}f)(x) = \frac{1}{\Gamma(\alpha)} \int_{-\infty}^{x} \frac{f(t)dt}{(x-t)^{1-\alpha}} \tag{1.3.61}$$

and

$$({}^{\mathsf{L}}I_{-}^{\alpha}f)(x) = \frac{1}{\Gamma(\alpha)}\int_{x}^{\infty}\frac{f(t)dt}{(t-x)^{1-\alpha}}, \tag{1.3.62}$$

where $x \in \mathbb{R}$ and $\Re e(\alpha) > 0$, while the fractional Liouville derivatives are defined as

$$\begin{aligned}({}^{\mathsf{L}}D_{+}^{\alpha}y)(x) &= \left(\frac{d}{dx}\right)^{n}({}^{\mathsf{L}}I_{+}^{n-\alpha}y)(x) \\ &= \frac{1}{\Gamma(n-\alpha)}\left(\frac{d}{dx}\right)^{n}\int_{-\infty}^{x}\frac{y(t)dt}{(x-t)^{\alpha-n+1}}\end{aligned} \tag{1.3.63}$$

and

$$\begin{aligned}({}^{\mathsf{L}}D_{-}^{\alpha}y)(x) &= \left(-\frac{d}{dx}\right)^{n}({}^{\mathsf{L}}I_{-}^{n-\alpha}y)(x) \\ &= \frac{1}{\Gamma(n-\alpha)}\left(-\frac{d}{dx}\right)^{n}\int_{x}^{\infty}\frac{y(t)dt}{(t-x)^{\alpha-n+1}},\end{aligned} \tag{1.3.64}$$

where $n = -[-\Re e(\alpha)]$, $\Re e(\alpha) \geq 0$ and $x \in \mathbb{R}$.

The expressions for ${}^{\mathsf{L}}I_{+}^{\alpha}f$ and ${}^{\mathsf{L}}I_{-}^{\alpha}f$ in (1.3.61) and (1.3.62), and for ${}^{\mathsf{L}}D_{+}^{\alpha}y$ and ${}^{\mathsf{L}}D_{-}^{\alpha}y$ in (1.3.63) and (1.3.64), are called Liouville left- and right-sided fractional integrals and fractional derivatives on the whole axis $\mathbb{R}$, respectively.

In particular, when $\alpha = n \in \mathbb{N}_0$, then

$$({}^{\mathsf{L}}D_{+}^{0}y)(x) = ({}^{\mathsf{L}}D_{-}^{0}y)(x) = y(x) \tag{1.3.65}$$

and

$$({}^{\mathsf{L}}D_{+}^{n}y)(x) = y^{(n)}(x) \quad \text{and} \quad ({}^{\mathsf{L}}D_{-}^{n}y)(x) = (-1)^{n}y^{(n)}(x) \quad (n \in \mathbb{N}), \tag{1.3.66}$$

where $y^{(n)}(x)$ is the usual derivative of $y(x)$ of order n.

If $0 < \Re e(\alpha) < 1$ and $x \in \mathbb{R}$, then

$$({}^{\mathsf{L}}D_{+}^{\alpha}y)(x) = \frac{1}{\Gamma(1-\alpha)}\frac{d}{dx}\int_{-\infty}^{x}\frac{y(t)dt}{(x-t)^{\alpha-[\Re e(\alpha)]}} \tag{1.3.67}$$

and

$$({}^{\mathsf{L}}D_{-}^{\alpha}y)(x) = -\frac{1}{\Gamma(1-\alpha)}\frac{d}{dx}\int_{x}^{\infty}\frac{y(t)dt}{(t-x)^{\alpha-[\Re e(\alpha)]}}. \tag{1.3.68}$$

The Liouville fractional operators ${}^{\mathsf{L}}I_{+}^{\alpha}$ and ${}^{\mathsf{L}}D_{+}^{\alpha}$ of the exponential function $e^{\lambda x}$ yield the same exponential function, both apart from a constant multiplication factor, *i. e.*, if $\lambda > 0$ and $\Re e(\alpha) \geq 0$,

$$({}^{\mathrm{L}}I_{+}^{\alpha}e^{\lambda t})(x) = \lambda^{-\alpha}e^{\lambda x}; \tag{1.3.69}$$

$$\left({}^{\mathrm{L}}I_{-}^{\alpha}e^{-\lambda t}\right)(x) = \lambda^{-\alpha}e^{-\lambda x}; \tag{1.3.70}$$

$$({}^{\mathrm{L}}D_{+}^{\alpha}e^{\lambda t})(x) = \lambda^{\alpha}e^{\lambda x}; \tag{1.3.71}$$

and

$$\left({}^{\mathrm{L}}D_{-}^{\alpha}e^{-\lambda t}\right)(x) = \lambda^{\alpha}e^{-\lambda x}. \tag{1.3.72}$$

On the other hand, if $\alpha > 0$, $\beta > 0$, then, for "sufficiently good" functions, we have

$$({}^{\mathrm{L}}I_{+}^{\alpha}\,{}^{\mathrm{L}}I_{+}^{\beta}f)(x) = ({}^{\mathrm{L}}I_{+}^{\alpha+\beta}f)(x) \text{ and } ({}^{\mathrm{L}}I_{-}^{\alpha}\,{}^{\mathrm{L}}I_{-}^{\beta}f)(x) = ({}^{\mathrm{L}}I_{-}^{\alpha+\beta}f)(x); \tag{1.3.73}$$

$$({}^{\mathrm{L}}D_{+}^{\alpha}\,{}^{\mathrm{L}}I_{+}^{\alpha}f)(x) = f(x), \text{ and } ({}^{\mathrm{L}}D_{-}^{\alpha}\,{}^{\mathrm{L}}I_{-}^{\alpha}f)(x) = f(x). \tag{1.3.74}$$

If in addition $\alpha > \beta > 0$, then the formulas

$$({}^{\mathrm{L}}D_{+}^{\beta}\,{}^{\mathrm{L}}I_{+}^{\alpha}f)(x) = ({}^{\mathrm{L}}I_{0+}^{\alpha-\beta}f)(x) \text{ and } ({}^{\mathrm{L}}D_{-}^{\beta}\,{}^{\mathrm{L}}I_{-}^{\alpha}f)(x) = ({}^{\mathrm{L}}I_{-}^{\alpha-\beta}f)(x) \tag{1.3.75}$$

hold. Furthermore, when $\beta = k \in \mathbb{N}$ and $\mathfrak{Re}(\alpha) > k$, then

$$(D^{k}\,{}^{\mathrm{L}}I_{+}^{\alpha}f)(x) = {}^{\mathrm{L}}I_{+}^{\alpha-k}f(x), \text{ and } (D^{k}\,{}^{\mathrm{L}}I_{-}^{\alpha}f)(x) = (-1)^{k}\,{}^{\mathrm{L}}I_{-}^{\alpha-k}f(x). \tag{1.3.76}$$

The Fourier transform (1.1.1) of the Liouville fractional integrals ${}^{\mathrm{L}}I_{+}^{\alpha}f$ and ${}^{\mathrm{L}}I_{-}^{\alpha}f$ is given for $0 < \mathfrak{Re}(\alpha) < 1$, by the following relations

$$(\mathcal{F}\,{}^{\mathrm{L}}I_{+}^{\alpha}f)(\kappa) = \frac{(\mathcal{F}f)(\kappa)}{(-i\kappa)^{\alpha}} \tag{1.3.77}$$

and

$$(\mathcal{F}\,{}^{\mathrm{L}}I_{-}^{\alpha}f)(\kappa) = \frac{(\mathcal{F}f)(\kappa)}{(i\kappa)^{\alpha}}. \tag{1.3.78}$$

Here $(\mp i\kappa)^{\alpha}$ means

$$(\mp i\kappa)^{\alpha} = |\kappa|^{\alpha}e^{\mp\alpha\pi i\ \mathrm{sgn}(\kappa)/2}. \tag{1.3.79}$$

Moreover, if $\mathfrak{Re}(\alpha) > 0$, then, for "sufficiently good" functions $f(x)$, the equations (1.3.77) and (1.3.78) are valid as well as the following corresponding relations for the Liouville fractional derivatives ${}^{\mathrm{L}}D_{+}^{\alpha}f$ and ${}^{\mathrm{L}}D_{-}^{\alpha}f$

$$(\mathcal{F}\,{}^{\mathtt{L}}D^{\alpha}_{+}f)(\kappa) = (-i\kappa)^{\alpha}(\mathcal{F}f)(\kappa) \tag{1.3.80}$$

and

$$(\mathcal{F}\,{}^{\mathtt{L}}D^{\alpha}_{-}f)(\kappa) = (i\kappa)^{\alpha}(\mathcal{F}f)(\kappa), \tag{1.3.81}$$

where $(\mp i\kappa)^{\alpha}$ is defined by (1.3.79).

The rules for fractional integration by parts, for $\alpha > 0$, and for "sufficiently good" functions, are given by

$$\int_{-\infty}^{\infty} \varphi(x) \left({}^{\mathtt{L}}I^{\alpha}_{+}\psi\right)(x)dx = \int_{-\infty}^{\infty} \psi(x) \left({}^{\mathtt{L}}I^{\alpha}_{-}\varphi\right)(x)dx. \tag{1.3.82}$$

$$\int_{-\infty}^{\infty} f(x) \left({}^{\mathtt{L}}D^{\alpha}_{+}g\right)(x)dx = \int_{-\infty}^{\infty} g(x) \left({}^{\mathtt{L}}D^{\alpha}_{-}f\right)(x)dx. \tag{1.3.83}$$

$$\int_{0}^{\infty} \varphi(x) \left({}^{\mathtt{RL}}I^{\alpha}_{0+}\psi\right)(x)dx = \int_{0}^{\infty} \psi(x) \left({}^{\mathtt{L}}I^{\alpha}_{-}\varphi\right)(x)dx. \tag{1.3.84}$$

$$\int_{0}^{\infty} f(x) \left({}^{\mathtt{RL}}D^{\alpha}_{0+}g\right)(x)dx = \int_{0}^{\infty} g(x) \left({}^{\mathtt{L}}D^{\alpha}_{-}f\right)(x)dx. \tag{1.3.85}$$

The Liouville fractional derivatives ${}^{\mathtt{L}}D^{\alpha}_{+}f$ and ${}^{\mathtt{L}}D^{\alpha}_{-}f$ exist for suitable functions f, but they are not defined, for example, for constant functions. Nevertheless, they can be reduced in general to more convenient forms which admit fractional differentiation of a constant function. In this way we come to the Marchaud fractional derivatives ${}^{\mathtt{M}}D^{\alpha}_{+}f$ and ${}^{\mathtt{M}}D^{\alpha}_{-}f$ of order $\alpha \in \mathbb{C}$, defined by

$$\left({}^{\mathtt{M}}D^{\alpha}_{+}f\right)(x) = \frac{1}{\kappa(\alpha,k)} \int_{0}^{\infty} \frac{(\Delta^{k}_{t}f)(x)}{t^{1+\alpha}}dt \quad (k > \Re e(\alpha) > 0) \tag{1.3.86}$$

and

$$\left({}^{\mathtt{M}}D^{\alpha}_{-}f\right)(x) = \frac{1}{\kappa(\alpha,k)} \int_{0}^{\infty} \frac{(\Delta^{k}_{-t}f)(x)}{t^{1+\alpha}}dt \quad (k > \Re e(\alpha) > 0), \tag{1.3.87}$$

respectively. Here $\kappa(\alpha,k)$ is the constant

$$\kappa(\alpha,k) = \int_{0}^{\infty} (1-e^{-u})^{k} \frac{du}{u^{1+\alpha}} \quad (k \in \mathbb{N},\ k > \Re e(\alpha) > 0), \tag{1.3.88}$$

and $(\Delta_h^k f)(x)$ is the finite difference of order k of a function $f(x)$ with increment h

$$\left(\Delta_h^k f\right)(x) = \sum_{j=0}^{k} (-1)^j \binom{k}{j} f(x - jh). \tag{1.3.89}$$

Note that

$$\left(\Delta_h^k 1\right)(x) = \sum_{j=0}^{k} (-1)^j \binom{k}{j} = (1-1)^k = 0 \quad (k \in \mathbb{N}). \tag{1.3.90}$$

In particular, when $h = -1$, (1.3.89) coincides with the difference (1.3.89) with $h = +1$ except for the factor $(-1)^k$, *i. e.*

$$\left(\Delta_{-1}^k f\right)(x) = (-1)^k (\Delta^k f)(x).$$

It is known that for a real number $\alpha > 0$ the right-hand sides of (1.3.86) and (1.3.87) do not depend on the choice of k $(k > \alpha)$. Similarly, for complex α the right-hand sides of (1.3.86) and (1.3.87) do not depend on the choice of k $(k > \Re e(\alpha))$.

The Marchaud fractional derivatives ${}^{\mathtt{M}}D_+^\alpha f$ and ${}^{\mathtt{M}}D_-^\alpha f$ are defined for the constant function $f = c \in \mathbb{C}$, and in accordance with (1.3.90)

$$\left({}^{\mathtt{M}}D_+^\alpha c\right)(x) = \left({}^{\mathtt{M}}D_-^\alpha c\right)(x) = 0 \quad (\alpha \in \mathbb{C},\ \Re e(\alpha) > 0). \tag{1.3.91}$$

For "suitable functions" f, the Marchaud fractional derivatives coincide with the Liouville fractional derivatives for same α

$${}^{\mathtt{M}}D_+^\alpha f = {}^{\mathtt{L}}D_+^\alpha f; \quad {}^{\mathtt{M}}D_-^\alpha f = {}^{\mathtt{L}}D_-^\alpha f.$$

In particular, they have the same properties as the Liouville derivatives over exponential functions e^{bt} and e^{-bt}, in the following sense

$$\left({}^{\mathtt{M}}D_+^\alpha e^{bt}\right)(x) = b^\alpha e^{bx}, \tag{1.3.92}$$

$$\left({}^{\mathtt{M}}D_-^\alpha e^{-bt}\right)(x) = b^\alpha e^{-bx}. \tag{1.3.93}$$

for $\alpha \in \mathbb{C}$ $(\mathrm{Re}(\alpha) > 0)$ and $b \in \mathbb{C}$ $(\Re e(b) > 0)$.

1.3.4 *Generalized exponential functions*

In this section we consider two special functions, which play a role as generalized exponentials. The first one is the Mittag-Leffler function

$$E_\alpha(\lambda z^\alpha) = \sum_{k=0}^{\infty} \frac{\lambda^k z^{\alpha k}}{\Gamma(\alpha k + 1)} \quad (z, \lambda \in \mathbb{C};\ \ \Re e(\alpha) > 0), \tag{1.3.94}$$

and the second one is defined in terms of the Mittag-Leffler type function by

$$e_\alpha^{\lambda z} = z^{\alpha-1} E_{\alpha,\alpha}(\lambda z^\alpha) = \sum_{k=0}^{\infty} \frac{\lambda^k z^{\alpha(k-1)}}{\Gamma(\alpha(k+1))} \quad (z, \lambda \in \mathbb{C},\ \ \Re e(\alpha) > 0). \tag{1.3.95}$$

The above functions satisfy the identity

$$E_1(\lambda z) = e_1^{\lambda z} = e^{\lambda z} \quad (z, \lambda \in \mathbb{C}), \tag{1.3.96}$$

and therefore they are generalizations of the classical exponential function.

Some others properties of the first of these functions, for $z \in \mathbb{C}\backslash\{0\}$, $\alpha, \lambda \in \mathbb{C}$ ($\Re e(\alpha) > 0$), and $n \in \mathbb{N}$, are the following

$$\lim_{z\to a+} E_\alpha(\lambda(z-a)^\alpha) = 1; \tag{1.3.97}$$

$$\left(\frac{\partial}{\partial z}\right)^n [E_\alpha(\lambda z^\alpha)] = z^{-n} E_{\alpha,1-n}(\lambda z^\alpha); \tag{1.3.98}$$

$$\left(\frac{\partial}{\partial \lambda}\right)^n [E_\alpha(\lambda z^\alpha)] = n! z^{\alpha n} E_{\alpha,\alpha n+1}^{n+1}(\lambda z^\alpha); \tag{1.3.99}$$

$$\mathfrak{L}[E_\alpha(\lambda z^\alpha)](s) = \frac{s^{\alpha-1}}{s^\alpha - \lambda} \quad (\Re e(s) > 0;\ \ |\lambda s^{-\alpha}| < 1); \tag{1.3.100}$$

and

$$\mathfrak{L}\left[t^{\alpha n}\left(\frac{\partial}{\partial \lambda}\right)^n E_\alpha(\lambda z^\alpha)\right](s) = \frac{n! s^{\alpha-1}}{(s^\alpha - \lambda)^{n+1}}. \tag{1.3.101}$$

On the other hand, for the function $e_\alpha^{\lambda z}$, we have the following properties

$$\lim_{z\to a+}\left[(z-a)^{1-\alpha} e_\alpha^{\lambda(z-a)}\right] = \frac{1}{\Gamma(\alpha)}; \tag{1.3.102}$$

$$\left(\frac{\partial}{\partial z}\right)^n [e_\alpha^{\lambda z}] = z^{\alpha-n-1} E_{\alpha,\alpha-n}(\lambda z^\alpha); \tag{1.3.103}$$

$$\left(\frac{\partial}{\partial\lambda}\right)^n \left[e_\alpha^{\lambda z}\right] = n! z^{\alpha n+\alpha-1} E_{\alpha,(n+1)\alpha}^{n+1}(\lambda z^\alpha); \tag{1.3.104}$$

$$\mathfrak{L}\left[e_\alpha^{\lambda z}\right](s) = \frac{1}{s^\alpha - \lambda} \quad \left(\Re e(s) > 0; \ \ |\lambda s^{-\alpha}| < 1\right); \tag{1.3.105}$$

and

$$\mathfrak{L}\left[\left(\frac{\partial}{\partial\lambda}\right)^n e_\alpha^{\lambda z}\right](s) = \frac{n!}{(s^\alpha - \lambda)^{n+1}}. \tag{1.3.106}$$

The generalized α-exponential functions do not have the index property, *i. e.*, in general

$$E_\alpha(\lambda z)E_\alpha(\mu z) \neq E_\alpha((\lambda+\mu)z); \quad e_\alpha^{\lambda z} e_\alpha^{\mu z} \neq e_\alpha^{(\lambda+\mu)z} \quad (\alpha \neq 1). \tag{1.3.107}$$

For example, if $\alpha = 2$ and $z = 1$, then in accordance with the second relation in (1.2.34), we have

$$e_2^\lambda = \sqrt{\lambda}\sinh(\sqrt{\lambda}), \tag{1.3.108}$$

but

$$[\sqrt{\lambda}\sinh(\sqrt{\lambda})][\sqrt{\mu}\sinh(\sqrt{\mu})] \neq \sqrt{(\lambda+\mu)}\sinh(\sqrt{(\lambda+\mu)}). \tag{1.3.109}$$

Let $M_n(\mathbb{R})$ ($n \in \mathbb{N}$) be the set of all matrices $\mathbf{A} = [a_{jk}]$ of order $n \times n$ with $a_{jk} \in \mathbb{R}$ ($j = 1, \ldots, n$). By analogy with (1.3.95), for $\alpha \in \mathbb{C}\backslash\{0\}$ ($\Re e(\alpha) > 0$), and $\mathbf{A} \in M_n(\mathbb{R})$, here we introduce a matrix α-exponential function defined by

$$e_\alpha^{\mathbf{A}z} = z^{\alpha-1}\sum_{k=0}^{\infty} \mathbf{A}^k \frac{z^{\alpha k}}{\Gamma((k+1)\alpha)}, \tag{1.3.110}$$

and also the function

$$E_\alpha(\mathbf{A}z^\alpha) = \sum_{k=0}^{\infty} \frac{\mathbf{A}^k z^{k\alpha}}{\Gamma(\alpha k+1)} \quad (z \in \mathbb{C}; \ \ \Re e(\alpha) > 0). \tag{1.3.111}$$

When $\alpha = 1$ we have

$$E_1(\mathbf{A}z) = e_1^{\mathbf{A}z} = e^{\mathbf{A}z} \quad (z \in \mathbb{C}). \tag{1.3.112}$$

Of course, in general, the semigroup property does not hold for the generalized matrix exponential functions

$$e_\alpha^{\mathbf{A}z} e_\alpha^{\mathbf{B}z} \neq e_\alpha^{(\mathbf{A}+\mathbf{B})z} \quad (z, \alpha \in \mathbb{C}; \ \mathbf{A}, \mathbf{B} \in M_n(\mathbb{R})). \tag{1.3.113}$$

Similarly, the inversion formula

$$(e^{\mathbf{A}z})^{-1} = e^{-\mathbf{A}z}, \tag{1.3.114}$$

valid for the matrix exponential function $e^{\mathbf{A}z}$, is not true, in general, for matrix α-exponential function $e_\alpha^{\mathbf{A}z}$,

$$(e_\alpha^{\mathbf{A}z})^{-1} \neq e_\alpha^{-\mathbf{A}z}. \tag{1.3.115}$$

If we define the norm $\|\mathbf{A}\|$ of the matrix $\mathbf{A}$ with elements $a_{jk} \in \mathbb{R}$ $(j, k = 1, \ldots, n)$ by

$$\|\mathbf{A}\| = \max_{j,k \in \mathbb{N}} |a_{jk}| \tag{1.3.116}$$

then, from (1.3.110), we derive the estimate for the norm of $e_\alpha^{\mathbf{A}z}$. For any fixed $z \in \mathbb{C}$, the following relation holds

$$\|e_\alpha^{\mathbf{A}z}\| \leq \sum_{k=0}^{\infty} \|\mathbf{A}\|^k \frac{|z|^{\Re e(\alpha)k}}{|\Gamma((k+1)\alpha)|}. \tag{1.3.117}$$

When $z = x > 0$ and $\alpha > 0$, the above formula takes the simpler form

$$\|e_\alpha^{\mathbf{A}x}\| \leq \sum_{k=0}^{\infty} \|\mathbf{A}\|^k \frac{x^{\alpha k}}{\Gamma((k+1)\alpha)}. \tag{1.3.118}$$

Corresponding properties can be proved for the other generalized matrix exponential functions $E_\alpha(\mathbf{A}z^\alpha)$.

Particularly important are the following properties of the functions $E_\alpha(\lambda(z-a)^\alpha)$ and $e_\alpha^{\lambda(z-a)}$:

$$\left({}^{\mathrm{C}}D_{a+}^{\alpha} E_\alpha(\lambda(z-a)^\alpha)\right)(x) = \lambda E_\alpha(\lambda(x-a)^\alpha) \tag{1.3.119}$$

and

$$\left({}^{\mathrm{RL}}D_{a+}^{\alpha} e_\alpha^{(\lambda(z-a))}\right)(x) = \lambda e_\alpha^{(\lambda(x-a))} \tag{1.3.120}$$

with $\alpha > 0$, $a \in \mathbb{R}$ and $\lambda \in \mathbb{C}$. Thus, the two generalized exponential functions are eigenfunctions of Caputo and Riemann-Liouville differential operators, respectively.

The behavior of $E_\alpha(-x)$ and e_α^{-x} for various values of α in the intervals $(0, 1)$ and $(1, 2)$ can be seen from the following figures.

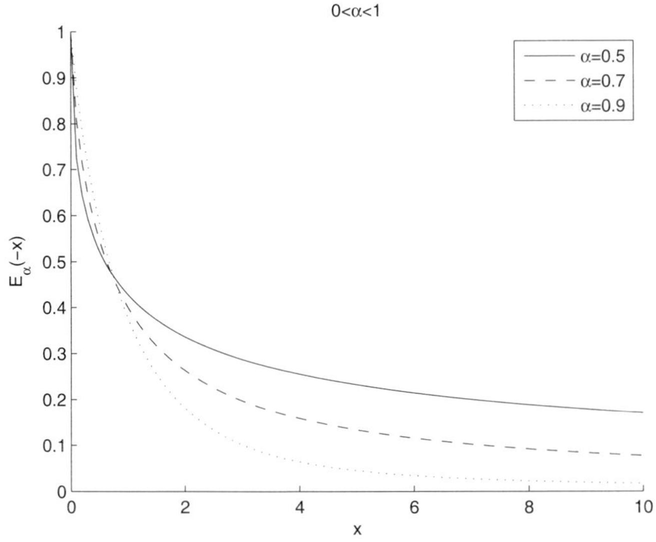

Fig. 1.1 Representation of $E_\alpha(-x)$ for some values of $\alpha \in (0, 1)$.

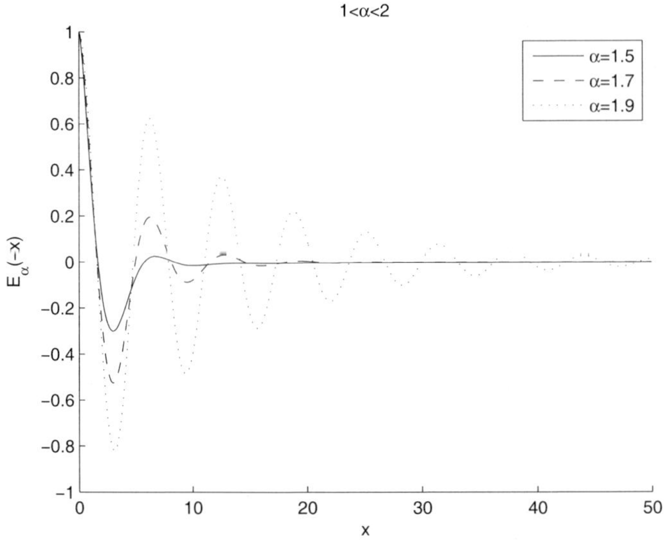

Fig. 1.2 Representation of $E_\alpha(-x)$ for some values of $\alpha \in (1, 2)$.

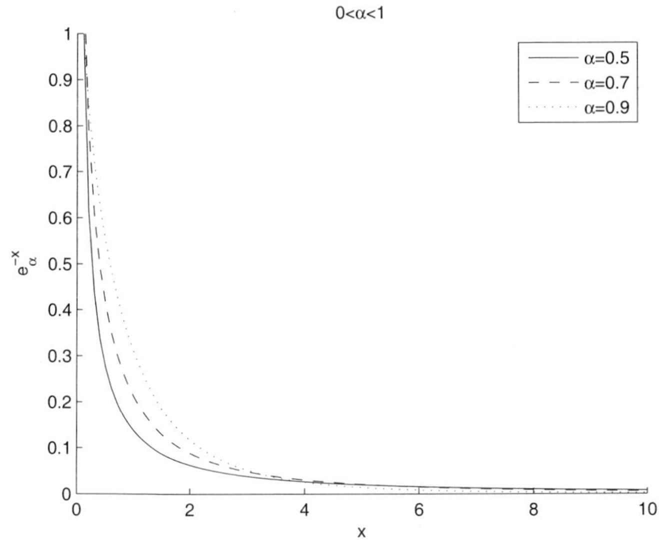

Fig. 1.3 Representation of e_α^{-x} for some values of $\alpha \in (0,1)$.

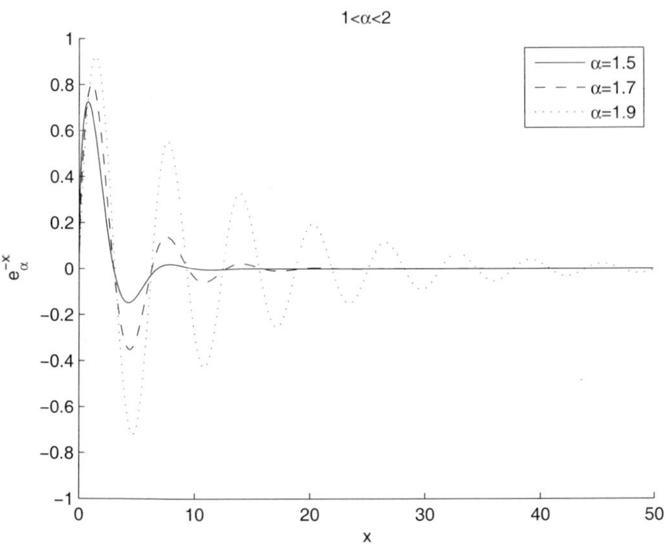

Fig. 1.4 Representation of e_α^{-x} for some values of $\alpha \in (1,2)$.

1.3.5 *Hadamard type fractional integrals and fractional derivatives*

In this section we present the definitions and some properties of the Hadamard type fractional integrals and fractional derivatives.

Let (a, b) $(0 \leq a < b \leq \infty)$ be a finite or infinite interval of the half-axis $\mathbb{R}^+$, and let $\Re e(\alpha) > 0$ and $\mu \in \mathbb{C}$. We consider the left-sided and right-sided integrals of fractional order $\alpha \in \mathbb{C}$ $(\Re e(\alpha) > 0)$ defined by

$$\left({}^{\mathrm{H}}I_{a+}^{\alpha} f\right)(x) = \frac{1}{\Gamma(\alpha)} \int_a^x \left(\log \frac{x}{t}\right)^{\alpha-1} \frac{f(t)dt}{t} \quad (a < x < b) \tag{1.3.121}$$

and

$$\left({}^{\mathrm{H}}I_{b-}^{\alpha} f\right)(x) = \frac{1}{\Gamma(\alpha)} \int_x^b \left(\log \frac{t}{x}\right)^{\alpha-1} \frac{f(t)dt}{t} \quad (a < x < b), \tag{1.3.122}$$

respectively. When $a = 0$ and $b = \infty$, these relations are given by

$$\left({}^{\mathrm{H}}I_{0+}^{\alpha} f\right)(x) = \frac{1}{\Gamma(\alpha)} \int_0^x \left(\log \frac{x}{t}\right)^{\alpha-1} \frac{f(t)dt}{t} \quad (x > 0) \tag{1.3.123}$$

and

$$\left({}^{\mathrm{H}}I_{-}^{\alpha} f\right)(x) = \frac{1}{\Gamma(\alpha)} \int_x^\infty \left(\log \frac{t}{x}\right)^{\alpha-1} \frac{f(t)dt}{t} \quad (x > 0). \tag{1.3.124}$$

More general fractional integrals than those in (1.3.123) and (1.3.124) are defined by

$$\left({}^{\mathrm{H}}I_{0+,\mu}^{\alpha} f\right)(x) = \frac{1}{\Gamma(\alpha)} \int_0^x \left(\frac{t}{x}\right)^{\mu} \left(\log \frac{x}{t}\right)^{\alpha-1} \frac{f(t)dt}{t} \quad (x > 0) \tag{1.3.125}$$

and

$$\left({}^{\mathrm{H}}I_{-,\mu}^{\alpha} f\right)(x) = \frac{1}{\Gamma(\alpha)} \int_x^\infty \left(\frac{x}{t}\right)^{\mu} \left(\log \frac{t}{x}\right)^{\alpha-1} \frac{f(t)dt}{t} \quad (x > 0) \tag{1.3.126}$$

with $\mu \in \mathbb{C}$.

The integral in (1.3.123) was introduced by Hadamard [277]. Therefore, the integrals (1.3.121), (1.3.122) and (1.3.123), (1.3.124) are often referred to as the Hadamard fractional integrals of order α. The more general integrals (1.3.125) and (1.3.126), introduced by Butzer *et al.* [126], are called the Hadamard type fractional integrals of order α.

The left- and right-sided Hadamard fractional derivatives of order $\alpha \in \mathbb{C}$ ($\Re e(\alpha) \geq 0$) on (a, b) are defined by

$$\begin{aligned}\left({}^{\mathrm{H}}D^{\alpha}_{a+}y\right)(x) &= \delta^n \left({}^{\mathrm{H}}I^{n-\alpha}_{a+}y\right)(x)\\ &= \left(x\frac{d}{dx}\right)^n \frac{1}{\Gamma(n-\alpha)} \int_a^x \left(\log\frac{x}{t}\right)^{n-\alpha+1} \frac{y(t)dt}{t}\end{aligned} \tag{1.3.127}$$

and

$$\begin{aligned}\left({}^{\mathrm{H}}D^{\alpha}_{b-}y\right)(x) &= (-\delta)^n \left({}^{\mathrm{H}}I^{n-\alpha}_{b-}y\right)(x)\\ &= \left(-x\frac{d}{dx}\right)^n \frac{1}{\Gamma(n-\alpha)} \int_x^b \left(\log\frac{t}{x}\right)^{n-\alpha+1} \frac{y(t)dt}{t}\end{aligned} \tag{1.3.128}$$

for $a < x < b$, respectively, where $n = -[-\Re e(\alpha)]$ and $\delta = xD$ ($D = d/dx$). When $a = 0$ and $b = \infty$, we have

$$\left({}^{\mathrm{H}}D^{\alpha}_{0+}y\right)(x) = \delta^n \left({}^{\mathrm{H}}I^{n-\alpha}_{0+}y\right)(x) \quad (x > 0)\,; \tag{1.3.129}$$

$$\left({}^{\mathrm{H}}D^{\alpha}_{-}y\right)(x) = (-\delta)^n \left({}^{\mathrm{H}}I^{n-\alpha}_{-}y\right)(x) \quad (x > 0)\,. \tag{1.3.130}$$

The Hadamard type fractional derivatives of order α with $\mu \in \mathbb{C}$, more general than those in (1.3.129) and (1.3.130), are defined for $\Re e(\alpha) \geq 0$ by

$$\left({}^{\mathrm{H}}D^{\alpha}_{0+,\mu}y\right)(x) = x^{-\mu}\delta^n x^{\mu} \left({}^{\mathrm{H}}I^{n-\alpha}_{0+,\mu}y\right)(x) \quad (x > 0)\,; \tag{1.3.131}$$

$$\left({}^{\mathrm{H}}D^{\alpha}_{-,\mu}y\right)(x) = x^{\mu}(-\delta)^n x^{-\mu} \left({}^{\mathrm{H}}I^{n-\alpha}_{-,\mu}y\right)(x) \quad (x > 0)\,. \tag{1.3.132}$$

The Hadamard type operators ${}^{\mathrm{H}}I^0_{0+,\mu}$, ${}^{\mathrm{H}}D^0_{0+,\mu}$ and ${}^{\mathrm{H}}I^0_{-,\mu}$, ${}^{\mathrm{H}}D^0_{-,\mu}$ can be defined as the identity operator

$${}^{\mathrm{H}}I^0_{0+,\mu}f \equiv {}^{\mathrm{H}}D^0_{0+,\mu}f = f, \qquad {}^{\mathrm{H}}I^0_{-,\mu}f \equiv {}^{\mathrm{H}}D^0_{-,\mu}f = f, \tag{1.3.133}$$

and in particular,

$${}^{\mathrm{H}}I^0_{0+}f \equiv {}^{\mathrm{H}}D^0_{0+}f = f, \qquad {}^{\mathrm{H}}I^0_{-}f \equiv {}^{\mathrm{H}}D^0_{-}f = f. \tag{1.3.134}$$

For $\Re e(\alpha) > 0$, $n = -[-\Re e(\alpha)]$, and $0 < a < b < \infty$, we have that $\left({}^{\mathrm{H}}D^{\alpha}_{a+}y\right)(x) = 0$ is valid if, and only if,

$$y(x) = \sum_{j=1}^{n} c_j \left(\log\frac{x}{a}\right)^{\alpha-j}, \tag{1.3.135}$$

where $c_j \in \mathbb{R}$ ($j = 1, \ldots, n$) are arbitrary constants. In particular, when $0 < \Re e(\alpha) \leq 1$, the relation $(\mathcal{D}^{\alpha}_{a+}y)(x) = 0$ holds if, and only if, $y(x) = c\left(\log\frac{x}{a}\right)^{\alpha-1}$ for any $c \in \mathbb{R}$.

On the other hand, the equality $\left({}^{\mathrm{H}}D_{b-}^{\alpha}y\right)(x)=0$ is valid if, and only if,

$$y(x)=\sum_{j=1}^{n}d_j\left(\log\frac{b}{x}\right)^{\alpha-j}, \tag{1.3.136}$$

where $d_j\in\mathbb{R}$ $(j=1,\ldots,n)$ are arbitrary constants. In particular, when $0<\Re e(\alpha)\le 1$, the relation $\left({}^{\mathrm{H}}D_{b-}^{\alpha}y\right)(x)=0$ holds if, and only if, $y(x)=d\left(\log\frac{b}{x}\right)^{\alpha-1}$ for any $d\in\mathbb{R}$.

It can also be directly verified that the Hadamard and Hadamard type fractional integrals and derivatives (1.3.123)–(1.3.126) and (1.3.129)–(1.3.132) of the power function x^{β} yield the same function, apart from a constant multiplication factor, that is, if $\Re e(\alpha)>0$, $\beta,\mu\in\mathbb{C}$, and $\Re e(\beta+\mu)>0$, then

$$\left({}^{\mathrm{H}}I_{0+,\mu}^{\alpha}t^{\beta}\right)(x)=(\mu+\beta)^{-\alpha}x^{\beta} \tag{1.3.137}$$

and

$$\left({}^{\mathrm{H}}D_{0+,\mu}^{\alpha}t^{\beta}\right)(x)=(\mu+\beta)^{\alpha}x^{\beta}. \tag{1.3.138}$$

On the other hand, if $\Re e(\beta-\mu)<0$, then

$$\left({}^{\mathrm{H}}I_{-,\mu}^{\alpha}t^{\beta}\right)(x)=(\mu-\beta)^{-\alpha}x^{\beta} \tag{1.3.139}$$

and

$$\left({}^{\mathrm{H}}D_{-,\mu}^{\alpha}t^{\beta}\right)(x)=(\mu-\beta)^{\alpha}x^{\beta}. \tag{1.3.140}$$

In particular, we have

$$\left({}^{\mathrm{H}}I_{0+}^{\alpha}t^{\beta}\right)(x)=\beta^{-\alpha}x^{\beta}\quad\text{and}\quad\left({}^{\mathrm{H}}D_{0+}^{\alpha}t^{\beta}\right)(x)=\beta^{\alpha}x^{\beta}\quad(\Re e(\beta)>0) \tag{1.3.141}$$

and

$$\left({}^{\mathrm{H}}I_{-}^{\alpha}t^{\beta}\right)(x)=(-\beta)^{-\alpha}x^{\beta}\quad\text{and}\quad\left({}^{\mathrm{H}}D_{-}^{\alpha}t^{\beta}\right)(x)=(-\beta)^{\alpha}x^{\beta}\quad(\Re e(\beta)<0). \tag{1.3.142}$$

The Hadamard and Hadamard type fractional integrals (1.3.121)–(1.3.126) satisfy the semigroup property

$${}^{\mathrm{H}}I_{a+}^{\alpha}\,{}^{\mathrm{H}}I_{a+}^{\beta}f={}^{\mathrm{H}}I_{a+}^{\alpha+\beta}f\quad\text{and}\quad{}^{\mathrm{H}}I_{b-}^{\alpha}\,{}^{\mathrm{H}}I_{b-}^{\beta}f={}^{\mathrm{H}}I_{b-}^{\alpha+\beta}f, \tag{1.3.143}$$

for $\Re e(\alpha)>0$, $\Re e(\beta)>0$ and $0<a<b<\infty$.

If $\mu\in\mathbb{C}$, $a=0$ and $b=\infty$, then

$${}^{\mathrm{H}}I_{0+,\mu}^{\alpha}\,{}^{\mathrm{H}}I_{0+,\mu}^{\beta}f={}^{\mathrm{H}}I_{0+,\mu}^{\alpha+\beta}f \tag{1.3.144}$$

and

$$ {}^{\mathrm{H}}I^{\alpha}_{-,\mu}\,{}^{\mathrm{H}}I^{\beta}_{-,\mu}f = {}^{\mathrm{H}}I^{\alpha+\beta}_{-,\mu}f. \tag{1.3.145} $$

In particular, when $\mu = 0$,

$$ {}^{\mathrm{H}}I^{\alpha}_{0+}\,{}^{\mathrm{H}}I^{\beta}_{0+}f = {}^{\mathrm{H}}I^{\alpha+\beta}_{0+}f; \qquad {}^{\mathrm{H}}I^{\alpha}_{-}\mathcal{J}^{\beta}_{-}f = {}^{\mathrm{H}}I^{\alpha+\beta}_{-}f. \tag{1.3.146} $$

Now we give the properties of compositions between the operators of fractional differentiation (1.3.127)–(1.3.132) and fractional integration (1.3.121)–(1.3.126).

If $\alpha \in \mathbb{C}$ and $\beta \in \mathbb{C}$ are such that $\mathfrak{Re}(\alpha) > \mathfrak{Re}(\beta) > 0$, then

$$ {}^{\mathrm{H}}D^{\beta}_{a+}\,{}^{\mathrm{H}}I^{\alpha}_{a+}f = {}^{\mathrm{H}}I^{\alpha-\beta}_{a+}f \ \text{ and } \ {}^{\mathrm{H}}D^{\beta}_{b-}\,{}^{\mathrm{H}}I^{\alpha}_{b-}f = {}^{\mathrm{H}}I^{\alpha-\beta}_{b-}f \tag{1.3.147} $$

when $0 < a < b < \infty$. In particular, if $\beta = m \in \mathbb{N}$, then

$$ {}^{\mathrm{H}}D^{m}_{a+}\,{}^{\mathrm{H}}I^{\alpha}_{a+}f = {}^{\mathrm{H}}I^{\alpha-m}_{a+}f \ \text{ and } \ {}^{\mathrm{H}}D^{m}_{b-}\,{}^{\mathrm{H}}I^{\alpha}_{b-}f = {}^{\mathrm{H}}I^{\alpha-m}_{b-}f. \tag{1.3.148} $$

If $\mu \in \mathbb{C}$, $a = 0$ and $b = \infty$, then

$$ {}^{\mathrm{H}}D^{\beta}_{0+,\mu}\,{}^{\mathrm{H}}I^{\alpha}_{0+,\mu}f = {}^{\mathrm{H}}I^{\alpha-\beta}_{0+,\mu}f \tag{1.3.149} $$

and

$$ {}^{\mathrm{H}}D^{\beta}_{-,\mu}\,{}^{\mathrm{H}}I^{\alpha}_{-,\mu}f = {}^{\mathrm{H}}I^{\alpha-\beta}_{-,\mu}f. \tag{1.3.150} $$

In particular, if $\beta = m \in \mathbb{N}$, then

$$ {}^{\mathrm{H}}D^{m}_{0+,\mu}\,{}^{\mathrm{H}}I^{\alpha}_{0+,\mu}f = {}^{\mathrm{H}}I^{\alpha-m}_{0+,\mu}f \tag{1.3.151} $$

and

$$ {}^{\mathrm{H}}D^{m}_{,\mu}\,{}^{\mathrm{H}}I^{\alpha}_{,\mu}f = {}^{\mathrm{H}}I^{\alpha-m}_{-,\mu}f. \tag{1.3.152} $$

while, when $\mu = 0$ and $m \in \mathbb{N}$,

$$ {}^{\mathrm{H}}D^{\beta}_{0+}\,{}^{\mathrm{H}}I^{\alpha}_{0+}f = {}^{\mathrm{H}}I^{\alpha-\beta}_{0+}f; \qquad {}^{\mathrm{H}}D^{\beta}_{-}\,{}^{\mathrm{H}}I^{\alpha}_{-}f = {}^{\mathrm{H}}I^{\alpha-\beta}_{-}f \tag{1.3.153} $$

and

$$ {}^{\mathrm{H}}D^{m}_{0+}\,{}^{\mathrm{H}}I^{\alpha}_{0+}f = {}^{\mathrm{H}}I^{\alpha-m}_{0+}f; \qquad {}^{\mathrm{H}}D^{m}_{-}\,{}^{\mathrm{H}}I^{\alpha}_{-}f = {}^{\mathrm{H}}I^{\alpha-m}_{-}f. \tag{1.3.154} $$

The Hadamard and Hadamard type fractional derivatives (1.3.127), (1.3.128) and (1.3.131), (1.3.132) are operators inverse to the corresponding fractional integrals (1.3.121), (1.3.122) and (1.3.125), (1.3.126), that is, if $\mathfrak{Re}(\alpha) > 0$ and $0 < a < b < \infty$ then

$$ {}^{\mathrm{H}}D^{\alpha}_{a+}\,{}^{\mathrm{H}}I^{\alpha}_{a+}f = f \ \text{ and } \ {}^{\mathrm{H}}D^{\alpha}_{b-}\,{}^{\mathrm{H}}I^{\alpha}_{b-}f = f, \tag{1.3.155} $$

whereas if $\Re e(\alpha) > 0$, $\mu \in \mathbb{C}$, $a = 0$ and $b = \infty$, then

$$ {}^{\mathrm{H}}D^{\alpha}_{0+,\mu}\,{}^{\mathrm{H}}I^{\alpha}_{0+,\mu} f = f \quad \text{and} \quad {}^{\mathrm{H}}D^{\alpha}_{-,\mu}\,{}^{\mathrm{H}}I^{\alpha}_{-,\mu} f = f. \tag{1.3.156} $$

In particular, if $\mu = 0$, then

$$ {}^{\mathrm{H}}D^{\alpha}_{0+}\,{}^{\mathrm{H}}I^{\alpha}_{0+} f = f \quad \text{and} \quad {}^{\mathrm{H}}D^{\alpha}_{-}\,{}^{\mathrm{H}}I^{\alpha}_{-} f = f. \tag{1.3.157} $$

The following property yields the formula for the composition of the fractional differentiation operator ${}^{\mathrm{H}}D^{\alpha}_{a+}$ with the fractional integration operator ${}^{\mathrm{H}}I^{\alpha}_{a+}$.

$$ ({}^{\mathrm{H}}I^{\alpha}_{a+}\,{}^{\mathrm{H}}D^{\alpha}_{a+} y)(x) = y(x) - \sum_{k=1}^{n} \frac{(\delta^{n-k}({}^{\mathrm{H}}I^{n-\alpha}_{a+} y))(a)}{\Gamma(\alpha - k + 1)} \left(\log \frac{x}{a}\right)^{\alpha - k}, \tag{1.3.158} $$

for $\Re e(\alpha) > 0$, $n = -[-\Re e(\alpha)]$ and $0 < a < b < \infty$.

It is known that function series admit a term-by-term Riemann-Liouville fractional integration and differentiation under certain conditions. Similar assertions are true for Hadamard-type fractional integration and differentiation operators ${}^{\mathrm{H}}I^{\alpha}_{0+,\mu}$ and ${}^{\mathrm{H}}D^{\alpha}_{0+,\mu}$.

Proposition 1.1. *Let* $\alpha \in \mathbb{C}$, $\mu > 0$, $l > 0$, *and let* $f(x) = \sum_{k=0}^{\infty} f_k(x)$, $f_k(x) \in C([0,l])$.

(1) *If* $\Re e(\alpha) > 0$ *and the series* $f(x) = \sum_{k=0}^{\infty} f_k(x)$ *is uniformly convergent on* $[0,l]$, *then its termwise Hadamard-type integration* ${}^{\mathrm{H}}I^{\alpha}_{0+,\mu}$ *is admissible*

$$ \left({}^{\mathrm{H}}I^{\alpha}_{0+,\mu} \sum_{k=0}^{\infty} f_k\right)(x) = \sum_{k=0}^{\infty} ({}^{\mathrm{H}}I^{\alpha}_{0+,\mu} f_k)(x) \quad (0 < x < l), \tag{1.3.159} $$

and the series $\sum_{k=0}^{\infty} ({}^{\mathrm{H}}I^{\alpha}_{0+,\mu} f_k)(x)$ *is also uniformly convergent on* $[0,l]$;

(2) *If* $\Re e(\alpha) \geq 0$ *and the series* $\sum_{k=0}^{\infty} f_k(x)$ *and* $\sum_{k=0}^{\infty} ({}^{\mathrm{H}}D^{\alpha}_{0+,\mu} f_k)(x)$ *are uniformly convergent on* $[\epsilon, l]$ $(\epsilon > 0)$, *then the former series admits termwise Hadamard-type fractional differentiation* ${}^{\mathrm{H}}D^{\alpha}_{0+,\mu}$ *by the formula*

$$ \left({}^{\mathrm{H}}D^{\alpha}_{0+,\mu} \sum_{k=0}^{\infty} f_k\right)(x) = \sum_{k=0}^{\infty} \left({}^{\mathrm{H}}D^{\alpha}_{0+,\mu} f_k\right)(x) \ (0 < x < l). \tag{1.3.160} $$

Proposition 1.2. *Let $\alpha \in \mathbb{C}$, $\mu > 0$, and let $f(x)$ be a convergent power series*

$$f(x) = \sum_{k=0}^{\infty} a_k x^k \quad (a_k \in \mathbb{C},\ k \in \mathbb{N}_0). \tag{1.3.161}$$

(1) *If $\Re(\alpha) > 0$, then the Hadamard-type integral ${}^{\mathrm{H}}I^{\alpha}_{0+,\mu}f$ is also represented by the convergent power series*

$$\left({}^{\mathrm{H}}I^{\alpha}_{0+,\mu}f\right)(x) = \sum_{k=0}^{\infty} (\mu + k)^{-\alpha} a_k x^k. \tag{1.3.162}$$

(2) *If $\Re(\alpha) \geq 0$, then the Hadamard-type derivative ${}^{\mathrm{H}}D^{\alpha}_{0+,\mu}f$ is also represented by the convergent power series*

$$\left({}^{\mathrm{H}}D^{\alpha}_{0+,\mu}f\right)(x) = \sum_{k=0}^{\infty} (\mu + k)^{\alpha} a_k x^k. \tag{1.3.163}$$

The radii of convergence of the series in (1.3.161), (1.3.162) *and* (1.3.163) *coincide.*

1.3.6 *Fractional integrals and fractional derivatives of a function with respect to another function*

In this section we present the definitions and some properties of the fractional integrals and fractional derivatives of a function f with respect to another function g. The Hadamard fractional integral and derivative are particular cases of these new operators.

Let (a, b) $(-\infty \leq a < b \leq \infty)$ be a finite or infinite interval of the real line $\mathbb{R}$ and $\Re(\alpha) > 0$. Also let $g(x)$ be an increasing and positive monotone function on $(a, b]$, having a continuous derivative $g'(x)$ on (a, b). The left- and right-sided fractional integrals of a function f with respect to another function g on $[a, b]$ are defined by

$$(I^{\alpha}_{a+;g}f)(x) = \frac{1}{\Gamma(\alpha)} \int_a^x \frac{g'(t)f(t)dt}{[g(x) - g(t)]^{1-\alpha}} \quad (x > a;\ \Re(\alpha) > 0) \tag{1.3.164}$$

and

$$(I^{\alpha}_{b-;g}f)(x) = \frac{1}{\Gamma(\alpha)} \int_x^b \frac{g'(t)f(t)dt}{[g(t) - g(x)]^{1-\alpha}} \quad (x < b;\ \Re(\alpha) > 0), \tag{1.3.165}$$

respectively.

When $a = 0$ and $b = \infty$, we shall use the following notations

$$(I^{\alpha}_{0+;g}f)(x) = \frac{1}{\Gamma(\alpha)} \int_0^x \frac{g'(t)f(t)dt}{[g(x)-g(t)]^{1-\alpha}} \quad (x > 0;\ \Re e(\alpha) > 0), \quad (1.3.166)$$

$$(I^{\alpha}_{-;g}f)(x) = \frac{1}{\Gamma(\alpha)} \int_x^{\infty} \frac{g'(t)f(t)dt}{[g(t)-g(x)]^{1-\alpha}} \quad (x > 0;\ \Re e(\alpha) > 0); \quad (1.3.167)$$

while, for $a = -\infty$ and $b = \infty$, we have

$$(I^{\alpha}_{+;g}f)(x) = \frac{1}{\Gamma(\alpha)} \int_{-\infty}^x \frac{g'(t)f(t)dt}{[g(x)-g(t)]^{1-\alpha}} \quad (x \in \mathbb{R};\ \Re e(\alpha) > 0), \quad (1.3.168)$$

$$(I^{\alpha}_{-;g}f)(x) = \frac{1}{\Gamma(\alpha)} \int_x^{\infty} \frac{g'(t)f(t)dt}{[g(t)-g(x)]^{1-\alpha}} \quad (x \in \mathbb{R};\ \Re e(\alpha) > 0). \quad (1.3.169)$$

Integrals (1.3.164) and (1.3.165) are called the g-Riemann-Liouville fractional integrals on a finite interval $[a,b]$, (1.3.166) and (1.3.167) the g-Liouville fractional integrals on a half-axis $\mathbb{R}^+$, while (1.3.168) and (1.3.169) are called the g-Liouville fractional integrals on the whole axis $\mathbb{R}$.

For $\Re e(\alpha) > 0$ and $\Re e(\beta) > 0$, we have

$$(I^{\alpha}_{a+;g}f_+)(x) = \frac{\Gamma(\beta)}{\Gamma(\alpha+\beta)}[g(x)-g(a)]^{\alpha+\beta-1}, \quad (1.3.170)$$

where $f_+(x) = [g(x)-g(a)]^{\beta-1}$. If $f_-(x) = [g(b)-g(x)]^{\beta-1}$, then

$$(I^{\alpha}_{b-;g}f_-)(x) = \frac{\Gamma(\beta)}{\Gamma(\alpha+\beta)}[g(b)-g(x)]^{\alpha+\beta-1}. \quad (1.3.171)$$

Moreover, if $\Re e(\alpha) > 0$ and $\lambda > 0$, then

$$(I^{\alpha}_{+;g}e^{\lambda g(t)})(x) = \lambda^{-\alpha}e^{\lambda g(x)} \quad (1.3.172)$$

and

$$(I^{\alpha}_{-;g}e^{-\lambda g(t)})(x) = \lambda^{-\alpha}e^{-\lambda g(x)}. \quad (1.3.173)$$

The semigroup property also holds, *i. e.*, if $\Re e(\alpha) > 0$ and $\Re e(\beta) > 0$, then the relations

$$(I^{\alpha}_{a+;g}I^{\beta}_{a+;g}f)(x) = (I^{\alpha+\beta}_{a+;g}f)(x); \quad (I^{\alpha}_{b-;g}I^{\beta}_{b-;g}f)(x) = (I^{\alpha+\beta}_{b-;g}f)(x) \quad (1.3.174)$$

and

$$(I^{\alpha}_{+;g}I^{\beta}_{+;g}f)(x) = (I^{\alpha+\beta}_{+;g}f)(x); \quad (I^{\alpha}_{-;g}I^{\beta}_{-;g}f)(x) = (I^{\alpha+\beta}_{-;g}f)(x) \quad (1.3.175)$$

hold for "sufficiently good" functions $f(x)$.

Let $g'(x) \neq 0$ ($-\infty \leq a < x < b \leq \infty$) and $\Re e(\alpha) \geq 0$ ($\alpha \neq 0$). Also let $n = -[-\Re e(\alpha)]$ and $D = d/dx$. The g-Riemann-Liouville and g-Liouville fractional derivatives of a function y with respect to g of order α ($\Re e(\alpha) \geq 0$; $\alpha \neq 0$), corresponding to the g-Riemann-Liouville and g-Liouville integrals in (1.3.164)–(1.3.165), (1.3.166)–(1.3.167), and (1.3.168)–(1.3.169), are defined by

$$(D^{\alpha}_{a+;g}y)(x) = \left(\frac{1}{g'(x)}D\right)^n (I^{n-\alpha}_{a+;g}y)(x) \quad (x < b), \tag{1.3.176}$$

$$(D^{\alpha}_{b-;g}y)(x) = \left(-\frac{1}{g'(x)}D\right)^n (I^{n-\alpha}_{b-;g}y)(x) \quad (x < b), \tag{1.3.177}$$

and

$$(D^{\alpha}_{+;g}y)(x) = \left(\frac{1}{g'(x)}D\right)^n (I^{n-\alpha}_{+;g}y)(x) \quad (x \in \mathbb{R}), \tag{1.3.178}$$

$$(D^{\alpha}_{-;g}y)(x) = \left(-\frac{1}{g'(x)}D\right)^n (I^{n-\alpha}_{-;g}y)(x) \quad (x \in \mathbb{R}), \tag{1.3.179}$$

respectively.

When $g(x) = x$, (1.3.176) and (1.3.177) coincide with the Riemann-Liouville fractional derivatives (1.3.3) and (1.3.4)

$$(D^{\alpha}_{a+;x}y)(x) = ({}^{\mathrm{RL}}D^{\alpha}_{a+}y)(x) \quad \text{and} \quad (D^{\alpha}_{b-;x}y)(x) = ({}^{\mathrm{RL}}D^{\alpha}_{b-}y)(x), \tag{1.3.180}$$

and (1.3.178) and (1.3.179) coincide with the Liouville fractional derivatives (1.3.63) and (1.3.64)

$$(D^{\alpha}_{+;x}y)(x) = ({}^{\mathrm{L}}D^{\alpha}_{+}y)(x) \quad \text{and} \quad (D^{\alpha}_{-;x}y)(x) = ({}^{\mathrm{L}}D^{\alpha}_{-}y)(x). \tag{1.3.181}$$

For $\Re e(\alpha) \geq 0$ ($\alpha \neq 0$) and $\Re e(\beta) > n-1$, the above derivatives have the properties

$$(D^{\alpha}_{a+;g}y_{+})(x) = \frac{\Gamma(\beta+1)}{\Gamma(\beta-\alpha+1)}[g(x)-g(a)]^{\beta-\alpha} \tag{1.3.182}$$

where $y_{+}(x) = [g(x)-g(a)]^{\beta}$, and

$$(D^{\alpha}_{b-;g}y_{-})(x) = \frac{\Gamma(\beta+1)}{\Gamma(\beta-\alpha+1)}[g(b)-g(x)]^{\beta-\alpha} \tag{1.3.183}$$

where $y_{-}(x) = [g(b)-g(x)]^{\beta}$.

For $\Re e(\alpha) \geq 0$ ($\alpha \neq 0$) and $\lambda > 0$, we have

$$(D^{\alpha}_{+;g}e^{\lambda g(t)})(x) = \lambda^{\alpha}e^{\lambda g(x)} \tag{1.3.184}$$

and

$$(D^{\alpha}_{-;g}e^{-\lambda g(t)})(x) = \lambda^{\alpha}e^{-\lambda g(x)}. \tag{1.3.185}$$

1.3.7 *Grünwald-Letnikov fractional derivatives*

In this section we give the definition of the Grünwald-Letnikov fractional derivatives and some of their properties. Such a fractional differentiation is based on a generalization of the classical differentiation of a function $y(x)$ of order $n \in \mathbb{N}$ via differential quotients,

$$y^{(n)}(x) = \lim_{h \to 0} \frac{(\Delta_h^n y)(x)}{h^n}. \tag{1.3.186}$$

Here $(\Delta_h^n y)(x)$ is a finite difference of order $n \in \mathbb{N}_0$ of a function $y(x)$ with a step $h \in \mathbb{R}$ and centered at the point $x \in \mathbb{R}$ defined by

$$(\Delta_h^n y)(x) = \sum_{k=0}^{n} (-1)^k \binom{n}{k} y(x - kh) \quad (x, h \in \mathbb{R};\ n \in \mathbb{N}) \tag{1.3.187}$$

and

$$(\Delta_h^0 f)(x) = f(x). \tag{1.3.188}$$

Property (1.3.186) is used to define a fractional derivative by directly replacing $n \in \mathbb{N}$ in (1.3.186) by $\alpha > 0$. For this, h^n is replaced by h^α, while the finite difference $(\Delta_h^n y)(x)$ is replaced by the difference $(\Delta_h^\alpha y)(x)$ of a fractional order $\alpha \in \mathbb{R}$ defined by the following series

$$(\Delta_h^\alpha y)(x) = \sum_{k=0}^{\infty} (-1)^k \binom{\alpha}{k} y(x - kh) \quad (x, h \in \mathbb{R};\ \alpha > 0), \tag{1.3.189}$$

where $\binom{\alpha}{k}$ are the binomial coefficients. When $h > 0$, the difference (1.3.189) is called left-sided difference, while for $h < 0$ it is called a right-sided difference .

The series in (1.3.189) converges absolutely and uniformly for each $\alpha > 0$ and for every bounded function $y(x)$.

The fractional difference $(\Delta_h^\alpha y)(x)$ has the following semigroup property

$$(\Delta_h^\alpha \Delta_h^\beta y)(x) = (\Delta_h^{\alpha+\beta} y)(x) \tag{1.3.190}$$

for $\alpha > 0$ and $\beta > 0$.

On the other hand, if $\alpha > 0$ and $y(x) \in L_1(\mathbb{R})$, then the Fourier transform (1.1.1) of Δ_h^α is given by

$$(\mathcal{F}\Delta_h^\alpha y)(\kappa) = \left(1 - e^{i\kappa h}\right)^\alpha (\mathcal{F}y)(\kappa). \tag{1.3.191}$$

Following (1.3.186), the left- and right-sided Grünwald-Letnikov derivatives $y_+^{\alpha)}(x)$ and $y_-^{\alpha)}(x)$ are defined by

$$y_+^{\alpha)}(x) = \lim_{h\to+0} \frac{(\Delta_h^\alpha y)(x)}{h^\alpha} \quad (\alpha > 0) \tag{1.3.192}$$

and

$$y_-^{\alpha)}(x) = \lim_{h\to+0} \frac{(\Delta_{-h}^\alpha y)(x)}{h^\alpha} \quad (\alpha > 0), \tag{1.3.193}$$

respectively. These constructions coincide with the Marchaud fractional derivatives for $y(x) \in L_p(\mathbb{R})$ $(1 \le p < \infty)$.

Then, by analogy with (1.3.192) and (1.3.193), the left- and right-sided Grünwald-Letnikov fractional derivatives of order $\alpha > 0$ on a finite interval $[a, b]$ are defined by

$$y_{a+}^{\alpha)}(x) = \lim_{h\to+0} \frac{(\Delta_{h,a+}^\alpha y)(x)}{h^\alpha} \tag{1.3.194}$$

and

$$y_{b-}^{\alpha)}(x) = \lim_{h\to+0} \frac{(\Delta_{h,b-}^\alpha y)(x)}{h^\alpha}, \tag{1.3.195}$$

respectively, where

$$(\Delta_{h,a+}^\alpha y)(x) = \sum_{k=0}^{[\frac{x-a}{h}]} (-1)^k \binom{\alpha}{k} y(x-kh) \quad (x \in \mathbb{R};\ \alpha, h > 0) \tag{1.3.196}$$

and

$$(\Delta_{h,b-}^\alpha y)(x) = \sum_{k=0}^{[\frac{b-x}{h}]} (-1)^k \binom{n}{k} y(x-kh) \quad (x, \in \mathbb{R};\ \alpha, h > 0). \tag{1.3.197}$$

Such Grünwald-Letnikov fractional derivatives coincide with the Marchaud fractional derivatives, for sufficiently well-behaved functions, and can be represented in the form

$$y_{a+}^{\alpha)}(x) = \frac{y(x)}{\Gamma(1-\alpha)(x-a)^\alpha} + \frac{\alpha}{\Gamma(1-\alpha)} \int_a^x \frac{y(x)-y(t)}{(x-t)^{1+\alpha}} dt \quad (0 < \alpha < 1) \tag{1.3.198}$$

and

$$y_{b-}^{\alpha)}(x) = \frac{y(x)}{\Gamma(1-\alpha)(b-x)^\alpha} + \frac{\alpha}{\Gamma(1-\alpha)} \int_x^b \frac{y(x)-y(t)}{(t-x)^{1+\alpha}} dt \quad (0 < \alpha < 1). \tag{1.3.199}$$

1.3.8 *Variable-order fractional derivatives*

All the various notions of fractional order operators that we have introduced so far share the property that the order of the operator is constant, *i. e.* it does not depend on the free variable. This behaviour is not always appropriate; in some applications one would like to have the feature of an order that depends on the free variable. There are a number of different possibilities to realize this. Here we mention the concept proposed by Coimbra [156]; alternative approaches are discussed in [616]:

The Coimbra fractional derivative of variable order $\alpha(x)$ is defined by

$$\begin{aligned} &{}^{\mathsf{Co}}D_a^{\alpha(x)} y(x) \\ &= \frac{1}{\Gamma(1-\alpha(x))} \int_a^x (x-t)^{-\alpha(x)} Dy(t)dt + \frac{y(a^+) - y(a^-)}{\Gamma(1-\alpha(x))} x^{-\alpha(x)} \end{aligned} \tag{1.3.200}$$

if $0 \le \alpha(x) < 1$ for all x in the interval of interest.

One should note that this definition permits arbitrary functions $\alpha(x)$ as long as their function values remain in the indicated interval; specifically, it is allowed (and has been discussed, *e. g.*, in Section 3 of [156]) to choose $\alpha(x) = \phi(y(x))$ with some function ϕ, so that the order of the differential operator depends on the function that is being differentiated.

Clearly, in the special case where α is a constant function and y is continuous at the point a, the Coimbra derivative reduces to the derivative of Caputo's type.

Chapter 2

A Survey of Numerical Methods for the Solution of Ordinary and Partial Fractional Differential Equations

When working with problems stemming from "real-world" applications, it is only rarely possible to evaluate the solution of a given fractional differential equation in closed form, and even if such an analytic solution is available, it is typically too complicated to be used in practice. Therefore it is indispensable to have a number of numerical algorithms at hand so that one is able to compute numerical solutions with a sufficient accuracy in reasonable time. Thus, this and the following chapter will be devoted to a study of such algorithms. To be precise, in this chapter we will give a survey of the standard numerical methods that should give the reader an impression of what he or she may expect from today's state-of-the-art algorithms. In Chapter 3 we will then look at the most important algorithms more closely, giving a detailed account of their respective strengths and weaknesses.

Since almost all numerical methods for fractional differential equations are in some sense based on the approximation of fractional differential or integral operators by appropriate formulas, we shall begin our presentation in Section 2.1 with a look at this problem. The subsequent sections will be devoted to numerical methods for fractional ordinary differential equations with an emphasis on initial value problems; specifically Section 2.2 will deal with what we call *direct methods*, *i. e.* methods where the numerical discretization scheme is applied directly to the differential operator appearing in the equation under consideration, whereas in Section 2.3 we look at *indirect methods* where we apply some analytic manipulation to the differential equation before the numerical work begins. Then, Section 2.4 gives an overview of a particularly important class of methods that are based on a generalization of classical methods for first-order equations, namely the so-called *linear multistep methods.* In a similar spirit, the extension of *spectral methods* to the fractional setting is discussed in Section 2.5. A brief survey of other methods is contained in Section 2.6. (It should be noted here that the classes of algorithms described in these sections are not disjoint; some methods fall into more than one category.) Section 2.7 is devoted to a study of numerical methods for a slightly different class of problems, the so-called terminal value problems. In Section 2.8 we then turn our attention towards formulas for the treatment of initial value problems for equations containing more than one differential operator. Finally, in Section 2.9, we will extend the concepts that we have established to the problem of fractional partial differential equations.

For the sake of simplicity, we shall always assume here and in Chapter 3 that the fractional operators are taken with starting point 0. This will allow us to drop the corresponding index from our notation, thus enhancing the readability of the formulas. Moreover, the parameter α that will be used to denote the order of the differential and integral operators under consideration will always satisfy the relation $\alpha > 0$ (and, unless stated otherwise, also $\alpha \notin \mathbb{N}$).

Finally we note that we shall only discuss the case of scalar problems explicitly. The extension of the numerical schemes described below to the vector-valued setting (*i. e.* to systems of differential equations) is straightforward and does not introduce any substantial difficulties. It is sufficient to interpret the scalar solutions as appropriate vectors.

2.1 The Approximation of Fractional Differential and Integral Operators

Let us recall from the introductory chapter the fractional integral in the sense of Riemann and Liouville, defined in Eqs. (1.3.8) and (1.3.1), *viz.*

$$J^{\alpha} f(x) = \frac{1}{\Gamma(\alpha)} \int_0^x (x-t)^{\alpha-1} f(t) dt. \qquad (2.1.1)$$

Obviously, J^{α} is an integral operator. The Riemann-Liouville fractional derivative from Eqs. (1.3.9), (1.3.3) and (1.3.1), *i. e.*

$$^{\mathrm{RL}}D^{\alpha} f(x) = \frac{d^n}{dx^n} J^{n-\alpha} f(x), \qquad n = \lceil \alpha \rceil, \qquad (2.1.2)$$

where $\lceil \cdot \rceil$ denotes the ceiling function that rounds up to the nearest integer, is at first sight a combination of a classical differential operator and an integral operator, but for suitable functions f it can also be written as a pure integral operator, namely

$$^{\mathrm{RL}}D^{\alpha} f(x) = \frac{1}{\Gamma(-\alpha)} \int_0^x (x-t)^{-\alpha-1} f(t) dt, \tag{2.1.3}$$

see, *e. g.*, [219]. Here, the integral has a strong singularity (its order being $1+\alpha > 1$), and it must be interpreted according to Hadamard's finite-part concept [278]. Note that formally the representation (2.1.3) of the Riemann-Liouville derivative can be obtained from the representation (2.1.1) of the Riemann-Liouville integral if we simply replace α by $-\alpha$. Additionally, we recall the definition of the Caputo derivative from Eq. (1.3.36), namely

$$^{\mathrm{C}}D^{\alpha} f(x) = J^{n-\alpha} D^n f(x), \qquad n = \lceil \alpha \rceil, \tag{2.1.4}$$

where D^n denotes the classical derivative of (integer) order n. As is well known, for a large class of functions this can be rewritten in the equivalent form

$$^{\mathrm{C}}D^{\alpha} f(x) = {}^{\mathrm{RL}}D^{\alpha}(f - T_{n-1}[f])(x) \tag{2.1.5}$$

that establishes the connection between the Riemann-Liouville and the Caputo derivatives. Here once again $n = \lceil \alpha \rceil$ and

$$T_{n-1}[f](u) = \sum_{k=0}^{n-1} \frac{u^k}{k!} f^{(k)}(0)$$

is the Taylor polynomial of degree $n-1$ for the function f, centered at the point 0. A useful consequence of Eq. (2.1.5) is that any numerical approximation method for Caputo derivatives immediately yields a corresponding algorithm for their Riemann-Liouville counterparts and vice versa. In the upcoming developments we therefore only need to describe the approaches for one of these two fractional derivatives explicitly; the other one can then be constructed easily with the help of Eq. (2.1.5)

2.1.1 *Methods based on quadrature theory*

It is evident from the representations (2.1.1), (2.1.3) and (2.1.4) that all fractional derivatives and integrals that we are interested in can be interpreted as integral operators in some sense. Therefore it seems to be very natural that the first approximation methods for such operators that we shall look at are based on principles from quadrature theory, *i. e.* the theory of numerical integration.

In the integrals appearing in Eqs. (2.1.1) and (2.1.3), the integrands always consist of two factors. The first factor is very easily described, it being the function $(x-t)^{\alpha-1}$ or $(x-t)^{-\alpha-1}$, respectively, and it has a singularity at one end point of the interval of integration. The other factor, $f(t)$, will typically have a much more complicated analytic representation, but is nonsingular in general (and, in many cases, possesses additional known smoothness properties). Under these circumstances, it is well known from the theory of numerical integration [174] that the idea of product integration is a useful approach. This means that we replace the smooth but complicated factor f by an approximation $\tilde{f}$, say, that is easily computed and for which it is also a simple matter to determine the quantity $J^\alpha \tilde{f}$ or ${}^{\mathrm{RL}}D^\alpha \tilde{f}$, respectively. Of course, in view of relation (2.1.5), the approach for the Riemann-Liouville derivative ${}^{\mathrm{RL}}D^\alpha$ can — as mentioned above — easily be modified to construct a corresponding method for the Caputo derivative ${}^{\mathrm{C}}D^\alpha$.

In practical applications, the idea of using piecewise polynomial interpolation with constant step size h to construct $\tilde{f}$ from f has proven to be very useful. We shall restrict our attention to the two most important special cases, namely piecewise polynomials of degree 0 (*i. e.* step functions) and degree 1 (polygons). For the discretization on the fundamental interval $[0, b]$ we will use the grid points $x_j = jh$, $j = 1, 2, \ldots, N$ where, for the sake of simplicity, we assume h to be chosen such that $b = x_N = Nh$ with some integer N.

For the Riemann-Liouville integral J^α, the discretization of f by step functions has been investigated in a detailed way in [198]. The main results were that the resulting approximation, the so-called *product rectangle formula*, can be written in the form

$$J^\alpha[f](x_k) \approx {}_hJ^\alpha_{\mathrm{Re}}[f](x_k) = h^\alpha \sum_{j=0}^{k-1} b_{j,k} f(x_j) \tag{2.1.6}$$

where

$$b_{j,k} = \frac{1}{\Gamma(1+\alpha)} \left((k-j)^\alpha - (k-1-j)^\alpha\right) \tag{2.1.7}$$

and that the approximation quality of this scheme can be described as follows (see Theorem 2.4 of [198] where, in particular, details of the proof may be found).

Theorem 2.1.

(a) *Let* $f \in C^1[0,b]$. *Then,*

$$|J^\alpha[f](x_k) - {}_hJ^\alpha_{\mathrm{Re}}[f](x_k)| \leq \frac{1}{\alpha} \sup_{x\in[0,b]} |f'(x)| x_k^\alpha h.$$

(b) *Let* $f(x) = x^p$ *for some* $p \in (0,1)$. *Then,*

$$|J^\alpha[f](x_k) - {}_hJ^\alpha_{\mathrm{Re}}[f](x_k)| \leq C^{\alpha,p}_{\mathrm{Re}} x_k^{\alpha+p-1} h$$

where $C^{\alpha,p}_{\mathrm{Re}}$ *is a constant that depends only on* α *and* p.

We thus conclude

Corollary 2.1.

(a) *Let* $f \in C^1[0,b]$. *Then,*

$$|J^\alpha[f](x_k) - {}_hJ^\alpha_{\mathrm{Re}}[f](x_k)| = O(h)$$

uniformly on $[0,b]$.

(b) *Let* $f(x) = x^p$ *for some* $p \in (0,1)$. *Then,*

$$|J^\alpha[f](x_k) - {}_hJ^\alpha_{\mathrm{Re}}[f](x_k)| = O(h)$$

pointwise on $[0,b]$. *The convergence is uniform if and only if* $\alpha + p \geq 1$.

For the engineer, physicist, *etc.* who wants to solve a concrete problem involving fractional operators numerically, an $O(h)$ convergence order is often not good enough. It is therefore useful to know that a slightly more sophisticated method can be employed to improve this rate of convergence. The most natural approach is then to use polygons (continuous piecewise polynomials of degree 1) instead of step functions which leads us to the *product trapezoidal method*

$$J^\alpha[f](x_k) \approx {}_hJ^\alpha_{\mathrm{Tr}}[f](x_k) = h^\alpha \sum_{j=0}^{k} a_{j,k} f(x_j) \tag{2.1.8}$$

where

$$a_{j,k} = \frac{1}{\Gamma(2+\alpha)} \times \begin{cases} \left((k-1)^{1+\alpha} - (k-\alpha-1)k^{\alpha}\right) & \text{if } j=0, \\ \left((k-j+1)^{1+\alpha} + (k-j-1)^{1+\alpha} \right. \\ \qquad \left. -2(k-j)^{1+\alpha}\right) & \text{if } 1 \le j \le k-1, \\ 1 & \text{if } j=k. \end{cases} \tag{2.1.9}$$

whose approximation quality is described in the following results taken from Theorem 2.5 of [198].

Theorem 2.2.

(a) *Let $f \in C^2[0,T]$. Then there is a constant C^{α}_{Tr} depending only on α such that*

$$|J^{\alpha}[f](x_k) - {}_hJ^{\alpha}_{\mathrm{Tr}}[f](x_k)| \le C^{\alpha}_{\mathrm{Tr}} \sup_{x\in[0,b]} |f''(x)| x_k^{\alpha} h^2.$$

(b) *Let $f \in C^1[0,T]$ and assume that f' fulfils a Lipschitz condition of order μ for some $\mu \in (0,1)$. Then, there exist positive constants $B^{\alpha,\mu}_{\mathrm{Tr}}$ (depending only on α and μ) and $M(f,\mu)$ (depending only on f and μ) such that*

$$|J^{\alpha}[f](x_k) - {}_hJ^{\alpha}_{\mathrm{Tr}}[f](x_k)| \le B^{\alpha,\mu}_{\mathrm{Tr}} M(f,\mu) x_k^{\alpha} h^{1+\mu}.$$

(c) *Let $f(x) = x^p$ for some $p \in (0,2)$ and $\varrho = \min(2, p+1)$. Then,*

$$|J^{\alpha}[f](x_k) - {}_hJ^{\alpha}_{\mathrm{Tr}}[f](x_k)| \le C^{\alpha,p}_{\mathrm{Tr}} x_k^{\alpha+p-\varrho} h^{\varrho}$$

where $C^{\alpha,p}_{\mathrm{Tr}}$ is a constant that depends only on α and p.

The idea of piecewise linear interpolation that is the background of the product trapezoidal method (2.1.8) for the Riemann-Liouville integral has also been used successfully for the Riemann-Liouville derivative [181]. Not surprisingly, the resulting formula has the form

$${}^{\mathrm{RL}}D^{\alpha}[f](x_k) \approx {}^{\mathrm{RL}}_{h}D^{\alpha}_{\mathrm{Tr}}[f](x_k) = h^{-\alpha} \sum_{j=0}^{k} A_{j,k} f(x_j) \tag{2.1.10}$$

with

$$A_{j,k} = \frac{1}{\Gamma(2-\alpha)} \times \begin{cases} \left((k-1)^{1-\alpha} - (k+\alpha-1)k^{-\alpha}\right) & \text{if } j = 0, \\ \left((k-j+1)^{1-\alpha} + (k-j-1)^{1-\alpha} \right. \\ \qquad \left. -2(k-j)^{1-\alpha}\right) & \text{if } 1 \le j \le k-1, \\ 1 & \text{if } j = k. \end{cases} \tag{2.1.11}$$

In other words, the product trapezoidal approximation for the Riemann-Liouville derivative, given by Eqs. (2.1.10) and (2.1.11), can be obtained from the corresponding expressions for the Riemann-Liouville integral, *viz.* Eqs. (2.1.8) and (2.1.9), by formally replacing α by $-\alpha$.

For the approximation quality of this latter method, we have from Lemma 2.2 of [181] (see also Theorem 2.3 of [182]):

Theorem 2.3. *Let $f \in C^2[0,T]$. Then there is a constant $\bar{C}^{\alpha}_{\mathrm{Tr}}$ depending only on α such that*

$$\left| {}^{\mathrm{RL}}D^{\alpha}[f](x_k) - {}^{\mathrm{RL}}_{h}D^{\alpha}_{\mathrm{Tr}}[f](x_k) \right| \le \bar{C}^{\alpha}_{\mathrm{Tr}} \sup_{x \subset [0,b]} |f''(x)| x_k^{-\alpha} h^2.$$

Since $x_k \ge x_1 = h$, we derive the uniform estimate

Corollary 2.2. *Let $f \in C^2[0,T]$. Then,*

$$\left| {}^{\mathrm{RL}}D^{\alpha}[f](x_k) - {}^{\mathrm{RL}}_{h}D^{\alpha}_{\mathrm{Tr}}[f](x_k) \right| = O(h^{2-\alpha})$$

uniformly for all k.

Thus we note that the quality of this approach deteriorates as α (the order of the differential operator) increases. In particular, we cannot expect any convergence at all if $\alpha \ge 2$. However, for $0 < \alpha < 1$ — the case that is relevant for the majority of the classical applications — we do observe quite good convergence properties.

Finally we remark that it is not difficult to generalize the concepts introduced above to the case that the mesh is not uniformly spaced.

2.1.2 *Grünwald-Letnikov methods*

Another very obvious approach for the discretization of fractional differential and integral operators is based on a straightforward generalization of concepts from classical calculus to the fractional case. Specifically, it is well known that an integer-order derivative can be written as a differential quotient, *viz.*

$$D^n f(x) = \lim_{h\to 0} \frac{1}{h^n} \sum_{k=0}^{n} (-1)^k \frac{\Gamma(n+1)}{\Gamma(k+1)\Gamma(n-k+1)} f(x-kh). \qquad (2.1.12)$$

Recalling that the Gamma function has a pole at the nonpositive integers, we deduce that the sum does not change if we replace the upper summation limit n by ∞: all terms that we add have the value zero, so the infinite series in fact collapses to the sum stated in (2.1.12). In the resulting formula, there is — at least from a formal standpoint — no obstacle that prevents us from replacing the integer number n by a real number $\alpha > 0$, thus giving

$$\lim_{h\to 0} \frac{1}{h^\alpha} \sum_{k=0}^{\infty} (-1)^k \frac{\Gamma(\alpha+1)}{\Gamma(k+1)\Gamma(\alpha-k+1)} f(x-kh), \qquad (2.1.13)$$

the classical *Grünwald-Letnikov derivative* [275, 365, 375] that we had already met in Eq. (1.3.192).

However there are now two substantial difficulties that we have introduced by going from the integer-order case (2.1.12) to the fractional case in (2.1.13): First, we now have a truly infinite series, and so we must discuss its convergence properties, and second we need to evaluate f at every point in $(-\infty, x)$ in order to compute the expression given in Eq. (2.1.13) whereas in the integer order case it was sufficient to deal with values of f in an arbitrarily small neighborhood of x only. The latter point is a major problem because when dealing with differential equations one usually wants to work with functions defined on a finite interval $[0, b]$, say, with the initial condition given at the point 0, and not on an interval that extends to $-\infty$. In order to remove this problem, we can use a very simple idea, namely we indeed restrict ourselves to functions defined on $[0, b]$ as desired and extend these functions to the complete interval $(-\infty, b]$ by setting $f(x) = 0$ for $x < 0$. Evidently, this means that once again the infinite series reduces to a finite sum as long as h is fixed and positive, but the number of summands grows in an unbounded way as $h \to 0$. This gives us the definition of the Grünwald-Letnikov differential operator that is appropriate in our context,

$$^{\mathtt{GL}}D^\alpha f(x) = \lim_{h\to 0} \frac{1}{h^\alpha} \sum_{k=0}^{\lfloor x/h \rfloor} (-1)^k \frac{\Gamma(\alpha+1)}{\Gamma(k+1)\Gamma(\alpha-k+1)} f(x-kh), \quad (2.1.14)$$

cf. Eq. (1.3.194) in combination with Eq. (1.3.196). It can be shown that this operator is closely related to the Riemann-Liouville derivative:

Theorem 2.4. *Let* $\alpha > 0$ *and* $f \in C^{\lceil \alpha \rceil}[0, b]$. *Then, for* $x \in (0, b]$,

$$^{\mathrm{GL}}D^{\alpha} f(x) = {}^{\mathrm{RL}}D^{\alpha} f(x).$$

For a proof of this result we refer to [219] or Theorem 20.6 and pp. 228–229 of [542].

From the representation (2.1.14), it is evident that the corresponding difference quotient with step length h,

$$^{\mathrm{GL}}_{\;h}D^{\alpha} f(x) = \frac{1}{h^{\alpha}} \sum_{k=0}^{\lfloor x/h \rfloor} (-1)^k \frac{\Gamma(\alpha+1)}{\Gamma(k+1)\Gamma(\alpha-k+1)} f(x-kh), \qquad (2.1.15)$$

is a candidate for an approximation of $^{\mathrm{RL}}D^{\alpha} f(x)$. We shall call this method the *Grünwald-Letnikov finite difference* of order α with step size h.

We note that, in the case $\alpha \in \mathbb{N}$, a formally simpler representation can be obtained because of the relation

$$\frac{\Gamma(\alpha+1)}{\Gamma(k+1)\Gamma(\alpha-k+1)} = \frac{\alpha!}{k!(\alpha-k)!} = \binom{\alpha}{k}.$$

For the sake of simplicity we shall use this binomial coefficient notation also in the case that $\alpha \notin \mathbb{N}$ in accordance with Eqs. (1.2.20) and (1.2.23).

It will turn out that the Grünwald-Letnikov method is a special case of the fractional linear multistep methods discussed in the next subsection. Therefore we will not discuss it any further here and refer to the general treatment below instead.

2.1.3 *Lubich's fractional linear multistep methods*

None of the methods described so far has exhibited very fast convergence. To overcome this difficulty we now describe a class of methods introduced by Lubich in a series of papers in the 1980s [384–387].

The starting point for Lubich's approach is a classical linear multistep method for first-order initial value problems of the form

$$y'(x) = f(x, y(x)), \qquad y(x_0) = y_0. \qquad (2.1.16)$$

We recall some of the background of these methods. This and more information may be found in many standard textbooks on this topic, *e. g.* in Chapter III of [280].

Given a uniform discretization

$$x_j = x_0 + jh, \quad j = 0, 1, 2, \ldots, N$$

of the basic interval $[x_0, b]$ with a prescribed value $N \in \mathbb{N}$ and $h = (b - x_0)/N$, a *linear multistep method* for the problem (2.1.16) takes the form

$$\sum_{k=-1}^{p} \alpha_k y_{m-k} = h \sum_{k=-1}^{p} \beta_k f(x_{m-k}, y_{m-k}) \tag{2.1.17}$$

with some real constants α_k and β_k, $k = -1, 0, 1, \ldots, p$. To be precise, it is called a $(p+1)$-step method if $\alpha_p \neq 0$ or $\beta_p \neq 0$. In the following discussion we shall always assume that at least one of these two conditions is satisfied. The method is explicit if $\beta_{-1} = 0$ and implicit otherwise. It should be noted that Eq. (2.1.17) requires the data $y_m, y_{m-1}, \ldots, y_{m-p}$ to compute the new value y_{m+1}. Thus one has to assume the starting values $y_0, y_1, \ldots y_p$ to be given (they would typically be computed by some other numerical scheme), and then the values y_{m+1}, $m = p, p+1, \ldots, N-1$, will be determined by solving Eq. (2.1.17).

In the investigation of such methods, two polynomials play a fundamental role, namely the *first characteristic polynomial*

$$\rho(\zeta) = \sum_{k=-1}^{p} \alpha_k \zeta^{p-k} \tag{2.1.18}$$

and the *second characteristic polynomial*

$$\sigma(\zeta) = \sum_{k=-1}^{p} \beta_k \zeta^{p-k}. \tag{2.1.19}$$

It is obvious that we may immediately compute the characteristic polynomials for any given linear multistep method, and we may also construct the multistep method if the characteristic polynomials are given. Thus, it is common to speak of *the linear multistep method* (ρ, σ) instead of using the representation (2.1.17). From ρ and σ we may compute another function that is important for the analysis, namely the *generating function*

$$\omega(\zeta) = \frac{\sigma(1/\zeta)}{\rho(1/\zeta)} \tag{2.1.20}$$

of the linear multistep method. For the two-step backward differentiation formula

$$\frac{3}{2}y_{m+1} - 2y_m + \frac{1}{2}y_{m-1} = hf(x_{m+1}, y_{m+1}),$$

for example, we obtain the characteristic polynomials

$$\rho(\zeta) = \frac{3}{2}\zeta^2 - 2\zeta + \frac{1}{2} \quad \text{and} \quad \sigma(\zeta) = \zeta^2$$

and the generating function

$$\omega(\zeta) = \frac{1}{\frac{3}{2} - 2\zeta + \frac{1}{2}\zeta^2} = \sum_{m=0}^{\infty} \omega_m \zeta^m$$

with

$$\omega_m = 1 - 3^{-m-1} \qquad (m = 0, 1, \ldots).$$

We want to apply such a method to the particularly simple initial value problem

$$y'(x) = f(x), \quad y(x_0) = 0,$$

whose solution obviously is

$$y(x) = \int_{x_0}^{x} f(\xi) d\xi, \quad x \in [x_0, b].$$

In this way we produce an approximation for y, *i. e.* for the primitive of f, on the given grid $\{x_j : j = 0, 1, \ldots, N\}$. Thus we are effectively constructing a quadrature formula. This quadrature formula can be written in a form that will be the basis for the extension of the concept to the fractional setting:

Lemma 2.1. *Consider the numerical solution of the initial value problem* $y'(x) = f(x)$, $y(x_0) = 0$, *on the interval* $[x_0, b]$ *with the linear* $(p+1)$*-step formula* (ρ, σ), *under the assumption that* $f(x) = 0$ *for* $x_0 \le x \le x_p$ *and using the exact initial values* $y_j = 0$ *for* $j = 0, 1, \ldots, p$. *This numerical solution can be written in the* convolution quadrature form

$$({}_hJf)(x_m) = y_m = h\sum_{j=0}^{m} \omega_{m-j} f(x_j), \quad m = p+1, p+2, \ldots, N,$$

where the convolution weights ω_m, $m = 0, 1, 2, \ldots$, *are the coefficients of the power series of the generating function* ω, *viz.*

$$\omega(\zeta) = \sum_{m=0}^{\infty} \omega_m \zeta^m.$$

This result follows from Lemma 2.1 of [382].

The generalization to the fractional setting is now a relatively simple matter. Indeed, to find an approximation for $J^\alpha f$ we proceed much as above; we only need to replace the coefficients of the power series of the generating function ω itself by the corresponding coefficients of the αth power of ω. To put it formally, we start from a classical multistep formula and use the function ω defined via the two characteristic polynomials ρ and σ as in Eq. (2.1.20). Then we write the power series expansion of ω^α, *viz.*

$$(\omega(\zeta))^\alpha = \sum_{m=0}^{\infty} \omega_m \zeta^m. \tag{2.1.21}$$

We stress that the coefficients ω_m in this representation depend on the order α of the fractional operator; however we have decided to follow the common standard of not mentioning this explicitly in our notation. Since α is kept fixed in most of the applications, this should not lead to any confusion. Based on this concept we can formulate a fractional generalization of Lemma 2.1.

Lemma 2.2. *For $0 < \alpha < 1$, consider the numerical solution of the initial value problem ${}^{\mathrm{C}}D^\alpha y(x) = f(x)$, $y(x_0) = 0$, on the interval $[x_0, b]$ with the fractional version of the linear $(p+1)$-step formula (ρ, σ), under the assumption that $f(x) = 0$ for $x_0 \le x \le x_p$ and using the exact initial values $y_j = 0$ for $j = 0, 1, \ldots, p$. This numerical solution can be written in the* convolution quadrature form

$$({}_hJ^\alpha f)(x_m) = y_m = h^\alpha \sum_{j=0}^{m} \omega_{m-j} f(x_j), \quad m = p+1, p+2, \ldots, N,$$

where the convolution weights ω_m, $m = 0, 1, 2, \ldots$, *are the coefficients of the power series of the αth power of the generating function ω as given in Eq. (2.1.21). If the underlying classical linear multistep method (ρ, σ) is of order r then the fractional version also satisfies*

$$y(x_m) - y_m = O(h^r).$$

The assumption that $f(x) = 0$ for $x_0 \le x \le x_p$ is somewhat strange. Its main purpose is to enforce a certain regular behavior of the exact solution near the point x_0. For such a regular function, this convolution quadrature is perfectly appropriate. However, as we shall see later, in a typical application we cannot expect this condition to be satisfied. Rather we have to deal with functions whose asymptotic behavior near x_0 is of the form

$$\sum_{j=0}^{J}\sum_{k-0}^{K} z_{jk}(x-x_0)^{j+k\alpha} + O((x-x_0)^S)$$

with certain positive integers J, K and S. (This follows from the general theory of the solutions of such equations developed by Miller and Feldstein [431] and extended by Lubich [383].) In such a case it turns out that the convolution quadrature cannot handle the terms with exponents $j+k\alpha \notin \mathbb{N}$ very accurately. Therefore we have to introduce additional terms to correct these flaws. Such an approach is possible; it leads to the following concept.

Definition 2.1. Let $0 < \alpha < 1$. An approximation to the fractional differential equation ${}^{\mathrm{C}}D^\alpha y(x) = f(x)$, $y(x_0) = 0$, in the form

$$({}_hJ^\alpha f)(x_m) = h^\alpha \sum_{j=0}^{m} \omega_{m-j} f(x_j) + h^\alpha \sum_{j=0}^{s} w_{mj} f(x_j), \qquad m = 0, 1, \ldots, N$$

with some fixed $s \in \mathbb{N}$ is called a *fractional convolution quadrature*. The weights ω_j are called the *convolution weights*, and the term

$${}_h\Omega^\alpha f(x_m) = h^\alpha \sum_{j=0}^{m} \omega_{m-j} f(x_j)$$

is the *convolution part* of ${}_hJ^\alpha$ with the corresponding *convolution quadrature error* ${}_hE^\alpha = J^\alpha - {}_h\Omega^\alpha$. The term

$${}_hS^\alpha f(x_m) = h^\alpha \sum_{j=0}^{s} w_{mj} f(x_j)$$

is called the *starting part* or *starting quadrature*, and the weights w_{mj} are the *starting weights*.

We have already seen above how we can find suitable convolution weights. The search for starting weights leading to highly accurate methods will be postponed until §2.4.

2.2 Direct Methods for Fractional Ordinary Differential Equations

We shall now apply the approximation methods introduced above in the context of the numerical solution of fractional differential equations. In this and the following two sections, we will look at ordinary differential equations of fractional order; Section 2.9 will then be devoted to an extension of the ideas to partial differential equations.

For the purpose of exposition of the methods for solving ordinary fractional differential equations, we believe a distinction of cases to be useful. Specifically, we shall classify the numerical methods under consideration into *direct* and *indirect* methods. The direct methods will be the topic of this section; we will discuss indirect methods in §2.3. Moreover we will look at *linear multistep methods* in §2.4. Many important special instances of the latter can actually be interpreted as either direct or indirect methods, but the general concept allows to describe a much larger class whose members may or may not fall into one of these categories.

2.2.1 *The basic idea*

So let us now start our investigations by introducing the direct methods. They are characterized by the fact that we take the given initial value problem

$$ {}^{\mathrm{C}}D^{\alpha} y(x) = f(x, y(x)), \qquad y^{(k)}(0) = y_0^{(k)} \quad (k = 0, 1, \ldots, \lceil \alpha \rceil - 1), \tag{2.2.1} $$

and apply one of the approximation algorithms of §2.1 directly to the operator ${}^{\mathrm{C}}D^{\alpha}$. Since we shall restrict our attention to linear schemes, *i. e.* methods of the form

$$ {}^{\mathrm{C}}D^{\alpha} y(x_n) \approx \sum_{k=0}^{n} a_{k,n} y(x_k) \tag{2.2.2} $$

(where, without loss of generality, the grid points are assumed to satisfy the relation $0 \le x_0 < x_1 < \cdots < x_n < T$), this gives rise to numerical approaches for the initial value problem (2.2.1) that have the form

$$ \sum_{k=0}^{n} a_{k,n} y_k = f(x_n, y_n), \qquad n = 0, 1, \ldots \tag{2.2.3} $$

which can be seen by combining Eqs. (2.2.1) and (2.2.2) and by replacing the unknown values $y(x_k)$ by their respective approximations y_k. In the next subsections, we shall study some important special cases of these algorithms. Unless explicitly stated otherwise, we will always assume that a solution of (2.2.1) is sought on some interval $[0, X]$, and that we have a uniform grid

$$x_j = \frac{j}{N} X \qquad (j = 0, 1, \dots, N) \tag{2.2.4}$$

on the basic interval $[0, X]$ with some preassigned integer N. The step size is denoted by $h = X/N$. Our job will then be to construct approximate solution values $y_k \approx y(x_k)$ at the grid points. Of course, we will use the given initial value y_0 at the point $x_0 = 0$, so actual work needs to be done for $k = 1, 2, \dots, N$.

2.2.2 *Quadrature-based direct methods*

We begin by looking at the method obtained by using a discretization of the differential operator based on quadrature theory, *i. e.* the formulas introduced in Eqs. (2.1.10) and (2.1.11). This leads to an algorithm of the form

$$h^{-\alpha} \sum_{j=0}^{k} A_{j,k} y_j = f(x_k, y_k) \qquad (k = 1, 2, \dots, N) \tag{2.2.5}$$

with the $A_{j,k}$ given in Eq. (2.1.11). Since $A_{k,k} = 1/\Gamma(2-\alpha)$, we may rewrite this identity in the form

$$y_k = \Gamma(2-\alpha) h^{\alpha} f(x_k, y_k) - \Gamma(2-\alpha) \sum_{j=0}^{k-1} A_{j,k} y_j. \tag{2.2.6}$$

The usual procedure for the computation of the numerical solution is to calculate y_1 first, then y_2, then y_3, *etc.*, until y_N is reached. In this procedure, the currently computed value is the only unknown in the equation; all other data have already been calculated in previous steps and hence are known. It is then natural to ask whether Eq. (2.2.6) can be solved for y_k. The following result states that the answer is positive provided that the step size is sufficiently small.

Theorem 2.5. *Assume that the function f on the right-hand side of the differential equation* (2.2.1) *is continuous and satisfies a Lipschitz condition* *with respect to its second variable with Lipschitz constant L. Moreover let $h < (\Gamma(2-\alpha)L)^{-1/\alpha}$. Then, Eq. (2.2.6) has a unique solution y_k for all k.*

The assumption on f is not unnatural; it is just the standard assumption imposed in the usual analytical theorems [192] that assert the existence and uniqueness of the solution of the initial value problem, which of course is a prerequisite of any numerical work.

Proof. For an arbitrary but fixed $k \in \{1, 2, \ldots, N\}$ let us define

$$\Phi(u) = \Gamma(2-\alpha)h^\alpha f(x_k, u) - \Gamma(2-\alpha)\sum_{j=0}^{k-1} A_{j,k}y_j.$$

From Eq. (2.2.6) it is then evident that y_k is a fixed point of the operator $\Phi : \mathbb{R} \to \mathbb{R}$. In view of the Lipschitz condition on f we deduce for every $u_1, u_2 \in \mathbb{R}$

$$\begin{aligned} |\Phi(u_1) - \Phi(u_2)| &= \Gamma(2-\alpha)h^\alpha|f(x_k, u_1) - f(x_k, u_2)| \\ &\leq \Gamma(2-\alpha)h^\alpha L|u_1 - u_2|. \end{aligned}$$

By our assumption on h we have $\Gamma(2-\alpha)h^\alpha L < 1$, *i. e.* the mapping Φ is contractive. Thus, by Banach's fixed point theorem, it has a unique fixed point, *i. e.* Eq. (2.2.6) has a unique solution y_k. □

As a consequence of this proof we see that we may actually find the required numerical solution y_k from Eq. (2.2.6) by means of the Picard iteration sequence

$$y_{k,\ell+1} = \Phi(y_k, \ell) = \Gamma(2-\alpha)\left(h^\alpha f(x_k, y_{k,\ell}) - \sum_{j=0}^{k-1} A_{j,k}y_j\right) \quad (\ell = 0, 1, \ldots)$$

that will converge to y_k as $\ell \to \infty$ for any starting value $y_{k,0}$. A default choice for this starting value that is useful in most cases is $y_{k,0} = y_{k-1}$.

Remark 2.1. In the limit case $\alpha \to 1$ this method reduces to the classical first-order backward differentiation formula (BDF1). Therefore it seems reasonable to call this method a fractional backward differentiation formula. However we shall see shortly that other generalizations of the classical BDF1 exist as well. Similar problems arise with almost every other numerical method that we shall encounter: Most of these can be interpreted as a generalization of a well known routine for first-order initial value problems, but these basic routines can be extended to the fractional setting in many different ways, and all these generalizations have a certain right to inherit the name of their classical ancestor. Thus, the reader must be very careful when looking at the literature on numerical methods for fractional differential equations because quite frequently one will find different methods being denoted by identical names.

The convergence behavior of this fractional backward differentiation formula can be summarized in the following way.

Theorem 2.6. *The fractional backward differentiation formula* (2.2.6), *with weights* A_{jk} *as in Eq.* (2.1.11), *produces an approximate solution for the initial value problem* (2.2.1) *that satisfies the inequality*

$$|y(x_j) - y_j| = O(h^{2-\alpha})$$

uniformly for all j *if the solution* y *satisfies the smoothness condition* $y \in C^2[0, X]$.

In the case of a linear equation this has been shown in Corollary 1.2 of [181]; the extension to nonlinear problems can be done using standard methods.

Once again, as in Theorem 2.3 and in Corollary 2.2, it turns out that the approximation is quite good if α is close to zero but deteriorates as α increases, up to the point that no convergence at all can be expected if $\alpha \geq 2$.

2.3 Indirect Methods for Fractional Ordinary Differential Equations

2.3.1 *The basic idea*

In contrast to the direct methods that we had introduced in the previous section, indirect methods are constructed not by applying a discretization directly to the fractional differential equation, but by first performing some analytical manipulation on the initial value problem and by then applying the numerical method to the equation obtained as the result of this analytical operation. The most common idea in this context is to apply the

Riemann-Liouville integral operator J^α to the initial value problem (2.2.1), thus creating the nonlinear and weakly singular Volterra integral equation of the second kind

$$y(x) = \sum_{k=0}^{\lceil\alpha\rceil-1} \frac{y_0^{(k)}}{k!} x^k + J^\alpha[f(\cdot, y(\cdot))](x). \tag{2.3.1}$$

The following result from Lemma 2.1 of [192] shows under very weak conditions on the given initial value problem that this operation does not change the exact solution.

Theorem 2.7. *If the function f is continuous, then the initial value problem* (2.2.1) *is equivalent to the Volterra integral equation* (2.3.1). *In other words, every continuous solution of* (2.3.1) *is also a solution of our original initial value problem* (2.2.1) *and vice versa.*

Thus, we can now use a numerical method to discretize the Riemann-Liouville integral operator J^α and not, as in the direct methods, the Caputo differential operator ${}^{\mathrm{C}}D^\alpha$. Once again we only consider linear schemes

$$J^\alpha z(x_n) \approx \sum_{k=0}^{n} a_{k,n} z(x_k) \tag{2.3.2}$$

hence producing numerical approaches for the Volterra equation (2.3.1) that have the form

$$y_n = \sum_{k=0}^{n} a_{k,n} f(x_k, y_k), \qquad n = 0, 1, \ldots. \tag{2.3.3}$$

In the next subsections, we shall again look at some special cases of this idea more closely, but before doing that we shall note one observation that catches the eye when comparing the direct method of Eq. (2.2.3) and the indirect method described in Eq. (2.3.3). Specifically, the computational complexity of the two methods appears to be somewhat different because each step of the direct method involves only one evaluation of the given function f, whereas the nth step of the indirect method requires $O(n)$ such evaluations. This can be an important aspect because the computation of f may, in some cases, be an extremely time-consuming job. Since in a practical application one would have to solve Eqs. (2.2.3) and Eq. (2.3.3), respectively, by some iterative procedure, the number of function calls would in both cases have to be multiplied by some factor. This feature is likely to make the difference even bigger. However, by using a careful implementation it is possible to store and re-use the values $f(x_j, y_j)$ in the indirect approach, thus obtaining a computational cost that effectively is almost identical to the cost of the direct method.

We shall once again assume that we are looking for an approximate solution of the fractional differential equation (2.2.1) or, equivalently, the Volterra integral equation (2.3.1), on the interval $[0, X]$ and that we have discretized this interval in a uniform way with grid points given in Eq. (2.2.4), *viz.*

$$x_j = \frac{j}{N} X \qquad (j = 0, 1, \ldots, N).$$

Thus the step size is again $h = X/N$. The approximation for $y(x_j)$ will be denoted by y_j as usual.

2.3.2 *An Adams-type predictor-corrector method*

The first method that we shall look at in this section has been introduced by Diethelm and Freed [200, 201] and analyzed in a detailed way in [197, 198]. It is based on the approximation of the integral operator in Eq. (2.3.1) by the product trapezoidal method introduced in §2.1.1. This leads to the formula

$$y_k = \sum_{j=0}^{\lceil \alpha \rceil - 1} \frac{x_k^j}{j!} y_0^{(j)} + h^\alpha \sum_{j=0}^{k} a_{jk} f(x_j, y_j) \tag{2.3.4}$$

for $k = 1, 2, \ldots, N$ where the weights a_{jk} are given in Eq. (2.1.9). In the limit case $\alpha \to 1$ this method reduces to the classical second-order Adams-Moulton formula (see, *e. g.*, §III.1 of [280]); hence we shall call our method a *fractional Adams-Moulton formula* (but we explicitly recall the note of caution mentioned in Remark 2.1 with respect to the terminology).

It is evident from Eq. (2.3.4) that this method is implicit because the unknown value of the kth step, *viz.* the number y_k, appears not only on its own on the left-hand side of the equation but also inside the function f on the right-hand side. Thus we need to consider the question of solving Eq. (2.3.4) for y_k.

The first idea in this context is to create an iterative method for the solution of Eq. (2.3.4) for each k and to run these iterations until the difference between two successive iterations is sufficiently close (typically, a small multiple of machine epsilon). From an applied standpoint, this essentially amounts to computing the exact solution of Eq. (2.3.4). Such a process is indeed possible, and it results in the fractional Adams-Moulton formula in its pure form. We summarize the essential properties of this idea as follows.

Theorem 2.8. *Assume that the function f on the right-hand side of the differential equation (2.2.1) is continuous and satisfies a Lipschitz condition with respect to its second variable with Lipschitz constant L. Moreover let $h < (\Gamma(2+\alpha)/L)^{1/\alpha}$. Then, Eq. (2.3.4) has a unique solution y_k for all k.*

Furthermore, the approximation computed in this way satisfies the error bound

$$|y(x_j) - y_j| = O(h^2)$$

uniformly for all j under either of the following conditions:

- ${}^{C}D^{\alpha}y \in C^2[0, X]$,
- $\alpha > 1$ *and* $f \in C^3([0, X] \times [y_0^{(0)} - K, y_0^{(0)} + K])$ *with some* $K > 0$.

The first part of this result corresponds to Theorem 2.5 for the first-order backward differentiation formula of Diethelm; it can be proved in exactly the same way. The second part is an analogue of Theorem 2.6. It can be deduced from Theorem 2.2.

It must be noted however that the solution of the nonlinear equation (2.3.4) by the method of iteration until convergence will frequently require a rather large number of iteration steps, and hence the entire process is likely to be very time-consuming. Fortunately, it is actually not necessary to go into this expensive procedure. The reason is that y_k is only an approximation for the desired value $y(x_k)$ with a certain accuracy. Thus, if we only solve Eq. (2.3.4) up to an error of the same order, then the total error will still be of this order. Our goal is thus to try to find such a sufficiently good approximate solution of Eq. (2.3.4) using only a small number of arithmetical operations. To this end we recall some facts from the classical theory of first-order equations (see, *e. g.*, Sections III.1 and III.8 of [280] and Section V.1 of [281]).

Specifically, a common procedure when dealing with such a method (that falls into the category of implicit multistep methods) whose order of convergence is $O(h^p)$ is to solve Eq. (2.3.4) by the following iterative procedure:

(1) Compute a first approximation of the solution y_k of Eq. (2.3.4), the so-called *predictor* $y_{k,0}$, by means of an explicit algorithm whose order of convergence is $O(h^{p-1})$.

(2) Perform one step of the iteration and take the result $y_{k,1}$ of this step, the so-called *corrector*, as the desired approximate solution, *i. e.* write

$$y_{k,1} = \sum_{j=0}^{\lceil\alpha\rceil-1} \frac{x_k^j}{j!} y_0^{(j)} + h^\alpha \sum_{j=0}^{k-1} a_{jk} f(x_j, y_j) + h^\alpha a_{kk} f(x_k, y_{k,0}). \quad (2.3.5)$$

and use this value in place of the true solution y_k in all future steps.

Such a procedure is known as a *predictor-corrector method* or, more precisely, as a PECE (predict-evaluate-correct-evaluate) method because we first compute the predictor $y_{k,0}$, then we evaluate the function f with arguments $(x_k, y_{k,0})$, then we use the result to compute the corrector $y_{k,1}$, and finally we evaluate f once again, this time with arguments $(x_k, y_{k,1})$, because we need this value for the process of computing y_{k+1}. The classical theory of numerical methods for first-order equation then tells us that the values $y_k = y_{k,1}$ approximate the true solution $y(x_k)$ of our initial value problem up to an error of magnitude $O(h^p)$.

Of course, it is possible to construct a similar process for the fractional setting. A natural candidate for a predictor for the fractional Adams-Moultion scheme is the *fractional Adams-Bashforth method*. This method is constructed in the same way as the Adams-Moulton method, except that we replace the product trapezoidal method by the product rectangle method that we had also introduced in §2.1.1. It is thus given by

$$y_k = \sum_{j=0}^{\lceil\alpha\rceil-1} \frac{x_k^j}{j!} y_0^{(j)} + h^\alpha \sum_{j=0}^{k-1} b_{jk} f(x_j, y_j) \quad (2.3.6)$$

for $k = 1, 2, \ldots, N$ where the weights b_{jk} are given in Eq. (2.1.7). In the limit case $\alpha \to 1$ this method reduces to the classical first-order Adams-Bashforth formula (see, *e. g.*, §III.1 of [280]) that happens to coincide with the forward Euler method.

The Adams-Bashforth method can of course be considered as a numerical scheme for solving fractional differential equations in its own right. It is an explicit method since y_k does not appear on the right-hand side of Eq. (2.3.6); therefore we do not need to worry about conditions under which

Eq. (2.3.6) can be solved for y_k. As a consequence of Corollary 2.1, the error can be estimated as follows:

Theorem 2.9. *The approximation computed by the Adams-Bashforth method satisfies the error bound*

$$|y(x_j) - y_j| = O(h)$$

uniformly for all j if ${}^{\mathrm{C}}D^{\alpha} y \in C^1[0, X]$.

However, our plan was not to use one of these methods on its own but rather to combine them to a predictor-corrector pair. Thus the complete representation of the Adams-Bashforth-Moulton algorithm requires us to choose a certain value of $m \in \mathbb{N}$ (the number of corrector iterations; it is very common to use $m = 1$), write

$$y_{k,0} = \sum_{j=0}^{\lceil\alpha\rceil-1} \frac{x_k^j}{j!} y_0^{(j)} + h^{\alpha} \sum_{j=0}^{k-1} b_{jk} f(x_j, y_j), \tag{2.3.7a}$$

$$\begin{aligned} y_{k,\mu} = {} & \sum_{j=0}^{\lceil\alpha\rceil-1} \frac{x_k^j}{j!} y_0^{(j)} + h^{\alpha} \sum_{j=0}^{k-1} a_{jk} f(x_j, y_j) \\ & + h^{\alpha} a_{kk} f(x_k, y_{k,\mu-1}) \qquad (\mu = 1, 2, \ldots, m) \end{aligned} \tag{2.3.7b}$$

$$y_k = y_{k,m} \tag{2.3.7c}$$

for $k = 1, 2, \ldots, N$ and use the values y_k as the desired approximations. The weights a_{jk} and b_{jk} must be taken from Eqs. (2.1.9) and (2.1.7), respectively.

Looking at the error of this predictor-corrector method, we find a slight peculiarity of the fractional case in comparison to first-order equations. For the latter we know that each application of a corrector iteration increases the order of the approximation by 1 until the full order of the corrector is reached. Thus, if the predictor is an $O(h^{p_1})$ method and the pure corrector is $O(h^{p_2})$ with $p_2 > p_1$, then the predictor-corrector method with m corrector steps as described in Eq. (2.3.7) has an order $O(h^{\min\{p_1+m,p_2\}})$. In the fractional case, we observe that we need more corrector steps to achieve such a goal. To be precise, each corrector iteration only increases the order by α (see, *e. g.*, Section 6 of [183]). Combining the results of [183] and [198], the formal result reads as follows.

Theorem 2.10. *The relation*

$$|y(x_k) - y_k| = O(h^{\min\{2,1+m\alpha\}})$$

holds uniformly for all $j = 1, 2, \ldots, N$ under either of the conditions mentioned in Theorem 2.8.

Other error estimates under slightly different assumptions may be found in Section 3 of [198]. A full algorithmic description of the method in a pseudocode form has been given in [199].

2.3.3 *Deng's modification of the fractional Adams-Bashforth-Moulton method*

A useful modification of this fundamental approach has been discovered by Deng [178]. Like our original method described above, it is based on a predictor-corrector strategy with the product trapezoidal formula as a corrector. However, it uses a different predictor. The development of this predictor was again motivated by the idea of splitting up the Riemann-Liouville integral $J^\alpha[f(\cdot, y(\cdot))](x_k)$ in the Volterra form of the initial value problem into subintervals $[x_j, x_{j+1}]$ $(j = 0, 1, \ldots, k-1)$ and finding a suitable approximation on each subinterval. While our Adams-Bashforth method used the product rectangle rule on all subintervals, Deng suggests to use this formula only on the last subinterval, *i. e.* for $j = k - 1$, and to employ the product trapezoidal formula on all other subintervals. Thus, Deng's method combines the predictor formula

$$y_{k,0} = \sum_{j=0}^{\lceil\alpha\rceil-1} \frac{x_k^j}{j!} y_0^{(j)} + h^\alpha \sum_{j=0}^{k-2} a_{jk} f(x_j, y_j) + h^\alpha c_k f(x_{k-1}, y_{k-1}) \tag{2.3.8a}$$

with, as already mentioned, an m-fold application of the corrector

$$\begin{aligned} y_{k,\mu} &= \sum_{j=0}^{\lceil\alpha\rceil-1} \frac{x_k^j}{j!} y_0^{(j)} + h^\alpha \sum_{j=0}^{k-1} a_{jk} f(x_j, y_j) \\ &\quad + h^\alpha a_{kk} f(x_k, y_{k,\mu-1}) \qquad (\mu = 1, 2, \ldots, m) \qquad (2.3.8\text{b}) \\ y_k &= y_{k,m} \qquad (2.3.8\text{c}) \end{aligned}$$

that we simply copied from Eqs. (2.3.7b) and (2.3.7c). In these equations, the coefficients a_{jk} are the same as in our original Adams-Moulton method, *i. e.* they are taken from Eq. (2.1.9), and the value c_k can be computed by a method that is analog to the one that lead to the a_{jk} and b_{jk}; this leads to

$$c_1 = \frac{h^{-\alpha}}{\Gamma(\alpha)} \int_{x_0}^{x_1} (x_1 - t)^{\alpha-1} dt = \frac{1}{\Gamma(1+\alpha)} \tag{2.3.9a}$$

and

$$\begin{aligned} c_k &= \frac{h^{-\alpha}}{\Gamma(\alpha)} \left(\int_{x_{k-1}}^{x_k} (x_k - t)^{\alpha-1} dt + \int_{x_{k-2}}^{x_{k-1}} (x_k - t)^{\alpha-1} \frac{t - x_{k-2}}{h} dt \right) \\ &= \frac{2^{\alpha+1} - 1}{\Gamma(2+\alpha)} \end{aligned} \tag{2.3.9b}$$

for $k \geq 2$.

A comparison of Deng's method (2.3.8) with the original Adams-Bashforth-Moulton method (2.3.7) that we constructed in the previous subsection reveals differences between the two approaches with respect to their accuracy and with respect to their computational cost.

Firstly, the computational cost of both algorithms is dominated by the arithmetic operations required for computing the sums that extend from $j = 0$ to $j = k-1$ in the predictor and the corrector of the kth discretization step (or, in the case of Deng's predictor in Eq. (2.3.8a), that extends from $j = 0$ to $j = k - 2$). In the original method (2.3.7), the sum for the predictor step needs to be computed, and the sum in the first corrector step needs to be computed as well because it differs from the predictor sum. In subsequent corrector steps, the sum from the first corrector step can be re-used, so no additional computations are necessary. In contrast, Deng's approach has the same sum appearing in the predictor and in the corrector (except for one single term that is present in the corrector only and that can be computed at a negligible cost in comparison to the remainder of the sum), and so it is sufficient to compute the sum only once. Therefore, in an asympotic sense, Deng's method comes with only half of the computational cost of the Adams-Bashforth-Moulton method.

Of course, the fact that Deng uses a different predictor than the Adams-Bashforth-Moulton method also leads to a change in the convergence properties. As demonstrated on p. 1520 of [178], the methods used in [198] for the Adams-Bashforth-Moulton method can be applied for Deng's algorithm as well; the outcome reads as follows.

Theorem 2.11. *If* ${}^{\mathrm{C}}D^\alpha y \in C^2[0, X]$ *then the relation*

$$|y(x_k) - y_k| = O(h^{\min\{2,1+(m+1)\alpha\}})$$

holds uniformly for all $j = 1, 2, \ldots, N$.

When we compare this error estimate with its counterpart for the classical Adams-Bashforth-Moulton method from Theorem 2.10, we see that Deng's method with m corrector iterations yields the same accuracy as the Adams method with $m + 1$ corrector iterations.

2.3.4 *The Cao-Burrage-Abdullah approach*

A substantially different indirect method has recently been proposed by Cao, Burrage and Abdullah [138]. Also starting from the initial value problem

$${}^{\mathrm{C}}D^\alpha y(x) = f(x, y(x)), \qquad y(0) = y_0, \tag{2.3.10}$$

with $0 < \alpha < 1$, they decided not to apply the *integral* operator of order α but a *differential* operator of order $1 - \alpha$ to the differential equation. Interestingly, it is necessary here not to use a Caputo operator but a Riemann-Liouville operator, thus — using the formal relation ${}^{\mathrm{RL}}D^{1-\alpha}\,{}^{\mathrm{C}}D^\alpha = (D^1 J^\alpha)(J^{1-\alpha} D^1) = D^1 J^1 D^1 = D^1$ — obtaining the problem

$$y'(x) = {}^{\mathrm{RL}}D^{1-\alpha}\,{}^{\mathrm{C}}D^\alpha y(x) = {}^{\mathrm{RL}}D^{1-\alpha}[f(\cdot, y(\cdot))](x), \tag{2.3.11}$$

still augmented by the originally given initial condition $y(0) = y_0$.

Now, in a numerical approach we need to approximate the two differential operators D^1 and ${}^{\mathrm{RL}}D^{1-\alpha}$ on the right-hand side and the left-hand side of Eq. (2.3.11), respectively. Since the integer-order derivative D^1 can be handled numerically by very simple standard methods, the additional amount of work (compared to the approaches of the previous sections where we only had to deal with one differential or integral operator) is negligible. Moreover, the nature of the methods discussed earlier essentially forced us to use an equispaced mesh. In many cases such a straightforward approach is perfectly appropriate, but there are situations where one would rather prefer to use a non-uniform mesh. Examples include cases where the given data is only provided in tabulated form at some irregularly spaced abscissas, or when one has reason to believe that the solution behaves in an unpleasant (*e. g.*, highly oscillatory) fashion in some subinterval, thus requiring small step sizes there, whereas the variation is small (and hence admits much more economical large stepsizes) elsewhere. It is precisely this setting of nonuniform grids that Cao, Burrage and Abdullah had in mind when they developed their concept.

Specifically, let $0 = x_0 < x_1 < x_2 < \cdots < x_N$ be the mesh points. Then, the discretization at the point x_j, $1 \le j \le N$, takes the form

$$\frac{y_j - y_{j-1}}{x_j - x_{j-1}} = \sum_{k=0}^{j} a_{kj} f(x_k, y_k) \tag{2.3.12}$$

where the weights a_{kj} are taken from a suitable discretization of the Riemann-Liouville derivative. A number of possible choices are available here, for example a non-equispaced version of the product trapezoidal method from §2.1.1. Of course there remains the question which choice is good in this context. Unfortunately, almost no analytical results seem to be known in this connection. In addition, since the method has only recently been published, there is not much numerical experience available either.

The price that we have to pay for the high degree of flexibility in the choice of the mesh introduced by this approach is on the analytical side: At present we are not aware of any thorough error analysis of the algorithm. Moreover, there is a substantial difference between a discretization of Eq. (2.3.11) and the discretization that we had to perform in the earlier parts of this chapter. This difference is due to the fact that Eq. (2.3.11) will typically be valid only on the half-open interval $(0, x_N]$. The reason is that, as is well known [431], the solution y of the originally given fractional differential equation (2.3.10) in most cases behaves as $y(x) = y_0 + cx^\alpha$ as $x \to 0$ with some constant c. Thus, since $0 < \alpha < 1$, we have (except for the very rare special case where $c = 0$) that y is not differentiable at 0, and so Eq. (2.3.11) is undefined there. Moreover it is unclear what the influence of this discontinuity of y' is on the numerical approximation method. Onc certainly needs to take into account the possibility that the approximation in (2.3.12) may be poor for $j = 1$, giving a possibly very inaccurate value for y_1, and it is unknown how this potentially large error is propagated to the values $y_2, y_3, \ldots$. This is not to say that the method will be doomed to failure (indeed the numerical results in [138] look promising at least for the examples considered there), but one has to be warned that a convergence proof or error analysis under reasonably general assumptions is lacking at the moment.

2.4 Linear Multistep Methods

A particularly well understood class of methods is based on using the *linear multistep methods* described in Subsection 2.1.3 above for the discretization of the fractional differential and integral operators arising in our equations. In particular, we recall the concept of fractional convolution quadratures from Definition 2.1 and introduce the following notions.

Definition 2.2. Consider the fractional convolution quadrature

$$({}_hJ^\alpha f)(x_m) = h^\alpha \sum_{j=0}^{m} \omega_{m-j} f(x_j) + h^\alpha \sum_{j=0}^{s} w_{mj} f(x_j), \quad m = 0, 1, \ldots, \tag{2.4.1}$$

for some $\alpha > 0$.

- The quadrature is called *stable* for the integral J^α if
$$\omega_n = O(n^{\alpha-1}).$$
- It is called *consistent* of order p for J^α if
$$h^\alpha \omega(\exp(-h)) = 1 + O(h^p).$$
- It is called *convergent* of order p to J^α if
$${}_h\Omega^\alpha \pi_{z-1}(1) - J^\alpha \pi_{z-1}(1) = O(h^z) + O(h^p)$$
holds for all $z \in \mathbb{C} \setminus \{0, -1, -2, \ldots\}$ where we have used the notation $\pi_k(x) = x^k$.

Recall that the function ω used in this definition is related to the coefficients of the fractional convolution quadrature via the identity

$$\omega(z) = \sum_{k=0}^{\infty} \omega_k z^k$$

and that, as stated in Definition 2.1, ${}_h\Omega^\alpha$ is the convolution part of ${}_hJ^\alpha$, *i. e.* the first of the two sums on the right-hand side of Eq. (2.4.1). It is important to note a special feature of Definition 2.2: The stability, consistency and convergence of a fractional convolution quadrature depend only on its convolution part. The starting quadrature does not play a role here. We shall come back to this observation in Theorem 2.14 below. First however we note one important result that generalizes Dahlquist's well known result for first-order problems [170, 171] to the fractional setting. For a proof of this result we refer to Lubich's classical paper [385] or the detailed discussion in the book of Brunner and van der Houwen [119] or in Section 4.3 of Weilbeer's thesis [624].

Theorem 2.12. *Let α be a non-integer number and r_1 and r_2 be rational functions. Then, a fractional convolution quadrature whose convolution weights are defined by the generating function $\omega(\zeta) = (r_1(\zeta))^\alpha r_2(\zeta)$ is convergent of order p if and only if it is stable and consistent of order p.*

Whereas this result allows us to determine whether a given fractional convolution quadrature is convergent or not, another important result by Dahlquist [172] that can be generalized to the fractional setting (see Theorem 4.3.11 of [624]) allows us to construct fractional linear multistep methods in such a way that the convergence is guaranteed a priori. This construction may be based on any classical linear multistep method for first order problems that is known to be convergent.

Theorem 2.13. *Let (ρ, σ) denote an implicit classical linear multistep method for first order equations which is stable and consistent of order p. Moreover assume that all zeros of σ lie inside or on the boundary of the unit disc. Denote the generating function of this linear multistep method by ω. Then, for $\alpha > 0$, the fractional linear multistep method with generating function ω^α is convergent of order p to J^α.*

Now let us come back to the observation mentioned above that the starting quadrature does not play a role in the context of stability, consistency and convergence of the fractional convolution quadrature. Specifically, if we consider a fractional convolution quadrature ${}_hJ^\alpha$ as introduced in Definition 2.1 that converges to J^α with order p (as defined in Definition 2.2, *i. e.* without taking the starting part into consideration) then we can always find a set of starting weights such that for a sufficiently well behaved function

f the error ${}_hJ^\alpha f - J^\alpha f$ is also of order p. The following theorem indicates what we mean by a sufficiently well behaved function in this context and shows how to obtain the required starting weights. It is based on a result given in [384].

Theorem 2.14. *Let (ρ, σ) be a convergent implicit linear multistep method of order $p \geq 1$ for first order differential equations, and let all zeros of σ lie inside or on the boundary of the unit disc. Moreover let $0 < \alpha < 1$ and*

$$\omega^\alpha(\zeta) = \left(\frac{\sigma(1/\zeta)}{\rho(1/\zeta)}\right)^\alpha.$$

Furthermore, set $A = \{\gamma = k + j\alpha : j, k \in \mathbb{N}_0, \gamma \leq p-1\}$ and let $s+1$ be the number of elements of A. Define the starting weights w_{nj} to be the solution of the linear system

$$\sum_{j=0}^{s} w_{nj} j^\gamma = \frac{\Gamma(\gamma+1)}{\Gamma(\gamma+\alpha+1)} n^{\gamma+\alpha} - \sum_{j=0}^{n} \omega_{n-j} j^\gamma, \quad \gamma \in A, \tag{2.4.2}$$

then we have $w_{nj} = O(n^{\alpha-1})$, $j = 0, 1, \ldots, s$.

Moreover, if y_n $(n = 0, 1, \ldots, N)$ is the numerical approximation of the solution y of the fractional initial value problem ${}^{C}D^\alpha y(x) = f(x, y(x))$, $y(0) = y_0$, at the point $x_n = nh$, defined by

$$y_n = y_0 + h^\alpha \sum_{j=0}^{n} \omega_{n-j} f(x_j, y_j) + h^\alpha \sum_{j=0}^{s} w_{n,j} f(x_j, y_j)$$

where f is sufficiently smooth, then this numerical solution satisfies

$$\max_{n=0,1,\ldots,N} |y_n - y(x_n)| = O(h^{p-\epsilon})$$

with some $\epsilon \in [0, 1-\alpha)$. In particular, $\epsilon = 0$ if $\alpha = q/(q+1)$ with some $q \in \mathbb{N}$.

This means that we can always find a set of starting weights such that the error of the fractional convolution quadrature behaves in the same way as it does in the classical (first-order) case.

It thus turns out that, in theory, these formulas can be used to compute highly accurate numerical solutions for fractional differential equations. In practice however things are not that simple. In particular, it has been observed [190] that the computation of the starting weights can be a very ill-conditioned problem for certain choices of the order α of the differential operator whereas it behaves much better for other values of α. This is due to the fact that the coefficient matrix of the linear system (2.4.2) that defines these weights may, depending on α, contain rows that almost coincide, *i. e.* the system may be very close to a singular system. In such a case, no accurate method for the solution of the system, *i. e.* for the computation of the starting weights, is available, and the use of a standard method for the solution of this system leads to highly inaccurate values for the starting weights and consequently to highly inaccurate numerical solutions of the fractional differential equations in question. We shall provide a detailed discussion about the cases where such problems may be expected and, on the other hand, where fractional linear multistep methods may be considered reliable in Subsection 3.1.3.

For the moment we conclude our discussion of this approach with a reference to the work of Ford and Connolly [239] who have provided a comparison of the performance of the linear multistep methods with the other algorithms described above. In particular, their goal was to display the accuracy that can be achieved by the methods in relation to their respective computational costs. The results of [239] indicate that the predictor-corrector method of Subsection 2.3.2 tends to be a very useful choice in many cases and that a third order backward differentiation scheme (which is a special case of a linear multistep method) might also be a good candidate, but the information presently known about the latter is not yet sufficient to fully decide on its usefulness.

2.5 Spectral Methods

The methods that we discussed so far have almost exclusively been of a nature that can be described as local in the sense that one essentially decomposes the interval of interest into a number of small subintervals and then constructs a solution on one subinterval after the other. Bearing in mind that fractional differential operators are actually non-local, the appropriateness of such a local approach is sometimes considered doubtful; *cf.*, *e. g.*, Subsection 1.1 of [645]. Based on this observation, the authors of that paper suggest to use a different class of numerical solution methods for fractional differential equations that is of a global type in the sense that an approximate solution for the complete interval is computed in one step.

Specifically, the class of methods proposed and investigated in [645] are so-called *spectral methods*, *i. e.* methods based on an expansion of the solution in terms of eigenfunctions of a certain operator. We shall now briefly present the basic ideas behind spectral methods and review some of their properties even though the entire development presently seems to be in an incomplete state which is in particular reflected by the fact that we shall restrict our attention to the rather simple differential equation

$$ {}^{\mathrm{C}}D^{\alpha} y(x) = f(x) \tag{2.5.1a} $$

whose right-hand side does not depend on the unknown function y, where $0 < \alpha < 1$, subject to the initial condition

$$ y(0) = y_0 \tag{2.5.1b} $$

(which is in fact the only case handled in the original work of Zayernouri and Karniadakis [645]).

Given the initial value problem (2.5.1) with $0 < \alpha < 1$, one first recalls the classical Jacobi polynomials with parameters $-\alpha/2$ and $\alpha/2$, *i. e.* the polynomials $P_k^{(-\alpha/2,\alpha/2)}$ of degree k that are orthogonal with respect to the weight function $(1-x)^{-\alpha/2}(1+x)^{\alpha/2}$ over the interval $[-1,1]$, *cf.*, *e. g.*, [584]. Assuming that we are interested in solving the initial value problem on the interval $[0, X]$, we use these orthogonal polynomials to define the basis functions

$$ \mathcal{P}_k^{(\alpha)}(x) := \left(\frac{2x}{X}\right)^{\alpha/2} P_{k-1}^{(-\alpha/2,\alpha/2)}\left(-1 + \frac{2x}{X}\right) \tag{2.5.2} $$

for $x \in [0, X]$ and $k \in \mathbb{N}$. It is immediately clear that

$$ \mathcal{P}_k^{(\alpha)}(0) = 0 \tag{2.5.3} $$

for all k; moreover, an explicit calculation reveals that

$$ {}^{\mathrm{C}}D^{\alpha/2} \mathcal{P}_k^{(\alpha)}(x) = {}^{\mathrm{RL}}D^{\alpha/2} \mathcal{P}_k^{(\alpha)}(x) = \left(\frac{2}{X}\right)^{\alpha/2} \frac{\Gamma(k+\alpha/2)}{\Gamma(k)} P_{k-1}\left(-1 + \frac{2x}{X}\right) \tag{2.5.4} $$

where P_{k-1} denotes the standard Legendre polynomial of degree $k-1$.

The goal of the method is to find an approximate solution to the initial value problem (2.5.1) in the form

$$y_n(x) = y_0 + \sum_{k=1}^{n} p_k \mathcal{P}_k^{(\alpha)}(x) \tag{2.5.5}$$

with suitable coefficients p_k. From Eq. (2.5.3) it is clear that such a function will always satisfy the initial condition. The determination of the coefficients will then be based on a Petrov-Galerkin approach, *i. e.* we choose a set $\{\mathcal{Q}_k^{(\alpha)} : k = 1, 2, \ldots, n\}$ of test functions and demand that all these test functions be orthogonal to the residual of the approximation y_n with respect to the standard L^2 inner product on the interval $[0, X]$, *i. e.*

$$\int_0^X \left({}^{\mathrm{C}}D^\alpha y_n(t) - f(t)\right) \mathcal{Q}_j^{(\alpha)}(t)dt = 0 \tag{2.5.6}$$

which is equivalent to requiring that

$$\sum_{k=1}^{n} p_k \int_0^X {}^{\mathrm{C}}D^\alpha \mathcal{P}_k^{(\alpha)}(t)\mathcal{Q}_j^{(\alpha)}(t)dt = \int_0^X f(t)\mathcal{Q}_j^{(\alpha)}(t)dt \tag{2.5.7}$$

for $j = 1, 2, \ldots, n$. It follows from the theory of fractional Sturm-Liouville eigenproblems [644] that a useful choice of the test functions is given by

$$\mathcal{Q}_k^{(\alpha)}(x) := \left(\frac{2(X-x)}{X}\right)^{\alpha/2} P_{k-1}^{(\alpha/2,-\alpha/2)}\left(-1 + \frac{2x}{X}\right); \tag{2.5.8}$$

in fact these functions have a property similar to (2.5.4), but with the right-sided instead of the left-sided differential operators, namely

$${}^{\mathrm{C}}D_{X-}^{\alpha/2}\mathcal{Q}_k^{(\alpha)}(x) = {}^{\mathrm{RL}}D_{X-}^{\alpha/2}\mathcal{Q}_k^{(\alpha)}(x) = \left(\frac{2}{X}\right)^{\alpha/2} \frac{\Gamma(k+\alpha/2)}{\Gamma(k)} P_{k-1}\left(-1 + \frac{2x}{X}\right) \tag{2.5.9}$$

which, in combination with Eq. (2.5.4) yields

$${}^{\mathrm{C}}D_{X-}^{\alpha/2}\mathcal{Q}_k^{(\alpha)}(x) = {}^{\mathrm{C}}D^{\alpha/2}\mathcal{P}_k^{(\alpha)}(x). \tag{2.5.10}$$

In order to compute the approximation y_n, we have to find the coefficients p_k, *i. e.* we must solve the linear system (2.5.7), and to this end we may recall from [370] the identity

$$\int_0^X {}^{\mathrm{C}}D^\alpha \mathcal{P}_k^{(\alpha)}(t)\mathcal{Q}_j^{(\alpha)}(t)dt = \int_0^X {}^{\mathrm{C}}D^{\alpha/2}\mathcal{P}_k^{(\alpha)}(t) \cdot {}^{\mathrm{C}}D_{X-}^{\alpha/2}\mathcal{Q}_j^{(\alpha)}(t)dt \tag{2.5.11}$$

which, in conjunction with Eqs. (2.5.4) and (2.5.9), can be simplified to

$$\begin{aligned}
&\int_0^X {}^{\mathrm{C}}D^\alpha \mathcal{P}_k^{(\alpha)}(t)\mathcal{Q}_j^{(\alpha)}(t)dt \\
&= \left(\frac{2}{X}\right)^\alpha \frac{\Gamma(k+\alpha/2)}{\Gamma(k)} \frac{\Gamma(j+\alpha/2)}{\Gamma(j)} \int_0^X P_{k-1}\left(-1+\frac{2t}{X}\right) P_{j-1}\left(-1+\frac{2t}{X}\right) dt \\
&= \left(\frac{2}{X}\right)^{\alpha-1} \frac{\Gamma(k+\alpha/2)}{\Gamma(k)} \frac{\Gamma(j+\alpha/2)}{\Gamma(j)} \int_{-1}^1 P_{k-1}(u)P_{j-1}(u)du \\
&= \left(\frac{2}{X}\right)^{\alpha-1} \left(\frac{\Gamma(k+\alpha/2)}{\Gamma(k)}\right)^2 \frac{2}{2k-1}\delta_{jk}. \qquad (2.5.12)
\end{aligned}$$

It thus turns out, in view of the orthogonality of the classical Legendre polynomials, that the equation system (2.5.7) possesses a diagonal coefficient matrix, so that its solution can be immediately computed; we obtain the spectral approximation to the solution of the initial value problem (2.5.1) as

$$y_n(x) = y_0 + \left(\frac{X}{2}\right)^{\alpha-1} \sum_{k=1}^n \left(\frac{\Gamma(k)}{\Gamma(k+\alpha/2)}\right)^2 \frac{2k-1}{2} \int_0^X f(t)\mathcal{Q}_k^{(\alpha)}(t)dt \cdot \mathcal{P}_k^{(\alpha)}(x). \qquad (2.5.13)$$

Zayernouri and Karniadakis [645] report the results of this approximation method for various special examples, including cases where f is smooth and y is non-smooth or vice versa; in all these experiments they observed rapid (typically exponential) convergence behaviour. However, it must be said that their examples are such that the integrals $\int_0^X f(t)\mathcal{Q}_k^{(\alpha)}(t)dt$ that arise in the formula (2.5.13) for the desired approximate solution can be computed exactly. In general, this will not be the case, and then one has to resort to a numerical integration method whose use introduces an additional error component that needs to be taken into account in a complete error analysis.

The construction of the spectral approximation that we have described so far is restricted to differential equations of the form (2.5.1a), *i. e.* to equations whose right-hand side is independent of the unknown solution y. A formal extension to the case of linear differential equations

$${}^{\mathrm{C}}D^\alpha y(x) = f(x) + g(x)y(x) \qquad (2.5.14a)$$

subject to the usual initial condition

$$y(0) = y_0 \qquad (2.5.14b)$$

and the restriction $0 < \alpha < 1$ is straightforward; it still leads to an approximate solution in the form (2.5.5) where now the coefficients p_k are the solutions of the linear system

$$p_j \left(\frac{2}{X}\right)^{\alpha-1} \left(\frac{\Gamma(j+\alpha/2)}{\Gamma(j)}\right)^2 \frac{2}{2j-1} - \sum_{k=1}^{n} p_k \int_0^X g(t)\mathcal{P}_k^{(\alpha)}(t)\mathcal{Q}_j^{(\alpha)}(t)dt$$
$$= \int_0^X (f(t) + y_0 g(t))\,\mathcal{Q}_k^{(\alpha)}(t)dt \qquad (j = 1, 2, \ldots, n).$$

We are presently not aware of an explicit convergence and stability analysis for this algorithm, but clearly the same comments as above apply with respect to the evaluation of the integrals in these equations.

A generalization of the approach to differential equations with arbitrary nonlinear right-hand sides does not seem feasible. This fact seriously restricts the applicability of the algorithm for ordinary differential equations. However, we believe that it may have a significant role to play as a building block for numerical methods for solving partial differential equations of fractional order because for problems of this type linear equations are much more commonly used for modeling various phenomena than in the case of ordinary differential equations.

2.6 Other Methods

2.6.1 *A decomposition technique*

Some authors have also suggested other approaches than those described so far. Most of these methods have not gained a substantial amount of attention. The primary exception is probably the so-called *Adomian decomposition* that is usually traced back to Adomian's books [7, 8] even though its roots can actually be found in a series of much older papers by Perron [479–482]. The idea of the method is to write the differential equation in the abstract form

$$M(x, y(x)) = g(x) \tag{2.6.1}$$

where g is a given function, y is the unknown solution and M is a suitable operator (in our case, usually a combination of a fractional differential operator and some other, often nonlinear, functions applied to y). One then decomposes M according to

$$M = L_1 + L_2 + N \tag{2.6.2}$$

where L_1 and L_2 are linear operators chosen such that the inverse of L_1 can be found easily, while N denotes the remaining part of M that includes, in particular, all the nonlinearities. The method is based on the assumption that the exact solution y can be decomposed into a convergent series,

$$y = \sum_{n=0}^{\infty} y_n, \tag{2.6.3}$$

and the goal of the approach is to provide a reasonably simple method to compute the summands y_n of this series. To this end, we also decompose the nonlinearity N in the special series form

$$Ny(x) = \sum_{n=0}^{\infty} A_n(x)$$

with the so-called *Adomian polynomials*

$$A_n(x) = \frac{1}{n!}\left[\frac{d^n}{d\lambda^n} N\left(x, \sum_{j=0}^{n} \lambda^j y_j\right)\right]_{\lambda=0}.$$

Apparently A_n depends only on $y_0, y_1, \ldots, y_n$. One then starts with

$$y_0 = L_1^{-1} g$$

(which can be computed easily in view of our assumption on L_1) and proceeds by the recurrence relation

$$y_n = -L_1^{-1}(L_2 y_{n-1} - A_{n-1}), \qquad n = 1, 2, \ldots,$$

that defines the remaining summands of the series expansion (2.6.3) of y.

It is clear from the description above that the user of such an approach has a large amount of freedom in a concrete application. For example, the exact choice of the operators L_1, L_2 and N in Eq. (2.6.2) can be done in a number of different ways. Whereas it is obvious that the precise choice of these features can have a significant influence on the behavior of the method, it seems that no general rules are available that give the user concrete hints as to which choice may be useful in the context of the specific problem at hand.

Moreover, as stated above, most authors dealing with this method *assume* that the series expansion of Eq. (2.6.3) converges, but they do not look for conditions under which this assumption is satisfied, *i. e.* they use a hypothesis in terms of the unknown solution rather than in terms of the given data. In addition they neither deal with the question for the appropriate notion of convergence in this context (Convergence with respect to which norm?) nor do they discuss whether the series converges rapidly or slowly. The values of the argument of y for which the series converges (*i. e.* the convergence radius of the series) is usually not investigated either, but there is ample evidence from results of numerous examples which indicates that the radius of convergence is rather small in many cases [216, 234, 515].

In addition, the computation of the Adomian polynomials A_n is possible only if N, and hence M, is sufficiently smooth, and even in this case it requires an enormous amount of work if it is done analytically and is subject to severe numerical difficulties if done numerically. Therefore we believe that the precisely described and well understood methods of the previous sections should be preferred at least until more information about the decomposition becomes available.

2.6.2 *The variational iteration method*

Alternatively one may use the *variational iteration method* proposed by He [289] (see also the recent survey [290]). Like the Adomian method it is based on the abstract form (2.6.1) of the problem. The operator M is now decomposed in a slightly different way, namely as

$$M = L + N \tag{2.6.4}$$

where L is a linear operator. Starting from an initial approximation y_0 that is usually chosen to satisfy the initial conditions (and, if present, also the boundary conditions), one then constructs a sequence $(y_n)_{n=0}^{\infty}$ of approximations in a recursive manner using the scheme

$$y_n(t) = y_{n-1}(t) + \int_0^t \lambda(\tau)\left(Ly_{n-1}(\tau) + N\tilde{y}_{n-1}(\tau) - g(\tau)\right) d\tau \quad (n = 1, 2, \ldots). \tag{2.6.5}$$

Here, λ is a Lagrange multiplier that is characterized with the help of methods from variational calculus, and $\tilde{y}_{n-1}$ is a restricted variation, *i. e.*

$\delta\tilde{y}_{n-1} = 0$. Under certain conditions (that may be difficult to verify in practice) one can then show that $y_n \to y$ as $n \to \infty$ where y denotes the exact solution of the problem under consideration. We refer to [593] for more details on a related question; the approach presented there can be extended to our problem. Since, in practice, it is usually very difficult to determine the integrals on the right-hand side of Eq. (2.6.5), this method is of a rather limited value in concrete applications.

Similar remarks apply to a few other methods that have been proposed in the past like the *homotopy analysis method* [373] (that actually can be interpreted as a generalization of the Adomian decomposition), the *homotopy perturbation method* [437], and the *generalized differential transform method* [463]. In particular, the convergence properties of many of these methods tend to be rather poor [372].

2.6.3 *Methods based on nonclassical representations of fractional differential operators*

Another completely different approach that is probably much more useful in practice is due to Yuan and Agrawal [640]. Their method is based on the observation that one can rewrite the Caputo derivative of the function y in the form

$$ {}^{\mathrm{C}}D^{\alpha} y(x) = \int_0^{\infty} \phi(w, x)\, dw \tag{2.6.6} $$

where

$$ \phi(w, x) = (-1)^{\lfloor \alpha \rfloor} \frac{2 \sin \pi\alpha}{\pi} w^{2\alpha - 2\lceil \alpha \rceil + 1} \int_0^x y^{(\lceil \alpha \rceil)}(\tau) e^{-(x-\tau)w^2}\, d\tau. \tag{2.6.7} $$

Then they note that this function ϕ is a solution of the *first-order* initial value problem

$$ \begin{aligned} \frac{\partial}{\partial x}\phi(w, x) &= (-1)^{\lfloor \alpha \rfloor} \frac{2 \sin \pi\alpha}{\pi} w^{2\alpha - 2\lceil \alpha \rceil + 1} y^{(\lceil \alpha \rceil)}(x) - w^2 \phi(w, x), \qquad (2.6.8) \\ \phi(w, 0) &= 0, \end{aligned} $$

for each fixed $w > 0$. We consider this to be an ordinary, not a partial, differential equation because w is assumed to be fixed, so that x is actually the only free variable. Based on these identities they then suggest to compute a numerical approximation for the Caputo derivative by replacing the integral on the right-hand side of Eq. (2.6.6) by a Gauss-Laguerre quadrature.

The required function values of ϕ in this quadrature are not computed by its definition as given in Eq. (2.6.7) but by solving the ordinary differential equation (2.6.8) numerically by a suitable algorithm.

Lu and Hanyga [381] and, in particular, Schmidt and Gaul [558] have criticized the poor approximation quality of this method. Later, Diethelm [184] has identified the mathematical background of this unsatisfactory behavior, it being the fact that the Gauss-Laguerre quadrature formula suggested by Yuan and Agrawal for the numerical approximation of the integral in Eq. (2.6.6) is really unsuitable because of the properties of the integrand. By replacing the Gauss-Laguerre method by a suitably transformed Gauss-Jacobi rule, much better results can be obtained. For details we refer to [184]. Birk and Song [103, 104] have successfully applied this modified version of the Yuan-Agrawal method in connection with the investigation of certain diffusion problems.

Alternatively, a very similar approach is due to Chatterjee and Singh [150, 566]. Instead of using the function ϕ given in Eq. (2.6.7), they define

$$\phi(w,x) = \frac{(-1)^{\lfloor\alpha\rfloor}\sin\pi\alpha}{\pi(\alpha-\lceil\alpha\rceil+1)}\int_0^x y^{(\lceil\alpha\rceil)}(\tau)\exp\left(-(x-\tau)w^{1/(\alpha-\lceil\alpha\rceil+1)}\right)d\tau \tag{2.6.9}$$

and show that this function also satisfies Eq. (2.6.6). Moreover, it can also be expressed as the solution of a first-order initial value problem similar to Eq. (2.6.8) that now takes the form

$$\begin{aligned}\frac{\partial}{\partial x}\phi(w,x) &= \frac{(-1)^{\lfloor\alpha\rfloor}\sin\pi\alpha}{\pi(\alpha-\lceil\alpha\rceil+1)}y^{(\lceil\alpha\rceil)}(x) - w^{1/(\alpha-\lceil\alpha\rceil-1)}\phi(w,x), \\ \phi(w,0) &= 0,\end{aligned} \tag{2.6.10}$$

for a fixed $w > 0$. Either of the numerical techniques used for the original Yuan-Agrawal approach above is applicable to this version too. First steps in this direction are described in [184].

2.6.4 *Collocation*

To conclude this section we want to mention the collocation method of Blank [108] that is much closer in spirit to the methods of the previous sections than the method of Adomian or the variational iteration method in the sense that the latter are really more analytical than numerical approximations. The collocation scheme is based on choosing an N-dimensional linear space L of ansatz functions, for example a space of polynomial splines [175], and a set of N distinct points $x_1, \ldots, x_N$ in the interval where the solution of the differential equations is sought. One then tries to find a function $\phi \in L$ that satisfies the initial value problem (2.2.1) or, more conveniently, its Volterra integral formulation (2.3.1), at the points x_j, *i. e.*

$$\phi(x_j) = \sum_{k=0}^{\lceil\alpha\rceil-1} \frac{y_0^{(k)}}{k!} x_j^k + J^\alpha[f(\cdot, \phi(\cdot))](x_j), \quad j = 1, 2, \ldots, N. \tag{2.6.11}$$

This function ϕ is then taken as an approximation to the exact solution y of the initial value problem. In practice one chooses a basis $\{\phi_j : j = 1, 2, \ldots, N\}$ of the space L, represents the as yet unknown function ϕ as a linear combination of these basis function, *viz.*

$$\phi(x) = \sum_{k=1}^{N} a_k \phi_k(x),$$

and inserts this relation into each of the equations of the system (2.6.11). In this way we obtain an N-dimensional nonlinear system of equations with N unknowns that typically can be solved using appropriate methods.

In the special case that the differential equation is linear, the system becomes linear too and the solution theory and numerics become rather simple. Under this additional linearity assumption, Blank [108] has demonstrated that unique solutions exist and has developed a method to effectively compute these solutions.

Instead of using splines as basis functions, it is also possible to use functions from other classes such as, *e. g.*, the functions $\mathcal{P}_k^{(\alpha)}$ from Eq. (2.5.2). In this case which, following the nomenclature of Section 2.5, is known as *spectral collocation*, a similar theory can be established.

No matter which basis functions we use, a crucial point is always the selection of the collocation points. An unsuitable choice can lead to the system of collocation equations (2.6.11) to become very ill-conditioned or even unsolvable.

2.7 Methods for Terminal Value Problems

Sometimes one is interested in the solution of a somewhat different class of problems that are occasionally known under the name *terminal value problems*, see, *e. g.*, pp. 107–109 of [186]. Specifically we still want to solve a differential equation of the form

$$^{\mathrm{C}}D^{\alpha}y(x) = f(x, y(x)) \tag{2.7.1a}$$

on some interval $[0, x^*]$, say. For the sake of simplicity we restrict our attention to the case $0 < \alpha < 1$. Now, in contrast to the previously discussed problem described in Eq. (2.2.1), our additional condition that asserts the existence and uniqueness of the solution now takes the modified form

$$y(x^*) = y^*, \tag{2.7.1b}$$

i. e. we prescribe the function value at the end point of the interval of interest (hence the name, terminal value problem) and not at the starting point of the differential operator. It has been shown in [196] that under reasonable assumptions this problem indeed is uniquely solvable; thus it is a natural question to ask for numerical methods to find an approximate solution.

A proposal for a potential numerical algorithm can be derived from the analysis provided in Theorem 6.18 of [186]. This result essentially states that the terminal value problem (2.7.1) can be rewritten as an equivalent integral equation of the form

$$y(x) = y^* + \frac{1}{\Gamma(\alpha)} \int_0^{x^*} G(x,t) f(t, y(t)) dt$$

where

$$G(x,t) = \begin{cases} -(x^* - t)^{\alpha-1} & \text{for } t > x, \\ (x-t)^{\alpha-1} - (x^* - t)^{\alpha-1} & \text{for } t \le x. \end{cases}$$

The significant difference to the initial value problems treated previously is that this integral equation is an equation of Fredholm and not Volterra type. Nevertheless it is a class of integral equations for which numerical methods are available (see, *e. g.*, [273, 276, 327, 615]), so an application of one of these methods to the Fredholm equation immediately yields a numerical scheme for the terminal value problem.

An alternative way to construct a numerical method for terminal value problems has been briefly described in [196]. It is based on replacing the terminal condition (2.7.1b) by a standard-type initial condition

$$y(0) = y_0 \tag{2.7.2}$$

with an as yet unknown value y_0. One then uses the fact (see Corollary 6.16 of [186]) that the graphs of two solutions for the differential equation (2.7.1a) subject to two different initial conditions will, under standard assumptions, never intersect (the restriction $0 < \alpha < 1$ is relevant in this context though). Thus one starts with an arbitrary guess for the value y_0 and solves the initial value problem that consists of the differential equation (2.7.1a) and the initial condition (2.7.2) using any of the numerical schemes described above. Then one takes a look at the approximate solution at the point x^*. If this value is greater than the value y^* given in the terminal condition (2.7.1b) then one decreases y_0 and starts again; if it is smaller than y^* then y_0 has to be increased. This process is repeated in an iterative way, for example using a bisection strategy once one has reached a stage where an approximation with a too large value and an approximation with a too small value have been found, until one has determined an y_0 that leads to a numerical solution that matches the terminal condition up to the required accuracy.

In order to illustrate this scheme we recall here the example considered in [196], *viz.* the terminal value problem

$$ {}^{\mathrm{C}}D^{1/2}y(x) = \sin y(x), \qquad y(1) = 2.5. \tag{2.7.3} $$

We start by choosing an arbitrary initial value, in our case $y_0 = 1$, and compute the numerical approximation for $y(1)$ using the Adams-Bashforth-Moulton method with a step size of $1/200$. It turns out (see Table 2.1) that this leads to a too small value. Thus we restart the Adams method with $y_0 = 2$, which leads to a too large value. From here on we employ a simple bisection technique to find a new initial value, $y_0 = 1.5$, and compute $y(1)$ again. Proceeding in an iterative manner we find the required successive values for y_0 indicated in Table 2.1 and see that we can get as close to the desired exact solution as we like.

Table 2.1 Results of numerical algorithm for the terminal value problem (2.7.3).

y_0	1	2	1.5	1.75	1.625	1.6875	1.71875
$y(1)$	2.0556	2.63485	2.37728	2.51106	2.44567	2.47871	2.49496

The fact that the algorithm whose results we report here uses a simple bisection technique to compute the successive starting values y_0 explains why the convergence towards the correct terminal value is rather slow; a more sophisticated approach could be used to improve the convergence speed.

2.8 Numerical Methods for Multi-Term Fractional Differential Equations and Multi-Order Fractional Differential Systems

Up to this point we have only discussed so-called *single-term* fractional differential equations, *i. e.* equations containing only one differential operator. Whereas these equations form appropriate models for many problems in physics and other sciences, they are insufficient in some other cases. Indeed it is sometimes necessary to use differential equations involving differential operators of more than one order. Among the most prominent examples of such a situation we mention the *Bagley-Torvik equation* [605]

$$AD^2 y(x) + B{}^{\mathrm{C}}D^{3/2} y(x) + Cy(x) = g(x), \quad y(0) = y_0, y'(0) = y_0', \quad (2.8.1)$$

where $A \neq 0$ and B and C are arbitrary real numbers, This equation can be used to model the motion of a rigid plate immersed in a Newtonian fluid; see, *e. g.*, [605] or Section 8.3.2 of [487].

Other applications arise in the work of Koeller [347] who has given a theoretical explanation of why differential equations of the form

$$p_0\sigma(t) + p_1{}^{\mathrm{C}}D^{\alpha_1}\sigma(t) + p_2{}^{\mathrm{C}}D^{\alpha_2}\sigma(t) = q_0\epsilon(t) + q_1{}^{\mathrm{C}}D^{\alpha_1}\epsilon(t) + q_2{}^{\mathrm{C}}D^{\alpha_2}\epsilon(t)$$

may be used to model the mechanical behavior of viscoelastic materials with two relaxation modes. Here ϵ is strain and σ is stress. One of these quantities (in most cases ϵ) is usually known whereas the other one is the unknown solution of the multi-term equation. Koeller was particularly interested in the case $\alpha_2 = 2\alpha_1$, but this is by no means a must. Rossikhin and Shitikova [532] have recently pointed out that Koeller's equation is just a special case of a more general model for viscoelastic behavior introduced in a formally different but equivalent way by Rabotnov [512].

Our final example for an application of multi-term equations is the *Basset equation*

$$D^1 y(x) + b{}^{\mathrm{C}}D^{\alpha} y(x) + cy(x) = f(x), \qquad y(0) = y_0,$$

where $0 < \alpha < 1$. This equation describes the forces that occur when a sphere sinks in a (relatively less dense) fluid; see, *e. g.*, [397].

All these equations can be written in the form

$$ {}^{\mathrm{C}}D^{\alpha_n} y(x) = f(x, y(x), {}^{\mathrm{C}}D^{\alpha_1} y(x), \ldots, {}^{\mathrm{C}}D^{\alpha_{n-1}} y(x)) \tag{2.8.2} $$

with initial conditions

$$ y^{(k)}(0) = y_0^{(k)}, \qquad k = 0, 1, \ldots, \lceil \alpha_n \rceil - 1, \tag{2.8.3} $$

where we assume that $0 < \alpha_1 < \alpha_2 < \cdots < \alpha_n$. An equation of the form (2.8.2) is called a *multi-term fractional differential equation*. Without loss of generality we shall assume that all the integers that are contained in the interval $(0, \alpha_n)$ are also members of the finite sequence $(\alpha_k)_{k=1}^n$. In other words, it is impossible for two consecutive elements of the finite sequence (α_k) to lie on opposite sides of an integer number. It is an immediate consequence of this assumption that we have $0 < \alpha_{j+1} - \alpha_j \leq 1$ for all $j = 1, 2, \ldots, n-1$.

In order to illustrate this assumption, we look at the Bagley-Torvik equation (2.8.1). Here we would choose $\alpha_1 = 1$, $\alpha_2 = 3/2$, $\alpha_3 = 2$, and $f(x, y(x), {}^{\mathrm{C}}D^{\alpha_1} y(x), {}^{\mathrm{C}}D^{\alpha_2} y(x)) = (g(x) - B {}^{\mathrm{C}}D^{\alpha_2} y(x) - Cy(x))/A$, thus bringing the equation into the form (2.8.2). An analytical theory for equations of the type (2.8.2), including existence and uniqueness theorems, has been given in [194]. That paper also contains first steps towards the development of a general-purpose numerical algorithm for multi-term equations. Additional information may be found in the survey article [185].

Apart from these multi-term equations, there is a different possibility to introduce more than one fractional derivative into a mathematical model. Specifically, we may use a system of fractional differential equations, each of which has an order that may or may not coincide with the orders of the other equations. To put it more formally, we consider models of the type

$$ \begin{aligned} {}^{\mathrm{C}}D^{\alpha_1} y_1(x) &= f_1(x, y_1(x), \ldots, y_n(x)), \\ \vdots \quad & \quad \vdots \\ {}^{\mathrm{C}}D^{\alpha_n} y_n(x) &= f_n(x, y_1(x), \ldots, y_n(x)). \end{aligned} \tag{2.8.4} $$

As we shall see it is sufficient for our purposes to assume that $0 < \alpha_k \leq 1$ for all k. This implies that the initial conditions for the differential equation system (2.8.4) are

$$\begin{aligned} y_1(0) &= y_{1,0}, \\ \vdots \quad & \quad \vdots \\ y_n(0) &= y_{n,0}. \end{aligned} \tag{2.8.5}$$

A system of this class will be called a *multi-order fractional differential system*. Such systems seem to be investigated less frequently than multi-term equations, but we will now reveal some close connections between the two concepts, and thus the former deserve some attention at least in view of the fact that they can be very useful tools for the numerical treatment of the latter. The connections may be constructed in two different ways.

Specifically, given the equation (2.8.2), we may first write $\beta_1 = \alpha_1$, $\beta_j = \alpha_j - \alpha_{j-1}$ $(j = 2, 3, \ldots, n)$, $y_1 = y$ and $y_j = {}^{\mathrm{C}}D^{\alpha_{j-1}} y$, $j = 2, 3, \ldots, n$. Note that under our assumptions on the α_j it is clear that $0 < \beta_j \leq 1$ for all j. Then we can conclude:

Theorem 2.15. *The multi-term equation* (2.8.2) *with initial conditions* (2.8.3) *is equivalent to the system*

$$\begin{aligned} {}^{\mathrm{C}}D^{\beta_1} y_1(x) &= y_2(x), \\ {}^{\mathrm{C}}D^{\beta_2} y_2(x) &= y_3(x), \\ \vdots \quad & \quad \vdots \\ {}^{\mathrm{C}}D^{\beta_{n-1}} y_{n-1}(x) &= y_n(x), \\ {}^{\mathrm{C}}D^{\beta_n} y_n(x) &= f(x, y_1(x), y_2(x), \ldots, y_n(x)) \end{aligned} \tag{2.8.6}$$

with the initial conditions

$$y_j(0) = \begin{cases} y_0^{(0)} & \text{if } j = 1, \\ y_0^{(k)} & \text{if } \alpha_{j-1} = k \in \mathbb{N}, \\ 0 & \text{else} \end{cases} \tag{2.8.7}$$

in the following sense:

(*a*) *Whenever the function* $y \in C^{\lceil \alpha_n \rceil}[0, X]$ *is a solution of the multi-term equation* (2.8.2) *with initial conditions* (2.8.3), *the vector-valued function* $Y = (y_1, \ldots, y_n)^{\mathrm{T}}$ *with*

$$y_j(x) = \begin{cases} y(x) & \text{if } j = 1, \\ {}^{\mathrm{C}}D^{\alpha_{j-1}} y(x) & \text{if } j \geq 2, \end{cases} \tag{2.8.8}$$

is a solution of the multi-order system (2.8.6) *with initial conditions* (2.8.7).

(b) *Whenever the vector-valued function* $Y = (y_1, \ldots, y_n)^{\mathrm{T}}$ *is a solution of the multi-order system* (2.8.6) *with initial conditions* (2.8.7), *the function* $y = y_1$ *is a solution of the multi-term equation* (2.8.2) *with initial conditions* (2.8.3).

A proof of this result may be found in [185] (see also [193]).

To describe the second possible method we assume, for the sake of simplicity, that the highest order differential operator is not an integer-order derivative. Otherwise some small formal modifications in the notation are necessary, but the basic concept and the main results remain unchanged. For a general formal description of this alternative approach it is advantageous to express the multi-term equation in the form

$$\begin{aligned} {}^{\mathrm{C}}D^{k+\delta_{k,n_k}} y(x) = f(x, {}^{\mathrm{C}}D^0 y(x), {}^{\mathrm{C}}D^{\delta_{0,1}} y(x), \ldots, {}^{\mathrm{C}}D^{\delta_{0,n_0}} y(x), \\ {}^{\mathrm{C}}D^1 y(x), {}^{\mathrm{C}}D^{1+\delta_{1,1}} y(x), \ldots, {}^{\mathrm{C}}D^{1+\delta_{1,n_1}} y(x), \ldots, \\ {}^{\mathrm{C}}D^k y(x) \ldots, {}^{\mathrm{C}}D^{k+\delta_{k,n_k-1}} y(x)), \end{aligned} \tag{2.8.9}$$

where $0 < \delta_{j,1} < \delta_{j,2} < \cdots < \delta_{j,n_j} < 1$ for all j. The corresponding initial conditions are then

$$y^{(j)}(0) = y_0^{(j)}, \qquad j = 0, 1, \ldots, k. \tag{2.8.10}$$

In order to achieve our goal, we define

$$s(\mu, \sigma) = \sigma + \mu + 1 + \sum_{j=0}^{\mu-1} n_j \quad \text{and} \quad N = s(k, n_k) - 1 = k + \sum_{j=0}^{k} n_j.$$

Using this notation, we come to the following statement.

Theorem 2.16. *The multi-term equation* (2.8.9) *with initial conditions* (2.8.10) *is equivalent to the* N*-dimensional system*

$$\begin{aligned} {}^{\mathrm{C}}D^{\delta_{\mu,\sigma}} y_{s(\mu,0)}(x) &= y_{s(\mu,\sigma)}(x), \quad \mu = 0, 1, \ldots, k, \quad \sigma = 1, 2, \ldots, \widehat{\sigma}_\mu, \\ {}^{\mathrm{C}}D^1 y_{s(\mu,0)}(x) &= y_{s(\mu+1,0)}(x), \quad \mu = 0, 1, \ldots, k-1, \\ {}^{\mathrm{C}}D^{\delta_{k,n_k}} y_{s(k,0)}(x) &= f(x, y_1(x), y_2(x), \ldots, y_N(x)) \end{aligned} \tag{2.8.11}$$

where $\widehat{\sigma}_\mu = n_\mu$ *if* $0 \le \mu < k$ *and* $\widehat{\sigma}_k = n_k - 1$, *with the initial conditions*

$$y_j(0) = \begin{cases} y_0^{(k)} & \text{if there exists } k \text{ such that } j = s(k,0), \\ 0 & \text{else} \end{cases} \tag{2.8.12}$$

in the following sense:

(a) *Whenever the function* $y \in C^{k+1}[0, X]$ *is a solution of the multi-term equation* (2.8.9) *with initial conditions* [2.8.10], *the vector-valued function* $Y = (y_1, \ldots, y_N)^{\mathrm{T}}$ *with*

$$y_{s(\mu,\sigma)}(x) = \begin{cases} {}^{\mathrm{C}}D^{\mu}y(x) & \text{for } \sigma = 0, \\ {}^{\mathrm{C}}D^{\mu+\delta_{\mu,\sigma}}y(x) & \text{for } \sigma = 1, 2, \ldots, \widehat{\sigma}_{\mu}, \end{cases} \qquad \mu = 0, 1, \ldots, k, \tag{2.8.13}$$

is a solution of the multi-order system (2.8.11) *with initial conditions* (2.8.12).

(b) *Whenever the vector-valued function* $Y = (y_1, \ldots, y_N)^{\mathrm{T}}$ *is a solution of the multi-order system* (2.8.11) *with initial conditions* (2.8.12), *the function* $y = y_1$ *is a solution of the multi-term equation* (2.8.9) *with initial conditions* (2.8.10).

This result has been stated without proof for a subset of the class of equations described in (2.8.9) in [215]. A possible method of proof has been outlined in [185].

The key difference between these two approaches is that the method of Theorem 2.15 will convert the given multi-term equation into an n-dimensional multi-order system where all orders of the differential operators are strictly fractional (unless we have some j with $\alpha_j - \alpha_{j-1} = 1$), while the method of Theorem 2.16 will produce an n-dimensional system with $\lfloor \alpha_n \rfloor$ equations of order 1 and $n - \lfloor \alpha_n \rfloor$ equations of strictly fractional order. The latter may be advantageous from a numerical point of view because the first order equations involve local differential operators that may be discretized with a smaller computational cost.

A third method has been proposed by Diethelm and Ford [194]. That method however requires some number-theoretic assumptions on the orders α_j of the differential operators to be satisfied. In such a case the method provides a system of first-order equations that is equivalent to the given multi-term equation too. The formal advantage of this system is that all the equations are of the same order. This simplifies the analysis and implementation of the numerical method. However, there are also two disadvantages associated with the approach of [194]. Firstly, if the assumptions mentioned above are not satisfied then the method is not directly applicable; it can nevertheless be modified to provide a system of single-order equations that is not exactly, but at least approximately, equivalent to the given multi-term system. This is not a major drawback because, in principle, the differences in the solutions can be made arbitrarily small, and since one normally can only approximately solve the system anyway, the additional error introduced in this way will not have significant effects. The second disadvantage may be more serious: The dimension of the system that will be constructed can be very large. This may lead to an unnecessarily high arithmetic complexity [183]. Therefore we shall not go into the details of this third approach here.

Ford and Simpson [241] have shown in passing that these techniques may also be used to reformulate a single-order fractional differential equation whose order is greater than 1 in the form of an equivalent multi-order system with orders less than or equal to 1. The latter is a somewhat easier object for numerical work as some algorithms tend to behave much worse when applied to equations of higher order. In addition, this concept may be considered an extension of the well known classical technique for the numerical solution of initial value problems of higher integer order which consists of rewriting the problem in the form of a first-order system and solving this system numerically with the help of an algorithm for first-order initial value problems.

An even more general problem is the question for numerical methods for so-called *distributed-order equations*, *i. e.* equations of the form

$$\int_0^m A(r, {}^{\mathsf{C}}D^r y(x))dr = f(x, y(x)) \tag{2.8.14}$$

with certain functions A and f and suitable initial conditions. Equations of this type arise in a number of applications, see, *e. g.*, [42, 48, 49, 142–144, 286, 573, 609]. From a numerical point of view, the present knowledge about this class of problems is very limited. Therefore we shall restrict ourselves to drawing the reader's attention towards the works of Diethelm and Ford [191, 195] and the references cited therein.

We remark at this point that we also do not intend to discuss *variable-order equations*, *i. e.* equations of the form

$${}^{\mathsf{Co}}D^{\alpha(x)}y(x) = f(x, y(x)) \tag{2.8.15}$$

that contain a Coimbra operator as defined in Subsection 1.3.8, *i. e.* where the order of the differential operator is not a constant but depends on the free variable x. For applications of models of this (or an even more general) type we refer, *e. g.*, to [156, 583, 616].

2.9 The Extension to Fractional Partial Differential Equations

2.9.1 *General formulation of the problem*

In this Chapter, we have so far only dealt with *ordinary* differential equations of fractional order. However, in the modeling of various phenomena in finance, engineering, physics and other areas the use of *partial* differential equations with fractional differential operators is becoming more and more popular. A particularly important class of applications arising in this context are the so-called *time-fractional diffusion-wave equations*

$$ {}^{\mathrm{C}}D_t^\alpha y(x,t) + \phi(x,t) \sum_{j=1}^{n} \frac{\partial^2}{\partial x_j^2} y(x,t) = f(x,t) \tag{2.9.1}$$

for $x = (x_1, \ldots, x_n) \in G \subset \mathbb{R}^n$ and $t \in [0,T]$. Here, the notation ${}^{\mathrm{C}}D_t^\alpha y(x,t)$ indicates the partial derivative of Caputo type of order α of the function y with respect to t. It is clear that this equation unifies the classical concepts of the n-dimensional diffusion equation (heat equation) which is obtained for $\alpha = 1$ and the n-dimensional wave equation ($\alpha = 2$). In order to obtain a well-posed problem, it is natural to combine Eq. (2.9.1) with two boundary conditions and one (if $\alpha \le 1$) or two (if $1 < \alpha \le 2$) initial conditions. These conditions are usually given in the same form as in the classical case, *viz.*

$$ A(x,t)y(x,t) + B(x,t)\frac{\partial}{\partial n} y(x,t) = g(x,t) \quad \text{for } t \ge 0 \text{ and } x \in \partial G \tag{2.9.2}$$

(with $\partial/\partial n$ denoting the partial derivative in the direction of the outer normal of the boundary at the point x) and

$$ y(x,0) = f_1(x), \quad \frac{\partial}{\partial t} y(x,0) = f_2(x) \quad \text{for } x \in G \tag{2.9.3}$$

(the second condition of Eq. (2.9.3) of course being applicable only in the case $1 < \alpha \le 2$). As the name *time-fractional diffusion-wave equations* indicates, we can obtain this class of equations by taking a classical diffusion or wave equation and replacing the first- or second-order derivative with respect to the time variable by a fractional derivative. From a formal point of view it would seem to be equally natural to replace the second-order derivatives with respect to the space variables by their fractional-order generalizations. This has indeed been done occasionally; however, some authors raise strong objections against this idea because it leads to certain phenomena that violate fundamental principles of physics. For a thorough and well written discussion of this issue we refer to Section 2.3.2 of Hilfer's work [305].

A straightforward way to construct efficient numerical methods for this class of equations can be described in a very simple way. Specifically we can discretize the second order derivatives with respect to space by a classical second order central difference quotient whereas one of the numerical schemes mentioned in the previous sections can be used to discretize the fractional derivative with respect to the time variable. We shall demonstrate the details of this method by means of an example in one space dimension,

$$ {}^{\mathrm{C}}D_t^{\alpha} u(x,t) + \phi(x,t)\frac{\partial^2}{\partial x^2}u(x,t) = f(x,t) \tag{2.9.4} $$

for $t \in [0,T]$ and $x \in [a,b]$, combined with appropriate initial and boundary conditions that will be discussed more precisely later.

Let $\Delta x = (b-a)/N$ and $\Delta t = T/M$ denote the step size of the discretization in the space and time axis respectively, where the values N and M are assumed to be given values defining the size of the discretization grid

$$ \{(x_i, t_j) : i = 0,1,\ldots,N, j = 0,1,\ldots,M\} $$

where $x_0 = a$ and $x_N - b$. Using the discretization on the space axis, the second derivative $u_{xx}(x,t)$ can be approximated by the central difference of second order

$$ \frac{1}{\Delta x^2}\begin{pmatrix} d_1 & o_1 & & & \\ 1 & -2 & 1 & & \\ & \ddots & \ddots & \ddots & \\ & & 1 & -2 & 1 \\ & & & o_2 & d_2 \end{pmatrix}\begin{pmatrix} u(a,t) \\ u(x_1,t) \\ \vdots \\ u(x_{N-1},t) \\ u(b,t) \end{pmatrix}, \tag{2.9.5} $$

where the values d_1, o_1, o_2, d_2 are determined by the initial and boundary conditions and will be given later in this section.

While the discretization in space is done as in the case of classical diffusion or wave equations, the discretization of the time component in the diffusion-wave equation (2.9.4) contains, as mentioned in the previous sections, a more complex structure because of the non-local character of fractional derivatives.

We had seen above that, in general, the discretization of ${}^{\mathrm{C}}D_t^{\alpha}u(x,t)$ can be defined by

$$\frac{1}{\Delta t^{\alpha}}\begin{pmatrix}\omega_{0,0} & & \\ \vdots & \ddots & \\ \omega_{M,0} & \dots & \omega_{M,M}\end{pmatrix}\begin{pmatrix}u(x,t_0)\\ \vdots \\ u(x,t_M)\end{pmatrix} \tag{2.9.6}$$

with some weights $\omega_{k,j}$ defined by the approximation method used. At the time step t_k, $k=1,\dots,M$, the values for $u(x_i,t_j)$ for $i=0,1,\dots,N$ and $j=0,\dots,k-1$ are known. Thus at the time step t_k the sum

$$\sum_{\nu=0}^{k-1}\omega_{k,\nu}u(x_i,t_\nu), \quad i=0,1,\dots,N$$

can be transferred to the right hand side of the discretization scheme. Therefore at the time step t_k a tridiagonal system has to be solved, where the coefficient matrix is given by

$$\phi(x_i,t_k)\mathbf{A}+\mathbf{B}, \quad i=0,1,\dots,N.$$

Here the matrix $\mathbf{A}$ is defined by the discretization (2.9.5) and the matrix $\mathbf{B}$ is the matrix containing only the main diagonal of the discretization (2.9.6). With the above discretization, the right hand side at time step t_k is given by

$$\begin{pmatrix} r_1 \\ f(x_1,t_k) \\ \vdots \\ f(x_{N-1},t_k) \\ r_2 \end{pmatrix} + \begin{pmatrix} 0 \\ {}^{\mathrm{C}}D_t^{\alpha}(T_{n-1}[u;0])(x_1,t_k) \\ \vdots \\ {}^{\mathrm{C}}D_t^{\alpha}(T_{n-1}[u;0])(x_{N-1},t_k) \\ 0 \end{pmatrix} - \begin{pmatrix} 0 \\ \sum_{\nu=0}^{k-1}\omega_{k,\nu}u(x_1,t_\nu) \\ \vdots \\ \sum_{\nu=0}^{k-1}\omega_{k,\nu}u(x_{N-1},t_\nu) \\ 0 \end{pmatrix}.$$

The matrix entries d_1, o_1, o_2, d_2 and the vector entries r_1, r_2 are determined by the initial and boundary conditions using Taylor approximation.

Denoting $\xi_1 = a$ and $\xi_2 = b$ they are given for $i = 1, 2$ at time step t_k by:

$$
\begin{aligned}
d_i &= \frac{g_i(t_k)}{\Delta x} + h_i(t_k)(-1)^i \left[\frac{1}{\Delta x^2} - \frac{\omega_{k,k}}{2\phi(\xi_i, t_k)\Delta t^\alpha}\right], \\
o_i &= (-1)^{i+1}\frac{h_i(t_k)}{\Delta x^2}, \\
r_i &= \frac{r_i(t_k)}{\Delta x} + \frac{(-1)^{i+1} h_i(t_k)}{2\phi(\xi_i, t_k)} \\
&\quad \times \left[f(\xi_i, t_k) - \frac{1}{\Delta t^\alpha}\sum_{\nu=0}^{k-1} \omega_{k,\nu} u(\xi_i, t_\nu) + D_t^\alpha(T_{n-1}[u;0])(\xi_i, t_k))\right].
\end{aligned}
$$

In order to complete the description, it remains to say explicitly what the coefficients in the discretization of the fractional differential operator, *i. e.* the entries in the matrix of Eq. (2.9.6), look like. In view of the experience available from the classical situation $\alpha = 1$ it seems most natural to use backward differentiation formulas (BDF). For the beginning, only the most simple instances, *i. e.* BDFs having only first order accuracy, will be considered. There are two main ways to construct this. The first one is the approach of Subsection 2.2.2, the second one is based on the ideas discussed in Section 2.4.

In the above numerical method those two schemes differ only in the weights $\omega_{k,j}$ of (2.9.6). For the method of Subsection 2.2.2 they are given by

$$
\omega_{k,k-j} = \frac{1}{\Gamma(2-\alpha)} \times \begin{cases} 1 & \text{for } j = 0, \\ (j-1)^{1-\alpha} - 2j^{1-\alpha} + (j+1)^{1-\alpha} & \text{for } 1 \le k \le j-1, \\ (k-1)^{1-\alpha} - (\alpha-1)k^{-\alpha} - k^{1-\alpha} & \text{for } k = j \ge 1 \end{cases} \tag{2.9.7}
$$

according to Eq. (2.1.11).

For the Grünwald-Letnikov approach the weights are identical on the main diagonal and any off-diagonal. Thus the first row in (2.9.6) determines the whole set of weights and is given by (see [385, 488])

$$
\omega_{k,0} = (-1)^k \binom{\alpha}{k}. \tag{2.9.8}
$$

This approach gives a first order approximation if the function $u(x, t)$ is a causal function with respect to time (*i. e.* $u(x, t) \equiv 0$ if $t < 0$), and if

$u(x,0) = 0$ (which usually is not the case unless a homogeneous initial condition is defined). For the cases where $u(x,0) \neq 0$, additional starting weights $w_{k,0}$ need to be added to the first row in (2.9.6), given by

$$\begin{aligned} w_{k,0} &= \frac{k^{-\alpha}}{\Gamma(1-\alpha)} - (-1)^k \frac{\Gamma(\alpha)}{\Gamma(\alpha-k)\Gamma(k+1)} \\ &= \frac{k^{-\alpha}}{\Gamma(k-\alpha)} - \sum_{j=0}^{k} \omega_{j,0}, \end{aligned} \tag{2.9.9}$$

while the rest of the matrix stays as defined in (2.9.8) above (see, *e. g.*, [385]).

While the convergence order for the time-discretization of Diethelm's method is $\mathcal{O}(\Delta t^{2-\alpha})$, the order of convergence for Lubich's method is $\mathcal{O}(\Delta t^{\alpha})$ (see, *e. g.*, [189]). Thus for fractional diffusion equations, Diethelm's method is superior, while for fractional wave equations Lubich's approach is advantageous. In the following example in each case the theoretically better suited method is used.

2.9.2 *Examples*

Consider the very simple time-fractional diffusion problem

$$^{\mathrm{C}}D_t^{1/2} u(x,t) - \frac{\partial^2}{\partial x^2} u(x,t) = 0 \tag{2.9.10}$$

with the initial condition

$$u(x,0) = u_0(x) = x$$

and boundary conditions

$$u(a,t) = 0, \quad u(b,t) = 1.$$

The solution on $[a,b] \times [0,T] = [0,1] \times [0,1]$ is given by $u(x,t) = x$. The algorithm described above reproduced the exact solution up to machine precision with grid parameters $N = 67$ and $M = 80$ both for the choice of weights indicated in Eq. (2.9.7) and for the weights given according to Eq. (2.9.8).

For the second example the differential equation

$$^{\mathrm{C}}D_t^{\alpha} u(x,t) - \frac{\partial^2}{\partial x^2} u(x,t) = 0 \tag{2.9.11a}$$

is solved on $[0, \pi] \times [0, 10]$ with the initial condition

$$u(x,0) = \sin(x), \quad u_t(x,0) = 0 \tag{2.9.11b}$$

and boundary conditions $u(0,t) = 0$, $u(\pi, t) = 0$. This equation can either be viewed as the problem of temperature distribution in a bar generated by a point heat source kept in the middle (if $\alpha \in (0,1)$), or (if $\alpha \in (1,2)$) as the problem of the deflection of a string subjected to a point load at the center of the string in a string vibration setting. Of course the second initial condition mentioned in Eq. (2.9.11b) must be ignored in the case $0 < \alpha < 1$. (See also [17] for related results).

In Fig. 2.1 the numerical solution is plotted for the case $\alpha = 1/2$. The number of nodes in time and space were 100 and 30 respectively. In Fig. 2.2 the solution for the classical version of the same diffusion problem (*i. e.* $\alpha = 1$) is shown for comparison. The obvious difference between those two cases is that the fractional case exhibits fast diffusion in the beginning and slow diffusion later on. If α is chosen smaller, the diffusion process over time becomes even slower. This result was to be expected by the analytical background described in Section 2 of [202].

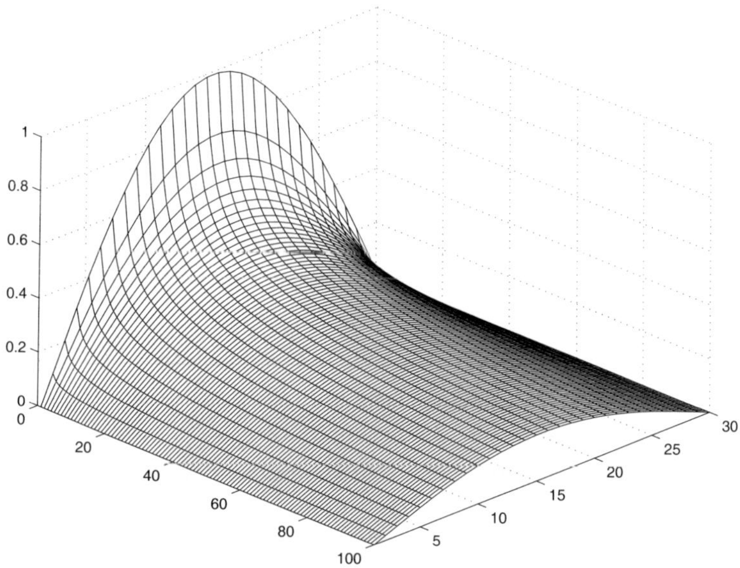

Fig. 2.1 Numerical solution of Eq. (2.9.11) for $\alpha = 1/2$.

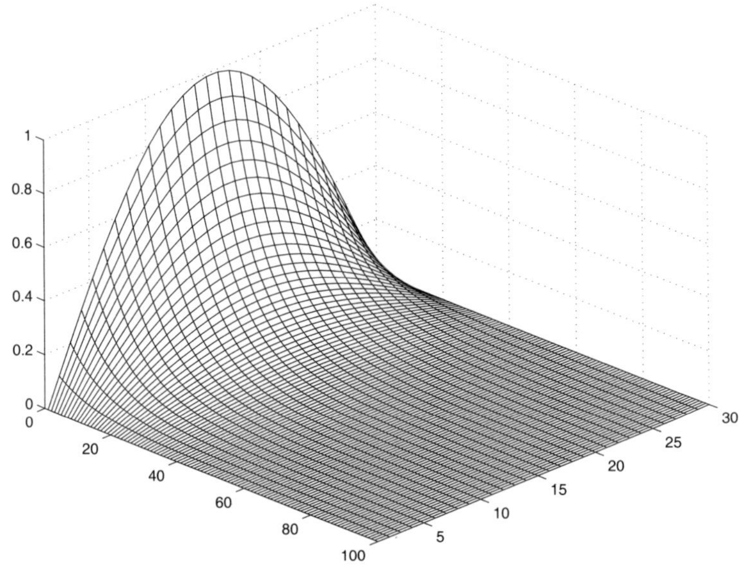

Fig. 2.2 Numerical solution of Eq. (2.9.11) for $\alpha = 1$.

In addition, in Fig. 2.3 the numerical solution for $\alpha = 3/2$ is plotted, which exhibits a behavior that can be said to be of an intermediate form between classical diffusion and classical waves, but with the diffusive part dominating the character. If we increase the parameter α then the qualitative behavior becomes closer to that of the wave equation. This is evident from Fig. 2.4, where $\alpha = 1.75$. The solution of the classical wave equation (*i. e.* $\alpha = 2$) is plotted in Fig. 2.5 for comparison.

The described behaviors were also reproduced by Agrawal [17] for a similar problem, but using a completely different method (namely, the exact solution was computed explicitly in the form of an infinite series, and a finite partial sum of this series was then used as an approximation to the exact solution).

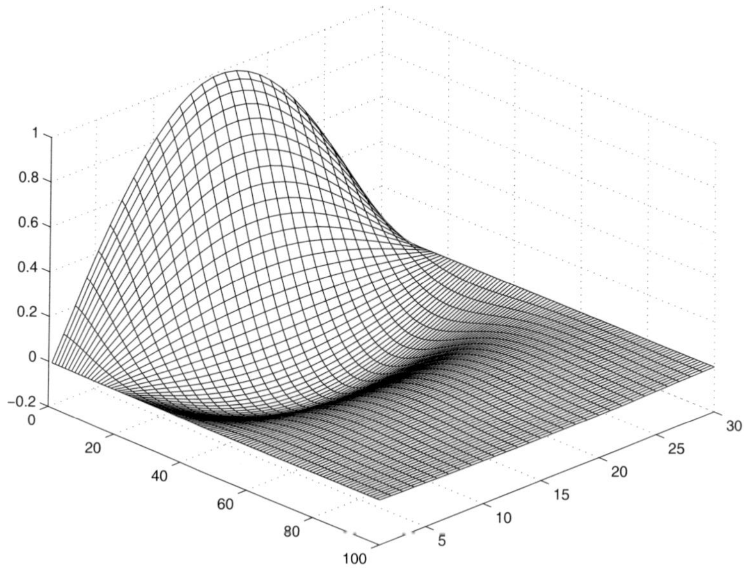

Fig. 2.3 Numerical solution of Eq. (2.9.11) for $\alpha = 3/2$.

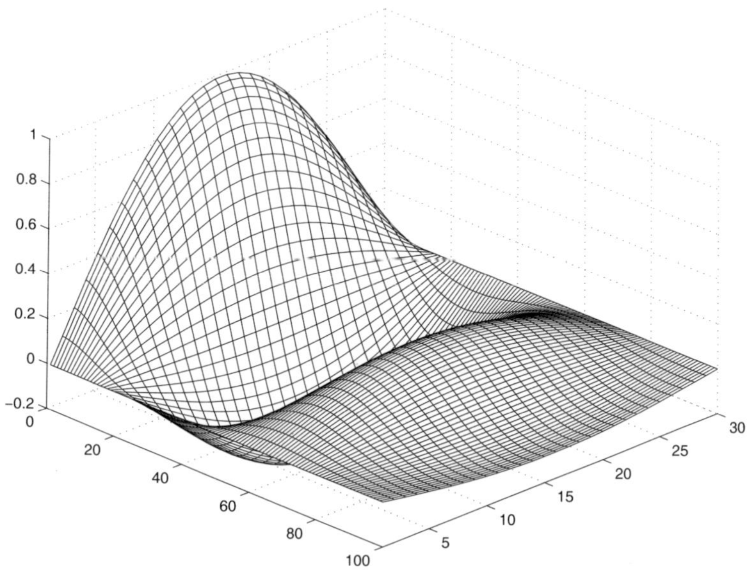

Fig. 2.4 Numerical solution of Eq. (2.9.11) for $\alpha = 7/4$.

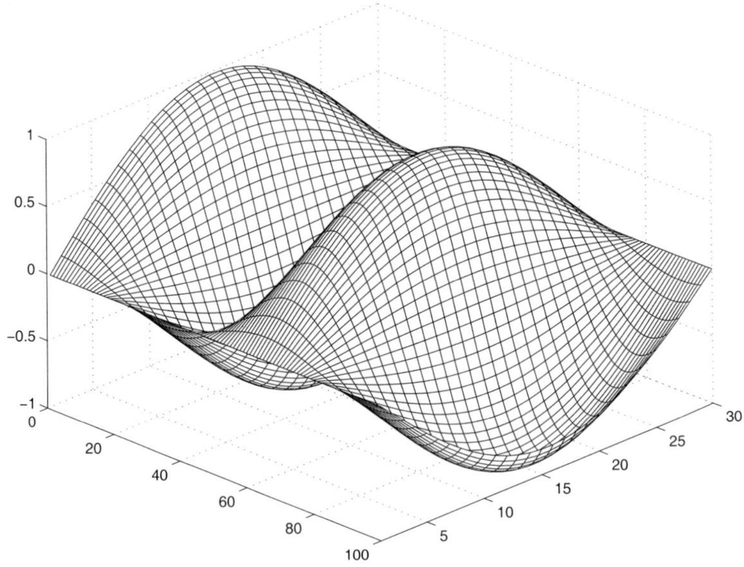

Fig. 2.5 Numerical solution of Eq. (2.9.11) for $\alpha = 2$.

Chapter 3

Specific Efficient Methods for the Solution of Ordinary and Partial Fractional Differential Equations

In the previous chapter we had introduced a number of numerical methods for (ordinary and partial) fractional differential equations. Our goal now is to look more closely at some especially important methods from this selection. In particular we want to provide the reader with some basic knowledge that allows to decide for a concrete algorithm for a specific problem at hand. Moreover we want to mention a few techniques that are generally applicable to many classes of algorithms and that allow to obtain slightly modified versions that retain the convergence behavior but that require less computational effort.

The source codes of some very significant special algorithms will be listed in Appendix A.

3.1 Methods for Ordinary Differential Equations

We shall begin with the investigation of algorithms for ordinary fractional differential equations; partial fractional differential equations will be treated later.

3.1.1 *Dealing with non-locality: The finite memory principle, nested meshes, and the approaches of Deng and Li*

A key problem in connection with the numerical solution of fractional differential equations is the fact that fractional differential operators are never local. This means that, whenever we want to compute the value of the expression ${}^{C}D^{\alpha}y(x)$, we need to take into consideration the entire history of the function y, *i. e.* the function values on the complete interval $[0, x]$. In contrast, for a derivative of integer order it would be sufficient to have information on an arbitrarily small neighborhood of x. This crucial difference implies, as can be seen from a close inspection of the numerical algorithms introduced in Chapter 2, that the computation of a numerical solution of an ordinary fractional differential equation on a fixed interval $[0, b]$, say, by one of the standard algorithms using a step size of h has an arithmetic complexity of $O(h^{-2})$ which is much more effort than the $O(h^{-1})$ operation count that we can observe for differential equations of first order. In particular if one is interested in a solution over a rather long interval, this can be a serious problem. Therefore a number of ideas has been proposed in order to overcome this difficulty. We shall look at the three best known and most frequently used concepts and discuss their merits.

The first of these three methods is the *finite memory principle* (sometimes also denoted as *short memory principle*). This very intuitive idea is described, *e. g.*, in Section 7.3 of [487]. Basically, one defines a fixed value $T > 0$, the so-called memory length, and modifies the Caputo derivative

$$ {}^{C}D^{\alpha}y(x) = \frac{1}{\Gamma(n-\alpha)} \int_0^x (x-t)^{n-\alpha-1} y^{(n)}(t)dt $$

(where $n = \lceil \alpha \rceil$) of the function y if $x > T$. (No changes are necessary for $x \leq T$.) In the case $x > T$ the length of the range of integration, *i. e.* the length of the actual memory used in the definition of the Caputo operator, is greater than the prescribed memory length T, and it grows even larger as x grows. The idea of the finite memory principle is then not to use the complete memory but only a portion of length T of it. Thus one uses

$$ \frac{1}{\Gamma(n-\alpha)} \int_{x-T}^x (x-t)^{n-\alpha-1} y^{(n)}(t)dt $$

instead of ${}^{C}D^{\alpha}y(x)$ if $x > T$. In mathematical terms, we increase the lower terminal point of the interval of integration, *i. e.* we choose to take into account the right part of $[0, x]$ and to ignore the contribution of the left part of this interval. This choice has been taken because of the monotonicity property of the kernel $(x-t)^{n-\alpha-1}$: the exponent of this expression is negative, and so the contribution of the left part is likely to be much smaller than the contribution of the right part, thus this choice minimizes the error introduced by the scheme. In this way one always needs to approximate an operator that only takes into account function values from the interval $[x-T, x]$ whose length never exceeds T. Hence, this very simple strategy allows to reduce the arithmetic complexity of standard numerical methods to the $O(h^{-1})$ count that we know from algorithms for first-order equations.

Of course, there is a price that we have to pay for this reduction in computational cost, and this is a loss of accuracy due to the fact that we simply ignore a part of the integration interval. Some authors claim that it is possible to keep this loss of accuracy so small that the error introduced by the numerical scheme for the differential equation itself still dominates the newly introduced contribution, thus allowing us to effectively neglect the latter. However, the careful inspection of Ford and Simpson [242] reveals that in order to really achieve this goal, one must choose the memory length T so large that the real improvement of the run time is negligible too. Thus it seems that this method is too simple to be successful in practice.

To overcome these problems, Ford and Simpson [242] have then derived a modified approach that allows to retain the order of convergence of the underlying basic numerical algorithm and at the same time reduce the arithmetic complexity to $O(h^{-1} \ln |h|)$ which is much better than the original $O(h^{-2})$ operation count and only marginally worse than the $O(h^{-1})$ that we know from the theory of first-order equations. Their idea is best explained by looking at a concrete example. For this purpose we shall choose the quadrature-based direct method developed in Subsection 2.2.2 and consider the practically most important case $0 < \alpha < 1$. We recall that it was based on a direct approximation of the differential operator in the given initial value problem

$$^{\mathrm{C}}D^{\alpha} y(x) = f(x, y(x)), \qquad y^{(k)}(0) = y_0^{(k)}, \quad k = 0, 1, \ldots, \lceil \alpha \rceil - 1. \quad (3.1.1)$$

with the help of a quadrature formula with equispaced nodes. Thus, at the grid point x_j we need to take into consideration function values of the given function f at all previously handled grid points $x_0, x_1, \ldots, x_{j-1}$ which gives rise to an $O(j^2)$ operation count after j steps and hence $O(N^2) = O(h^{-2})$ operations at the end point of the interval of interest. Now we recall the observation that Podlubny had already used in his finite memory principle but use it in a different way. To be precise, in the analysis of the finite memory approach we had seen that we cannot afford to ignore the contribution coming from the parts that are a long way away from x completely. But it is possible to approximate the corresponding part of the integral with a more economical formula, *i. e.* a formula that uses a smaller number of nodes. This will evidently reduce the arithmetic complexity and, if this new quadrature formula is chosen carefully, then we can still retain the order of convergence.

Specifically the idea is to choose two parameters $w > 0$ and $T > 0$, compute (for each x) the smallest integer m that satisfies $x < w^{m+1}T$ (*i. e.* $m = \lceil(\ln x - \ln T)/\ln w\rceil - 1$) and then to decompose the interval of integration according to

$$[0, x] = [0, x - w^m T] \cup [x - w^m T, x - w^{m-1}T] \cup \cdots \cup [x - wT, x - T] \cup [x - T, x].$$

One can then distribute the computational effort over the history of the function in a logarithmic instead of a uniform way. This means that we choose a basic step size $h > 0$ but use this step size only on the rightmost subinterval of the partition mentioned above, *i. e.* on $[x - T, x]$. As we move farther and farther away from the point x, we then increase the step size according to a simple strategy: On an interval of the form $[x - w^j T, x - w^{j-1}T]$ we use a step size of $w^{j-1}h$. The leftmost interval may not be amenable to this approach because it is possible that its length is not an integer multiple of the required step size; one can simply use the basic step size h here without sacrificing too much run time because this interval is rather small. An example of the distribution of the quadrature nodes is given in Figure 3.1 where we have chosen $w = 2$. The scheme requires only 91 nodes for this particular choice of the parameters whereas a uniform grid would use 201 nodes. It is clear that the advantage of the nested mesh approach grows as x becomes larger; to be precise the number of nodes used at the grid point $x = kh$ reduces from $k + 1$ to $O(\ln k)$. This means that the overall complexity of a differential equation solver with step size h that uses this approach is only $O(h^{-1}|\ln h|)$ and not $O(h^{-2})$ as a standard approach would have. Nevertheless it is possible to prove a very pleasant result about the error of the scheme (see Theorem 1 of [242]):

Theorem 3.1. *The nested mesh scheme preserves the order of the underlying quadrature rule on which it is based.*

Yet another promising approach (that can actually be combined with the

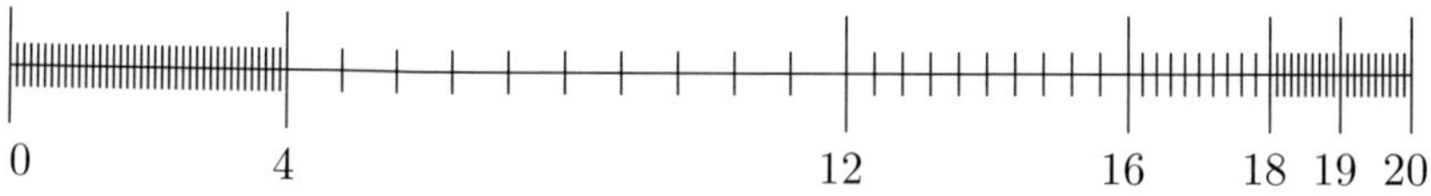

Fig. 3.1 Distribution of quadrature nodes according to the nested mesh principle of Ford and Simpson for $x = 20$, $w = 2$, $h = 1/10$ and $T = 1$. The step sizes are $h = 1/10$ in the leftmost subinterval $[0, 4]$ and in the two rightmost subintervals $[18, 19]$ and $[19, 20]$, $h = 2/10$ in $[16, 18]$, $h = 4/10$ in $[12, 16]$ and $h = 8/10$ in $[4, 12]$.

nested mesh idea to produce an even stronger method) has been suggested by Deng [179]. It is based on an analytic observation that has its origin in a well known property of first-order equations. In order to explain the idea we shall once again start from the usual initial value problem

$$ {}^{C}D^{\alpha} y(x) = f(x, y(x)), \qquad y^{(k)}(0) = y_0^{(k)}, \quad k = 0, 1, \ldots, \lceil \alpha \rceil - 1, \quad (3.1.2) $$

and recall that it can be rewritten in the equivalent form of a Volterra integral equation

$$ y(x) = \sum_{k=0}^{\lceil \alpha \rceil - 1} \frac{y_0^{(k)}}{k!} x^k + \frac{1}{\Gamma(\alpha)} \int_0^x (x - t)^{\alpha - 1} f(t, y(t)) dt \qquad (3.1.3) $$

(see, *e. g.*, Lemma 6.2 of [186]). Now we look at this relation at two consecutive grid points, *i. e.* for $x = x_j$ and $x = x_{j+1}$, and subtract the equations from each other. This yields

$$ \begin{aligned} & y(x_{j+1}) - y(x_j) \\ & = \sum_{k=0}^{\lceil \alpha \rceil - 1} \frac{y_0^{(k)}}{k!} (x_{j+1}^k - x_j^k) + \frac{1}{\Gamma(\alpha)} \int_{x_j}^{x_{j+1}} (x_{j+1} - t)^{\alpha - 1} f(t, y(t)) dt \\ & \quad + \frac{1}{\Gamma(\alpha)} \int_0^{x_j} \left((x_{j+1} - t)^{\alpha - 1} - (x_j - t)^{\alpha - 1} \right) f(t, y(t)) dt \end{aligned} $$

or, equivalently,

$$ \begin{aligned} y(x_{j+1}) = y(x_j) & + \sum_{k=0}^{\lceil \alpha \rceil - 1} \frac{y_0^{(k)}}{k!} (x_{j+1}^k - x_j^k) \\ & + \frac{1}{\Gamma(\alpha)} \int_{x_j}^{x_{j+1}} (x_{j+1} - t)^{\alpha - 1} f(t, y(t)) dt \qquad (3.1.4) \\ & + \frac{1}{\Gamma(\alpha)} \int_0^{x_j} \left[(x_{j+1} - t)^{\alpha - 1} - (x_j - t)^{\alpha - 1} \right] f(t, y(t)) dt. \end{aligned} $$

Now when we want to approximately compute $y(x_{j+1})$ then we have already computed the expression $y(x_j)$ that appears on the right-hand side so we only need to look this value up which can be achieved in constant time. In addition, the sum on the right-hand side of Eq. (3.1.4) is zero for $0 < \alpha \leq 1$ and it can be easily computed in $O(1)$ time, independent of j, if $\alpha > 1$. Moreover the first integral on the right-hand side extends over an interval of length h, also independent of j, and therefore it can be computed by an appropriate quadrature formula in $O(1)$ time as well. Thus all that remains to be done is to construct an efficient approximation for the second integral on the right-hand side, *i. e.* the integral that contains the history of the function. It is well known that this integral is zero for the classical (non-fractional) case $\alpha = 1$ (this is clear because the factor in brackets vanishes). Unfortunately no such simplification is available when $\alpha \neq 1$, but at least we may still observe that $(x_{j+1} - t)^{\alpha-1} \approx (x_j - t)^{\alpha-1}$ if $x_j - t \gg x_{j+1} - x_j$. In other words, the factor in brackets is very small at least for those t that are relatively far away from x_j and thus it is unproblematic to use a less accurate but computationally cheap quadrature formula (*e. g.* a formula with a rather coarse mesh) in this range.

It is sometimes also helpful to use the following alternative representation discussed by Li [369] for the Riemann-Liouville integral operator J_a^n.

Theorem 3.2. *Let $f \in C[0, b]$, $\alpha > 0$ and $x \in [0, b]$. Then,*

$$J^\alpha f(x) = \frac{1}{\Gamma(\alpha)\Gamma(1-\alpha)} \int_0^\infty g(\xi, x)\xi^{-\alpha} d\xi$$

where

$$g(\xi, x) = \int_0^x \exp(-\xi(x-t))f(t)dt.$$

Remark 3.1. Note that the function g is actually independent of α. Thus, if one needs to compute $J^\alpha f$ for various different values of α then it suffices to determine g once; only the integration of $g(\cdot, x)(\cdot)^{-\alpha}$ needs to be performed a multiple number of times.

The main application of Theorem 3.2 is based on the following property of the function g which is an immediate consequence of its definition.

Lemma 3.1. *Let $f \in C[0, b]$, $\alpha > 0$, $h > 0$ and $x, x+h \in [0, b]$. Then, the function g introduced in Theorem 3.2 satisfies*

$$g(\xi, x+h) = \exp(-\xi h)g(\xi, x) + \tilde{g}(\xi, x, h)$$

for all $\xi \geq 0$, *where*

$$\tilde{g}(\xi, x, h) = \int_x^{x+h} \exp(-\xi(x + h - t)) f(t) dt.$$

Lemma 3.1 in conjunction with the representation of Theorem 3.2 allows us to establish a useful way of computing $J^\alpha f(x + h)$ for some $h > 0$ if $J^\alpha f(x)$ is already known. Specifically, while a direct application of the definition of J^α would require us to recompute the entire integral from 0 to x, and thus to re-evaluate the complete history of f, we are now in a position to invoke the auxiliary function g. Assuming that we have already computed $g(\cdot, x)$, the computation of $g(\cdot, x + h)$ requires only the multiplication of $g(\cdot, x)$ by a constant and the addition of a locally defined integral, *i. e.* an integral over the short interval $[x, x + h]$, followed by the evaluation of the integral $\int_0^\infty g(\xi, x)\xi^{-n} d\xi$ by a suitable method. Efficient numerical techniques for the latter problem have been discussed by Li [369].

Remark 3.2. For the sake of completeness we remark that the function g from Theorem 3.2 solves the first-order initial value problem

$$\frac{\partial}{\partial x} g(\xi, x) = -\xi g(\xi, x) + f(x), \quad g(\xi, 0) = 0$$

which can be easily verified by inserting the definition of g. Note that, since ξ can be considered to be a constant parameter in this differential equation, we effectively deal with a linear inhomogeneous ordinary differential equation with constant coefficients. Thus, this approach for the Riemann-Liouville integral is very similar to the techniques for Caputo type derivatives mentioned in Subsection 2.6.3.

3.1.2 *Parallelization of algorithms*

Another straightforward approach to tackle the computational complexity introduced by the non-locality of the fractional differential operators is based on using parallel computers. We shall explain this approach, first discussed in [187], on the basis of the Adams-Bashforth-Moulton method in its standard form. However, the basic principle can also be applied to modifications of this algorithm like the nested mesh concept described above or even to many completely different classes of numerical methods.

To this end, we shall look at the usual initial value problem

$$ {}^{\mathrm{C}}D^{\alpha}y(x) = f(x, y(x)), \qquad y^{(k)}(0) = y_0^{(k)}, \quad k = 0, 1, \ldots, \lceil \alpha \rceil - 1 \quad (3.1.5) $$

and recall the standard Adams-Bashforth-Moulton method from Subsection 2.3.2. Our first observation is that the computation of the weights a_{jk} and b_{jk} of the method as given by Eqs. (2.1.7) and (2.1.9), respectively, is computationally extremely cheap compared to the effort required for the evaluation of the approximate solution according to Eqs. (2.3.5) and (2.3.6), respectively. Thus we shall concentrate on the parallelization of the algorithm for the latter equations.

For this purpose, we assume to have a hardware platform with p compute cores. Our idea is then to divide the set $y_1, y_2, \ldots, y_N$ of values to be computed into blocks of size p in such a way that the ℓth block contains the variables $y_{(\ell-1)p+1}, \ldots, y_{\ell p}$ ($\ell = 1, 2, \ldots, \lceil N/p \rceil - 1$), and the $\lceil N/p \rceil$th block contains the remaining variables $y_{(\lceil N/p \rceil - 1)p+1}, \ldots, y_N$. (This last block contains p elements if and only if p is a divisor of N; otherwise it will contain less than p elements. No problems will result from this fact.) The processing will then happen by handling one block after the other, and during the handling of each block, each processor will be assigned to exactly one of the block's variables. After the computations for each block have been finished, all processors communicate their results to the other processors. This implies that, at the beginning of the ℓth block, all y_j computed in the previous blocks are known to all processors. Based on this structure, the exact procedure used in the computations for the ℓth block can now be given.

The key to an efficient parallelization of the Adams method is that we exploit the structure of the sums in Eqs. (2.3.6) and (2.3.5). Specifically we rewrite these two sums in the forms

$$ y_{j+1,0} = I_{j+1} + h^{\alpha} H_{j,\ell}^{(0)} + h^{\alpha} L_{j,\ell}^{(0)} \qquad (3.1.6) $$

and

$$ y_{j+1} = I_{j+1} + h^{\alpha} H_{j,\ell} + h^{\alpha} L_{j,\ell}, \qquad (3.1.7) $$

respectively. Here,

$$ I_{j+1} = \sum_{k=0}^{\lceil \alpha \rceil - 1} \frac{x_{j+1}^k}{k!} y_0^{(k)} \qquad (3.1.8) $$

is a sum with a fixed and typically very small number of summands that appears in both sums. Moreover, the other quantities appearing in Eqs. (3.1.6) and (3.1.7) are defined by

$$H_{j,\ell}^{(0)} = \sum_{k=0}^{(\ell-1)p} b_{j-k} f(x_k, y_k),$$

$$L_{j,\ell}^{(0)} = \sum_{k=(\ell-1)p+1}^{j} b_{j-k} f(x_k, y_k),$$

$$H_{j,\ell} = c_j f(x_0, y_0) + \sum_{k=1}^{(\ell-1)p} a_{j-k} f(x_k, y_k),$$

and

$$L_{j,\ell} = \sum_{k=(\ell-1)p+1}^{j} a_{j-k} f(x_k, y_k) + \frac{f(x_{j+1}, y_{j+1,0})}{\Gamma(\alpha+2)}. \tag{3.1.9}$$

We need to recall here that, in the block ℓ presently under consideration, these expressions must be calculated for $j = (\ell-1)p, (\ell-1)p+1, \ldots, \ell p-1$. Each of our p processors will then be assigned with the task of computing $y_{j+1,0}$ and y_{j+1} for exactly one value of j. To this end, each processor first computes the sums I_{j+1}, $H_{j,\ell}^{(0)}$ and $H_{j,\ell}$ for its value of j. With respect to these computations, we remark a few observations:

(1) For these computations, one needs to know only the initial values (which are given as part of the initial value problem), the weights of the algorithm (that have been computed in advance) and the y_k for $k \le (\ell-1)p$, *i. e.* approximate solution values that have already been computed in previous blocks. Thus, at the beginning of the ℓth block, all required data are available.
(2) The computations to be performed by any of the processors are completely independent of the computations of the others processors. In particular, no communication is required between the processors in this part of the block.
(3) Each processor has the task of computing I_j, *i. e.* a sum of $\lceil\alpha\rceil$ terms, and two sums of $(\ell-1)p+1$ terms each, namely $H_{j,\ell}^{(0)}$ and $H_{\ell,j}$. Thus, the work load is distributed across the processors in a uniform way; we hence have a very good load balancing, and it may be expected that all processors require similar amounts of time for this part of the task.

(4) The amount of time required for this part of the block depends on ℓ in a linear way. When ℓ is small, it can be completed very quickly, but as ℓ increases it will take longer and longer.

In the second part of the block we will deal with the remaining sums $L_{j,\ell}^{(0)}$ and $L_{j,\ell}$. In this context we note that the sum $L_{j,\ell}^{(0)}$ is empty for $j = (\ell-1)p$ (i.e. the smallest of the indices under consideration in the current block). Thus, for this value of j the value $y_{j+1,0}$ can be computed by the appropriate processor. Moreover, the expression $L_{j,\ell} = L_{(\ell-1)p,\ell}$ in this case reduces to $f(x_{(\ell-1)p+1}, y_{(\ell-1)p+1,0})/\Gamma(\alpha+2)$ which can now also be computed, and so we can complete the calculation of $y_{(\ell-1)p+1}$. The processor associated with this task then additionally computes $f(x_{(\ell-1)p+1}, y_{(\ell-1)p+1})$ and sends all its results to all other processors. Its job is finished for this block, and so it becomes idle. (Strictly speaking, it would not have been necessary to compute and store the value $f(x_j, y_j)$, but since this value will be used again and again in future steps, it is more efficient to do so than to call the function f with the same set of arguments for a large number of times.) The other processors have had to wait for this information about $y_{(\ell-1)p+1}$, but once they have received it, they are in a position to compute the first summands of their respective values $L_{j,\ell}^{(0)}$ and $L_{j,\ell}$. This sum only consists of this one summand for the next value of j, *viz.* $j = (\ell-1)p+1$, so the associated processor can complete the calculation of $y_{j+1,0} = y_{(\ell-1)p+2,0}$ and subsequently also the computation of $y_{(\ell-1)p+2}$ and $f(x_{(\ell-1)p+2}, y_{(\ell-1)p+2})$ itself since all the data required for the evaluation of $L_{(\ell-1)p+1,\ell}$ are now present. This processor then also passes the result of the computation to all other processors (including the first one that has already completed its task and that needs this value in the next block). In this fashion, we work our way through all y_j so that, at the end of this process, all processors have computed their y_j and $f(x_j, y_j)$ and passed the results to all other processors. This concludes the ℓth block, and we now have the data of y_j, $j = 1, 2, \ldots, \ell p$, available at all processors. With respect to this part of the block, none of our four observations above applies. Instead we find:

(1) Only the first value $y_{(\ell-1)p+1}$ could be immediately computed. All other y_j needed to have input from earlier values computed in this block and had to wait for the other processors to provide these values.

(2) As a consequence of the first comment, a certain amount of communication between the processors is required. Specifically, each processor computes its assigned value y_j in the indicated way and then passes the result to all other processors.

(3) The work load is small for the first processor and monotonically increases up to the last processor, so in this part of the block we do not have a good load balance. On average, each processor remains idle for half of the time spent in the second part of the block.

(4) The absolute amount of time required for this part of the block is independent of ℓ. Thus, in view of our previous observation on the first part of the block, the relative amount of time that this part of the block takes becomes smaller as ℓ increases, and hence, as the algorithm proceeds, the inefficiency caused by the load imbalance becomes less and less severe.

The procedure described above for the ℓth block then needs to be repeated in an iterative manner for $\ell = 1, 2, \ldots, \lceil N/p \rceil - 1$. This will give us the numerical solution at the points $x = x_j$, $j = 1, 2, \ldots, p(\lceil N/p \rceil - 1)$. For the remaining values x_j with $j = p(\lceil N/p \rceil - 1) + 1, \ldots, N$ we proceed in an almost identical way, except that now the number of values that need to be computed may be smaller than the number p of processors. (As mentioned above, the numbers will be equal if and only if p is a divisor of N.) Thus, for this last block we only employ as many processors as we have approximate function values that need to be computed, and these processors perform their computations in the same way as they did in the previous blocks; the other processors may remain idle.

For a detailed investigation of the performance of this parallelization of the Adams scheme, we refer to [187]. In particular it can be seen from the results reported there that this approach works very well for shared memory multiprocessor systems.

3.1.3 *When and when not to use fractional linear multistep formulas*

Traditionally, the fractional linear multistep formulas (in particular, the fractional backward differentiation formulas) proposed by Lubich that we had introduced in Subsection 2.1.3 and in Section 2.4 have been considered to be very good methods for the numerical solution of fractional differential equations due to the fact that they combine a conceptual simplicity, a straightforward method to compute their coefficients from the corresponding coefficients of well known methods for first-order differential equations, and a high order of convergence. This opinion has been supported by many numerical examples in the early literature on numerical methods for fractional differential equations where these methods have been used very successfully [279, 384]. However, in those early days of this field of research, almost all fractional differential equations that arose in real-world application were equations of order 1/2, and so this was the case that was usually tested.

More recently it has turned out that one needs to be somewhat more careful in judging the quality of fractional linear multistep formulas [190]. Specifically, while it is indeed true that they work very well for certain values of the order α of the differential equation, they can seriously fail for other values of α. To explain the background of this phenomenon, it is useful to recall the construction of these methods. As seen in Definition 2.1, a fractional linear multistep method consists of two components, the convolution quadrature and the starting quadrature. In particular, the latter is responsible for eliminating all lower order terms in the error expansion, and hence for making sure that the convergence order is sufficiently high. Theorem 2.14 gives us a representation for the weights of both parts. The concept for the convolution part is explicit and rather simple to implement; in particular this can easily be done in a numerically stable way. The difficulty lies in the the starting quadrature. The weights w_{nj} of this formula are only given implicitly as the solution of the linear equation system (2.4.2). This system reads

$$\sum_{j=0}^{s} w_{nj} j^{\gamma} = \frac{\Gamma(\gamma+1)}{\Gamma(\gamma+\alpha+1)} n^{\gamma+\alpha} - \sum_{j=0}^{n} \omega_{n-j} j^{\gamma}, \quad \gamma \in A, \tag{3.1.10}$$

where $A = \{\gamma = k + j\alpha : j, k \in \mathbb{N}_0, \gamma \leq p-1\}$, $s+1$ is the number of elements of A, and p is the order of convergence of the formula under consideration. The computation of the expressions on the right-hand side of this system requires the knowledge of the weights ω_{n-j} of the convolution quadrature, but as mentioned above this is not a major problem. Rather, the main obstacle lies in the structure of the coefficient matrix V of the system that is evident from the left-hand side. This matrix depends on α in a somewhat irregular way. If, for example, $p = 2$ and $\alpha = 1/2$ then we have $A = \{0, 1/2, 1\}$, and hence $s = 2$ and the matrix V has the form

$$V = \begin{pmatrix} 1 & 1 & 1 \\ 0 & 1^{1/2} & 2^{1/2} \\ 0 & 1^1 & 2^1 \end{pmatrix}.$$

This is essentially a Vandermode matrix which is known to be mildly ill-conditioned. A number of reliable algorithms for the accurate solution of systems with coefficient matrices of this type are known [107, 303]. These methods can, in principle, be used whenever $\alpha = 1/k$ with some integer k but, due to the structure of the equation system, they tend to become more and more unstable as k increases [190]. What is more, they cannot be applied directly at all if α is not the reciprocal of an integer.

As a particular example, let $p = 2$ and $\alpha = 0.499$ (*i. e.* we only perturb α by a rather small amount in comparison to the case described above). Then we have $A = \{0, 0.499, 0.998, 1\}$, hence $s = 3$ and the matrix V has the form

$$V = \begin{pmatrix} 1 & 1 & 1 & 1 \\ 0 & 1^{0.499} & 2^{0.499} & 3^{0.499} \\ 0 & 1^{0.998} & 2^{0.998} & 3^{0.998} \\ 0 & 1^1 & 2^1 & 3^1 \end{pmatrix} \approx \begin{pmatrix} 1 & 1 & 1.00000 & 1.00000 \\ 0 & 1 & 1.41323 & 1.73015 \\ 0 & 1 & 1.99723 & 2.99342 \\ 0 & 1 & 2.00000 & 3.00000 \end{pmatrix}.$$

Thus we observe that the small perturbation of α has resulted in a change of the dimension of the matrix and, what is much worse, the last two rows of the matrix are almost identical. In other words, the matrix is very close to a singular matrix; therefore its condition is extremely poor. Diethelm *et al.* [190] have reported results of a detailed comparison of various numerical approaches for handling this ill-conditioned system. The result was that the standard GMRES method [534, 535] and its modification using the Householder transform [621] usually gave the best results, but it was rather slow. In contrast, a classical LU decomposition was much faster and produced only slightly worse results. The accuracy of the Björck-Pereyra method was comparable only for $\alpha = 1/2$, somewhat worse for $\alpha = 1/3$ and extremely poor for $\alpha = 1/k$ with $k > 3$; as mentioned above, this algorithm is not applicable at all for other values of α.

Thus the conclusion is that high order fractional linear multistep methods are a very useful concept in theory but as long as the problem of finding an accurate scheme for the numerical computation of the starting weights is unsolved, they cannot be reliably computed in practice except for a small set of values for α and therefore one should avoid using them unless the order of the differential equation under consideration happens to be in this set.

3.1.4 *The use of series expansions*

An essential feature of fractional multistep formulas is that they consist of two components, the convolution quadrature and the starting quadrature. The main problem of these formulas that we had found in the previous subsection was that it is extremely difficult to compute the starting quadrature with a sufficient accuracy. No such problems are associated with the convolution quadrature. Thus one might be tempted to find some approach that avoids the use of the starting quadrature. In order to construct such an approach it is useful to recall the background that has lead us to introducing the starting quadrature in the first place. In fact the reason was that we wanted to have a high order of convergence for a most general class of equations, and it is known that the solutions of such equations are typically not smooth at the origin; rather they have an asymptotic expansion in powers of the form $x^{j+k\alpha}$ with $j, k \in \mathbb{N}$ (see [431] or Section 6.4 of [186]). The introduction of the starting quadratures was an attempt to make the complete formula exact for linear combinations of such powers with small j and k, thus eliminating all low order terms in the asymptotic expansion of the error with respect to the step size.

From this derivation we can see that there is an alternative to the introduction of the notoriously difficult starting quadrature. Specifically we can try to modify the initial value problem

$$^{\mathrm{C}}D^{\alpha}y(x) = f(x, y(x)), \qquad y^{(k)}(0) = y_0^{(k)}, \quad k = 0, 1, \ldots, \lceil\alpha\rceil - 1, \quad (3.1.11)$$

so that the initial value problem newly obtained in this way has two important properties:

(a) Near the starting point 0, the solution $\tilde{y}$, say, of the modified problem has an asymptotic expansion of the form

$$\tilde{y}(x) = \sum_{j=0}^{\mu} \sum_{k=0}^{\nu} \tilde{c}_{jk} x^{j+k\alpha} + o(x^\rho)$$

with $\rho > \mu + \nu\alpha$ and

$$\tilde{c}_{jk} = 0 \quad \text{for } k \neq 0. \tag{3.1.12}$$

(b) The solution of the original problem can be computed from the solution of the modified problem in an efficient and numerically stable manner.

Condition (3.1.12) then asserts that those terms of the asymptotic expansion of the original solution that required the use of starting quadratures to obtain the desired order of convergence are absent. Thus, as a consequence of this condition, we can dispense with the starting quadrature completely and use only the convolution quadrature part of the fractional multistep method without losing the order of convergence.

The key point in this approach is, of course, that we need to make sure that (3.1.12) is satisfied. This is by no means trivial but may be possible in certain situations. In particular, if we know the coefficients c_{jk} in the asymptotic expansion

$$y(x) = \sum_{j=0}^{\mu} \sum_{k=0}^{\nu} c_{jk} x^{j+k\alpha} + o(x^\rho) \tag{3.1.13}$$

of the exact solution of the initial value problem (3.1.11) then we can define

$$\tilde{y}(x) = y(x) - \sum_{j=0}^{\mu} \sum_{k=1}^{\nu} c_{jk} x^{j+k\alpha}$$

which immediately implies our required condition (3.1.12). Moreover, as a consequence of this definition and the original initial value problem (3.1.11) we can see that $\tilde{y}$ solves the modified differential equation

$$\begin{aligned} {}^{\mathrm{C}}D^\alpha \tilde{y}(x) &= {}^{\mathrm{C}}D^\alpha y(x) - \sum_{j=0}^{\mu} \sum_{k=1}^{\nu} \frac{c_{jk}\Gamma(j+1+k\alpha)}{\Gamma(j+1+(k-1)\alpha)} x^{j+(k-1)\alpha} \\ &= f(x, \tilde{y}(x) + \sum_{j=0}^{\mu} \sum_{k=1}^{\nu} c_{jk} x^{j+k\alpha}) \qquad \text{(3.1.14a)} \\ &\quad - \sum_{j=0}^{\mu} \sum_{k=1}^{\nu} \frac{c_{jk}\Gamma(j+1+k\alpha)}{\Gamma(j+1+(k-1)\alpha)} x^{j+(k-1)\alpha} \end{aligned}$$

whereas the initial conditions can remain unchanged, *i. e.*

$$\tilde{y}^{(k)}(0) = y_0^{(k)}, \quad k = 0, 1, \ldots, \lceil \alpha \rceil - 1. \tag{3.1.14b}$$

Thus the remaining problem is that we need to find the coefficients of the first terms of the asymptotic expansion (3.1.13). In a number of special cases it may be possible to achieve this via a differential transform method [463] or an Adomian decomposition [8]. Note that we do not need to find more than the first few coefficients of the expansion, so the poor convergence properties usually associated with Adomian's method (see our comments in Section 2.6) may be irrelevant in this context.

3.1.5 *The generalized Adams methods as an efficient tool for multi-order fractional differential equations*

In Section 2.8 we had dealt with the basics of multi-term equations, *i. e.* equations involving more than one fractional derivative. In particular we had constructed a number of possible methods to convert such equations into systems of single-order equations. In some of these approaches, all differential equations of the resulting system were of fractional order (see, *e. g.*, Theorem 2.15); in other approaches (Theorem 2.16) we had a mixture of fractional-order and integer-order equations. The detailed investigations of Ford and Connolly [240] indicate that in most examples the former approach will lead to a computationally more efficient algorithm than the latter.

Of course, the reformulation of the given multi-term equation in the form of a system of single-term equations alone is not sufficient to obtain a numerical solution. Rather we need to solve the system that we have created by a suitable numerical method. According to [215] and [239] one can say that the Adams method described in Subsection 2.3.2 will usually lead to an algorithm with satisfactory convergence and stability properties.

As an example we consider the initial value problem

$$^{\mathrm{C}}D^{1.8}y(x) + 3x^2 \, ^{\mathrm{C}}D^{0.4}y(x) + 5(\sin x)y(x) = \exp x, \tag{3.1.15a}$$

$$y(0) = 1, \quad y'(0) = -7. \tag{3.1.15b}$$

Using the approach of Theorem 2.15, we may rewrite this problem in the form

$$
\begin{aligned}
&{}^{\mathrm{C}}D^{0.4}y_1(x) = y_2(x),\\
&{}^{\mathrm{C}}D^{0.6}y_2(x) = y_3(x),\\
&{}^{\mathrm{C}}D^{0.8}y_3(x) = -3x^2y_2(x) - 5(\sin x)y_1(x) + \exp x,\\
&y_1(0) = 1, \quad y_2(0) = 0, \quad y_3(0) = -7.
\end{aligned}
\tag{3.1.16}
$$

The first component y_1 of the solution vector $(y_1, y_2, y_3)^{\mathrm{T}}$ of this system is then identical to the required solution y of our multi-term equation (3.1.15). A numerical computation using the Adams method of the corresponding order for each of the three constituent equations of the system (3.1.16) with step sizes $h = 1/10$, $h = 1/20$ and $h = 1/40$, respectively, produced the results depicted in Fig. 3.2.

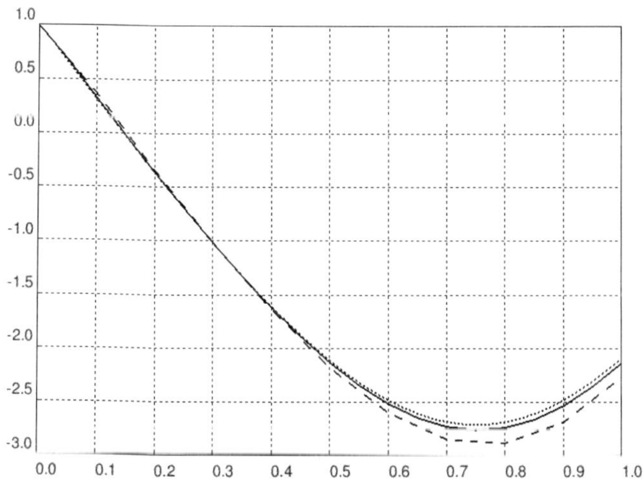

Fig. 3.2 Numerical solutions of Eq. (3.1.15) for $h = 1/10$ (dashed line), $h = 1/20$ (continuous line) and $h = 1/40$ (dotted line).

Of course, an explicit expression for the exact solution of this system is not known, but the plots of the approximate solutions (and the underlying numerical data) clearly indicate that the method converges. The error analysis of Edwards *et al.* [215] confirms this observation; indeed we know the convergence order of each single equation from the theory of Subsection

2.3.2, and using the methods from [215] one can see that the convergence order of the overall method is just the minimum of the convergence orders of each individual equation.

Now let us look at a slight modification of the problem (3.1.15), namely the equation

$$0.01\ {}^{\mathrm{C}}D^{1.8}y(x) + 3x^2\ {}^{\mathrm{C}}D^{0.4}y(x) + 5(\sin x)y(x) = \exp x, \quad (3.1.17\mathrm{a})$$

$$y(0) = 1, \quad y'(0) = -700 \quad (3.1.17\mathrm{b})$$

that is obtained by multiplying the highest order derivative in the original equation by a small coefficient and by multiplying the highest order initial value by the reciprocal of this coefficient. Using the same approach as above, we may rewrite this problem in the form

$$\begin{aligned} {}^{\mathrm{C}}D^{0.4}y_1(x) &= y_2(x), \\ {}^{\mathrm{C}}D^{0.6}y_2(x) &= y_3(x), \\ {}^{\mathrm{C}}D^{0.8}y_3(x) &= -300x^2y_2(x) - 500(\sin x)y_1(x) + 100\exp x, \\ y_1(0) = 1, \quad & y_2(0) = 0, \quad y_3(0) = -700. \end{aligned} \quad (3.1.18)$$

In order to obtain a reliable approximation of the exact solution we have first computed the solution numerically with a number of very small step sizes until the differences between consecutive approximate solutions were negligible. It was sufficient to use a step size of $h = 1/2000$; the resulting graph is shown in Fig. 3.3 as a continuous line. Clearly, the total variation of the solution is much larger than in the previous example. This is not surprising in view of the modification of the initial values.

In particular, we find that $y(1) \approx -1.502667$. If we use step sizes in the approximation that are similar to those of the previous example then we arrive at completely meaningless results as indicated in Table 3.1. Only significantly smaller step sizes yield a useful numerical result.

The reason for this apparently strange behavior becomes evident by looking at the graph of the "exact" solution in Fig. 3.3. Specifically, the gradient of the solution is negative and very large in modulus near the starting point $x = 0$ (notice the scale on the y-axis of Fig. 3.3 in comparison to the scale of the y-axis in Fig. 3.2), then it becomes large and positive very quickly near $x = 0.15$, becomes negative again near $x = 0.35$ and only then decreases in modulus. The Adams method (or, indeed, most other

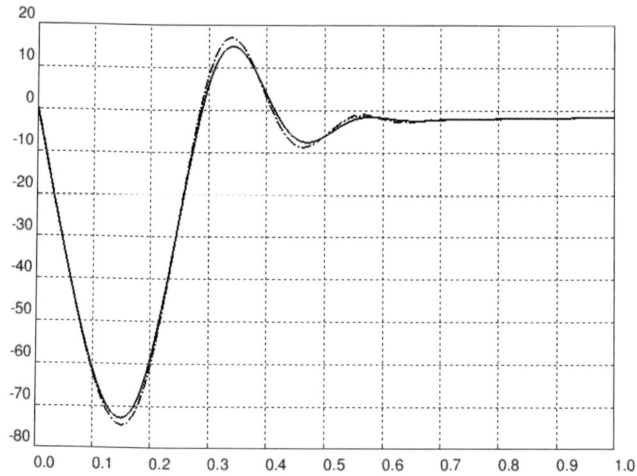

Fig. 3.3 Numerical solutions of Eq. (3.1.15) for uniform mesh with $h = 1/2000$ (continuous line; "exact" solution), uniform mesh with $h = 1/100$ (dotted line) and non-uniform mesh with $h = 1/100$ in the left half and $h = 1/60$ in the right half (dashed line).

Table 3.1 Numerical approximation for $y(1)$ from Eq. (3.1.17) using the Adams method and various step sizes.

step size h	approx. value of $y(1)$
1/10	984034.14
1/20	209443.96
1/40	−6.914879
1/100	−1.515596

methods) cannot follow these abrupt changes of the slope of y if the step size is too large, and so the numerical solution strongly deviates from the exact solution. However, it is apparent from Fig. 3.3 that this problem does not persist throughout the entire interval $[0, 1]$ where the solution is sought. For $x > 0.5$ we observe only very mild variations in the slope of y, and therefore one could use a coarser mesh in this range while retaining the fine mesh in the left half of the interval. It is one of the advantages of the Adams method that such a non-uniform mesh can be implemented very easily. To be precise, denoting the grid points by t_j, the resulting formula for differential equations of order α with initial point 0 becomes (see, *e. g.*, pp. 198–200 of [186])

$$y_{k+1} = \sum_{j=0}^{\lceil\alpha\rceil-1} \frac{t_{k+1}^j}{j!} y_0^{(j)} + \frac{1}{\Gamma(\alpha)} \left(\sum_{j=0}^{k} a_{j,k+1} f(t_j, y_j) + a_{k+1,k+1} f(t_{k+1}, y_{k+1}^{\mathrm{P}}) \right) \tag{3.1.19}$$

(this is the Adams-Moulton formula) with weights

$$a_{0,k+1} = \frac{(t_{k+1} - t_1)^{\alpha+1} + t_{k+1}^{\alpha}(\alpha t_1 + t_1 - t_{k+1})}{t_1 \alpha(\alpha+1)}, \tag{3.1.20a}$$

$$\begin{aligned} a_{j,k+1} &= \frac{(t_{k+1} - t_{j-1})^{\alpha+1} + (t_{k+1} - t_j)^{\alpha}(\alpha(t_{j-1} - t_j) + t_{j-1} - t_{k+1})}{(t_j - t_{j-1})\alpha(\alpha+1)} \\ &\quad + \frac{(t_{k+1} - t_{j+1})^{\alpha+1} - (t_{k+1} - t_j)^{\alpha}(\alpha(t_j - t_{j+1}) - t_{j+1} + t_{k+1})}{(t_{j+1} - t_j)\alpha(\alpha+1)} \end{aligned} \tag{3.1.20b}$$

if $1 \le j \le k$, and

$$a_{k+1,k+1} = \frac{(t_{k+1} - t_k)^{\alpha}}{\alpha(\alpha+1)}. \tag{3.1.20c}$$

The Adams-Bashforth predictor here is given by

$$y_{k+1}^{\mathrm{P}} = \sum_{j=0}^{\lceil\alpha\rceil-1} \frac{t_{k+1}^j}{j!} y_0^{(j)} + \frac{1}{\Gamma(\alpha)} \sum_{j=0}^{k} b_{j,k+1} f(t_j, y_j) \tag{3.1.21}$$

with

$$b_{j,k+1} = \frac{(t_{k+1} - t_j)^{\alpha} - (t_{k+1} - t_{j+1})^{\alpha}}{\alpha}. \tag{3.1.22}$$

We have applied this concept to the problem given in Eq. (3.1.17) with a mesh size of $1/100$ on $[0, 0.5]$ and $1/60$ on $[0.5, 1]$. The result is plotted in Fig. 3.3 as a dashed line. Specifically we obtained a value $y(1) \approx -1.514199$, so the accuracy was roughly comparable to the accuracy obtained by using a uniform grid with a mesh size $h = 1/100$ (the dotted line in Fig. 3.3; the dashed and dotted lines can hardly be distinguished from each other). However, the uniform grid has $N = 100$ mesh points whereas the non-uniform grid has only $N = 80$ mesh points. Since the computational cost is proportional to N^2, the use of the uniform grid is more than 50% more expensive than the non-uniform grid.

Another point worth mentioning in connection with the application of the Adams method to multi- or single-order systems arising from the reformulation of multi-term equations is connected to the structure of the initial value vector of the system. Specifically, a look at the equivalence results (Theorems 2.15 and 2.16) reveals that the initial conditions, given in Eqs. (2.8.7) and (2.8.12), respectively, may contain a large number of zeros. To be precise, only the coefficients that correspond to integer-order derivatives of the solution of the given multi-term equation may be non-zero values; all other components must vanish. Let us briefly look at an example to demonstrate why this may be a problem and what can be done to solve it. The simple linear problem

$$\sum_{k=0}^{14}(k+1)\,{}^{\mathrm{C}}D^{k/10}y(x) = \sin x, \qquad y(0) = 1, \quad y'(0) = 2, \tag{3.1.23}$$

will be sufficient to demonstrate the relevant effects. According to Theorem 2.15, the equivalent system of equations reads

$$\begin{aligned} {}^{\mathrm{C}}D^{1/10}y_j &= y_{j+1}, \qquad j = 1, 2, \ldots, 13, \\ {}^{\mathrm{C}}D^{1/10}y_{14} &= \frac{1}{15}\left(\sin x - \sum_{k=1}^{14} k y_k(x)\right) \end{aligned} \tag{3.1.24a}$$

subject to the initial conditions

$$y_k(0) = \begin{cases} 1 & \text{for } k = 1, \\ 2 & \text{for } k = 11, \\ 0 & \text{else.} \end{cases} \tag{3.1.24b}$$

Figure 3.4 shows the approximate solution of this system on the interval $[0, 1]$ that we have obtained using the Adams-Bashforth-Moulton method with a uniform step size of $h = 1/20$ (dotted line). It is evident (in particular from the inset picture in the bottom left corner that shows an enlarged version of the graphs in the range $[0, 0.2]$) that this numerical solution is constant on the subinterval $[0, 0.2]$, *i. e.* on the first four steps of the algorithm. A comparison with a numerical solution with a much smaller step size ($h = 1/2000$; the continuous line in Fig. 3.4) that we once again consider to be very close to the exact solution reveals that this means that the exact solution moves away from the numerical solution for $h = 1/20$ (which, as stated above, remains constant) on a relatively large initial subinterval. Only when we have left this subinterval, the numerical solution for $h = 1/20$ can start to move towards the exact solution and diminish the approximation error. Of course, this task is much more difficult than a similar problem where no such initial interval problem exists.

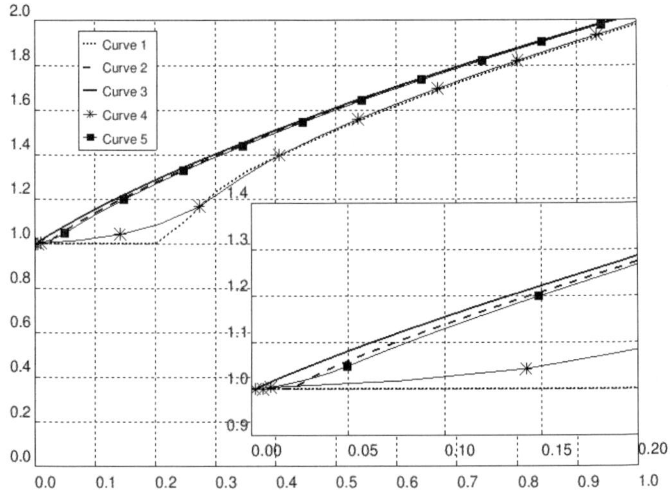

Fig. 3.4 Numerical solutions of Eq. (3.1.24) on $[0, 1]$ for uniform mesh with $h = 1/20$ (dotted line, Curve 1), uniform mesh with $h = 1/200$ (dashed line, Curve 2), uniform mesh with $h = 1/2000$ (continuous line, Curve 3; "exact" solution), non-uniform mesh with 5 steps of length $h = 1/500$ followed by 15 steps of length $h = 66/1000$ (continuous line marked with stars, Curve 4) and non-uniform mesh with 5 steps of length $h = 1/500$ followed by 100 steps of length $h = 99/10000$ (continuous line marked with squares, Curve 5). The inset in the bottom right corner is an enlarged view of the range $[0, 0.2]$ of the complete graph.

This behavior of the numerical solution does not seem to be justified by the analytical properties of the initial value problem. However, a close inspection of numerical values for other choices of the step size reveals that we can *always* observe this phenomenon: The approximate values for $y(x_1), \ldots, y(x_4)$ coincide with the initial value $y(0)$ in all these cases. The reason behind this observation becomes apparent when we look at how the Adams-Bashforth-Moulton method interacts with the system of equations (3.1.24) and, in particular, with the initial conditions (3.1.24b). To be precise, we note that the vector containing the initial conditions consists of the entry $y_0^{(0)}$ in the first component, followed by nine zeros. Only then the next non-zero entry may appear; its value comes from the given initial values of the original problem (3.1.23). The Predict-Evaluate-Correct-Evaluate (PECE) form of the Adams algorithm means that the non-zero elements are propagated by two rows in each step, and hence the algorithm needs to take five steps before it can produce an approximation that differs from the initial value.

Two possible strategies are immediately obvious in order to overcome these difficulties. The first approach uses the fact that we always need the same number of steps to get away from the initial value, no matter what step size we have chosen. Thus we can decide to choose a very small step size for the first few (in this example: four) steps, so that the exact solution will not move away too far from the initial value (*i. e.* from the numerical solution) on the interval covered by these first steps. Hence it will be much easier for the algorithm to bridge the gap and return to the neighborhood of the graph of the exact solution. Once the next (here: the fifth) step is reached, it is possible to return to a much coarser mesh in order to avoid excessive computational costs. The numerical results depicted in Fig. 3.4 (Curves 4 and, in particular, 5) demonstrate that this concept indeed allows us to obtain much better approximate solutions without increasing the complexity too much.

Alternatively we can also follow a completely different path originally proposed in [183]. As we had noted above, the problem that we are facing is due to the fact that there are many (in this case: nine) zeros in the initial value vector, and only two components are propagated at each step, so that it here takes five steps to eliminate all the zeros. The propagation by two components takes place in two parts of the algorithm, one in the predictor (the Adams-Bashforth method) and one in the corrector (the Adams-Moulton method). Thus we can also solve the problem by introducing additional corrector iterations, hence moving from the plain PECE structure to a P(EC)^{m}E form of the algorithm as indicated in Eq. (2.3.7). In our case, $m = 8$ would be an appropriate choice. According to Theorem 2.10 this approach has the pleasant side effect of increasing the convergence order. Moreover, a close inspection of the formula (2.3.7b) reveals that the corrector may be written in the form

$$y_{k,\mu} = \gamma_k + h^\alpha a_{kk} f(x_k, y_{k,\mu-1})$$

where

$$\gamma_k = \sum_{j=0}^{\lceil\alpha\rceil-1} \frac{x_k^j}{j!} y_0^{(j)} + h^\alpha \sum_{j=0}^{k-1} a_{jk} f(x_j, y_j)$$

is independent of μ (the index of the corrector iteration). Thus the total arithmetic complexity of the corrector part of the kth step (taking us from x_{k-1} to x_k) is $O(k)$ for the calculation of γ_k plus $O(m)$ for the m corrector steps, the sum of which (since m is constant) is asymptotically the same as the complexity in the case $m = 1$. We hence conclude that the computational cost of an Adams-Bashforth-Moulton predictor-corrector method with a large but fixed number of corrector iterations is asymptotically identical to the computational cost of the same method with only one application of the corrector. In summary, the use of a P(EC)^{m}E approach simultaneously solves our problem that was induced by the large number of zeros in the initial value vector (3.1.24b) and increases the convergence order of the algorithm and yet it only introduces a negligible amount of additional computational complexity. We have plotted some graphs of approximate solutions obtained in this way in Fig. 3.5. The good quality of the numerical solutions is evident. In particular, the zoom of the subinterval $[0, 0.2]$ clearly shows that we have completely succeeded in removing the unpleasant constant behavior of the approximate solution near the initial point (except for the dotted curve which corresponds to only three corrector iterations where we can confirm the theoretical expectation that the subinterval with a constant approximate solution decreases in size by a half).

For the purpose of comparison we have also built up Table 3.2 that displays the absolute errors of the various methods that we have tried at the point $x = 1$. The P(EC)^{m}E method is seen to give the best relation between computational effort (essentially proportional to N^2 where N is the total number of grid points) and accuracy.

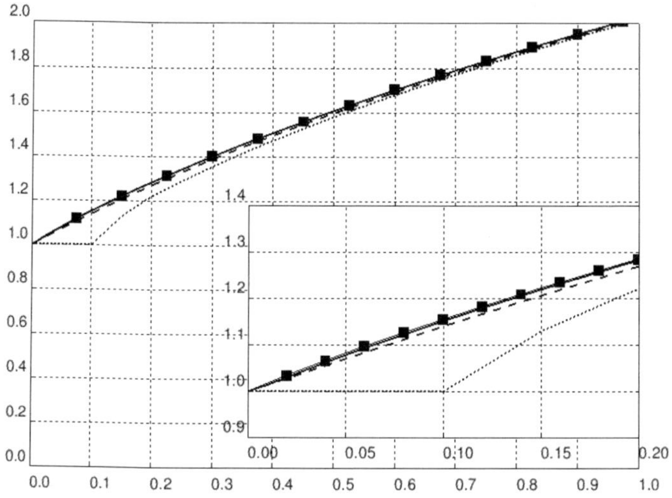

Fig. 3.5 Numerical solutions of Eq. (3.1.24) on $[0, 1]$ for uniform mesh with $h = 1/20$ and three corrector iterations (dotted line), $h = 1/20$ and nine corrector iterations (dashed line), $h = 1/100$ and nine corrector iterations (continuous line), and $h = 1/2000$ and nine corrector iterations (continuous line marked with star symbols; "exact" solution). The inset in the bottom right corner is an enlarged view of the range $[0, 0.2]$ of the complete graph.

Table 3.2 Absolute errors of numerical approximations for $y(1)$ from Eq. (3.1.23) using the Adams method with various step sizes and various numbers of corrector iterations.

grid	total no. of points	corrector iterations	abs. error at $x = 1$	visualization
uniform, $h = 1/20$	20	1	$5 \cdot 10^{-2}$	Fig. 3.4, Curve 1
uniform, $h = 1/200$	200	1	$4 \cdot 10^{-3}$	Fig. 3.4, Curve 2
uniform, $h = 1/2000$	2000	1	$2 \cdot 10^{-4}$	Fig. 3.4, Curve 3
$h = 1/500$ (5 steps), $h = 66/1000$ (15 steps)	20	1	$4 \cdot 10^{-2}$	Fig. 3.4, Curve 4
$h = 1/500$ (5 steps), $h = 99/10000$ (100 steps)	105	1	$5 \cdot 10^{-3}$	Fig. 3.4, Curve 5
uniform, $h = 1/20$	20	3	$1 \cdot 10^{-2}$	Fig. 3.5, dotted
uniform, $h = 1/20$	20	9	$9 \cdot 10^{-3}$	Fig. 3.5, dashed
uniform, $h = 1/100$	100	9	$2 \cdot 10^{-3}$	Fig. 3.5, contin.

3.1.6 *Two classes of singular equations as application examples*

We conclude this section with a subsection dealing with two particularly difficult types of applications. The purpose of the first of these examples, taken from [203], is to demonstrate that certain algorithms can be successfully applied even in singular situations where the standard assumptions are not satisfied. Specifically we shall look at the initial value problem

$$ {}^{\mathrm{C}}D^{1/2}y(x) = \ln y(x) + E\frac{q(x)}{y(x)}, \qquad y(0) = 0. \tag{3.1.25} $$

This equation has been used [321] to model the propagation of a flame in the context of a thermo-diffusive model with high activation energies using a gaseous mixture with simple chemistry $A \to B$. Specifically, $y(x)$ denotes the radius of the flame at the time instant x; the given function q models the energy input into the system via a point source.

The model described by Eq. (3.1.25) — which can be justified in a mathematically rigorous way [362] — has some rather natural important questions associated with it. Apart from the most obvious one for an (exact or approximate) solution for a specific choice of the parameters E and q, one is often strongly interested in the bifurcation behavior of the equation. Analytically, we know from [46] for a large class of functions q with $q(x) = 0$ for all $x > x_0$ with some x_0 that there exists a value $E_{\mathrm{crit}}(q)$ such that

- if $E > E_{\mathrm{crit}}(q)$ then y is defined on $[0, \infty)$ and $\lim_{x\to\infty} y(x) = \infty$,
- if $E = E_{\mathrm{crit}}(q)$ then y is defined on $[0, \infty)$ and $\lim_{y\to\infty} y(x) = 1$,
- if $E < E_{\mathrm{crit}}(q)$ then there exists some finite $x_{\max} > x_0$ such that y is defined on $[0, x_{\max}]$ and $\lim_{x\to x_{\max}} y(x) = 0$.

Stated explicitly, it says that the flame will quench in finite time if the energy added to the system is smaller than the critical level E_{crit}, and it will burn persistently if the energy is above E_{crit}. For safety considerations it is therefore very important to find out the value of E_{crit} (or at least lower bounds for it) if one is interested in keeping the fire under control. On the other hand, sometimes one is interested in constructing a permanently burning flame, and then one needs to know E_{crit} (or at least upper bounds for it) in order to find an efficient process that uses as little energy as possible.

For the numerical solution of Eq. (3.1.25) one must take into account that the right-hand side of the differential equation is singular at the initial point. Thus, since the right-hand side is undefined at $(0, y(0))$, we cannot use a numerical method that needs to evaluate the right-hand side at this point. It is possible, however, to use the backward differentiation method of Subsection 2.2.2 or the fractional generalization of the backward Euler method according to Lubich's approach (see Subsection 2.1.3 and Section 2.4). Both approaches prove to be successful [203]; we concentrate on the former here.

A classical test case frequently considered in the literature (see the references cited in [203]) is

$$q(x) = \begin{cases} x^{0.3}(1-x) & \text{if } 0 \le x \le 1, \\ 0 & \text{else.} \end{cases}$$

For this function q, we have attempted to numerically calculate the critical value $E_{\text{crit}}(q)$ with the backward differentiation method of Subsection 2.2.2. To this end we have solved the initial value problem (3.1.25) with various values of E and a very small step size $h = 1/3200$. The results are depicted in Fig. 3.6, and they indicate that the critical value is $E_{\text{crit}}(q) \approx 7.6655$.

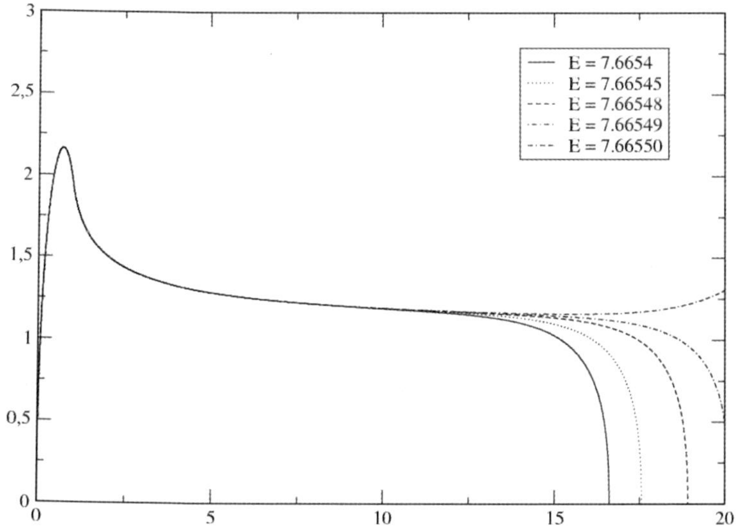

Fig. 3.6 Numerical solutions of Eq. (3.1.25) for $h = 1/3200$ and various values of E.

The second type of examples that we shall discuss in this subsection arises in a completely different context. Specifically, these equations originate from problems connected to the fractional calculus of variations as discussed in detail in Chapter 5. To be precise, we first deal with the system of equations

$$ {}^{\mathrm{C}}D_{a+}^{\alpha} q_2(t) = \lambda, \tag{3.1.26a} $$

$$ {}^{\mathrm{C}}D_{a+}^{\alpha} q_1(t) = p_1(t) - \lambda, \tag{3.1.26b} $$

$$ {}^{\mathrm{RL}}D_{b-}^{\alpha} p_1(t) = 0, \tag{3.1.26c} $$

$$ {}^{\mathrm{RL}}D_{b-}^{\alpha} p_2(t) = 0, \tag{3.1.26d} $$

that (in slightly more complicated notation whose use is not necessary here) forms the canonical equations of a fractional Euler-Lagrange problem, see Eq. (5.2.45) and its derivation. The interesting features of this system of equations are (a) that it contains both Caputo and Riemann-Liouville differential operators, (b) that it contains both left-sided and right-sided operators, and (c) that it is usually connected with initial conditions that lead to unbounded solutions.

Specifically, we shall consider the system (3.1.26) on the interval $[a, b] = [0, 1]$ with $\alpha = 0.85$, $\lambda = 3$ and the initial conditions

$$ q_2(0) = 0.5, \tag{3.1.27a} $$

$$ q_1(0) = 1.0, \tag{3.1.27b} $$

$$ J_{b-}^{\alpha} p_1(b) = \Gamma(\alpha), \tag{3.1.27c} $$

$$ J_{b-}^{\alpha} p_2(b) = \Gamma(\alpha). \tag{3.1.27d} $$

We first introduce the auxiliary functions $\tilde{p}_j(t) = p_j(t) - (b-t)^{\alpha-1}$ $(j = 1, 2)$, and notice that we can then replace equations (3.1.26c) and (3.1.26d) and the corresponding initial conditions (3.1.27c) and (3.1.27d), respectively, by

$$ {}^{\mathrm{C}}D_{b-}^{\alpha} \tilde{p}_1(t) = 0, \tag{3.1.28a} $$

$$ {}^{\mathrm{C}}D_{b-}^{\alpha} \tilde{p}_2(t) = 0, \tag{3.1.28b} $$

$$ \tilde{p}_1(b) = 0, \tag{3.1.28c} $$

$$ \tilde{p}_2(b) = 0. \tag{3.1.28d} $$

It is then clear that the system (3.1.28) has the solutions $\tilde{p}_1(t) = \tilde{p}_2(t) = 0$. This can be seen by an immediate analytical discussion, and the application of an arbitrary numerical method will lead to the same result. Thus, returning to our original unknowns, we derive that

$$p_1(t) = p_2(t) = (b - t)^{\alpha-1}.$$

We now have to solve the differential equations (3.1.26a) and (3.1.26b) together with the corresponding initial conditions (3.1.27a) and (3.1.27b) in order to determine the remaining unknowns q_1 and q_2. To this end we must take into consideration that the function p_1 that appears on the right-hand side of Eq. (3.1.26b) is singular at $t = b$, *i. e.* at the right end point of the interval of interest. Thus, following the concept known in the theory of numerical integration as "avoiding the singularity" [118, 511], we use a numerical method that does not evaluate the given function at the singular point $t = b$. An obvious choice here is the Adams-Bashforth (or forward Euler) method introduced in Subsection 2.3.2 above, *i. e.* the method obtained by only computing the predictor and using its value as the final approximation without employing the corrector step.

For the example with the parameters mentioned above, we have computed the numerical solution via this method with a step size of $h = 1/200$. Some numerical values for the solutions are given in Table 3.3, and plots of the approximate solutions are provided in Fig. 3.7.

Table 3.3 Numerical values for the components q_1, q_2, p_1 and p_2 of the solution of the system (3.1.26) subject to the initial conditions (3.1.27), with parameters $a = 0$, $b = 1$, $\lambda = 3$ and $\alpha = 0.85$, computed with the Adams-Bashforth method with step size $h = 1/200$.

t	$q_1(t)$	$q_2(t)$	$p_1(t)$	$p_2(t)$
0.00	1.000000	0.500000	1.000000	1.000000
0.10	0.702446	0.948135	1.015930	1.015930
0.25	0.356250	1.476467	1.044097	1.044097
0.50	−0.143113	2.260083	1.109569	1.109569
0.75	−0.577773	2.984338	1.231144	1.231144
0.95	−0.861913	3.537202	1.567309	1.567309
0.99	−0.896892	3.645564	1.995262	1.995262
1.00	−0.898149	3.672551	∞	∞

The equation system

$$^{\mathrm{C}}D^{\alpha}_{a+} q_2(t) = p_2(t), \tag{3.1.29a}$$

$$^{\mathrm{C}}D^{\alpha}_{a+} q_1(t) = p_1(t), \tag{3.1.29b}$$

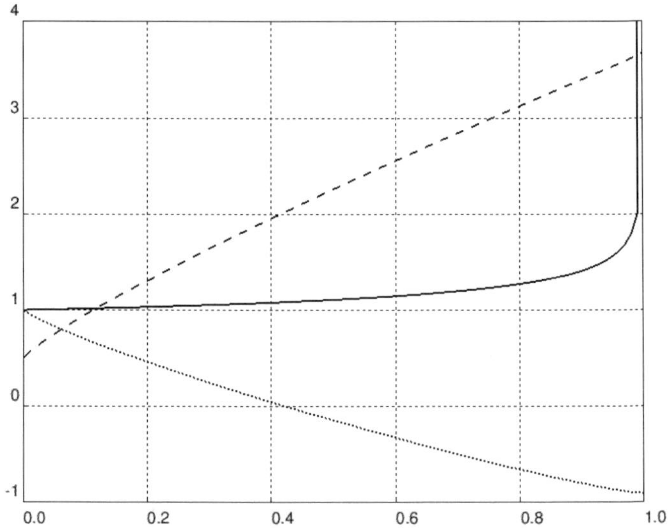

Fig. 3.7 Numerical values for the components q_1 (dotted), q_2 (dashed) and p_1 (continuous line) of the solution of the system (3.1.26) subject to the initial conditions (3.1.27), with parameters $a = 0$, $b = 1$, $\lambda = 3$ and $\alpha = 0.85$, computed with the Adams-Bashforth method with step size $h = 1/200$. The solution component p_2 coincides with p_1.

$$^{\mathrm{RL}}D_{b-}^{\alpha} p_1(t) = 0, \tag{3.1.29c}$$

$$^{\mathrm{RL}}D_{b-}^{\alpha} p_2(t) = 0 \tag{3.1.29d}$$

with which we shall conclude this section has a similar background; it arises in the context of fractional canonical equations in the extended phase space, see Eq. (5.2.54). For our numerical experiments we use $\alpha = 0.8$ and look for a solution on $[a, b] = [0, 10]$ with the initial conditions

$$q_2(0) = -1.0, \tag{3.1.30a}$$

$$q_1(0) = 1.0, \tag{3.1.30b}$$

$$J_{b-}^{\alpha} p_1(b) = \Gamma(\alpha), \tag{3.1.30c}$$

$$J_{b-}^{\alpha} p_2(b) = \Gamma(\alpha). \tag{3.1.30d}$$

Proceeding much as in the previous example and, in particular, using the same numerical method (except for the fact that we have chosen a step size of $h = 1/20$ to reflect the fact that the basic interval is now ten times as large as before), we can compute the approximate solution for the system (3.1.29) subject to the initial conditions (3.1.30). The results are indicated in Table 3.4 and Fig. 3.8.

Table 3.4 Numerical values for the components q_1, q_2, p_1 and p_2 of the solution of the system (3.1.29) subject to the initial conditions (3.1.30), with parameters $a = 0$, $b = 10$ and $\alpha = 0.8$, computed with the Adams-Bashforth method with step size $h = 1/20$.

t	$q_1(t)$	$q_2(t)$	$p_1(t)$	$p_2(t)$
0.00	1.000000	−1.000000	1.000000	1.000000
0.50	1.391091	−0.608909	0.637463	0.637463
1.00	1.684942	−0.315058	0.644394	0.644394
2.50	2.453250	0.453250	0.668325	0.668325
5.00	3.631618	1.631618	0.724780	0.724780
7.00	4.595544	2.595544	0.802742	0.802742
9.00	5.712011	3.712011	1.000000	1.000000
9.50	6.069920	4.069920	1.148698	1.148698
9.95	6.522876	4.522876	1.820564	1.820564
10.00	6.610243	4.610243	∞	∞

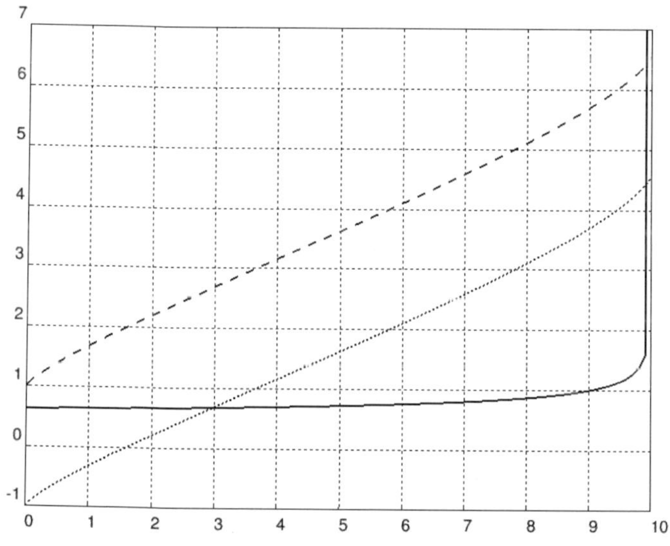

Fig. 3.8 Numerical values for the components q_1 (dashed), q_2 (dotted) and p_1 (continuous line) of the solution of the system (3.1.29) subject to the initial conditions (3.1.30), with parameters $a = 0$, $b = 10$ and $\alpha = 0.8$, computed with the Adams-Bashforth method with step size $h = 1/20$. The solution component p_2 coincides with p_1.

3.2 Methods for Partial Differential Equations

In our discussion of numerical methods for partial fractional differential equations, we shall first concentrate on generalizations of classical problems like diffusion or wave equations obtained by replacing the integer-order time derivatives with a differential operator of fractional order. In particular, the first three subsections of this section will be devoted to numerical algorithms for problems of this type. Mainly we shall describe how classical approaches can be extended to the fractional setting. In Subsection 3.2.4 we shall then briefly deal with numerical methods for equations that are of fractional order with respect to the space variables. Problems containing fractional operators in both space and time will not be explicitly discussed; however, corresponding numerical schemes can be constructed by combining the individual approaches.

Roughly speaking, we can classify the numerical approaches that we shall discuss into two categories: fully discrete schemes and semi-discrete schemes. In the former approach, we immediately discretize with respect to all variables and directly obtain a procedure to compute the required approximate solutions. In the latter, we discretize with respect to some of the variables (usually, either with respect to the space variables only or with respect to the time variable only) and thus obtain a system of problems of a simpler structure as an intermediate result. These simpler problems can then sometimes be solved exactly to obtain the required approximate solutions; more frequently they will also be solved by numerical schemes tailored for this simpler class of problems.

3.2.1 *The method of lines*

The first method that we shall look at explicitly, the *vertical method of lines*, belongs to the class of semi-discrete methods. This is a well known concept in the theory of classical partial differential equations of integer order where it is frequently employed for parabolic problems, see, *e. g.*, [556] or Section 7.2 of [345]. It is best illustrated by using a simple example. To this end, we shall look at the time-fractional diffusion equation in one space dimension,

$$ {}^{\mathrm{C}}D_t^{\alpha} y(x,t) = K(x,t)\frac{\partial^2}{\partial x^2} y(x,t) + f(x,t) \tag{3.2.1a} $$

for $t \in [0,T]$ and $x \in [a,b]$, say, with $0 < \alpha < 1$ and a strictly positive function K, subject to the initial condition

$$ y(x,0) = y_0(x) \quad \text{for } x \in [a,b] \tag{3.2.1b} $$

and the boundary conditions

$$ y(a,t) = r_1(t) \quad \text{and} \quad y(b,t) = r_2(t) \qquad \text{for } t \in [0,T] \tag{3.2.1c} $$

with certain given functions K, f, y_0, r_1 and r_2. From the description that we shall now give it will be evident that the method can easily be modified to handle other related problems like, for example, equations in more than one space dimension, equations with other types of boundary conditions or time-fractional wave equations, *i. e.* equations like (3.2.1) but with $1 < \alpha < 2$ and then, of course, with a second initial condition.

The idea is, as already indicated, to discretize the differential equation with respect to only one variable at first. In the case of interest here, the vertical method of lines, this is the space variable x. There is a number of possibilities to implement such a discretization. One might choose, for example, a finite element approach or a finite difference method. For our example, we will go for the latter choice, but it is no problem for the vertical method of lines to use a different scheme (see, *e. g.*, [243] for such an example). Thus, we select a uniform grid in the space variable that is given by the grid points $x_j = a + j(b-a)/N$, $j = 0, 1, \dots, N$, with a certain value $N \in \mathbb{N}$. For these grid points we then discretize the differential operator with respect to the space variables via a standard central difference formula,

$$ \frac{\partial^2}{\partial x^2} y(x_j,t) \approx \frac{1}{h^2}(y(x_{j+1},t) - 2y(x_j,t) + y(x_{j-1},t)) $$

for $j = 1, 2, \dots, N-1$ where we have introduced the parameter $h = (b-a)/N$. Notice that we do not need to consider $j = 0$ and $j = N$ as the solution $y(x_j,t)$ for these two values of j is already known from the boundary condition (3.2.1c). Combining this discretization with the originally given differential equation (3.2.1a) leads to the system of equations

$$ {}^{\mathrm{C}}D_t^{\alpha} y(x_j,t) = \frac{K(x_j,t)}{h^2}(y(x_{j+1},t) - 2y(x_j,t) + y(x_{j-1},t)) + f(x_j,t) \tag{3.2.2} $$

for $j = 1, 2, \ldots, N-1$. Clearly, this is a system of $N-1$ ordinary fractional differential equations of order α for the unknown functions $y(x_j, \cdot)$ that can be combined with the initial conditions

$$y(x_j, 0) = y_0(x_j), \qquad j = 1, 2, \ldots, N-1,$$

and the boundary conditions

$$y(x_0, t) = r_1(t) \quad \text{and} \quad y(x_N, t) = r_2(t) \qquad \text{for } t \in [0, T]$$

to obtain a system that can be uniquely solved. Depending on the nature of the functions f, K, r_1 and r_2, it may be possible to solve these equations analytically in closed form, thus producing the required approximation of the solution $y(x_j, t)$ to the fractional diffusion problem for the points of our grid with respect to the space variable and all t. In the more common case where an exact solution of the system of ordinary fractional differential equations cannot be found, one needs to apply a numerical scheme. In Chapter 2 we have presented a number of possible choices from which one may select here.

Remark 3.3. Using vector notation, we may reformulate the $N-1$ equations (3.2.2) in the form

$${}^{\mathrm{C}}D^{\alpha}Y(t) = \frac{1}{h^2}F(t, Y(t)) + \begin{pmatrix} f(x_1, t) \\ \vdots \\ f(x_{N-1}, t) \end{pmatrix}$$

with a suitably chosen function F. It is then clear that the Lipschitz constant of the right-hand side with respect to the variable Y may be very large if h is small, *i. e.* if the space variable has been discretized on a fine grid. This feature may lead to a problematic behavior of some explicit solvers for ordinary fractional differential equations and may force us either to use very small step sizes in the time variable or to revert to implicit algorithms. A corresponding stability analysis has been provided in [641].

As the name *vertical* method of lines indicates, there is also a *horizontal method of lines*, also known as *Rothe's method* (see Setion 7.2 of [345]). In this approach, one starts by discretizing with respect to the time variable. We can also do this here. Let us use a fractional linear multistep method as described in Section 2.4 for this purpose. Thus, using the time variable discretization $t_j = j\tau$ with $\tau = T/M$ for some $M \in \mathbb{N}$ and invoking Theorem 2.14 this means that our fractional initial-boundary value problem (3.2.1) is approximated by the system of elliptic differential equations

$$y(x, t_n) = y_0(x) + \tau^\alpha \sum_{j=0}^{n} \omega_{n-j} \left[K(x, t_j) \frac{\partial^2}{\partial x^2} y(x, t_j) + f(x, t_j) \right]$$
$$+ \tau^\alpha \sum_{j=0}^{s} w_{n,j} \left[K(x, t_j) \frac{\partial^2}{\partial x^2} y(x, t_j) + f(x, t_j) \right] \qquad (3.2.3a)$$

for $n = 1, 2, \ldots, M$ that does not contain any differential operators of non-integer order. This system needs to be augmented by the boundary conditions

$$y(a, t_n) = r_1(t_n) \quad \text{and} \quad y(b, t_n) = r_2(t_n) \qquad (3.2.3b)$$

for all $n = 1, 2, \ldots, M$ in order to be uniquely solvable, and its solution can once again be computed exactly if the given data permit or numerically otherwise. In either case, the solution procedure needs to take into account the fact that the equations of which the system (3.2.3a) is made up are coupled with each other in a specific manner: The starting weights $w_{n,j}$ typically cause a full coupling between all equations with $n = 1, 2, \ldots, s$, so the computation of the solution at the times $t_1, t_2, \ldots, t_s$ requires the handling of this s-dimensional subsystem. Once this has been accomplished, the situation becomes much simpler because for $n > s$ the differential equation contains the functions $y(\cdot, t_j)$ for $j = 0, 1, \ldots, n-1$ that are now known, which leaves $y(\cdot, t_n)$ as the only unknown function, so there is no more coupling with any other unknown that needs to be taken into consideration.

3.2.2 *Backward difference formulas for time-fractional parabolic and hyperbolic equations*

For non-fractional partial differential equations of parabolic type such as, *e. g.*, equations of the form (3.2.1) with $\alpha = 1$, finite difference methods are a standard choice (see, *e. g.*, Section 9.2 of [357]). It is therefore not surprising that the fractional generalizations of these methods are very popular tools when it comes to the numerical solution of fractional differential equations of the form (3.2.1), at least with $0 < \alpha < 1$. In this context it is particularly helpful to know that many well known properties of these algorithms can be carried over from the integer-order to the fractional case. Thus we shall now look at this class of methods in more detail.

Specifically, a finite difference method is characterized by the fact that it discretizes the given differential equation by using finite differences to replace the derivative operators. For derivatives of second order like those appearing on the right-hand side of Eq. (3.2.1a), it is natural to choose a central difference of second order for this purpose. If a step size of h is used, then this yields an $O(h^2)$ approximation with respect to the space variable. This approach to discretizing in the space direction can be followed no matter whether or not the time derivative is of integer order. In other words, there is no problem in transferring this idea from the case $\alpha = 1$ to the case $0 < \alpha < 1$. It remains to find a suitable discretization with respect to time.

The first order derivative with respect to time that appears in a classical parabolic equation can be discretized in a number of ways. Denoting the step size by k, the most important versions are the forward difference

$$z'(t) \approx \frac{z(t+k) - z(t)}{k}$$

and the backward difference

$$z'(t) \approx \frac{z(t) - z(t-k)}{k}$$

that lead to numerical methods of the forward Euler form

$$\frac{z_\ell - z_{\ell-1}}{k} = f(t_{\ell-1}, z_{\ell-1})$$

and the backward Euler form

$$\frac{z_\ell - z_{\ell-1}}{k} = f(t_\ell, z_\ell),$$

respectively, for the first-order equation $z'(t) = f(t, z(t))$ with respect to the time variable. It is well known that both these formulas have a convergence order of $O(k)$. But of course there is not only this similarity; there are also major differences between these approaches. Evidently, the forward Euler method is explicit, *i. e.* we can directly compute z_ℓ from $z_{\ell-1}$. The backward Euler method, on the other hand, is implicit because the unknown value z_ℓ also appears inside the given function f on the right-hand side. This means that typically the computational effort required for a backward

Euler method is significantly larger than the effort for a forward Euler method. However, this advantage for the forward scheme comes with the price of stability problems. Specifically, the forward method will become unstable if the step sizes are too large, whereas the backward method is unconditionally stable and therefore does not exhibit such problems. This is a very important aspect when one uses such methods for the approximation of the time derivative in our parabolic partial differential equation. As a matter of fact it implies that a finite difference method with a forward Euler method for the time integration can only converge if the step sizes in space and time are related by the inequality

$$k \leq ch^2$$

where c is a constant that depends on the coefficient function K of the differential equation (3.2.1a). Thus, if a small step size h is used for the space discretization then a very much smaller step size must be used for the time discretization. No such condition is required for the backward Euler scheme.

A sort of a compromise between these two methods is the Crank-Nicolson approximation

$$\frac{z_\ell - z_{\ell-1}}{k} = \frac{1}{2}(f(t_{\ell-1}, z_{\ell-1}) + f(t_\ell, z_\ell))$$

that is formally obtained by taking the arithemtic mean of the two equations defining the forward Euler and the backward Euler method, respectively. Like the backward Euler method it is implicit, and it shares some, but not all, of its favorable stability properties (see Chapter 7 of [345]). The advantage of this method is that it yields a better convergence order, namely $O(k^2)$, than the two methods described above.

When it comes to extending these concepts to time-fractional differential equations, it is rather simple to generalize the forward Euler method and the backward Euler method. All that needs to be done is that, as indicated in Section 2.9, one must replace the first-order (forward or backward) difference by an appropriately chosen fractional counterpart. We have seen in Chapter 2 that there are various different possible fractional versions of the finite difference; any of these can be used. It is known that the stability properties of the resulting fractional methods are essentially identical to those of the underlying classical (integer-order) methods. Specifically, a fractional forward difference scheme will only be stable under the restriction

$$k^{\alpha} \leq ch^2$$

where once again the constant c depends on the given function K. Since $0 < \alpha < 1$, this restriction is even more severe than its integer-order counterpart ($\alpha = 1$), and so we refrain from looking at fractional forward difference methods for time-fractional diffusion equations.

For the Crank-Nicolson method we also run into certain difficulties. To be precise, the background behind the Crank-Nicolson method is that, under standard smoothness assumptions, the forward and backward Euler methods have an error expansion at the point t_ℓ that has the form

$$z(t_\ell) - z_\ell = c_1 k + c_2 k^2 + \text{higher order terms}$$

where the constants c_1 and c_2 depend on the given differential equation and on the chosen method. However, for the constant c_1 one can show that this constant has the same absolute value for both algorithms, but different signs. Thus, the construction of the Crank-Nicolson scheme via an arithmetic mean of these two formulas implies that the coefficient of k in the error expansion of the Crank-Nicolson method vanishes, and the dominating term is of the order k^2 as mentioned above. For the fractional analog of the Euler schemes we have two problems in this connection. Firstly, no information about the relation of the constants c_1 for the two methods is available. Thus, even if it may be possible to eliminate the leading term of the error by a suitably weighted mean of forward and backward method, it is not known which values one should choose for the weights. And moreover, even if one would be able to find the correct values for these weights then one would typically still have to deal with the situation that the second term in the asymptotic expansion, *i. e.* the first term that would not be eliminated, is only going to be of the order $k^{1+\alpha}$ and not k^2, so the gain in accuracy would be less substantial than in the non-fractional case. An alternative path to a fractional version of the Crank-Nicolson scheme with the same order of convergence as in the classical case has been provided by McLean and Mustapha [414] whose approach however requires the use of a special non-uniform discretization for the time variable that may not always be desired. Thus we shall not pursue the Crank-Nicolson idea in the fractional setting any further either and concentrate only on the backward Euler scheme.

For the backward Euler method we may proceed in a similar way. Of course we need to use a backward difference instead of the forward difference with respect to time now. We may choose the formula from Eqs. (2.1.10) and (2.1.11) which then yields, after a suitable rearrangement of terms, the overall algorithm

$$\begin{aligned}
y_{0m} &= r_1(t_m),\\
y_{jm} &= \Gamma(2-\alpha)\frac{k^\alpha}{h^2}K(x_j,t_m)(y_{j+1,m} - 2y_{jm} + y_{j-1.m})\\
&\quad + (1-\alpha)m^{-\alpha}y_0(x_j) + \Gamma(2-\alpha)k^\alpha f(x_j,t_m) \qquad\qquad (3.2.4)\\
&\quad - \sum_{\mu=0}^{m-1} B_{\mu m} y_{j\mu} \qquad\qquad (1 \le j < N),\\
y_{Nm} &= r_2(t_m)
\end{aligned}$$

for $m = 0, 1, 2, \ldots, M$. In this algorithm, N is the discretization parameter for the mesh with respect to the space variables, $h = (b-a)/N$ is the corresponding step size, M is the discretization parameter for the time mesh and $k = T/M$ is the time step where T is the upper bound of the time interval on which we are looking for a solution. The grid points are then $t_\mu = \mu k$ and $x_j = a + jh$, and the weights are given by

$$B_{\mu m} = \begin{cases} (m-1)^{1-\alpha} - (m-1+\alpha)m^{-\alpha} & \text{for } \mu = 0,\\ (m-\mu+1)^{1-\alpha} - 2(m-\mu)^{1-\alpha} + (m-\mu-1)^{1-\alpha} & \text{else} \end{cases}$$

(which follows from Eqs. (2.1.10) and (2.1.11)). Finally the functions y_0, r_1 and r_2 are those given in the initial condition (3.2.1b) and the boundary conditions (3.2.1c), respectively. The values y_{jm} that we compute in this way can then be interpreted as approximations to the exact solution values $y(x_j, t_m)$. Clearly we have an implicit method because the values y_{jm}, $j = 0, 1, \ldots, N$, *i. e.* the unknowns of the system at the mth time step, appear not only on the left-hand side of the equation system (3.2.4) but also on the right-hand side. It is convenient to rewrite the system in matrix-vector form, thus obtaining

$$\begin{pmatrix} y_{0m} \\ y_{1m} \\ y_{2m} \\ \vdots \\ y_{N-1,m} \\ y_{Nm} \end{pmatrix} = \Gamma(2-\alpha)\frac{k^\alpha}{h^2} G_m \begin{pmatrix} y_{0m} \\ y_{1m} \\ y_{2m} \\ \vdots \\ y_{N-1,m} \\ y_{Nm} \end{pmatrix} + H_m \tag{3.2.5}$$

with the vector

$$H_m = \begin{pmatrix} r_1(t_m) \\ (1-\alpha)m^{-\alpha} y_0(x_1) + \Gamma(2-\alpha)k^\alpha f(x_1, t_m) - \sum_{\mu=0}^{m-1} B_{\mu m} y_{1\mu} \\ (1-\alpha)m^{-\alpha} y_0(x_2) + \Gamma(2-\alpha)k^\alpha f(x_2, t_m) - \sum_{\mu=0}^{m-1} B_{\mu m} y_{2\mu} \\ \vdots \\ (1-\alpha)m^{-\alpha} y_0(x_{N-1}) + \Gamma(2-\alpha)k^\alpha f(x_{N-1}, t_m) - \sum_{\mu=0}^{m-1} B_{\mu m} y_{N-1,\mu} \\ r_2(t_m) \end{pmatrix}$$

and the matrix $G_m = (g_{\rho\sigma}^{(m)})_{\rho,\sigma=0}^N$ with

$$g_{\rho\sigma}^{(m)} = \begin{cases} -2K(x_\rho, t_m) & \text{for } 1 \le \rho = \sigma \le N-1, \\ K(x_\rho, t_m) & \text{for } 1 \le \rho \le N-1 \text{ and } |\rho - \sigma| = 1, \\ 0 & \text{else} \end{cases}$$

(note that, for the sake of notational convenience, we have started the row and column indices at 0, not at 1). Hence our solution vector $Y_m = (y_{0m}, y_{1m}, \ldots y_{Nm})^{\mathrm{T}}$ at the mth time step can be obtained as the solution of the linear system

$$\left(I - \Gamma(2-\alpha)\frac{k^\alpha}{h^2} G_m \right) Y_m = H_m$$

where I is the unit matrix.

We now demonstrate the performance of this algorithm by means of an example problem. To this end we look at the initial-boundary value problem

$${}^{\mathrm{C}}D_t^{0.7} y(x,t) = \left(2 + \frac{1 + x(2\pi - x)}{5 + 5t} \right) \frac{\partial^2}{\partial x^2} y(x,t) + \cos x \cos t \tag{3.2.6a}$$

for $0 \le t \le 10$ and $0 \le x \le 2\pi$ with initial condition

$$y(x, 0) = \cos \frac{x}{2} \tag{3.2.6b}$$

and boundary conditions

$$y(0, t) = \exp(-t) \quad \text{and} \quad y(2\pi, t) = -\exp(-t). \tag{3.2.6c}$$

The resulting approximate solutions for various choices of the discretization parameters are shown in Figures 3.9 and 3.10. It is evident from these figures that the algorithm indeed behaves in a stable manner in the sense that an arbitrary combination of step sizes with respect to time and with respect to space yields useful values.

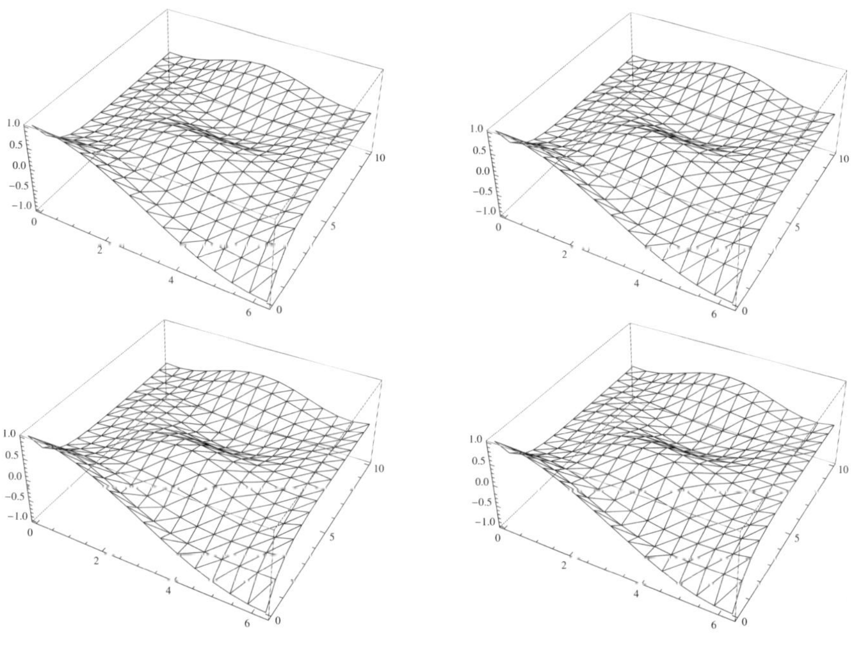

Fig. 3.9 Numerical solutions of Eq. (3.2.6) on $[0, 2\pi] \times [0, 10]$ for backward Euler method with space mesh size h and time mesh size k, where $h = \pi/5$, $k = 1$ (top left), $h = \pi/10$, $k = 0.5$ (top right), $h = \pi/50$, $k = 0.5$ (bottom left), $h = \pi/500$, $k = 0.5$ (bottom right).

For classical hyperbolic equations, *i. e.* equations of the form (3.2.1) with $\alpha = 2$ (and two initial conditions in Eq. (3.2.1b) instead of just one), two principal versions of finite difference approaches are frequently used. The first of these approaches is formally identical to the approach presented

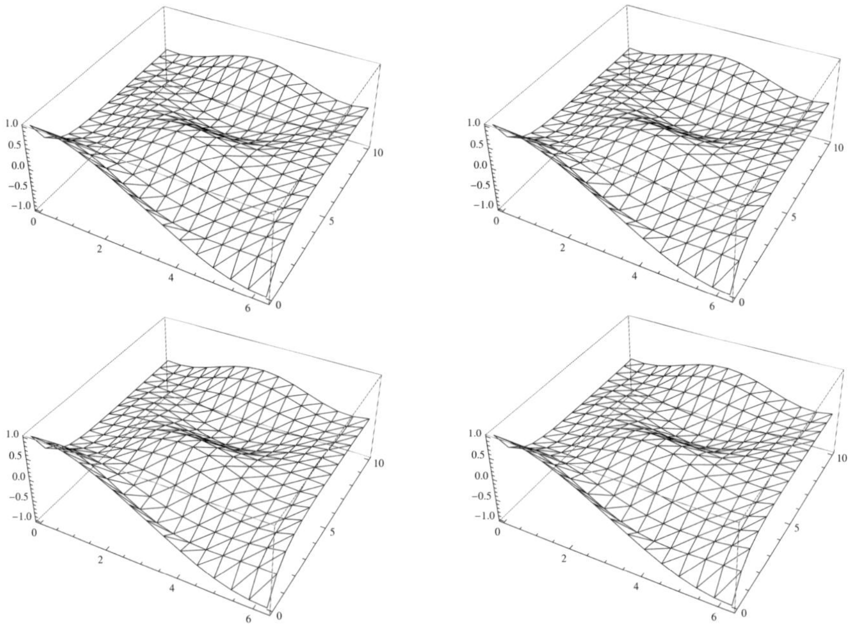

Fig. 3.10 Numerical solutions of Eq. (3.2.6) on $[0, 2\pi] \times [0, 10]$ for backward Euler method with space mesh size h and time mesh size k, where $h = \pi/50$, $k = 0.1$ (top left), $h = \pi/500$, $k = 0.1$ (top right), $h = \pi/10$, $k = 0.01$ (bottom left), $h = \pi/50$, $k = 0.01$ (bottom right).

above for parabolic problems, where we only have to set $\alpha = 2$ now. Of course one here needs to use second order differences with respect to time; centered differences are the usual method of choice. It is then possible to extend the previous theory and to prove convergence and stability under similar conditions as above. In particular, for appropriately chosen differences with respect to the space variable the stability condition then reads $k^2 \leq c'h^2$ or, more simply,

$$k \leq ch$$

where once again the constant c', and hence also c, depends on the coefficient function K, and once again no such restriction is required in the backward Euler method. For details we refer to [35] or Chapter 14 of [310] where one may also find a description of the very rich class of phenomena that may arise in connection with hyperbolic equations (in particular in the nonlinear case). This concept can be directly generalized to the fractional setting $1 < \alpha < 2$, thus immediately extending the results for the case $0 < \alpha < 1$.

The second commonly used option to approach the classical hyperbolic equation is to introduce a new two-dimensional function Y via the definition

$$Y = \begin{pmatrix} \partial y/\partial t \\ \partial y/\partial x \end{pmatrix}.$$

Using this new variable, the given equation (3.2.1a) with $\alpha = 2$ can be rewritten in the form of a two-dimensional system of partial differential equations of first order,

$$\begin{aligned} \frac{\partial}{\partial t} Y_1(x,t) \qquad\qquad - \frac{\partial}{\partial x} Y_2(x,t) &= 0, \\ \frac{\partial}{\partial t} Y_2(x,t) - K(x,t) \frac{\partial}{\partial x} Y_1(x,t) &= f(x,t), \end{aligned}$$

that we can write in matrix-vector form as

$$\frac{\partial}{\partial t} Y(x,t) + A(x,t) \frac{\partial}{\partial x} Y(x,t) = \begin{pmatrix} 0 \\ f(x,t) \end{pmatrix}$$

with

$$A = \begin{pmatrix} 0 & -1 \\ -K(x,t) & 0 \end{pmatrix}.$$

This approach can be generalized to the fractional situation too. Indeed, using the same new variable Y as above, we come to the system

$$\begin{aligned} \frac{\partial}{\partial t} Y_1(x,t) \qquad\qquad - \frac{\partial}{\partial x} Y_2(x,t) &= 0, \\ {}^{\mathrm{C}}D_t^{\alpha-1} Y_2(x,t) - K(x,t) \frac{\partial}{\partial x} Y_1(x,t) &= f(x,t), \end{aligned}$$

that evidently is a system of two differential equations of different orders. As we had seen above, it is in principle possible to deal with such a system numerically. However, as demonstrated in Section 14.4 of [310] (see also Chapter 12 of [357]), the special case $\alpha = 2$ already needs to be handled very carefully and tediously in order to obtain practically useful results, and the implications of the additional complications introduced by the fractional generalization seem to be unclear at the moment. We therefore do not follow this path any further here.

3.2.3 *Other methods*

It seems indeed that methods based on finite differences with respect to both the space and the time variables are the most popular choices in concrete applications. However there are also some other possibilities. A brief survey of finite difference and other methods for fractional partial differential equations has been given in Section 1 of [419]; see also the references cited therein. We shall now briefly describe a few potential alternatives to the finite difference concept.

We first deal with *collocation methods*. The essence of these methods is that one looks for an approximate solution in a finite-dimensional linear space. Given a basis of this space, the coefficients are then computed in such a way that the given differential equation is exactly fulfilled at a certain set of preassigned points. Of course the number of elements of this set must coincide with the dimension of the space of functions, and frequently additional conditions on the location of the points are required in order to obtain unique solutions with a good approximation quality. For the sake of illustration, we consider the problem (3.2.1) with $K \equiv 1$, $f \equiv 0$ and $r_1 \equiv r_2 \equiv 0$. Note, in particular, that the latter condition is not a restriction of generality as it can always be achieved by a suitable transformation.

In the construction of an exemplary special case of the method we follow Section 14.1 of [357] and discretize with respect to the space variable, *i. e.* we work according to the spirit of a vertical method of lines. Introducing a discretization parameter $N \in \mathbb{N}$, we set $h = (b-a)/N$ and use a piecewise polynomial space

$$S_N = \{v \in C^1[a,b] : v(0) = v(1) = 0, v|_{[a+jh,a+(j+1)h]} \in \mathcal{P}_{r-1} \text{ for all } j\}$$

with some $r \geq 4$, where $\mathcal{P}_{r-1}$ is the set of all polynomials of degree at most $r-1$. Now in each subinterval $[a+jh, a+(j+1)h]$ we define the collocations points $x_{j,1}, x_{j,2}, \ldots, x_{j,r-2}$ by $x_{j,\nu} = a+jh+h(\eta_\nu+1)/2$ where η_ν $(\nu = 1, 2, \ldots, r-2)$ is the νth zero of the Legendre polynomial P_{r-2}. Our problem then is to find a solution of the system of differential equations

$$^{\mathrm{C}}D_t^\alpha y_N(x_{j,\nu}, t) = \frac{\partial^2}{\partial x^2} y_N(x_{j,\nu}, t), \qquad 1 \leq j \leq M, \quad 1 \leq \nu \leq r-2, \quad (3.2.7)$$

subject to the initial condition $y_N(\cdot, 0) = y_{0,N}$ where $y_{0,N} \in S_N$ is a suitable approximation of the initial datum y_0 given in Eq. (3.2.1b).

In comparison to other methods, collocation methods frequently require more restrictive conditions on the regularity of the functions involved in order to converge sufficiently fast.

Our next objects of study in this subsection are the *spectral methods* (see, *e. g.*, Section 14.2 of [357] or [371]). Like collocation methods, spectral approximation is often employed with respect to the space variables in a partial differential equation. One possible way of describing the basic idea is once again to use a vertical method of lines approach. For the description it is again convenient to assume homogeneous boundary conditions $r_1 \equiv r_2 \equiv 0$ to be present in the original partial differential equation (3.2.1c) and hence also in the boundary conditions (3.2.3b). As mentioned above, this is not a loss of generality. Moreover, we once again assume $K \equiv 1$ as in our description of the collocation method. However, we do not impose any restriction on the inhomogeneity f of the given partial differential equation (3.2.1). Then we let $\{\phi_j : j = 1, 2, \ldots\}$ be a set of linearly independent and sufficiently smooth functions that span $L^2(a, b)$ and define S_N to be the linear span of $\{\phi_j : 1 \le j \le N\}$.

As in the collocation approach, we attempt to find an approximate solution in the set S_N (that now has a different meaning than above). To this end we invoke a Galerkin idea and require, instead of Eq. (3.2.7), that

$$({}^{\mathrm{C}}D_t^{\alpha} y_N, z) = (\frac{\partial^2}{\partial x^2} y_N, z) + (f, z) \qquad \text{for all } z \in S_N \text{ and } t > 0 \qquad (3.2.8)$$

where $(\cdot, \cdot)$ denotes the standard L^2 inner product on (a, b). Once again the resulting differential equations with respect to t are equipped with the initial condition $y_N(\cdot, 0) = y_{0,N}$ where $y_{0,N} \in S_N$ is a suitable approximation of the initial datum y_0 given in Eq. (3.2.1b).

If Π_N is the orthogonal projection from $L^2(a, b)$ to S_N then we may rewrite Eq. (3.2.7) as

$$ {}^{\mathrm{C}}D_t^{\alpha} y_N = A_N y_N + \Pi_N f$$

where $A_N = \Pi_N L \Pi_N$ and $L = \partial^2/\partial x^2$. With the basis representation $y_N(x, t) = \sum_{j=1}^{N} a_j(t)\phi_j(x)$ this amounts to a matrix-vector equation of the form

$$B\ {}^{\mathrm{C}}D_t^{\alpha} a(t) + Aa(t) = b(t), \qquad t > 0,$$

with the elements of the $(N \times N)$ matrices A and B being given by $(L\phi_i, \phi_j)$ and (ϕ_i, ϕ_j), respectively. In particular, the matrix B becomes the unit matrix if the basis function ϕ_j are orthonormal. A number of methods have been proposed in this book for solving this system of fractional differential equations with respect to t; any of these methods can be used to compute the function a and hence the final solution of our given partial fractional differential equation.

3.2.4 *Methods for equations with space-fractional operators*

It is mathematically no problem to replace the derivatives with respect to the space variables in a partial differential equation by their fractional-order generalizations. However, as indicated in the previous chapter, the resulting models have some features that are inconsistent with certain fundamental physical principles (see Section 2.3.2 of [305]). For certain other physical models, the corresponding authors argue in favor of space-fractional equations also based on ideas taken from physical principles [93]. Therefore the application of such space-fractional equations (or, even more generally, partial differential equations with fractional derivatives with respect to both space and time) needs to be done very carefully in order to obtain valid models. Nevertheless there may be situations where these models are appropriate, and therefore we shall devote a short section to their numerical solution too.

In this description, we discuss the fully general case of a partial differential equation that is fractional with respect to space and time. The case of an equation that has fractional derivatives only with respect to the space variable but not with respect to the time variable is contained *a fortiori* as a special case in this general setting.

The model problem that we shall numerically solve here has the convection-diffusion form

$$ {}^{\mathrm{C}}D_t^{\alpha} y(x,t) = -b(x)\frac{\partial}{\partial x} y(x,t) + a(x)\, {}^{\mathrm{RL}}D_x^{\beta} y(x,t) + q(x,t) \tag{3.2.9a} $$

for $x \in [0, L]$ and $t \in [0, T]$ with $0 < \alpha \leq 1$, $1 < \beta \leq 2$, continuous functions $a > 0$, $b > 0$ and q, the initial condition

$$ y(x, 0) = f(x) \tag{3.2.9b} $$

for all $x \in [0, L]$ and the homogeneous boundary conditions

$$ y(0, t) = y(L, t) = 0 \tag{3.2.9c} $$

for all $t \in [0, T]$. Obviously, the case of inhomogeneous boundary conditions can be reduced to this problem via the classical techniques.

The numerical method for the solution of the problem (3.2.9) that we shall propose here is due to Zhang [649]. It is based on approximating the three differential operators appearing in the differential equation (3.2.9a) by appropriate finite differences. Specifically, we introduce equispaced grids with respect to both variables, *i. e.* we define a time step $\tau > 0$ such that $T = N\tau$ with some $n \in \mathbb{N}$ and a space step $h = L/M$ with some $M \in \mathbb{N}$ and denote the grid points by $x_i = ih$ $(i = 0, 1, \ldots, M)$ and $t_j = j\tau$ $(j = 0, 1, \ldots, N)$. Then, the Caputo derivative with respect to the time variable is approximated by

$$\begin{aligned} &{}^{\mathrm{C}}D_t^{\alpha} y(x_i, t_{k+1}) \\ &\approx \frac{\tau^{-\alpha}}{\Gamma(2-\alpha)} \Big(y(x_i, t_{k+1}) \\ &\qquad + \sum_{j=0}^{k} y(x_i, t_{k-j}) \left[(j+2)^{1-\alpha} - 2(j+1)^{1-\alpha} + j^{1-\alpha} \right] \\ &\qquad - y(x_i, t_0) \left[(k+1)^{1-\alpha} - k^{1-\alpha} \right] \Big). \end{aligned} \tag{3.2.10}$$

This discretization can essentially be obtained by representing the Caputo differential operators in terms of their Riemann-Liouville counterparts using the well known relation (2.1.5) in combination with an approximation of the Riemann-Liouville derivatives via the formula developed in Eqs. (2.1.10) and (2.1.11). For the first-order space derivative we use the classical first difference

$$\frac{\partial}{\partial x} y(x_i, t_{k+1}) \approx h^{-1} \left(y(x_i, t_{k+1}) - y(x_{i-1}, t_{k+1}) \right). \tag{3.2.11}$$

And finally the fractional derivative with respect to the space variable x is approximated by the shifted Grünwald formula

$${}^{\mathrm{RL}}D_x^{\beta} y(x_i, t_{k+1}) \approx h^{-\beta} \sum_{j=0}^{i+1} g_j y(x_{i-j+1}, t_{k+1}) \tag{3.2.12a}$$

with

$$g_j = \frac{(-1)^j}{j!} \prod_{\mu=1}^{j} (\beta + 1 - \mu). \tag{3.2.12b}$$

The shifted Grünwald formula (3.2.12) evaluates the function in question at the point x_{i+1} in order to compute an approximation for a fractional derivative with starting point x_0 at the point x_i. The use of such a method may at first appear strange because the sampling point x_{i+1} is located outside of the interval $[x_0, x_i]$ over which the range of integration in the definition of the fractional derivative extends. However, Meerschaert and Tadjeran noted in §2 of their work [420] that such an approach is in general necessary to avoid severe instabilies in the constructed algorithms when dealing with space-fractional partial differential equations.

Approximations for $y(x_0, t)$ and $y(x_M, t)$ are not required because these values are already known exactly from the boundary conditions (3.2.9c). Thus, applying the discretizations (3.2.10), (3.2.11) and (3.2.12) to our problem (3.2.9), we construct a fully discretized scheme for the remaining unknowns $Y_k = (y_{1,k}, y_{2,k}, \ldots y_{M-1,k})^{\mathrm{T}}$ $(k = 1, 2, \ldots, N)$, *i. e.* the vectors of approximations for the exact solutions $(y(x_1, t_k), \ldots, y(x_{M-1}, t_k))^{\mathrm{T}}$. As described in detail in [649], this system can be written in the form

$$Y_0 = (f(x_1), f(x_2), \ldots, f(x_{M-1}))^{\mathrm{T}}, \tag{3.2.13a}$$

$$ZY_{k+1} = c_k Y_k - \sum_{\mu=1}^{k} d_\mu Y_{k+1-\mu} + \tau^\alpha Q_{k+1} \quad (0 \le k < N), \tag{3.2.13b}$$

where

$$Q_\mu = \Gamma(2-\alpha)(q(x_1, t_\mu), \ldots, q(x_{M-1}, t_\mu))^{\mathrm{T}},$$
$$d_\mu = (\mu+1)^{1-\alpha} - 2\mu^{1-\alpha} + (\mu-1)^{1-\alpha}$$

and

$$c_\mu = (\mu+1)^{1-\alpha} - \mu^{1-\alpha}.$$

Moreover, the matrix $Z = (z_{ij})_{i,j=1}^{M-1}$ in Eq. (3.2.13b) has entries of the form

$$z_{ij} = \begin{cases} 0 & \text{for } i < j-1, \\ -a(x_i)\tau^\alpha h^{-\beta}\Gamma(2-\alpha) & \text{for } i = j-1, \\ 1 + (b(x_i)h^{-1} - a(x_i)h^{-\beta}g_1)\tau^\alpha\Gamma(2-\alpha) & \text{for } i = j, \\ -(b(x_i)h^{-1} + a(x_i)h^{-\beta}g_2)\tau^\alpha\Gamma(2-\alpha) & \text{for } i = j+1, \\ -a(x_i)\tau^\alpha h^{-\beta}g_{i-j+1}\Gamma(2-\alpha) & \text{for } i > j+1 \end{cases}$$

which completes the description of the method.

The fact that this algorithm has useful stability and convergence properties has been demonstrated in [649].

Chapter 4

Generalized Stirling Numbers and Applications

In this Chapter, we will present the latest and most complete extension of the Stirling numbers using the framework of the fractional calculus that we are presently aware of.

4.1 Introduction

The classical Stirling numbers of the first kind $s(n, k)$, introduced by James Stirling in his famous manuscript *Methodus Differentialis* of 1730 [581], play, together with the Stirling numbers of second kind $S(n, k)$, an important role in the calculus of finite differences, in combinatorial problems, in numerical analysis, interpolation theory and number theory.

The classical Stirling numbers of the first kind $s(n, k)$ and of the second kind $S(n, k)$ can be defined via different generating functions, for example by

$$[x]_n = \sum_{k=0}^{n} s(n, k)x^k \quad (x \in \mathbb{R};\ n \in \mathbb{N}_0), \tag{4.1.1}$$

or

$$(\log(1 + x))^k = k! \sum_{n=k}^{\infty} s(n, k)\frac{x^n}{n!} \quad (|x| < 1;\ k \in \mathbb{N}_0) \tag{4.1.2}$$

and

$$x^n = \sum_{k=0}^{n} S(n, k)[x]_k \tag{4.1.3}$$

or

$$(e^x - 1)^k = k! \sum_{n=k}^{\infty} S(n, k)\frac{x^n}{n!} \quad (x \in \mathbb{R};\ k \in \mathbb{N}_0), \tag{4.1.4}$$

respectively. Equivalently, by

$$s(n,k) = \frac{1}{k!} \lim_{x \to 0} D^k [x]_n, \; D = \frac{d}{dx} \quad (x \in \mathbb{R}; \; n, k \in \mathbb{N}_0) \tag{4.1.5}$$

and

$$S(n,k) = \frac{1}{k!} \lim_{x \to 0} \Delta^k (x^n) \quad (x \in \mathbb{R}; \; n, k \in \mathbb{N}_0). \tag{4.1.6}$$

Here $[x]_n$ is the factorial polynomial defined for $x \in \mathbb{R}$ by

$$[x]_0 = 1, \; [x]_n = x(x-1)\cdots(x-n+1) \quad (n \in \mathbb{N}), \tag{4.1.7}$$

and Δ^k is the difference of order $k \in \mathbb{N}_0$ defined by

$$\Delta^0 f(x) = f(x), \; \Delta^1 f(x) = f(x+1) - f(x),$$

$$\Delta^{k+1} f(x) = \Delta \left(\Delta^k f \right)(x) \quad (k \in \mathbb{N}) \tag{4.1.8}$$

having the representation

$$\Delta^k f(x) = \sum_{j=0}^{k} (-1)^{k-j} \binom{k}{j} f(x+j) \quad (x \in \mathbb{R}; \; k \in \mathbb{N}_0) \tag{4.1.9}$$

in terms of the binomial coefficients

$$\binom{k}{j} = \frac{k!}{j!(k-j)!} \quad (k, j \in \mathbb{N}_0; \; j \le k). \tag{4.1.10}$$

The main definitions and properties of the classical Stirling numbers are given, for instance, in the textbooks by Riordan [523], Comtet [160], and on pp. 824–825 of the handbook by Abramowitz and Stegun [4]. For interesting applications of Stirling numbers in the setting of difference calculus, discrete mathematics and combinatorics one may consult the books by Jordan [320], by Graham *et al.* [274] and by Aigner [31].

Butzer *et al.* [123] introduced generalizations of $s(n,k)$ and $S(n,k)$ for the first parameter n, via Eqs. (4.1.5) and (4.1.6), for $\alpha \in \mathbb{R}$

$$s(\alpha,k) = \frac{1}{k!} \lim_{x \to 0} D^k [x]_\alpha \quad (-x \notin \mathbb{N}; \; \alpha \in \mathbb{R}, \; k \in \mathbb{N}_0) \tag{4.1.11}$$

with

$$[x]_\alpha = \frac{\Gamma(x+1)}{\Gamma(x-\alpha+1)},$$

and

$$S(\alpha,k) = \frac{1}{k!} \lim_{x \to 0} \Delta^k (x^\alpha) \quad (\alpha \ge 0, \; k \in \mathbb{N}_0), \tag{4.1.12}$$

the limit being taken in the sense

$$\lim_{x\to 0}\Delta^k(x^\alpha)=\lim_{\epsilon\to 0}\lim_{x\to 0}\Delta^k((x+\epsilon)^\alpha). \tag{4.1.13}$$

The mentioned authors and Butzer and Hauss [120, 121] investigated properties of the above generalized Stirling numbers which they finally called "Stirling function", that is, functions of the "continuous" parameter α. In particular, for such Stirling functions of the first kind they established the recurrence relation

$$s(\alpha+1,k)=s(\alpha,k-1)-\alpha s(\alpha,k);\quad s(\alpha,0)=\frac{1}{\Gamma(1-\alpha)} \tag{4.1.14}$$

with $\alpha\in\mathbb{R}$, $k\in\mathbb{N}$, and the integral representation

$$s(\alpha,k)=\frac{1}{2\pi i}\int_{\mathcal{L}}\frac{\Gamma(z+1)}{\Gamma(z-\alpha+1)}\frac{dz}{z^{n+1}}\quad(\alpha\in\mathbb{R};\ k\in\mathbb{N}_0), \tag{4.1.15}$$

with a special contour $\mathcal{L}$, while for Stirling functions of the second kind they established the closed form representation

$$S(\alpha,k)=\frac{1}{k!}\sum_{j=1}^{k}(-1)^{k-j}\binom{k}{j}j^\alpha\quad(\alpha>0;\ k\in\mathbb{N}) \tag{4.1.16}$$

and the recurrence relations for $k\in\mathbb{N}$

$$S(\alpha+1,k)=kS(\alpha,k)+S(\alpha,k-1)\quad(\alpha>0); \tag{4.1.17}$$

$$S(\alpha,0)=0,\ S(0,k)=0,\ S(0,0)=1\quad(\alpha>1).$$

Moreover Butzer *et al.* [122] extended $s(\alpha,k)$ to complex $\alpha\in\mathbb{C}$ ($\Re e(\alpha)>0$), and Loeb [378] followed up the matter of [120, 121], studying some properties of such generalized Stirling functions of the first kind with real α.

Generalizations of Stirling numbers of the first and the second kind to more general constructions containing more than two parameters were considered by several authors. For example, Rucinski and Voigt [533] gave a generalization of Stirling numbers of the second kind which are defined if in the formula (4.1.3) one replaces $[x]_k$ by the product $(x-a_0)\cdots(x-a_{k-1})$ containing a given sequence $(a_j)_{j=0}^{\infty}$. Butzer and Jansche [124] introduced a generalization $S_c(n,k)$ $(n,k\in\mathbb{N};\ n\geq k\geq 0)$ of Stirling numbers $S(n,k)$ with the property

$$\Theta^n f(x)=\sum_{k=0}^{n}S_c(n,k)x^k f^{(k)}(x), \tag{4.1.18}$$

where Θ^n is the differentiation operator of the form

$$(\Theta^n f)(x) = (\delta + c)^n f(x), \ \delta = x\frac{d}{dx}. \tag{4.1.19}$$

Also we must mention in this respect the generalized Stirling numbers of the second kind defined by Hauss [287] in his thesis, who developed a theory based on a very general approach, including representations in terms of various fractional integrals and derivatives, as well as the fine works by Platonov [486] and by Zhang [647].

Uniform asymptotic expansions for certain generalized Stirling functions were considered by Chelluri *et al.* [152].

As far as we know, a number of authors have introduced several generalizations of both Stirling numbers using different approaches, but without using the fractional calculus framework. We refer in this context, *e. g.*, to [5, 162, 235, 308, 355, 415, 517, 567–569, 601, 620], *etc.*

The aim of this chapter is to develop a global theory, within the framework of the fractional calculus, which allows us to introduce a natural extension of the classical concepts of $s(n,k)$ and $S(n,k)$ to Stirling functions with both parameters, n and k, complex numbers. Moreover, it is proved that such Stirling functions conserve the well known properties of the classical Stirling numbers. See, *e. g.*, the papers by Butzer *et al.* [125–129, 131].

4.2 Stirling Functions of the First Kind $s(\alpha, k)$ with Complex First Argument α

The Stirling numbers of first kind, the $s(n,k)$, can be defined in terms of their (horizontal) generating function

$$[z]_n = \sum_{k=0}^{n} s(n,k) z^k \quad (z \in \mathbb{C}; \ n \in \mathbb{N}_0), \tag{4.2.1}$$

thus, equivalently by (4.1.5),

$$s(n,k) = \frac{1}{k!} \lim_{x \to 0} D^k [x]_n, \ D = \frac{d}{dx} \quad (x \in \mathbb{R}; \ n, k \in \mathbb{N}_0).$$

A further equivalent approach is via their exponential generating function (4.1.2)

$$(\log(1+x))^k = k! \sum_{n=k}^{\infty} s(n,k) \frac{x^n}{n!} \quad (|x| < 1; \ k \in \mathbb{N}_0)$$

thus, in view of the Taylor expansion, also by

$$s(n,k) = \frac{1}{k!} \lim_{x\to 0} \left(\frac{d}{dx}\right)^n [\log(1+x)]^k \quad (x \in \mathbb{R}). \tag{4.2.2}$$

4.2.1 *Equivalent definitions*

The Stirling functions of the first kind, $s(\alpha, k)$, where $n \in \mathbb{N}$ is extended to real $\alpha \in \mathbb{R}$ as well as to complex $\alpha \in \mathbb{C}$, first studied from 1989 in [120–123], can be defined in terms of the infinite sum

$$[z]_\alpha = \sum_{k=0}^{\infty} s(\alpha, k) z^k \quad (|z| < 1; \;\; \alpha \in \mathbb{C}) \tag{4.2.3}$$

since $[z]_\alpha$ is holomorphic for $|z| < 1$, where $[z]_\alpha = \Gamma(z+1)/\Gamma(z+1-\alpha)$ $(\alpha \in \mathbb{C}\backslash\mathbb{Z}^-)$, therefore equivalently by

$$s(\alpha, k) = \frac{1}{k!} \lim_{x\to 0} \left(\frac{d}{dx}\right)^k [x]_\alpha \quad (\alpha \in \mathbb{C}; \; k \in \mathbb{N}_0). \tag{4.2.4}$$

The following representation for $s(\alpha, k)$ has been obtained by Butzer *et al.* [122].

Theorem 4.1 (Representation theorem). *For $\alpha \in \mathbb{C}$ and $k > \Re e(\alpha)$ $(k \in \mathbb{N})$ there holds*

$$\begin{aligned} s(\alpha, k) &= \frac{1}{\Gamma(-\alpha) k!} \int_{0+}^{1} \frac{[\log u]^k}{(1-u)^{\alpha+1}} du \\ &= (-1)^{k+1} \frac{\sin(\alpha\pi)}{\pi} \sum_{j=1}^{\infty} \frac{\Gamma(\alpha + j)}{(j-1)! \, j^{k+1}}. \end{aligned} \tag{4.2.5}$$

The Stirling functions of first kind, $s(\alpha, k)$, were also defined by the fractional counterpart of Eq. (4.2.2), namely

$$s(\alpha, k) = \frac{1}{k!} \lim_{x\to 1} \left({}^{\mathrm{RL}}D_{0+}^{\alpha} [\log(t)]^k\right)(x) \quad (\Re e(\alpha) > 0; \;\; k \in \mathbb{N}_0), \tag{4.2.6}$$

where ${}^{\mathrm{RL}}D_{0+}^{\alpha}$ is the Riemann-Liouville fractional derivative that we had described in Eq. (1.3.3). In particular, when $\alpha = n \in \mathbb{N}_0$, then the definition given in Eq. (4.2.6) coincides with that of Eq. (4.2.2) since the Riemann-Liouville derivative then reduces to the classical integer-order derivative, *i. e.*

$$\left({}^{\mathrm{RL}}D_{0+}^{n} [\log(t)]^k\right)(x) = \left(\frac{d}{dx}\right)^n [\log(x)]^k \quad (n \in \mathbb{N}_0).$$

With this approach we can also define the $s(\alpha,k)$ for $\Re e(\alpha)<0$ by

$$s(\alpha,k)=\frac{1}{k!}\lim_{x\to 1}\left({}^{\mathrm{RL}}I_{0+}^{-\alpha}[\log(t)]^k\right)(x)\quad (k\in\mathbb{Z}). \tag{4.2.7}$$

The first important result that we shall prove here is that the Stirling functions $s(\alpha,k)$, defined by Eqs. (4.2.6) and (4.2.7), coincide with those given by the definition given in Eq. (4.2.4). Thus these definitions are equivalent.

Theorem 4.2. *Let* $\alpha\in\mathbb{C}$ *and* $k\in\mathbb{N}_0$.

(a) *If* $\Re e(\alpha)\geq 0$, *then*

$$\begin{aligned}s(\alpha,k)&=\frac{1}{\Gamma(k+1)}\lim_{x\to 1}\left({}^{\mathrm{RL}}D_{0+}^{\alpha}[\log(t)]^k\right)(x) \qquad (4.2.8)\\ &=\frac{1}{k!}\lim_{x\to 0}\left(\frac{d}{dx}\right)^k\left[\frac{\Gamma(x+1)}{\Gamma(x+1-\alpha)}\right]=\frac{1}{k!}\lim_{x\to 0}\left(\frac{d}{dx}\right)^k[x]_\alpha.\end{aligned}$$

(b) *If* $\Re e(\alpha)<0$, *then*

$$\begin{aligned}s(\alpha,k)&=\frac{1}{\Gamma(k+1)}\lim_{x\to 1}\left({}^{\mathrm{RL}}I_{0+}^{-\alpha}[\log(t)]^k\right)(x)\\ &=\frac{1}{k!}\lim_{x\to 0}\left(\frac{d}{dx}\right)^k\left[\frac{\Gamma(x+1)}{\Gamma(x+1-\alpha)}\right]. \qquad (4.2.9)\end{aligned}$$

In particular,

$$s(\alpha,0)=\frac{1}{\Gamma(1-\alpha)}\quad (\alpha\in\mathbb{C}). \tag{4.2.10}$$

Proof.

(a) Let $\Re e(\alpha)\geq 0$, $\alpha\neq 0$, and $n=[\Re e(\alpha)]+1$. By (4.2.6) and (1.3.3),

$$s(\alpha,k)=\frac{1}{k!}\lim_{x\to 1}\left(\frac{d}{dx}\right)^n\left({}^{\mathrm{RL}}I_{0+}^{n-\alpha}[\log(t)]^k\right)(x). \tag{4.2.11}$$

Now, using the known property for the fractional Riemann-Liouville integral operator

$$\left({}^{\mathrm{RL}}I_{0+}^{n-\alpha}t^\gamma\right)(x)=\frac{\Gamma(\gamma+1)}{\Gamma(\gamma+1+n-\alpha)}x^{\gamma+n-\alpha}\quad (\gamma>-1),$$

in combination with a k-fold differentiation with respect to γ, yields for $\gamma > -1$, after an interchange of integration and differentiation,

$$\begin{aligned}
&\left({}^{\mathrm{RL}}I_{0+}^{n-\alpha} t^{\gamma}[\log(t)]^k\right)(x) \\
&= \left({}^{\mathrm{RL}}I_{0+}^{n-\alpha}\left(\frac{\partial}{\partial\gamma}\right)^k t^{\gamma}\right)(x) = \sum_{j=0}^{\infty}\left(\sum_{k=j}^{\infty} s(\alpha,k)S(k,j)\right) x^j f^{(j)}(x) \\
&= \Gamma(\alpha+1)\sum_{j=0}^{\infty}\frac{\sin[(\alpha-j)\pi]}{(\alpha-j)\pi}\frac{x^j f^{(j)}(x)}{j!}. \qquad (4.2.12)
\end{aligned}$$

Moreover,

$$\begin{aligned}
\left({}^{\mathrm{RL}}I_{0+}^{n-\alpha} t^{\gamma}[\log(t)]^k\right)(x) &= \left({}^{\mathrm{RL}}I_{0+}^{n-\alpha}\left(\frac{\partial}{\partial\gamma}\right)^k t^{\gamma}\right)(x) \\
&= \left(\frac{\partial}{\partial\gamma}\right)^k \left({}^{\mathrm{RL}}I_{0+}^{n-\alpha} t^{\gamma}\right)(x) \\
&= \left(\frac{\partial}{\partial\gamma}\right)^k \left[\frac{\Gamma(\gamma+1)}{\Gamma(\gamma+1+n-\alpha)}x^{\gamma+n-\alpha}\right].
\end{aligned}$$

Differentiating this expression k times with respect to x, and interchanging the order of differentiation, one has for $\gamma > -1$

$$\left(\frac{d}{dx}\right)^n \left({}^{\mathrm{RL}}I_{0+}^{n-\alpha} t^{\gamma}[\log(t)]^k\right)(x) = \left(\frac{\partial}{\partial\gamma}\right)^k \left[\frac{\Gamma(\gamma+1)}{\Gamma(\gamma+1-\alpha)}x^{\gamma-\alpha}\right]. \qquad (4.2.13)$$

Thus for $\gamma > -1$,

$$\frac{1}{k!}\left({}^{\mathrm{RL}}D_{0+}^{\alpha} t^{\gamma}[\log(t)]^k\right)(x) = \frac{1}{k!}\left(\frac{\partial}{\partial\gamma}\right)^k \left[\frac{\Gamma(\gamma+1)}{\Gamma(\gamma+1-\alpha)}x^{\gamma-\alpha}\right]. \quad (4.2.14)$$

Taking the limit for $\gamma \to 0$, one has by (4.2.8) and (4.2.14)

$$\begin{aligned}
s(\alpha,k) &= \frac{1}{k!}\lim_{x\to 1}\left({}^{\mathrm{RL}}D_{0+}^{\alpha}[\log(t)]^k\right)(x) \\
&= \frac{1}{k!}\lim_{x\to 1}\lim_{\gamma\to 0}\left({}^{\mathrm{RL}}D_{0+}^{\alpha} t^{\gamma}[\log(t)]^k\right)(x) \\
&= \frac{1}{k!}\lim_{\gamma\to 0}\lim_{x\to 1}\left(\frac{\partial}{\partial\gamma}\right)^k \left[\frac{\Gamma(\gamma+1)}{\Gamma(\gamma+1-\alpha)}x^{\gamma-\alpha}\right] \\
&= \frac{1}{k!}\lim_{\gamma\to 0}\left(\frac{\partial}{\partial\gamma}\right)^k \left[\frac{\Gamma(\gamma+1)}{\Gamma(\gamma+1-\alpha)}\right], \qquad (4.2.15)
\end{aligned}$$

establishing part (a) for $\mathfrak{Re}(\alpha) \geq 0,\ \alpha \neq 0$. If $\alpha = 0$, then by (4.2.6),

$$s(0,k) = \frac{1}{\Gamma(k+1)} \lim_{x\to 1} \left({}^{\mathrm{RL}}D^0_{0+}[\log(t)]^k\right)(x)$$
$$= \frac{1}{k!} \lim_{x\to 1} [\log(x)]^k = \begin{cases} 1 & \text{for } k = 0, \\ 0 & \text{for } k \in \mathbb{N}. \end{cases}$$

Thus $s(0,0) = 1$, and $s(0,k) = 0$ for $k \in \mathbb{N}$.

(b) In case $\mathfrak{Re}(\alpha) < 0$ one applies (4.2.7), using arguments similar to the above, but using (4.2.6) and (1.3.3) with $n - \alpha$ replaced by $-\alpha$, giving

$$s(\alpha,k) = \frac{1}{k!} \lim_{x\to 1} \left({}^{\mathrm{RL}}I^{-\alpha}_{0+}[\log(t)]^k\right)(x)$$
$$= \frac{1}{k!} \lim_{x\to 1} \lim_{\gamma\to 0} \left({}^{\mathrm{RL}}I^{-\alpha}_{0+} t^{\gamma}[\log(t)]^k\right)(x)$$
$$= \frac{1}{k!} \lim_{\gamma\to 0} \lim_{x\to 1} \left(\frac{\partial}{\partial\gamma}\right)^k \left[\frac{\Gamma(\gamma+1)}{\Gamma(\gamma+1-\alpha)}\, x^{\gamma-\alpha}\right]$$
$$= \frac{1}{k!} \lim_{\gamma\to 0} \left(\frac{\partial}{\partial\gamma}\right)^k \left[\frac{\Gamma(\gamma+1)}{\Gamma(\gamma+1-\alpha)}\right].$$

When $k = 0$, then, in accordance with (4.2.4), for any $\alpha \in \mathbb{C}$ we have

$$s(\alpha,0) = \frac{1}{0!} \lim_{x\to 0} [x_\alpha] = \lim_{x\to 0} \frac{\Gamma(x+1)}{\Gamma(x+1-\alpha)} = \frac{1}{\Gamma(1-\alpha)}.$$

This yields (4.2.10), and thus the theorem is proved. □

Now we obtain a recursion relation for the Stirling function $s(\alpha,k)$ in terms of the well known polygamma function (see, *e. g.*, Section 1.16 of [226]).

Theorem 4.3 (Recursion formula). *If $\alpha \in \mathbb{C}$ and $k \in \mathbb{N}$, then*

$$s(\alpha,k+1) = \frac{1}{k+1} \sum_{j=0}^{k} \frac{\psi^{(k-j)}(1) - \psi^{(k-j)}(1-\alpha)}{(k-j)!}\, s(\alpha,j) \quad (\alpha \notin \mathbb{N}), \tag{4.2.16}$$

where $\psi^{(m)}$ is the m-th polygamma function, that is,

$$\psi^{(m)}(z) = \left(\frac{d}{dz}\right)^m \psi(z) \quad (z \in \mathbb{C}\backslash\mathbb{Z}_0), \tag{4.2.17}$$

and $\psi(z) = \Gamma'(z)/\Gamma(z)$ is the digamma function.

Proof. The function $\Phi(x,\alpha) = \psi(x+1) - \psi(x+1-\alpha)$, defined for any $x \in V = \mathbb{C}\backslash\{x \in \mathbb{R}; x - \alpha \in \mathbb{Z}^-\}$, is holomorphic in V (see p. 261 of [454]), and thus in particular in $|x| < \epsilon$, for ϵ small, $\alpha \notin \mathbb{N}$. Hence it can be expanded as a power series about $x_0 = 0$: $\Phi(x,\alpha) = \sum_{k=0}^{\infty} \Psi_k(\alpha)x^k$ for $|x| < \epsilon$, $\alpha \notin \mathbb{N}$, where

$$\Psi_k(\alpha) = \frac{\psi^{(k)}(1) - \psi^{(k)}(1-\alpha)}{k!} \qquad (k \in \mathbb{N}_0),$$

noting that the function $\Psi_k(\alpha)$ equals

$$\Psi_k(\alpha) = \frac{1}{k!} \lim_{x\to 0} \left(\frac{d}{dx}\right)^k \Phi(x,\alpha).$$

Further, differentiating the series (4.2.3) for $|x| < 1$ yields

$$\frac{d}{dx}[x]_\alpha = \sum_{k=0}^{\infty} (k+1)s(\alpha, k+1)x^k. \tag{4.2.18}$$

But by definition the left-hand derivative equals

$$\begin{aligned} &\frac{d}{dx}\left[\frac{\Gamma(x+\alpha)}{\Gamma(x+1-\alpha)}\right] \\ &= \frac{\Gamma'(x+1)\Gamma(x+1-\alpha) - \Gamma(x+1)\Gamma'(x+1-\alpha)}{[\Gamma(x+1-\alpha)]^2} \\ &= \frac{\Gamma'(x+1)}{\Gamma(x+1)}\left\{\frac{\Gamma(x+1)}{\Gamma(x+1-\alpha)}\right\} - \left\{\frac{\Gamma(x+1)}{\Gamma(x+1-\alpha)}\right\}\frac{\Gamma'(x+1-\alpha)}{\Gamma(x+1-\alpha)}, \end{aligned} \tag{4.2.19}$$

thus, by power series multiplication, for $|x| < \epsilon$,

$$\frac{d}{dx}\left[\frac{\Gamma(x+\alpha)}{\Gamma(x+1-\alpha)}\right] = [x]_\alpha \Phi(x,\alpha) = \sum_{k=0}^{\infty}\left(\sum_{j=0}^{k} s(\alpha,j)\Psi_{k-j}(\alpha)\right)x^k. \tag{4.2.20}$$

A comparison of the coefficients of the series (4.2.18) and (4.2.20) yields the theorem. □

As a corollary of Theorem 4.2 and Theorem 4.3 we have

Corollary 4.1. *There hold for* $\alpha \in \mathbb{C}\backslash\mathbb{Z}$,

$$s(\alpha,1) = \frac{\psi(1)-\psi(1-\alpha)}{\Gamma(1-\alpha)} = \frac{\Phi(\alpha)}{\Gamma(1-\alpha)},$$

$$s(\alpha,2) = \frac{D^1\Phi(\alpha)+\Phi^2(\alpha)}{2\Gamma(1-\alpha)},$$

$$s(\alpha,3) = \frac{D^2\Phi(\alpha)+3\Phi(\alpha)D^1\Phi(\alpha)+\Phi^3(\alpha)}{6\;\Gamma(1-\alpha)},$$

$$s(\alpha,4) = \frac{1}{24\;\Gamma(1-\alpha)}\left[D^3\Phi(\alpha)+4\Phi(\alpha)D^2\Phi(\alpha)+6\Phi^2(\alpha)D^1\Phi(\alpha)\right.$$
$$\left.+3[D^1\Phi(\alpha)]^2+3\Phi^4(\alpha)\right],$$

where

$$D^k\Phi(\alpha) = \left(\frac{\partial}{\partial x}\right)^k [\psi(x+1)-\psi(x+1-\alpha)]\big|_{x=0} \quad (k\in\mathbb{N}_0).$$

Proof. Recalling the proof of (4.2.19) and (4.2.20), noting $[0]_\alpha = 1$,

$$s(\alpha,1) = \lim_{x\to 0}\frac{d}{dx}\left[\frac{\Gamma(x+1)}{\Gamma(x+1-\alpha)}\right] = \lim_{x\to 0}[x]_\alpha\Phi(x,\alpha) = \frac{\psi(1)-\psi(1-\alpha)}{\Gamma(1-\alpha)}.$$

As this procedure is somewhat cumbersome for $s(\alpha,2)$ *etc.*, we apply the recursion formula (4.2.16). Indeed, it easily follows that $s(\alpha,2) = \frac{1}{2}\{\Psi_1(\alpha)s(\alpha,0)+\Psi_0(\alpha)s(\alpha,1)\}$.

Let us also consider $s(\alpha,3)$; in fact,

$$s(\alpha,3) = \frac{1}{3}\{\Psi_2(\alpha)s(\alpha,0)+\Psi_1(\alpha)s(\alpha,1)+\Psi_0(\alpha)s(\alpha,2)\}.$$

The further $s(\alpha,k)$, $k=4,5,\ldots$, follow similarly. □

4.2.2 *Multiple sum representations. The Riemann Zeta function*

First let us recall the known expression of the classical $s(k,m)$ in terms of a multiple sum (see [130, 406]). In this way, for $2\le m\le k$ we have

$$s(k,m) = (-1)^{k+m}(k-1)!\sum_{j=m-1}^{k-1}(-1)^j s(j,m-1), \tag{4.2.21}$$

$$s(k,m) = (-1)^{k+m}(k-1)! \tag{4.2.22}$$
$$\times\sum_{j_{m-1}=m-1}^{k-1}\left(\frac{1}{j_{m-1}}\right)\sum_{j_{m-2}=m-2}^{j_{m-1}-1}\left(\frac{1}{j_{m-2}}\right)\cdots\sum_{j_1=1}^{j_2-1}\left(\frac{1}{j_1}\right).$$

In particular, for $m = 1, 2, 3$ and $k \geq m$

$$s(k,1) = (-1)^{k+1}(k-1)!, \tag{4.2.23}$$

$$s(k,2) = (-1)^{k+2}(k-1)! \sum_{j_1=1}^{k-1} \left(\frac{1}{j_1}\right), \tag{4.2.24}$$

$$s(k,3) = (-1)^{k+3}(k-1)! \sum_{j_2=2}^{k-1} \left(\frac{1}{j_2}\right) \sum_{j_1=1}^{j_2-1} \left(\frac{1}{j_1}\right). \tag{4.2.25}$$

These results can be generalized for the function $s(\alpha, m)$.

Theorem 4.4. *For $\alpha \in \mathbb{C}$ and $m > \Re e(\alpha)$, $m \in \mathbb{N}$, one has*

$$s(\alpha, m) = \frac{1}{\Gamma(-\alpha)} \sum_{k=m}^{\infty} \frac{(-1)^k s(k,m)}{k!(k-\alpha)}. \tag{4.2.26}$$

Proof. Replacing z by $-v$ in (4.2.1), dividing by $v^{\alpha+1}$ and integrating this power series, using Abel's limit theorem and Raabe's convergence criterion, it follows that

$$\begin{aligned}\int_{0+}^{1-} \frac{[\log(1-v)]^m dv}{v^{1+\alpha}} &= m! \sum_{k=m}^{\infty} \frac{(-1)^k s(k,m)}{k!} \int_{0+}^{1-} v^{k-1-\alpha} dv \\ &= m! \sum_{k=m}^{\infty} \frac{(-1)^k s(k,m)}{k!(k-\alpha)}.\end{aligned}$$

Comparing this result with the integral representation (4.2.5) of $s(\alpha, m)$ it immediately results (4.2.26). □

In view of (4.2.23) and (4.2.26), we have

$$s(\alpha, 1) = \frac{1}{\Gamma(-\alpha)} \sum_{k=1}^{\infty} \frac{(-1)^k (-1)^{k+1} (k-1)!}{k!(k-\alpha)} = -\frac{1}{\Gamma(-\alpha)} \sum_{k=1}^{\infty} \frac{1}{k(k-\alpha)}. \tag{4.2.27}$$

Further,

$$\begin{aligned}s(\alpha, 2) &= \frac{1}{\Gamma(-\alpha)} \sum_{k=2}^{\infty} \frac{(-1)^k}{k!(k-\alpha)} \left\{ (-1)^k (k-1)! \sum_{j=1}^{k-1} \frac{1}{j} \right\} \\ &= \frac{1}{\Gamma(-\alpha)} \sum_{k=2}^{\infty} \frac{1}{k(k-\alpha)} \sum_{j=1}^{k-1} \frac{1}{j}.\end{aligned} \tag{4.2.28}$$

Iterating this process yields the general multiple sum

$$s(\alpha, m) = \frac{(-1)^m}{\Gamma(-\alpha)} \sum_{j_m=m}^{\infty} \left(\frac{1}{j_m(j_m - \alpha)} \right) \sum_{j_{m-1}=m-1}^{j_m - 1} \left(\frac{1}{j_{m-1}} \right) \cdots \sum_{j_1=1}^{j_2-1} \left(\frac{1}{j_1} \right). \tag{4.2.29}$$

Observe that (4.2.26) can be considered as the counterpart of a classical result of Stirling on the connections between his numbers $s(k, m)$ and the famous Riemann zeta function

$$\zeta(z) = \sum_{j=0}^{\infty} \frac{1}{(j+1)^z} \qquad (\Re e(z) > 1)$$

for $z \in \mathbb{N}_0$, namely (see, *e. g.*, pp. 166 and 195 of [320]),

$$\zeta(m+1) = \sum_{k=m}^{\infty} \frac{(-1)^{k+m}\; s(k, m)}{k!\; k}. \tag{4.2.30}$$

The counterpart of the multiple sum formula (4.2.29) in the zeta function setting, first established in Butzer *et al.* [130] (see Adamchik [5]), reads

$$\zeta(m+1) = \sum_{j_m=m}^{\infty} \left(\frac{1}{j_m^2} \right) \sum_{j_{m-1}=m-1}^{j_m-1} \left(\frac{1}{j_{m-1}} \right) \cdots \sum_{j_1=1}^{j_2-1} \left(\frac{1}{j_1} \right), \tag{4.2.31}$$

valid for any $m > 0$. The proof follows by inserting (4.2.22) into (4.2.30).

4.3 General Stirling Functions of the First Kind $s(\alpha, \beta)$ with Complex Arguments

The purpose of this section is to generalize the Stirling functions $s(\alpha, k)$ with $\alpha \in \mathbb{C}$, $k \in \mathbb{N}_0$ to functions $s(\alpha, \beta)$ where *both* α and β are complex.

4.3.1 *Definition and main result*

As a natural generalization of the $s(\alpha, k)$, given in (4.2.6), to the $s(\alpha, \beta)$, we introduce the following definition.

Definition 4.1. Let α and β belong to $\mathbb{C}$. Then,

$$s(\alpha,\beta) = \frac{1}{\Gamma(\beta+1)} \lim_{x\to 1} \left({}^{\mathrm{RL}}D_{0+}^{\alpha}[\log(t)]^{\beta}\right)(x) \quad (\Re e(\alpha) \geq 0, \alpha \neq 0), \tag{4.3.1}$$

$$s(0,\beta) = \frac{1}{\Gamma(\beta+1)} \lim_{x\to 1} \left({}^{\mathrm{RL}}D_{0+}^{0}[\log(t)]^{\beta}\right)(x) \quad (\Re e(\beta) > 0), \tag{4.3.2}$$

$$s(\alpha,\beta) = \frac{1}{\Gamma(\beta+1)} \lim_{x\to 1} \left({}^{\mathrm{RL}}I_{0+}^{-\alpha}[\log(t)]^{\beta}\right)(x) \quad (\Re e(\alpha) < 0). \tag{4.3.3}$$

The main result of this section is Theorem 4.6. In order to explain better the motivation of our approach, it is advisable to write this main result at this stage somewhat differently, in two parts.

Theorem 4.5.

(a) *Let* $\alpha,\beta \in \mathbb{C}$ *such that* $\Re e(\alpha) < 0$, $\Re e(\beta) \geq 0$. *Then,*

$$\begin{aligned} s(\alpha,\beta) &= \frac{1}{\Gamma(\beta+1)\Gamma(-\alpha)} \int_{0+}^{1-} (1-t)^{-\alpha-1}[\log(t)]^{\beta} dt \\ &= \frac{e^{\beta\pi i}}{\Gamma(-\alpha)} \sum_{j=0}^{\infty} (-1)^j \binom{-\alpha-1}{j} \frac{1}{(j+1)^{\beta+1}}. \end{aligned} \tag{4.3.4}$$

(b) *Let* $\alpha,\beta \in \mathbb{C}$ *such that* $\Re e(\beta) > \Re e(\alpha) \geq 0$, *with* $n = [\Re e(\alpha)] + 1$. *Then,*

$$\begin{aligned} s(\alpha,\beta) &= \frac{1}{\Gamma(\beta+1)} \lim_{x\to 1} \left(\frac{\partial}{\partial x}\right)^{n} \frac{1}{\Gamma(n-\alpha)} \int_{0}^{x} (x-t)^{n-\alpha}[\log(t)]^{\beta} dt \\ &= \frac{e^{\beta\pi i}}{\Gamma(-\alpha)} \sum_{j=0}^{\infty} (-1)^j \binom{-\alpha-1}{j} \frac{1}{(j+1)^{\beta+1}}. \end{aligned} \tag{4.3.5}$$

The series in (4.2.5) coincides with the series in (4.3.5) in case $\beta \equiv k > \Re e(\alpha)$, noting the relations

$$\Gamma(z)\Gamma(1-z) = \frac{\sin(\pi z)}{\pi}, \qquad (z)_k = \frac{\Gamma(z+k)}{z}$$

with $z \in \mathbb{C}$ and $k \in \mathbb{N}$.

In the proof of Theorem 4.6 we will use the expansion

$$(1-z)^{-\mu} = \sum_{j=0}^{\infty} (\mu)_k \frac{z^k}{k!} \quad (|z| < 1,\ \mu \in \mathbb{C}), \tag{4.3.6}$$

where $z \in \mathbb{C}$, $j \in \mathbb{N}_0$ and $(z)_j$ is the Pochhammer symbol (see Section 2.1.1 of [226])

$$(z)_0 = 1, \quad (z)_j = z(z+1)...(z+j-1) \qquad (j \in \mathbb{N}). \tag{4.3.7}$$

We also need the following three preliminary results.

Lemma 4.1. *For $\beta \in \mathbb{C}$ and $m \in \mathbb{N}_0$ there holds the relation*

$$\lim_{x\to 1}\left(\frac{\partial}{\partial x}\right)^m [\log(x)+\log(t)]^\beta = \sum_{j=0}^{m} \frac{\Gamma(\beta+1)}{\Gamma(\beta-j+1)}\, c_{m,j}[\log(t)]^{\beta-j} \quad (4.3.8)$$

where

$$c_{m,m}=1, \quad c_{m,0}=0 \quad (m \in \mathbb{N}), \quad (4.3.9)$$

and

$$c_{m,j}=c_{m-1,j-1}-(m-1)c_{m-1,j} \quad (m \in \mathbb{N}; \quad j=1,\ldots,m-1). \quad (4.3.10)$$

In particular,

$$c_{0,0}=1; \quad c_{1,1}=1, \quad c_{1,0}=0; \quad c_{2,2}=1, \quad c_{2,1}=-1, \quad c_{2,0}=0;$$

$$c_{3,3}=1, \; c_{3,2}=-3, \quad c_{3,1}=2, \quad c_{3,0}=0; \quad (4.3.11)$$

$$c_{m,1}=(-1)^{m-1}(m-1)!, \quad c_{m,m-1}=-\frac{(m-1)m}{2} \quad (m \in \mathbb{N}).$$

Proof. For $m=0$, the identity (4.3.8) is clear. If $m=1$ and $m=2$, then

$$\frac{\partial}{\partial x}[\log(x)+\log(t)]^\beta = \beta\,[\log(xt)]^{\beta-1}\,\frac{1}{x}, \quad (4.3.12)$$

$$\left(\frac{\partial}{\partial x}\right)^2 [\log(xt)]^\beta = \frac{\beta(\beta-1)}{x^2}[\log(xt)]^{\beta-2} - \frac{\beta}{x^2}[\log(xt)]^{\beta-1}. \quad (4.3.13)$$

Taking the limit as $x \to 1$, we have

$$\lim_{x\to 1}\frac{\partial}{\partial x}[\log(xt)]^\beta = \beta[\log(t)]^{\beta-1},$$

$$\lim_{x\to 1}\left(\frac{\partial}{\partial x}\right)^2 [\log(xt)]^\beta = \beta(\beta-1)[\log(t)]^{\beta-2} - \beta[\log(t)]^{\beta-1},$$

and hence (4.3.8) follows for $m=1$ and $m=2$, respectively.

We may generalize (4.3.12) and (4.3.13) to

$$x^m\left(\frac{\partial}{\partial x}\right)^m [\log(xt)]^\beta = \sum_{j=1}^{m}\frac{\Gamma(\beta+1)}{\Gamma(\beta-j+1)}\, c_{m,j}[\log(xt)]^{\beta-j}, \quad (4.3.14)$$

where $m \in \mathbb{N}$ and $c_{m,j}$ are defined by (4.3.9)–(4.3.11). Then (4.3.8) will follow for $m \in \mathbb{N}_0$ from (4.3.14) by taking the limit $x \to 1$.

Formula (4.3.14) is proved by induction. Indeed, it has the forms (4.3.12) and (4.3.13) for $m = 1$ and $m = 2$. Suppose that it is valid for $m \in \mathbb{N}$. Using (4.3.14) we have

$$x^{m+1}\left(\frac{\partial}{\partial x}\right)^{m+1}[\log(xt)]^{\beta}$$

$$= \frac{\partial}{\partial x}\left\{\sum_{j=1}^{m}\frac{\Gamma(\beta+1)}{\Gamma(\beta-j+1)}c_{m,j}[\log(xt)]^{\beta-j}\right\}$$

$$= \sum_{j=1}^{m}\frac{\Gamma(\beta+1)}{\Gamma(\beta-j)}c_{m,j}[\log(xt)]^{\beta-j-1} - m\sum_{j=0}^{m}\frac{\Gamma(\beta+1)}{\Gamma(\beta-j+1)}c_{m,j}[\log(xt)]^{\beta-j}$$

$$= \frac{\Gamma(\beta+1)}{\Gamma(\beta-m-1)}c_{m,m}[\log(xt)]^{\beta-m-1} + \sum_{j=2}^{m}\frac{\Gamma(\beta+1)}{\Gamma(\beta-j+1)}c_{m,j-1}[\log(xt)]^{\beta-j}$$

$$- m\sum_{j=2}^{m}\frac{\Gamma(\beta+1)}{\Gamma(\beta-j+1)}c_{m,j}[\log(xt)]^{\beta-j} - m\frac{\Gamma(\beta+1)}{\Gamma(\beta)}\,c_{m,1}\,[\log(xt)]^{\beta-1}.$$

By (4.3.9), noting $c_{m+1,m+1} = 1$, so $c_{m,m} = 1 = c_{m+1,m+1}$, $c_{m+1,0} = 0$, and $mc_m = (-1)^{m+1}c_{m+1,1}$. Therefore,

$$x^{m+1}\left(\frac{\partial}{\partial x}\right)^{m+1}[\log(x)+\log(t)]^{\beta}$$

$$= \frac{\Gamma(\beta+1)}{\Gamma(\beta-m-1)}c_{m+1,m+1}\,[\log(xt)]^{\beta-m-1}$$

$$+ \sum_{j=2}^{m}\frac{\Gamma(\beta+1)}{\Gamma(\beta-j+1)}[c_{m,j-1} - mc_{m,j}][\log(xt)]^{\beta-j}$$

$$+ \frac{\Gamma(\beta+1)}{\Gamma(\beta)}\,c_{m+1,1}\,[\log(xt)]^{\beta-1} + c_{m+1,0}[\log(xt)]^{\beta}.$$

This yields (4.3.14) with m being replaced by $m+1$, if we take (4.3.10) into account. □

Lemma 4.2. *Let $\alpha \in \mathbb{C}$, $n \in \mathbb{N}$, $k \in \mathbb{N}_0$. Then $(-\alpha-k)_n$ has the representation*

$$(-\alpha-k)_n = \sum_{j=0}^{n}(-1)^j A_{n,j}(k+1)^j, \tag{4.3.15}$$

where the constants $A_{n,j} = A_{n,j}(\alpha)$ have the forms

$$A_{n,0} = (1-\alpha)(2-\alpha)\cdots(n-\alpha) = \frac{\Gamma(n+1-\alpha)}{\Gamma(1-\alpha)}; \tag{4.3.16}$$

$$\begin{aligned} A_{n,1} &= (1-\alpha)(2-\alpha)\cdots(n-1-\alpha) + (1-\alpha)(3-\alpha)\cdots(n-\alpha) + \cdots \\ &\quad +(2-\alpha)(3-\alpha)\cdots(n-\alpha) \\ &= \frac{\Gamma(n+1-\alpha)}{\Gamma(1-\alpha)} \sum_{i=1}^{n} \frac{1}{i-\alpha}; \end{aligned} \tag{4.3.17}$$

$$\begin{aligned} A_{n,2} &= (1-\alpha)(2-\alpha)\cdots(n-2-\alpha) + \cdots + (3-\alpha)(4-\alpha)\cdots(n-\alpha) \\ &= \frac{\Gamma(n+1-\alpha)}{\Gamma(1-\alpha)} \sum_{\substack{i_1,\, i_2 = 1 \\ (i_1 \neq i_2)}}^{n} \frac{1}{(i_1-\alpha)(i_2-\alpha)}; \ \cdots \end{aligned} \tag{4.3.18}$$

$$\begin{aligned} A_{n,j} &= (1-\alpha)(2-\alpha)\cdots(n-j-\alpha) + \cdots + (j+1-\alpha)\cdots(n-\alpha) \\ &= \frac{\Gamma(n+1-\alpha)}{\Gamma(1-\alpha)} \sum_{\substack{i_1, \cdots, i_j = 1 \\ (i_k \neq i_j)}}^{n} \frac{1}{(i_1-\alpha)\cdots(i_j-\alpha)} \\ &= \sum_{\substack{i_{j+1}, \cdots, i_n = 1 \\ (i_k \neq i_j)}}^{n} (i_{j+1}-\alpha)\cdots(i_n-\alpha); \end{aligned} \tag{4.3.19}$$

$$\begin{aligned} A_{n,n-2} &= (1-\alpha)(2-\alpha) + (1-\alpha)(3-\alpha) + \cdots + (1-\alpha)(n-\alpha) + \cdots \\ &\quad +(n-1-\alpha)(n-\alpha) = \sum_{\substack{i_1,\, i_2 = 1 \\ (i_1 \neq i_2)}}^{n} (i_1-\alpha)(i_2-\alpha); \end{aligned} \tag{4.3.20}$$

$$A_{n,n-1} = (1-\alpha) + (2-\alpha) + \cdots + (n-\alpha) = \sum_{i=1}^{n} (i-\alpha); \tag{4.3.21}$$

$$A_{n,n} = 1. \tag{4.3.22}$$

Proof. By (4.3.7) one has

$$\begin{aligned} (-\alpha-k)_n &= (-\alpha-k)(-\alpha-k+1)\cdots(-\alpha-k+n-1) \\ &= [(1-\alpha)-(k+1)][(2-\alpha)-(k+1)]\cdots[(n-\alpha)-(k+1)] \\ &= \sum_{j=0}^{n} (-1)^j A_{n,j} (k+1)^j, \end{aligned}$$

which yields (4.3.15). Since $(-\alpha-k)_n$ is a polynomial of degree n with respect to $(k+1)$, then (4.3.16)–(4.3.22) follow from known results from algebra. □

Lemma 4.3. *Let $n \in \mathbb{N}_0$ and $j = 0, 1, \ldots, n$. There hold the following relations*

$$\sum_{m=j}^{n} \binom{n}{m} \frac{\Gamma(n+1-\alpha)}{\Gamma(m+1-\alpha)} c_{m,j} = A_{n,j}, \tag{4.3.23}$$

where $c_{m,j}$ and $A_{n,j}$ are given by (4.3.9)–(4.3.10) and (4.3.16)–(4.3.22), respectively.

Proof. For $j = 0$ or $j = n$ the proof of (4.3.23) is simple. If $j = 0$, then in accordance with (4.3.11) and (4.3.9), $c_{0,0} = 1$ and $c_{m,j} = 0$ $(j = 1, \ldots, m)$. Using these relations we have

$$\sum_{m=0}^{n} \binom{n}{m} \frac{\Gamma(n+1-\alpha)}{\Gamma(m+1-\alpha)} c_{m,0} = \frac{\Gamma(n+1-\alpha)}{\Gamma(1-\alpha)} c_{0,0} = \frac{\Gamma(n+1-\alpha)}{\Gamma(1-\alpha)},$$

which proves (4.3.23) for $j = 0$, if we take (4.3.16) into account. If $j = n$, then (4.3.23) takes the form $c_{n,n} = A_{n,n}$, which is clear because according to (4.3.9) and (4.3.22) $c_{n,n} = A_{n,n} = 1$. If $j = n-1$, then, since $c_{n-1,n-1} = 1$, relation (4.3.23) takes the form

$$n(n-\alpha) + c_{n,n-1} = A_{n,n-1} \quad (n \in \mathbb{N}). \tag{4.3.24}$$

This can be seen to be valid because, according to (4.3.11), $n(n-\alpha) + c_{n,n-1} = \frac{n(n+1)}{2} - n\alpha$, while by (4.3.21),

$$A_{n,n-1} = (1 + 2 + \cdots + n) - n\alpha = \frac{n(n+1)}{2} - n\alpha.$$

The proofs of (4.3.23) in the cases $j = 1, \ldots, n-2$ can be carried out by direct applications of (4.3.9)–(4.3.11) and (4.3.16)–(4.3.22). They are cumbersome and therefore are omitted. □

Now to our main result for this section, Theorem 4.6, which was phrased in two parts in Theorem 4.5 for better understanding.

Theorem 4.6. *Let $\alpha, \beta \in \mathbb{C}$ such that $\Re e(\alpha) < \Re e(\beta)$. Then*

$$s(\alpha, \beta) = \frac{e^{\beta\pi i}}{\Gamma(-\alpha)} \sum_{j=0}^{\infty} \frac{(\alpha+1)_j}{j!(j+1)^{\beta+1}}$$

$$= \frac{e^{\beta\pi i}}{\Gamma(-\alpha)} \sum_{j=0}^{\infty} (-1)^j \binom{-\alpha-1}{j} \frac{1}{(j+1)^{\beta+1}}, \tag{4.3.25}$$

both series being absolutely convergent, where the power function x^β is defined for $x \in \mathbb{R}$, $x \neq 0$, and $\beta \in \mathbb{C}$ by the usual convention

$$x^\beta = \exp\{\beta[\log|x| + i \arg x]\} \quad (-\pi \leqq \arg x < \pi)$$

Proof. We first establish the result for $\mathfrak{Re}(\alpha) < \mathfrak{Re}(\beta)$, $\mathfrak{Re}(\beta) \geq 0$. By definition (4.3.3), a change of variables $u = e^{-t}$ and an interchange of the order of integration and summation yields that

$$\begin{aligned}
s(\alpha, \beta) &= \frac{1}{\Gamma(\beta+1)\Gamma(-\alpha)} \int_0^1 \frac{[\log(u)]^\beta du}{(1-u)^{\alpha+1}} \\
&= \frac{1}{\Gamma(\beta+1)\Gamma(-\alpha)} \int_0^\infty \left(1-e^{-t}\right)^{-\alpha-1} (-t)^\beta e^{-t} dt \\
&= \frac{e^{\beta\pi i}}{\Gamma(\beta+1)\Gamma(-\alpha)} \lim_{\epsilon\to 0+} \left[\int_\epsilon^\infty \left(1-e^{-t}\right)^{-\alpha-1} t^\beta e^{-t} dt\right] \\
&= \frac{e^{\beta\pi i}}{\Gamma(\beta+1)\Gamma(-\alpha)} \left[\sum_{j=1}^\infty \frac{(\alpha+1)_j}{j!} \lim_{\epsilon\to 0+} \int_\epsilon^\infty e^{-(j+1)t} t^\beta dt\right] \\
&= \frac{e^{\beta\pi i}}{\Gamma(\beta+1)\Gamma(-\alpha)} \left[\sum_{j=1}^\infty \frac{(\alpha+1)_j}{j!(j+1)^{\beta+1}} \lim_{\epsilon\to 0+} \Gamma\left(\beta+1, \epsilon(j+1)\right)\right],
\end{aligned}$$

where $\Gamma(z, w) = \int_w^\infty t^{z-1} e^{-t} dt$ is the incomplete Gamma function (see formula 6.9(21) in [226]). Since $\lim_{w\to 0+} \Gamma(z, w) = \Gamma(z)$, the last relation equals the first sum in (4.3.25).

The second sum in (4.3.25) follows from the first by noting the property

$$\frac{(\alpha+1)_j}{j!} = (-1)^j \binom{-\alpha-1}{j} \quad (\alpha \in \mathbb{C},\ j \in \mathbb{N}_0). \tag{4.3.26}$$

As to the convergence of the two series in (4.3.25), consider the general term d_j of the series

$$\sum_{j=0}^\infty d_j; \quad d_j = \frac{e^{\beta\pi i}}{\Gamma(-\alpha)} (-1)^j \binom{-\alpha-1}{j} \frac{1}{(j+1)^{\beta+1}}. \tag{4.3.27}$$

In view of the estimate for binomial coefficients, namely

$$\left|\binom{a}{b}\right| \leq \frac{c}{b^{1+\mathfrak{Re}(a)}} \quad (a, b \in \mathbb{C},\ a \neq -1, -2, \ldots), \tag{4.3.28}$$

for a certain constant $c > 0$, one has for d_j the estimate

$$|d_j| \leq \frac{c}{j^{\mathfrak{Re}(\beta)+1} j^{\mathfrak{Re}(-\alpha-1)+1}} = \frac{c}{j^{\mathfrak{Re}(\beta)+1-\mathfrak{Re}(\alpha)}}. \tag{4.3.29}$$

This estimate establishes the assertions of Theorem 4.6 only for $\Re e(\alpha) < 0$ and $\Re e(\beta) \geq 0$ since $\Re e(\beta) + 1 - \Re e(\alpha) > 1$.

If $\Re e(\alpha) \geq 0,\ \alpha \neq 0$, and $\Re e(\beta) > \Re e(\alpha)$, then instead of applying the definition (4.3.3) we work with (4.3.1) and (1.3.1). Then $s(\alpha, \beta)$ takes on the form for $n = [\Re e(\alpha)] + 1$,

$$\begin{aligned} s(\alpha,\beta) &= \frac{1}{\Gamma(\beta+1)} \lim_{x\to 1} \left(\frac{\partial}{\partial x}\right)^n \frac{1}{\Gamma(n-\alpha)} \int_0^x (x-u)^{n-\alpha-1} [\log(u)]^\beta du \\ &= \frac{1}{\Gamma(\beta+1)\Gamma(n-\alpha)} \lim_{x\to 1} \left(\frac{\partial}{\partial x}\right)^n \int_0^1 (1-t)^{n-\alpha} \left[x^{n-\alpha}[\log xt]^\beta\right] dt \\ &= \frac{1}{\Gamma(\beta+1)\Gamma(n-\alpha)} \int_0^1 (1-t)^{n-\alpha} \lim_{x\to 1} \left(\frac{\partial}{\partial x}\right)^n \left[x^{n-\alpha}[\log xt]^\beta\right] dt, \end{aligned}$$

where the change of variables $u = xt$ was made. Applying now the Leibniz rule for the derivative of a product, taking into account the known property

$$\left({}^{\mathrm{RL}}D^{\alpha}_{0+} t^{\gamma-1}\right)(x) = \frac{\Gamma(\gamma)\ x^{\gamma-\alpha-1}}{\Gamma(\gamma-\alpha)} \qquad (\Re e(\alpha) > 0,\quad \Re e(\gamma) > 0), \tag{4.3.30}$$

with $\alpha = n - m$ and $\gamma = n - \alpha + 1$, as well as Lemma 4.1, we find

$$\begin{aligned} &s(\alpha,\beta) \\ &= \frac{1}{\Gamma(\beta+1)\Gamma(n-\alpha)} \\ &\quad \times \sum_{m=0}^{n} \binom{n}{m} \int_0^1 (1-t)^{n-\alpha} \lim_{x\to 1} \left\{ \left(\frac{d}{dx}\right)^{n-m} \left[x^{n-\alpha}\right] \right. \\ &\qquad\qquad \left. \times \left(\frac{\partial}{\partial x}\right)^m \left[\log(x) + \log(t)]^\beta\right] \right\} dt \\ &= \frac{1}{\Gamma(\beta+1)\Gamma(n-\alpha)} \\ &\quad \times \sum_{m=0}^{n} \binom{n}{m} \frac{\Gamma(n+1-\alpha)}{\Gamma(m+1-\alpha)} \int_0^1 (1-t)^{n-\alpha} \lim_{x\to 1} \left(\frac{\partial}{\partial x}\right)^m \left[\log xt]^\beta\right] dt \\ &= \frac{1}{\Gamma(\beta+1)} \sum_{m=0}^{n} \binom{n}{m} \frac{\Gamma(n+1-\alpha)}{\Gamma(m+1-\alpha)} \\ &\quad \times \sum_{j=0}^{m} \frac{\Gamma(\beta+1)}{\Gamma(\beta-j+1)} \frac{c_{m,j}}{\Gamma(n-\alpha)} \int_0^1 (1-t)^{n-\alpha} [\log(t)]^{\beta-j} dt. \end{aligned} \tag{4.3.31}$$

Hence there follows

$$s(\alpha,\beta) = \sum_{m=0}^{n} \binom{n}{m} \frac{\Gamma(n+1-\alpha)}{\Gamma(m+1-\alpha)} \times \sum_{j=0}^{m} \frac{c_{m,j}}{\Gamma(\beta-j+1)} \lim_{x\to 1} \left({}^{\mathrm{RL}}I_{0+}^{n-\alpha}[\log(t)]^{\beta-j}\right)(x). \quad (4.3.32)$$

To evaluate the limit term for $x \to 1$, we apply the first part of the proof. Indeed, by (4.3.3) and (4.3.25) for $-\alpha$ and β replaced by $n-\alpha$ and $\beta - j$, respectively, we have since $\Re e(\alpha) < n$,

$$\lim_{x\to 1} \left({}^{\mathrm{RL}}I_{0+}^{n-\alpha}[\log(t)]^{\beta-j}\right)(x) = \Gamma(\beta-j+1)s(\alpha-n,\beta-j) = \frac{e^{(\beta-j)\pi i}\Gamma(\beta-j+1)}{\Gamma(n-\alpha)} \sum_{k=0}^{\infty} \frac{(\alpha-n+1)_k}{k!(k+1)^{\beta+1-j}}.$$

Thus (4.3.32) takes the form

$$\begin{aligned} s(\alpha,\beta) &= \frac{e^{\beta\pi i}}{\Gamma(n-\alpha)} \sum_{m=0}^{n} \binom{n}{m} \frac{\Gamma(n+1-\alpha)}{\Gamma(m+1-\alpha)} \sum_{j=0}^{m} (-1)^j c_{m,j} \times \sum_{k=0}^{\infty} \frac{(\alpha-n+1)_k}{k!(k+1)^{\beta+1-j}} \\ &= \frac{e^{\beta\pi i}}{\Gamma(n-\alpha)} \sum_{k=0}^{\infty} \frac{(\alpha-n+1)_k}{k!(k+1)^{\beta+1}} \sum_{m=0}^{n} \binom{n}{m} \frac{\Gamma(n+1-\alpha)}{\Gamma(m+1-\alpha)} \times \sum_{j=0}^{m} (-1)^j c_{m,j} (k+1)^j \\ &= \frac{e^{\beta\pi i}}{\Gamma(n-\alpha)} \sum_{k=0}^{\infty} \frac{(\alpha-n+1)_k}{k!(k+1)^{\beta+1}} \sum_{j=0}^{n} (-k-1)^j \times \sum_{m=j}^{n} \binom{n}{m} \frac{\Gamma(n+1-\alpha)}{\Gamma(m+1-\alpha)} c_{m,j}. \end{aligned} \quad (4.3.33)$$

According to Lemma 4.2 and Lemma 4.3,

$$\sum_{j=0}^{n} (-1)^j (k+1)^j \sum_{m=j}^{n} \binom{n}{m} \frac{\Gamma(n+1-\alpha)}{\Gamma(m+1-\alpha)} c_{m,j} = \sum_{j=0}^{n} (-1)^j A_{n,j} (k+1)^j = (-\alpha-k)_n, \quad (4.3.34)$$

and hence from (4.3.33) we obtain

$$s(\alpha,\beta) = e^{\beta\pi i}\sum_{k=0}^{\infty}\frac{(\alpha-n+1)_k(-\alpha-k)_n}{\Gamma(n-\alpha)}\frac{1}{k!(k+1)^{\beta+1}}, \tag{4.3.35}$$

but $(\alpha-n+1)_k(-\alpha-k)_n/\Gamma(n-\alpha) = (\alpha+1)_k/\Gamma(-\alpha)$, and thus (4.3.35) yields the first series in (4.3.25).

The second series in (4.3.25), clearly following from the first one, has the form of (4.3.27) with the d_j term again having the estimate (4.3.29); thus it is convergent for $\Re e(\alpha) < \Re e(\beta)$.

The relations in (4.3.25) remain valid also for $\alpha = 0$, $\Re e(\beta) > 0$. Indeed, according to definition (4.3.2) for $\alpha = 0$, $s(0,\beta) = \frac{1}{\Gamma(\beta+1)}\lim_{x\to 1}[\log x]^{\beta} = 0$, and

$$\lim_{\alpha\to 0+}\left[\frac{e^{\beta\pi i}}{\Gamma(-\alpha)}\sum_{j=0}^{\infty}\frac{(\alpha+1)_j}{j!(j+1)^{\beta+1}}\right] = 0.$$

This completes the proof of Theorem 4.6. □

It follows from Theorem 4.6 that if $\alpha,\beta\in\mathbb{C}$ such that $\Re e(\alpha) < \Re e(\beta)$, then the Stirling functions of the first kind $s(\alpha,\beta)$, defined by (4.3.1)–(4.3.3), have the same representations, namely

$$\begin{aligned} s(\alpha,\beta) &= \frac{e^{\beta\pi i}}{\Gamma(-\alpha)}\sum_{j=0}^{\infty}\frac{(\alpha+1)_j}{j!(j+1)^{\beta+1}} \\ &= \frac{e^{\beta\pi i}}{\Gamma(-\alpha)}\sum_{j=0}^{\infty}(-1)^j\binom{-\alpha-1}{j}\frac{1}{(j+1)^{\beta+1}} \end{aligned}$$

From Theorem 4.6 we also deduce the following result.

Corollary 4.2. *If $\alpha\in\mathbb{C}$ and $k\in\mathbb{Z}$ with $\Re e(\alpha) < k$, then*

$$\begin{aligned} s(\alpha,k) &= \frac{(-1)^k}{\Gamma(-\alpha)}\sum_{j=0}^{\infty}\frac{(\alpha+1)_j}{j!(j+1)^{k+1}} \\ &= \frac{(-1)^k}{\Gamma(-\alpha)}\sum_{j=0}^{\infty}(-1)^j\binom{-\alpha-1}{j}\frac{1}{(j+1)^{k+1}}. \end{aligned} \tag{4.3.36}$$

Corollary 4.3.

(a) *If* $n \in \mathbb{N}_0$ *and* $\beta \in \mathbb{C}$ *with* $\Re e(\beta) > n$, *then*

$$s(n, \beta) = 0, \quad s(0, \beta) = 0. \tag{4.3.37}$$

(b) *If* $n \in \mathbb{N}$ *and* $\beta \in \mathbb{C}$ *such that* $\Re e(\beta) > -n$, *then*

$$\begin{aligned} s(-n, \beta) &= \frac{e^{\beta\pi i}}{(n-1)!} \sum_{j=0}^{n-1} (-1)^j \binom{n-1}{j} \frac{1}{(j+1)^{\beta+1}} \\ &= \frac{e^{\beta\pi i}}{n!} \sum_{j=1}^{n} (-1)^{j-1} \binom{n}{j} \frac{1}{j^\beta}. \end{aligned} \tag{4.3.38}$$

In particular, for $\beta \in C$,

$$s(-1, \beta) = e^{\beta\pi i}, \quad s(-2, \beta) = e^{\beta\pi i}\left[1 - \frac{1}{2^{\beta+1}}\right].$$

Proof. As far as (4.3.38) is concerned, its left hand side exists under the given conditions, according to (4.3.3) with $\alpha = -n$ can be expressed by

$$s(-n, \beta) = \frac{1}{\Gamma(\beta+1)(n-1)!} \int_0^1 (1-t)^{n-1} [\log t]^\beta dt,$$

which is a convergent integral for $\Re e(\beta) > -n$. The right hand side turns out to be a finite sum since for $\alpha = -n$, $(-n+1)_j = 0$ for $j = n, n+1, \ldots$, so that it exists for any $n \in \mathbb{N}$ and $\beta \in \mathbb{C}$. To establish the right hand side of (4.3.38) one replaces $j + 1$ by k, and observes that

$$n \binom{n-1}{k-1} \frac{1}{k} = \binom{n}{k} \quad (n \in \mathbb{N},\ 1 \leqq k \leqq n-1). \tag{4.3.39}$$

□

4.3.2 *Differentiability of the* $s(\alpha, \beta)$*; The zeta function encore*

In Section 4.2.2 we indicated that the Stirling functions $s(\alpha, m)$ are connected to the zeta function $\zeta(m+1)$. Such a connection is indeed true also for the most general $s(\alpha, \beta)$ with $\alpha, \beta \in \mathbb{C}$.

For this purpose we first need a result on the continuity and differentiability of the $s(\alpha, \beta)$ with respect to α; the corresponding result for β is also given.

Theorem 4.7. *Let* $\alpha, \beta \in \mathbb{C}$ *be complex numbers such that* $\Re e(\alpha) < \Re e(\beta)$. *There hold the following assertions:*

(a) $s(\alpha, \beta)$ *as a function of* α *is continuously differentiable for* $\alpha \in \mathbb{C}, \alpha \neq 0$, *and*

$$\frac{\partial}{\partial \alpha} s(\alpha, \beta) = [\psi(-\alpha) - \psi(\alpha+1)] s(\alpha, \beta) + \frac{e^{\beta \pi i}}{\Gamma(-\alpha)} \sum_{j=0}^{\infty} \frac{(\alpha+1)_j}{j!(j+1)^{\beta+1}} \psi(\alpha+1+j). \quad (4.3.40)$$

(b) $s(\alpha, \beta)$ *as function of* β *is continuously differentiable for* $\beta \in \mathbb{C}$, *and for* $m \in \mathbb{N}$,

$$\left(\frac{\partial}{\partial \beta}\right)^m s(\alpha, \beta) = \frac{e^{\beta \pi i}}{\Gamma(-\alpha)} \sum_{j=0}^{\infty} \frac{(\alpha+1)_j}{j!(j+1)^{\beta+1}} \times \sum_{k=0}^{m} (-1)^k \binom{m}{k} (i\pi)^{m-k} [\log(j+1)]^k. \quad (4.3.41)$$

Proof. The continuity of $s(\alpha, \beta)$ as functions of α and β follow from the first formula of (4.3.25). The relations (4.3.40) and (4.3.41) are deduced by differentiation with respect to α and β of $s(\alpha, \beta)$, respectively. For the former, one makes use of the fact that

$$\frac{d}{dx} \frac{1}{\Gamma(-\alpha)} = \frac{\psi(-\alpha)}{\Gamma(-\alpha)}, \quad \frac{d}{d\alpha}(\alpha+1)_j = (\alpha+1)_j [\psi(\alpha+1+j) - \psi(\alpha+1)]$$

and for the latter, noting Leibniz' rule,

$$\left(\frac{\partial}{\partial \beta}\right)^m \left[e^{\beta \pi i} \frac{1}{(j+1)^{\beta+1}} \right] = \sum_{k=0}^{m} (-1)^k \binom{m}{k} \frac{(i\pi)^{m-k} e^{\beta \pi i}}{(j+1)^{\beta+1}} [\log(j+1)]^k.$$

The convergence of the series on the right sides of (4.3.40) and (4.3.41) follow by applying the relations (4.3.26), (4.3.28) and the asymptotic formulas for the gamma and psi-functions (see formulas 1.18(4) and 1.18(7) in [226]),

$$\frac{\Gamma(z+a)}{\Gamma(z+b)} = z^{a-b}[1 + O(z^{-1})], \quad \psi(z) = \log(z) + O(z^{-1}) \quad (z \to \infty). \quad (4.3.42)$$

□

We can now formulate the indicated connection between the Stirling functions and Riemann's zeta function.

Theorem 4.8. *Let* $\alpha, \beta \in \mathbb{C}$ *such that* $\Re e(\beta) > 0$ *and* $\Re e(\alpha) < \Re e(\beta)$. *Then*

$$\lim_{\alpha \to 0} \Gamma(-\alpha) s(\alpha, \beta) = e^{\beta \pi i} \zeta(\beta+1), \quad \lim_{\alpha \to 0} \frac{\partial}{\partial \alpha} s(\alpha, \beta) = -e^{\beta \pi i} \zeta(\beta+1).$$

Proof. In view of Theorem 4.6 we have, for $\Re e(\beta) > 0$ with $\Re e(\alpha) < \Re e(\beta)$,

$$\lim_{\alpha\to 0}\Gamma(-\alpha)s(\alpha,\beta) = \lim_{\alpha\to 0} e^{\beta\pi i}\sum_{j=0}^{\infty}\frac{(\alpha+1)_j}{j!(j+1)^{\beta+1}}. \tag{4.3.43}$$

Since $(1)_j = j!$ for $j \in \mathbb{N}_0$, the right-hand side of (4.3.43) is $e^{\beta\pi i}\zeta(\beta+1)$, establishing the first formula of Theorem 4.8.

Since

$$\psi(-\alpha)-\psi(\alpha+1) = \pi\cot(\alpha\pi),\quad \Gamma(-\alpha) = -\frac{1}{\alpha}\left[1+O\left(\frac{1}{z}\right)\right] \quad (\alpha\to 0), \tag{4.3.44}$$

and $\pi\cot(\pi\alpha) \sim 1/\alpha$ $(\alpha \to 0)$, we find

$$\lim_{\alpha\to 0}\frac{\psi(-\alpha)-\psi(\alpha+1)}{\Gamma(-\alpha)} = -1, \tag{4.3.45}$$

cf., *e. g.*, formula 1.7(11) of [226] and formula 6.1.3 of [4]. Finally, taking the limit as $\alpha \to 0$ in (4.3.40) and using (4.3.44) and (4.3.45), we deduce the second formula of the theorem. □

4.3.3 *Recurrence relations for* $s(\alpha, \beta)$

Theorem 4.9. *Let $\alpha, \beta \in \mathbb{C}$ such that $\Re e(\alpha) < \Re e(\beta) - 1$. Then the function $s(\alpha,\beta)$ satisfies the recurrence formula*

$$s(\alpha+1,\beta) = s(\alpha,\beta-1) - \alpha\ s(\alpha,\beta), \tag{4.3.46}$$

and for $n \in \mathbb{N}_0$ in addition

$$s(\alpha+1,\beta) = \sum_{j=0}^{n}(-\alpha)_j s(\alpha-j,\beta-1) + (-\alpha)_{n+1}s(\alpha-n,\beta). \tag{4.3.47}$$

Proof. If $\Re e(\alpha) < \Re e(\beta) - 1$, then according to Theorem 4.6 the Stirling functions $s(\alpha+1,\beta)$, $s(\alpha,\beta-1)$ and $s(\alpha,\beta)$ are well-defined. For $\alpha = 0$ relation (4.3.46) takes the form $s(1,\beta) = s(0,\beta-1)$ which is obvious since $s(1,\beta) = s(0,\beta-1) = 0$ by (4.3.37).

Let $\alpha \neq 0$. In view of Theorem 4.6 and since both series on the second side of (4.3.46) are absolutely convergent, then

$$\begin{aligned}
s(\alpha, \beta - 1) - \alpha s(\alpha, \beta) &= \frac{e^{(\beta-1)\pi i}}{\Gamma(-\alpha)} \sum_{j=0}^{\infty} \frac{(\alpha+1)_j}{j!(j+1)^{\beta}} \\
&\quad - \frac{\alpha e^{\beta \pi i}}{\Gamma(-\alpha)} \sum_{j=0}^{\infty} \frac{(\alpha+1)_j}{j!(j+1)^{\beta+1}} \\
&= \frac{e^{\beta \pi i}}{\Gamma(-\alpha)} \sum_{j=0}^{\infty} \frac{(\alpha+1)_j}{j!(j+1)^{\beta+1}} [-(\alpha+j+1)] \\
&= \frac{e^{\beta \pi i}(-\alpha-1)}{\Gamma(-\alpha)} \sum_{j=0}^{\infty} \frac{(\alpha+2)_j}{j!(j+1)^{\beta+1}},
\end{aligned}$$

where the relation $(\alpha+1)_j(\alpha+1+j) = (\alpha+1)(\alpha+2)_j$, $j \in \mathbb{N}_0$, was used. This will yield (4.3.46).

Relation (4.3.47) will be established by induction. Supposing it is valid for $n \in \mathbb{N}$, then using (4.3.46),

$$\begin{aligned}
&\sum_{j=0}^{n+1} (-\alpha)_j s(\alpha - j, \beta - 1) \\
&\quad = s(\alpha+1, \beta) + (-\alpha)_{n+1}[s(\alpha - n - 1, \beta - 1) - s(\alpha - n, \beta)] \\
&\quad = s(\alpha+1, \beta) - (-\alpha)_{n+1}(\alpha - n - 1)s(\alpha - n - 1, \beta),
\end{aligned}$$

that is

$$\sum_{j=0}^{n+1} (-\alpha)_j s(\alpha - j, \beta - 1) = s(\alpha+1, \beta) + (-\alpha)_{n+2}(\alpha - n - 1)s(\alpha - n - 1, \beta),$$

noting that $-(-\alpha)_{n+1}(\alpha - n - 1) = (-\alpha)_{n+2}$. This yields (4.3.47), n being replaced by $n+1$. Thus the proof is complete. □

Theorem 4.10. *For $m \in \mathbb{N}$ and $\beta \in \mathbb{C}$ the Stirling functions $s(-m, \beta)$ satisfy the recurrence*

$$s(-m, \beta) = s(-m-1, \beta-1) + (m+1)s(-m-1, \beta), \tag{4.3.48}$$

and for $m \in \mathbb{N}_0$ in addition

$$\begin{aligned}
s(-m, \beta) = \sum_{j=0}^{n} (m+1)_j \; s(-m-1-j, \beta-1) \\
+ (m+1)_{n+1} s(-m-1-n, \beta).
\end{aligned} \tag{4.3.49}$$

Proof. Taking into account (4.3.46), we have

$$s(-m-1,\beta-1)+(m+1)s(-m-1,\beta)$$

$$= \frac{e^{(\beta-1)\pi i}}{m!}\sum_{j=0}^{m}(-1)^j\binom{m}{j}\frac{1}{(j+1)^{\beta}} + \frac{e^{\beta\pi i}}{m!}\sum_{j=0}^{m}(-1)^j\binom{m}{j}\frac{1}{(j+1)^{\beta+1}}$$

$$= \frac{e^{\beta\pi i}}{m!}\sum_{j=0}^{m}(-1)^j\binom{m}{j}\left[\frac{m+1}{(j+1)^{\beta+1}} - \frac{1}{(j+1)^{\beta}}\right]$$

$$= \frac{e^{\beta\pi i}}{m!}\sum_{j=0}^{m-1}(-1)^j\binom{m}{j}\frac{(m-j)}{(j+1)^{\beta+1}},$$

which yields (4.3.48), since

$$\frac{1}{m}\binom{m}{j}(m-j) = \binom{m-1}{j} \qquad (m \in \mathbb{N};\ 1 \leq j \leq m-1),$$

and

$$s(-m-1,\beta-1)+(m+1)s(-m-1,\beta)$$

$$= \frac{e^{\beta\pi i}}{(m-1)!}\sum_{j=0}^{m-1}(-1)^j\binom{m-1}{j}\frac{1}{(j+1)^{\beta+1}}.$$

Relation (4.3.49) follows from (4.3.48) by induction. □

Let m and k both belong to $\mathbb{N}$ such that $m > k$. In particular, it is easy to proof the following formulas

$$s(-m,-k) = 0 \quad (m,k \in \mathbb{N};\ \ m > k). \tag{4.3.50}$$

and

$$s(\alpha+1,k) = s(\alpha,k-1) - \alpha s(\alpha,k) \quad (\alpha \in \mathbb{C},\ \ \Re e(\alpha) < k-1). \tag{4.3.51}$$

4.4 Stirling Functions of the Second Kind $S(\alpha, k)$ with Complex First Argument α

As we mentioned above, in the introduction of this chapter, Butzer *et al.* [123] have introduced the Stirling function $S(\alpha,k)$ defined by

$$S(\alpha,k) = \frac{1}{k!}\lim_{x\to 0}\Delta^k(x^{\alpha}) \quad (x \in \mathbb{R};\ \alpha \geq 0,\ k \in \mathbb{N}_0), \tag{4.4.1}$$

where the limit being taken in the sense

$$\lim_{x\to 0} \Delta^k (x^\alpha) = \lim_{\epsilon\to 0} \lim_{x\to 0} \Delta^k ((x+\epsilon)^\alpha). \qquad (4.4.2)$$

The aim of this section is to extend this definition of $S(\alpha, k)$ to complex values of the first parameter, into the fractional calculus framework, by employing differences of fractional order and fractional differentiation operators. Moreover, we will obtain several properties for such Stirling functions.

4.4.1 *Stirling functions $S(\alpha, k)$, $\alpha \geq 0$, and their representations by Liouville and Marchaud fractional derivatives*

In this section we consider Stirling functions of the second kind $S(\alpha, k)$ with nonnegative $\alpha \geq 0$.

Theorem 4.11. *Let $\alpha \geq 0$ and $k \in \mathbb{N}_0$. Then the Stirling function of the second kind $S(\alpha, k)$ has the following explicit representation*

$$S(\alpha, k) = \frac{1}{k!} \sum_{j=1}^{k} (-1)^{k-j} \binom{k}{j} j^\alpha \quad (\alpha > 0;\ k \in \mathbb{N}). \qquad (4.4.3)$$

In particular, we have

$$S(\alpha, 0) = 0 \quad (\alpha > 0); \quad S(0, k) = 0 \quad (k \in \mathbb{N}); \quad S(0,0) = 1. \qquad (4.4.4)$$

Proof. Using (4.4.1), (4.4.2) and applying (4.1.9) with $f(x) = x^\alpha$, we have

$$S(\alpha, k) = \frac{1}{k!} \sum_{j=0}^{k} (-1)^{k-j} \binom{k}{j} \lim_{\epsilon\to 0} \lim_{x\to 0} (x+\epsilon+j)^\alpha. \qquad (4.4.5)$$

If $\alpha > 0$, then

$$S(\alpha, k) = \frac{1}{k!} \sum_{j=0}^{k} (-1)^{k-j} \binom{k}{j} j^\alpha \qquad (4.4.6)$$

which yields (4.4.3) and the first relation in (4.4.4), when $k \in \mathbb{N}$ and $k = 0$, respectively. If $\alpha = 0$, then (4.4.5) takes on the form

$$S(0, k) = \frac{1}{k!} \sum_{j=0}^{k} (-1)^{k-j} \binom{k}{j},$$

since $S(0, k) = (1-1)^k / k! = 0$ for $k > 0$. Also, it is easy to see that $S(0,0) = 1$. □

Now we shall give representations of the Stirling functions $S(\alpha, k)$ with positive $\alpha > 0$ in terms of the Liouville fractional derivatives (1.3.63) and (1.3.64).

Theorem 4.12. *Let $\alpha > 0$, $k \in \mathbb{N}_0$ and let ${}^{\mathtt{L}}D_+^\alpha$ and ${}^{\mathtt{L}}D_-^\alpha$ be the operators of Liouville fractional differentiation* (1.3.63) *and* (1.3.64), *respectively. Then the Stirling functions of the second kind $S(\alpha, k)$ have the representations*

$$S(\alpha, k) = \frac{(-1)^k}{k!} \lim_{x \to 0} \left({}^{\mathtt{L}}D_+^\alpha \left((1 - e^t)^k - 1\right)\right)(x) \tag{4.4.7}$$

and

$$S(\alpha, k) = \frac{(-1)^k}{k!} \lim_{x \to 0} \left({}^{\mathtt{L}}D_-^\alpha \left((1 - e^{-t})^k - 1\right)\right)(x). \tag{4.4.8}$$

Proof. Using (1.3.63) and (1.3.71) and taking into account the relation

$$(1 - e^{-u})^k = \sum_{j=0}^{k} (-1)^j \binom{k}{j} e^{-ju}, \tag{4.4.9}$$

we have

$$\frac{(-1)^k}{k!} \left({}^{\mathtt{L}}D_+^\alpha \left((1 - e^t)^k - 1\right)\right)(x) = \frac{(-1)^k}{k!} \sum_{j=1}^{k} (-1)^j \binom{k}{j} \left({}^{\mathtt{L}}D_+^\alpha e^{jt}\right)(x)$$

$$= \frac{1}{k!} \sum_{j=1}^{k} (-1)^{k-j} \binom{k}{j} j^\alpha e^{jx}. \tag{4.4.10}$$

Taking the limit $x \to 0$, we obtain

$$\frac{(-1)^k}{k!} \lim_{x \to 0} \left({}^{\mathtt{L}}D_+^\alpha \left((1 - e^t)^k - 1\right)\right)(x) = \frac{1}{k!} \sum_{j=1}^{k} (-1)^{k-j} \binom{k}{j} j^\alpha = S(\alpha, k).$$

Relation (4.4.8) follows similarly by using Eqs. (1.3.64), (4.4.9) and (1.3.72). □

Stirling functions $S(\alpha, k)$ with positive $\alpha > 0$ can also be represented in terms of the Marchaud fractional derivatives (1.3.86) and (1.3.87).

Theorem 4.13. *Let $\alpha > 0$, $k \in \mathbb{N}_0$ and let ${}^{\mathtt{M}}D_+^\alpha$ and ${}^{\mathtt{M}}D_-^\alpha$ be the operators of Marchaud fractional differentiation* (1.3.86) *and* (1.3.87), *respectively. Then the Stirling functions of the second kind $S(\alpha, k)$ have the following representations*

$$S(\alpha, k) = \frac{1}{k!} \lim_{x \to 0} \left({}^{\mathtt{M}}D_+^\alpha \left((e^t - 1)^k\right)\right)(x) \tag{4.4.11}$$

and

$$S(\alpha, k) = \frac{1}{k!} \lim_{x \to 0} \left({}^{\mathtt{M}}D_-^\alpha \left((e^{-t} - 1)^k\right)\right)(x). \tag{4.4.12}$$

Proof. Using an analogue of Eq. (4.4.9), *viz.*

$$(e^t - 1)^k = \sum_{j=0}^{k} (-1)^{k-j} \binom{k}{j} e^{jt}, \tag{4.4.13}$$

and applying term-by-term differentiation, we have

$$\left({}^{\mathrm{M}}D_+^{\alpha}\left((e^t-1)^k\right)\right)(x) = \left({}^{\mathrm{M}}D_+^{\alpha}1\right)(x) + \sum_{j=1}^{k}(-1)^{k-j}\binom{k}{j}\left({}^{\mathrm{M}}D_+^{\alpha}e^{jt}\right)(x).$$

Using the first relation in (1.3.91) and (1.3.92) (with $b = j \in \mathbb{N}$), we deduce

$$\frac{1}{k!}\left({}^{\mathrm{M}}D_+^{\alpha}\left((e^t-1)^k\right)\right)(x) = \frac{1}{k!}\sum_{j=1}^{k}(-1)^{k-j}\binom{k}{j} j^{\alpha} e^{jx}. \tag{4.4.14}$$

Taking the limit $x \to 0$ in (4.4.14), we obtain (4.4.11). The representation (4.4.12) is proved similarly by applying (4.4.13), the second relation in (1.3.91), and (1.3.93). □

In the case $k > \alpha > 0$ there holds the following integral representation for $S(\alpha, k)$

Theorem 4.14. *Let $\alpha > 0$ and $k \in \mathbb{N}$ be such that $k > \alpha$. Then the Stirling functions of the second kind $S(\alpha, k)$ have the integral representation*

$$S(\alpha, k) = \frac{(-1)^k}{k!\Gamma(-\alpha)} \int_0^{\infty} (1 - e^{-t})^k \frac{dt}{t^{1+\alpha}}. \tag{4.4.15}$$

Proof. If $k > \alpha > 0$, the constant $\kappa(\alpha, k)$ given by (1.3.88) has the form

$$\kappa(\alpha, k) = -\Gamma(-\alpha)A_k(\alpha), \quad A_k(\alpha) = \sum_{j=0}^{k}(-1)^{j-1}\binom{k}{j} j^{\alpha},$$

see Eq. (5.81) of [542]. In accordance with (4.4.3),

$$A_k(\alpha) = (-1)^{k+1} k! S(\alpha, k) \tag{4.4.16}$$

and hence

$$S(\alpha, k) = \frac{(-1)^k}{k!\Gamma(-\alpha)}\kappa(\alpha, k) \quad (k \in \mathbb{N},\ k > \alpha > 0), \tag{4.4.17}$$

and then the theorem is proved. □

Corollary 4.4. *If $n \in \mathbb{N}$ and $k \in \mathbb{N}$ is such that $n < k$, then*

$$S(n, k) = 0 \quad (1 \le n \le k - 1). \tag{4.4.18}$$

Proof. In Theorem 4.14 let $\alpha = n \in \mathbb{N}$ and $k \in \mathbb{N}$ be such that $n < k$. Then in accordance with Eq. (5.74) of [542],

$$A_k(n) = 0 \ (n = 1, 2, \cdots, k - 1), \tag{4.4.19}$$

and (4.4.18) follows from (4.4.16). □

4.4.2 *Stirling functions $S(\alpha, k)$, $\alpha < 0$, and their representations by Liouville fractional integrals*

In this section we consider Stirling functions of the second kind $S(\alpha, k)$ with negative $\alpha < 0$ and $k \in \mathbb{N}$. For this purpose we use an approach similar to (4.4.1) in the form

$$S(\alpha, k) = \frac{1}{k!} \lim_{x\to 0} \Delta_*^k (x^\alpha) \quad (x \in \mathbb{R};\ \alpha < 0,\ k \in \mathbb{N}), \tag{4.4.20}$$

with the *"cut" finite difference*

$$\Delta_*^k f(x) = \sum_{j=1}^{k} (-1)^{k-j} \binom{k}{j} f(x+j) \quad (x \in \mathbb{R};\ k \in \mathbb{N}), \tag{4.4.21}$$

where the limit is taken in the sense of (4.4.2),

$$\lim_{x\to 0} \Delta_*^k (x^\alpha) = \lim_{\epsilon\to 0} \lim_{x\to 0} \Delta_*^k ((x+\epsilon)^\alpha) \quad (\alpha < 0). \tag{4.4.22}$$

Theorem 4.15. *Let $\alpha < 0$ and $k \in \mathbb{N}$. The Stirling functions of the second kind $S(\alpha, k)$ have the explicit representation*

$$S(\alpha, k) = \frac{1}{k!} \sum_{j=1}^{k} (-1)^{k-j} \binom{k}{j} j^\alpha. \tag{4.4.23}$$

Proof. Using (4.4.22) and applying (4.4.21) with $f(x) = x^\alpha$, we have

$$\begin{aligned} S(\alpha, k) &= \frac{1}{k!} \sum_{j=1}^{k} (-1)^{k-j} \binom{k}{j} \lim_{\epsilon\to 0} \lim_{x\to 0} (x+\epsilon+j)^\alpha \\ &= \frac{1}{k!} \sum_{j=1}^{k} (-1)^{k-j} \binom{k}{j} j^\alpha, \end{aligned} \tag{4.4.24}$$

which yields (4.4.23). □

Now we give the representation of $S(\alpha, k)$ $(\alpha < 0)$ in terms of the Liouville fractional integrals (1.3.61) and (1.3.62).

Theorem 4.16. *Let $\alpha < 0$, $k \in \mathbb{N}$ and let ${}^{\mathrm{L}}I_+^{-\alpha}$ and ${}^{\mathrm{L}}I_-^{-\alpha}$ be the operators of Liouville fractional integration* (1.3.61) *and* (1.3.62), *respectively. Then the Stirling functions of the second kind $S(\alpha, k)$ have the integral representations*

$$S(\alpha, k) = \frac{(-1)^k}{k!} \lim_{x\to 0} \left({}^{\mathrm{L}}I_+^{-\alpha} \left((1-e^t)^k - 1\right)\right)(x) \tag{4.4.25}$$

and

$$S(\alpha, k) = \frac{(-1)^k}{k!} \lim_{x\to 0} \left({}^{\mathrm{L}}I_-^{-\alpha} \left((1-e^{-t})^k - 1\right)\right)(x). \tag{4.4.26}$$

Proof. The theorem is proved in a similar way as Theorem 4.12, taking account the known relations (4.4.9) and (1.3.70). □

We point out that if $\alpha > 0$ and $k \in \mathbb{N}$, then the "cut" difference $\Delta_*^k(x^\alpha)$ coincides with the usual difference $\Delta^k(x^\alpha)$. According to this fact and (4.1.12) and (4.4.20), we can give the following unified representation to Stirling functions of the second kind $S(\alpha, k)$, for $x \in \mathbb{R}$, $\alpha \in \mathbb{R}$, $\alpha \neq 0$, and $k \in \mathbb{N}$, by

$$S(\alpha,k) = \frac{1}{k!} \lim_{x\to 0} \Delta_*^k(x^\alpha) \tag{4.4.27}$$

or

$$S(\alpha,k) = \frac{1}{k!} \sum_{j=1}^{k} (-1)^{k-j} \binom{k}{j} j^\alpha. \tag{4.4.28}$$

Finally, we must remark that in the above development for the Stirling numbers $S(\alpha, k)$, only the case $\alpha < 0$ and $k = 0$ has not yet been defined.

4.4.3 *Stirling functions $S(\alpha, k)$, $\alpha \in \mathbb{C}$, and their representations*

In this section we extend the results established in Sections 4.4.1 and 4.4.2 to Stirling functions of the second kind $S(\alpha, k)$ with $\alpha \in \mathbb{C}$, except when $\Re e(\alpha) < 0$ and $k = 0$.

First we use the extensions of the approaches (4.4.1) and (4.4.20). Thus we define the Stirling functions $S(\alpha, k)$ by

$$S(\alpha,k) = \frac{1}{k!} \lim_{x\to 0} \Delta^k(x^\alpha) \quad (x \in \mathbb{R};\ \Re e(\alpha) > 0,\ k \in \mathbb{N}_0), \tag{4.4.29}$$

$$S(\alpha,k) = \frac{1}{k!} \lim_{x\to 0} \Delta_*^k(x^\alpha) \quad (x \in \mathbb{R};\ \Re e(\alpha) \leq 0;\ \ k \in \mathbb{N}), \tag{4.4.30}$$

and $S(0, 0) = 1$. Here the limits are to be understood in the sense of (4.4.2) and (4.4.22).

Explicit representations for such Stirling functions are given by the following statement.

Theorem 4.17. *Let $\alpha \in \mathbb{C}$ and $k \in \mathbb{N}_0$. The Stirling function of the second kind $S(\alpha, k)$ has the explicit representation*

$$S(\alpha,k) = \frac{1}{k!} \sum_{j=1}^{k} (-1)^{k-j} \binom{k}{j} j^\alpha \quad (\alpha \neq 0;\ \ k \in \mathbb{N}), \tag{4.4.31}$$

$$S(\alpha,0) = 0\ (\Re e(\alpha) > 0);\quad S(0,0) = 1;\quad S(0,k) = 0\ (k \in \mathbb{N}). \tag{4.4.32}$$

Proof. The proof of this Theorem is the same as that of Theorems 4.11 and 4.15. □

Corollary 4.5. *If $k = 1$, then*

$$S(\alpha, 1) = 1 \quad (\alpha \in \mathbb{C},\ \alpha \neq 0), \tag{4.4.33}$$

and in particular

$$S(i\theta, 1) = 1 \quad (\theta \in \mathbb{R},\ \theta \neq 0). \tag{4.4.34}$$

If $n, k \in \mathbb{N}$ and $1 \leq n \leq k-1$ then

$$S(n, k) = 0 \tag{4.4.35}$$

The next result, the proof of which is similar to Theorem 4.12, yields the representations of the Stirling functions $S(\alpha, k)$ with $\Re e(\alpha) \geq 0$ in terms of the Liouville fractional derivatives (1.3.63) and (1.3.64).

Theorem 4.18. *Let $\Re e(\alpha) > 0$, $k \in \mathbb{N}_0$ or $\Re e(\alpha) = 0$ $(\alpha \neq 0)$, $k \in \mathbb{N}$, and let ${}^{\mathtt{L}}D_+^\alpha$ and ${}^{\mathtt{L}}D_-^\alpha$ be the operators of Liouville fractional differentiation* (1.3.63) *and (1.3.64), respectively. The Stirling functions $S(\alpha, k)$ have the representations* (4.4.7) *and* (4.4.8) *of Theorem 4.12.*

The representations of $S(\alpha, k)$ with $\mathrm{Re}(\alpha) \geq 0$ in terms of the Marchaud fractional derivatives (1.3.86) and (1.3.87) are given by the following statement whose proof is similar to that of Theorem 4.13.

Theorem 4.19. *Let $\Re e(\alpha) > 0$, $k \in \mathbb{N}_0$ and let ${}^{\mathtt{M}}D_+^\alpha$ and ${}^{\mathtt{M}}D_-^\alpha$ be the operators of Marchaud fractional differentiation* (1.3.86) *and* (1.3.87), *respectively. The Stirling functions of the second kind $S(\alpha, k)$ have the representations* (4.4.11) *and* (4.4.12) *of Theorem 4.13.*

When $k > \Re e(\alpha) > 0$, for $S(\alpha, k)$ there also holds the integral representation of the form (4.4.15). To prove such a representation we extend relation (4.4.17) to complex $\alpha \in \mathbb{C}$ $(\Re e(\alpha) > 0)$.

Lemma 4.4. *Let $\alpha \in \mathbb{C}$ and $k \in \mathbb{N}$ be such that $k > \Re e(\alpha) > 0$. The Stirling function $S(\alpha, k)$ is expressed via the constant $\kappa(\alpha, k)$, given in* (1.3.88), *by*

$$S(\alpha, k) = \frac{(-1)^{k+1}\alpha}{k!\Gamma(1-\alpha)}\kappa(\alpha, k) \quad (k \in \mathbb{N}, k > \mathrm{Re}(\alpha) > 0). \tag{4.4.36}$$

Proof. Let $\alpha > 0$ and $k > \alpha > 0$. Using (4.4.17) and the functional equation for the Gamma function $\Gamma(-\alpha) = -\Gamma(1-\alpha)/\alpha$, we have

$$S(\alpha,k) = \frac{(-1)^k}{k!\Gamma(-\alpha)}\kappa(\alpha,k)) = \frac{(-1)^{k+1}\alpha}{k!\Gamma(1-\alpha)}\kappa(\alpha,k),$$

which proves (4.4.36) for $\alpha > 0$. When $\alpha \in \mathbb{C}$, $k \in \mathbb{N}$ and $k > \Re e(\alpha) > 0$, the left- and right-hand sides of (4.4.36) are analytic functions of α in accordance with (4.4.31) and (1.3.88), respectively. Therefore (4.4.36) is true for such $\alpha \in \mathbb{C}$ by analytic continuation. □

Using Lemma 4.4 and (1.3.88) we deduce the integral representation for $S(\alpha,k)$ when $k > \text{Re}(\alpha) > 0$.

Theorem 4.20. *Let $\alpha \in \mathbb{C}$ and $k \in \mathbb{N}$ be such that $k > \Re e(\alpha) > 0$. The Stirling functions $S(\alpha,k)$ have the integral representation*

$$S(\alpha,k) = \frac{(-1)^{k+1}\alpha}{k!\Gamma(1-\alpha)}\int_0^\infty (1-e^{-t})^k \frac{dt}{t^{1+\alpha}}. \tag{4.4.37}$$

The integral representation of $S(\alpha,k)$ with $\Re e(\alpha) < 0$ in terms of the Liouville fractional integrals is given by the following result which is proved similarly to Theorem 4.16.

Theorem 4.21. *Let $\alpha \in \mathbb{C}$ (Re$(\alpha < 0)$), $k \in \mathbb{N}$ and let ${}^{\mathrm{L}}I_+^{-\alpha}$ and ${}^{\mathrm{L}}I_-^{-\alpha}$ be the operators of Liouville fractional integration given in* (1.3.61) *and* (1.3.62), *respectively. The Stirling functions of the second kind $S(\alpha,k)$ have the integral representations* (4.4.25) *and* (4.4.26) *of Theorem 4.16.*

4.4.4 *Stirling functions $S(\alpha,k)$, $\alpha \in \mathbb{C}$, and recurrence relations*

In the previous section we defined Stirling functions of the second kind $S(\alpha,k)$ for $\alpha \in \mathbb{C}$, except when $\Re e(\alpha) < 0$ and $k = 0$. In this section we will give a definition of $S(\alpha,k)$ for such α and k using the following recurrence relations.

Lemma 4.5. *For $\alpha \in \mathbb{C}$ ($\alpha \neq -1$) and $k \in \mathbb{N}_2 = \{2,3,\ldots\}$, the Stirling functions $S(\alpha,k)$ satisfy the recurrence relation*

$$S(\alpha+1,k) = kS(\alpha,k) + S(\alpha,k-1). \tag{4.4.38}$$

Proof. In accordance with (4.4.31) all terms of both sides of (4.4.38) are defined for the range of parameters α and k considered. Using (4.4.31) and the obvious property

$$\binom{k}{j} = k\binom{k-1}{j-1}\frac{1}{j},$$

we have, for $\alpha \neq 0$

$$\begin{aligned} S(\alpha+1,k) &= \frac{1}{k!}\sum_{j=1}^{k-1}(-1)^{k-j}\binom{k}{j}j^{\alpha+1} + \frac{1}{k!}k^{\alpha+1} \\ &= \frac{1}{(k-1)!}\sum_{j=1}^{k-1}(-1)^{k-j}\binom{k-1}{j-1}j^{\alpha} + \frac{1}{k!}k^{\alpha+1}. \quad (4.4.39) \end{aligned}$$

Applying the well known combinatorial formula

$$\binom{k}{j} = \binom{k-1}{j} + \binom{k-1}{j-1} \qquad (k, j \in \mathbb{N},\ 1 \leq j < k), \qquad (4.4.40)$$

we deduce

$$S(\alpha+1,k)$$

$$= \frac{1}{(k-1)!}\sum_{j=1}^{k-1}(-1)^{k-j}\binom{k}{j}j^{\alpha} + \frac{k}{k!}k^{\alpha} - \frac{1}{(k-1)!}\sum_{j=1}^{k-1}(-1)^{k-j}\binom{k-1}{j}j^{\alpha}$$

$$= k\frac{1}{k!}\sum_{j=1}^{k}(-1)^{k-j}\binom{k}{j}j^{\alpha} + \frac{1}{(k-1)!}\sum_{j=1}^{k-1}(-1)^{k-1-j}\binom{k-1}{j}j^{\alpha},$$

which, combined with (4.4.31), yields (4.4.38). When $\alpha = 0$, relation (4.4.38) follows from (4.4.32)–(4.4.35). □

If $k = 1$ and $\Re e(\alpha) > 0$, then $S(\alpha+1,1) = S(\alpha,1) = 1$ from (4.4.33), while $S(\alpha,0) = 0$ in accordance with (4.4.32); therefore the relation (4.4.38) is valid for $\Re e(\alpha) > 0$ and $k = 1$. It is also true for $k = 1$ and $\alpha = 0$ because $S(1,1) = 1$, $S(0,1) = 0$ and $S(0,0) = 1$ by (4.4.33) and (4.4.32).

Now we put $k = 1$ in (4.4.38) and rewrite it in the form

$$S(\alpha,0) = S(\alpha+1,1) - S(\alpha,1) \qquad (\alpha \neq -1). \qquad (4.4.41)$$

But in accordance with Theorem 4.17 the right-hand side of (4.4.41) is defined for $\alpha \in \mathbb{C}$ when $\Re e(\alpha) < 0$ and $\Re e(\alpha) = 0$ $(\alpha \neq 0)$. Therefore we can define the Stirling functions $S(\alpha, 0)$ for such α by (4.4.41). Then $S(\alpha, 0) = 0$ for $\Re e(\alpha) < 0$ $(\alpha \neq -1)$ and $\Re e(\alpha) = 0$, taking in account (4.4.33). Hence, in accordance with (4.4.32) and (4.4.33) we obtain the property

$$S(\alpha, 0) = 0 \quad (\alpha \in \mathbb{C};\ \alpha \neq 0, -1). \tag{4.4.42}$$

Therefore, from the first relation (4.4.33) and (4.4.42) we deduce that (4.4.38) holds for $k = 1$ when $\Re e(\alpha) < 0$ and $\Re e(\alpha) = 0$ $(\alpha \neq 0)$. Formula (4.4.38) can also be valid for $k = 0$ if we define the $S(\alpha, -1)$ by

$$S(\alpha, -1) = S(\alpha + 1, 0) \quad (\alpha \in \mathbb{C}), \tag{4.4.43}$$

and in accordance with (4.4.32) we obtain

$$S(-1, -1) = 1,\ S(\alpha, -1) = 0 \quad (\alpha \in \mathbb{C},\ \alpha \neq -1). \tag{4.4.44}$$

From Lemma 4.5 and the above arguments, relation (4.4.38) is valid for any $\alpha \in \mathbb{C}$ $(\alpha \neq -1)$ and any $k \in \mathbb{N}_0$. This yields the following result.

Theorem 4.22. *Let $\alpha \in \mathbb{C}$ $(\alpha \neq -1)$, $k \in \mathbb{N}_0$, let the Stirling functions $S(\alpha, k)$ be given by (4.4.31) for $\alpha \neq 0$ and by (4.4.32) for $\alpha = 0$, and let $S(\alpha, -1)$ be given by (4.4.44). Then the $S(\alpha, k)$ satisfy the recurrence relation (4.4.38).*

Corollary 4.6. *If $k \in \mathbb{N}_0$, then*

$$S(k, k) = 1. \tag{4.4.45}$$

Remark 4.1. Relation (4.4.38) can be rewritten in the form

$$S(\alpha, k) = S(\alpha + 1, k + 1) - (k + 1)S(\alpha, k + 1). \tag{4.4.46}$$

This formula together with (4.4.43), (4.4.44) can be used to define the Stirling functions $S(\alpha, k)$ for negative $k = -2, -3, \ldots$. For example, we define $S(\alpha, -2)$ by taking $k = -2$ in (4.4.46)

$$S(\alpha, -2) = S(\alpha + 1, -1) + S(\alpha, -1),$$

and an application of (4.4.44) yields

$$S(-2, -2) = 1,\ S(-1, -2) = 1,\ S(\alpha, -2) = 0 \quad (\alpha \in \mathbb{C},\ \alpha \neq -1, -2).$$

On the other hand, whereas $S(n, k) = 0$ for $k > n$, it should be observed that $S(-n, k)$ is in general not equal to zero for $n, k \in \mathbb{N}$. For example, as can be checked directly on the basis of (4.4.23), $S(-1, 2) = -3/4$ and $S(-1, 3) = 11/36$.

From Theorem 4.22 we deduce the following recurrence relations.

Proposition 4.1. *Let $\alpha \in \mathbb{C}$ and $n, k \in \mathbb{N}_0$. The Stirling functions $S(\alpha, k)$ satisfy the recurrence relations*

$$S(\alpha+n+1, k+n) = \sum_{j=0}^{n}(j+k)S(\alpha+j, k+j)+S(\alpha, k-1) \quad (k \in \mathbb{N}), \tag{4.4.47}$$

when $\alpha \neq -m$ $(m = 1, 2, \ldots, n, n+1)$, and

$$S(\alpha + n + 1, n) = \sum_{j=1}^{n} jS(\alpha + j, j) + S(\alpha + 1, 0), \tag{4.4.48}$$

when $\alpha \neq -m$ $(m = 2, \ldots, n, n + 1)$.

Proof. Relations (4.4.47) and (4.4.48) follow by induction on $n \in \mathbb{N}_0$, using (4.4.38), similar to the case $\alpha > 0$. □

The next result is also deduced similarly to the cases $\alpha > 0$, $\beta > 0$.

Proposition 4.2. *Let $\alpha \in \mathbb{C}$ and $\beta \in \mathbb{C}$ be such that $\alpha \neq 0$, $\beta \neq 0$ and $\alpha + \beta \neq 0$, and let $k \in \mathbb{N}$. There holds the relation*

$$S(\alpha + \beta, k) = \sum_{j=0}^{k} \sum_{i=0}^{k-j} \frac{(i + j)!}{j!} \binom{k - j}{i} S(\alpha, i + j)S(\beta, k - j) \quad (k \in \mathbb{N}). \tag{4.4.49}$$

In particular, the duplication formula is given by

$$S(2\alpha, k) = \sum_{j=0}^{k} \sum_{i=0}^{k-j} \frac{(i + j)!}{j!} \binom{k - j}{i} S(\alpha, i + j)S(\alpha, k - j) \quad (k \in \mathbb{N}). \tag{4.4.50}$$

4.4.5 *Further properties and first applications of Stirling functions $S(\alpha, k)$, $\alpha \in \mathbb{C}$*

In this section we present further properties of the Stirling functions $S(\alpha, k)$ of complex order α. First of all we present a representation for $S(\alpha, k)$, on the basis of (4.4.29), in terms of a complex integral, whose proof is similar to the proof of Proposition 7 in [123].

Proposition 4.3. *For $\alpha \in \mathbb{C}$ ($\alpha \neq 0$) there holds the integral representation*

$$S(\alpha, k) = \frac{1}{2\pi i}\int_{\mathcal{L}} \frac{(z-1)^\alpha}{(z-1)(z-2)\dots(z-k-1)}dz, \tag{4.4.51}$$

where $\mathcal{L}$ is an arbitrary rectifiable Jordan curve in the half-plane $\Re(z) > 1$ containing $n = 2, 3, \dots, k+1$.

Similarly to the proof of Theorem 6 in [123] there also holds the property

Proposition 4.4. *For $z \in \mathbb{C}$ and $\alpha \in \mathbb{C}$ ($\alpha \neq 0$) there holds the Newton series expansion*

$$z^\alpha = \sum_{k=1}^{\infty} S(\alpha, k)[z]_k, \tag{4.4.52}$$

$[z]_k$ being given by (4.1.7), the abscissa of convergence of the series on the right-hand side of (4.4.52) being $\lambda \leq \max\{0, \Re(\alpha) - 1/2\}$. This series converges absolutely for $|z| \geq \lambda + 1$ and uniformly on all compact sets of the half-plane $\Re(z) \geq \lambda + \epsilon$ with arbitrarily small ϵ.

In particular, for $n \in \mathbb{N}$ and $x > 0$ there holds the Newton series expansion

$$x^{-n} = \sum_{k=1}^{\infty} S(-n, k)[x]_k. \tag{4.4.53}$$

The next result deals with the asymptotic behavior of $S(\alpha, k)$ as $a \to \infty$.

Proposition 4.5. *For $k \in \mathbb{N}$ there hold the asymptotic estimates*

$$\lim_{|\alpha|\to\infty,\ -\pi/2+\epsilon\leq\arg(\alpha)\leq\pi/2-\epsilon} \left(k^{-\alpha}S(\alpha, k)\right) = \frac{1}{k!}, \tag{4.4.54}$$

$$\lim_{|\alpha|\to\infty,\ \pi/2+\epsilon\leq\arg(\alpha)\leq 3\pi/2-\epsilon} S(\alpha, k) = \frac{(-1)^k}{(k-1)!}, \tag{4.4.55}$$

where ϵ ($0 < \epsilon < \pi/2$) is a fixed number.

Proof. Let $0 < \epsilon < \pi/2$ and $\alpha \in \mathbb{C}$ ($\alpha \neq 0$). If $\Re(\alpha) > 0$, then by (4.4.31) we have

$$k^{-\alpha}S(\alpha, k) = \frac{1}{k!}\left\{\sum_{j=1}^{k-1}(-1)^{k-j}\binom{k}{j}\left(\frac{j}{k}\right)^\alpha + 1\right\},$$

and taking the limit $|\alpha| \to \infty$ with $-\pi/2+\epsilon \le \arg(\alpha) \le \pi/2-\epsilon$, we obtain (4.4.54).

Let $\Re e(\alpha) < 0$. When $k = 1$, $S(\alpha, 1) = 1$ by (4.4.33), and (4.4.54) is clear. When $k \ge 2$, then by (4.4.31)

$$S(\alpha,k) = \frac{(-1)^{k-1}}{(k-1)!} + \frac{1}{k!}\sum_{j=2}^{k}(-1)^{k-j}\binom{k}{j}\frac{1}{j^{-\alpha}},$$

and (4.4.55) is deduced by taking the limit $|\alpha| \to \infty$ with $\pi/2+\epsilon \le \arg(\alpha) \le 3\pi/2-\epsilon$. □

The "horizontal" generating function for $S(\alpha, k)$ is given by the following assertion which is proved similarly to the case of real positive $\alpha > 0$, if we take (4.4.31) into account (see Eqs. (6.12) and (6.13) of [123]).

Proposition 4.6. *For $\alpha \in \mathbb{C}$ ($\alpha \neq 0$) and $x \in \mathbb{R}$ the Stirling functions $S(\alpha, k)$ have the "horizontal" generating function in the form*

$$\Phi_\alpha(x) = e^{-x}\sum_{k=1}^{\infty}\frac{k^\alpha}{k!}x^k = \sum_{k=1}^{\infty}S(\alpha,k)x^k. \qquad (4.4.56)$$

This "generating" function satisfies a Rodriguez formula, the difference-differential equation

$$\Phi_\alpha(x) = x\left(\Phi_{\alpha-1}(x) + \Phi'_{\alpha-1}(x)\right).$$

In particular, for $n \in \mathbb{N}$ and $x \in \mathbb{R}$, there holds

$$\Phi_{-n}(x) = e^{-x}\sum_{k=1}^{\infty}\frac{x^k}{k!k^n} = \sum_{k=1}^{\infty}S(-n,k)x^k.$$

Remark 4.2. The Bell function $B(\alpha)$ of order $\alpha \ge 0$ can be defined by (see p. 40 in [123])

$$B(\alpha) = \sum_{k=1}^{\infty}S(\alpha,k). \qquad (4.4.57)$$

The extension to $\alpha \in \mathbb{C}$ is obvious. If $\alpha \neq 0$, then, in accordance with (4.4.32), (4.4.56) yields

$$B(\alpha) = \frac{1}{e}\sum_{k=1}^{\infty}\frac{k^\alpha}{k!} = \sum_{k=1}^{\infty}S(\alpha,k). \qquad (4.4.58)$$

In particular, the Bell number $B(\alpha)$ of order $\alpha = -n$ has the representation

$$B(-n) = e^{-1}\sum_{k=1}^{\infty}\frac{1}{k!k^n} = \sum_{k=1}^{\infty}S(-n,k). \qquad (4.4.59)$$

Bell numbers of negative order, which do not have a finite series, do not seem to have been considered so far; it should be of interest to interpret them in the context of combinatorics. Likewise it would of course be satisfying to view the new Stirling numbers $S(-n, k)$ in the setting of combinatorial analysis.

The function $\Phi_\alpha(x)$ can be applied to obtain a certain convolution theorem for $S(\alpha, k)$ and other recurrence relations than those given in Propositions 4.1 and 4.2.

Proposition 4.7. *For $\alpha \in \mathbb{C}$ ($\alpha \neq 0$), $\beta \in \mathbb{C}$ ($\beta \neq 0$) there holds for any $m = 2, 3, \ldots$ the convolution relation*

$$\sum_{k=1}^{m-1} S(\alpha, k) S(\beta, m-k) = \sum_{k=2}^{m} \sum_{j=1}^{k-1} (-2)^{m-k} \frac{j^\alpha (k-j)^\beta}{j!(k-j)!(m-k)!}. \tag{4.4.60}$$

Proposition 4.8. *For $\alpha \in \mathbb{C}$ ($\alpha \neq -1, 0$) one has*

$$\sum_{k=0}^{\infty} (k+1) S(\alpha, k) = \sum_{k=0}^{m} S(\alpha+1, k). \tag{4.4.61}$$

The following theorem, the proof of which also depends on $\Phi_\alpha(x)$, is basic for the next one.

Theorem 4.23. *For $\alpha \in \mathbb{C}$ ($\alpha \neq 0$) and $m \in \mathbb{N}$ there holds*

$$m^\alpha = \sum_{k=1}^{m} \binom{m}{k} k! S(\alpha, k). \tag{4.4.62}$$

In particular, for $\alpha = -n$ ($n \in \mathbb{N}$),

$$\frac{1}{m^n} = \sum_{k=1}^{m} \binom{m}{k} k! S(-n, k). \tag{4.4.63}$$

Proof. Using the first and the second formulas in (4.4.56) and taking into account the convergence of the series $\sum_{k=1}^{\infty} S(\alpha, k) x^k$ for any $x \in \mathbb{R}$, we have

$$m^\alpha = D^m [e^x \Phi_\alpha(x)]|_{x=0} = D^m \left[\left(\sum_{j=0}^{\infty} \frac{x^j}{j!} \right) \left(\sum_{k=1}^{\infty} S(\alpha, k) x^k \right) \right] \Bigg|_{x=0}$$

$$= D^m \left[\left(\sum_{j=0}^{\infty} \frac{x^j}{j!} \right) \left(\sum_{k=0}^{\infty} S(\alpha, k+1) x^{k+1} \right) \right] \Bigg|_{x=0}.$$

Applying the identity theorem for power series and the multiplication theorem, we deduce

$$
\begin{aligned}
m^\alpha &= D^m[e^x \Phi_\alpha(x)]|_{x=0} \\
&= D^m \left[\sum_{j=0}^{\infty} \left(\sum_{k=0}^{j} \frac{S(\alpha, k+1)}{(j-k)!} \right) x^{j+1} \right] \Bigg|_{x=0} \\
&= D^m \left[\sum_{j=1}^{\infty} \left(\sum_{k=0}^{j-1} \frac{S(\alpha, k+1)}{(j-k-1)!} \right) x^{j} \right] \Bigg|_{x=0} \\
&= D^m \left[\sum_{j=m}^{\infty} \left(\sum_{k=0}^{j-1} \frac{S(\alpha, k+1)}{(j-k-1)!} \right) \frac{j!}{(j-m)!} x^{j-m} \right] \Bigg|_{x=0} \\
&= \sum_{k=0}^{m-1} S(\alpha, k+1) \frac{m!}{(m-k-1)!} = \sum_{k=1}^{m} \frac{m!}{(m-k)!} S(\alpha, k), \quad (4.4.64)
\end{aligned}
$$

which proves (4.4.62). The relation (4.4.62) with $\alpha = -n$ ($n \in \mathbb{N}$) yields (4.4.63). Thus the proof is complete. □

The following result is based on Theorem 4.23.

Theorem 4.24. *For $\alpha \in \mathbb{C}$ ($\alpha \neq -1$) and $m \in \mathbb{N}$ there holds*

$$\sum_{k=1}^{m} k^\alpha = \sum_{k=1}^{m} \binom{m}{k} (k-1)! S(\alpha+1, k). \tag{4.4.65}$$

In particular, for $\alpha = -n$ ($n \in \mathbb{N}$, $n \neq 1$), we have

$$\sum_{k=1}^{m} \frac{1}{k^n} = \sum_{k=1}^{m} \binom{m}{k} (k-1)! S(-n+1, k). \tag{4.4.66}$$

Proof. The proof follows by induction. The case $m = 1$ is valid since $1 = S(\alpha+1, 1)$ holds for $\alpha \neq -1$ by (4.4.33). Let (4.4.65) be valid for $m = n \in \mathbb{N}$. By (4.4.40) we have for $m = n+1$

$$
\begin{aligned}
&\sum_{k=1}^{n+1} \binom{n+1}{k} (k-1)!\, S(\alpha+1, k) \qquad (4.4.67) \\
&= n! S(\alpha+1, n+1) + \sum_{k=1}^{n} \left[\binom{n}{k} + \binom{n}{k-1} \right] (k-1)!\, S(\alpha+1, k) \\
&= \sum_{k=1}^{n} \binom{n}{k} (k-1)!\, S(\alpha+1, k) + \sum_{k=1}^{n+1} \binom{n}{k-1} (k-1)!\, S(\alpha+1, k).
\end{aligned}
$$

Since, by induction, (4.4.65) holds for $m = n$,

$$\sum_{k=1}^{n} \binom{n}{k} (k-1)!\, S(\alpha+1,k) = \sum_{k=1}^{n} k^{\alpha}.$$

According to (4.1.10) and Theorem 4.23,

$$\sum_{k=1}^{n+1} \binom{n}{k-1} (k-1)!\, S(\alpha+1,k) = \sum_{k=1}^{n+1} \frac{n!}{(n-k+1)!} S(\alpha+1,k)$$

$$= \sum_{k=1}^{n+1} \binom{n+1}{k} k!\, \frac{1}{n+1} S(\alpha+1,k) = \frac{1}{n+1}(n+1)^{\alpha+1} = (n+1)^{\alpha}.$$

Substituting two last relations into (4.4.67), we obtain (4.4.65). The relation (4.4.66) follows from (4.4.65) for $\alpha = -n$ ($n \in \mathbb{N}$), which completes the proof of this theorem. □

Remark 4.3. The evaluation of $\sum_{k=1}^{m} k^{\alpha}$ for the classical case $\alpha = n \in \mathbb{N}$ is due to Bernoulli (who introduced the numbers named after him in its determination), and for real $\alpha > -1$ to [123]. The new formula (4.4.65) in the instance $\alpha \in \mathbb{C}$ with $\alpha \neq -1$, and in the particular case $\alpha = -n$ with $n \in \mathbb{N}$ ($n \neq 1$) of (4.4.66) emphasize the important role that the Stirling functions $S(\alpha+1,k)$, in particular $S(-n+1,k)$, play even in classical combinatorics and discrete mathematics. Formula (4.4.66) fails for $n = 1$ since the right-hand side would then be zero, noting $S(-n+1,k) = 0$ for $n = 1$ and all $k \in \mathbb{N}$.

Normally, $\sum_{k=1}^{m} k^{-n}$ is summed up for $n > 1$ by employing the Euler-Poisson summation formula, giving an infinite series involving the Bernoulli numbers. But (4.4.66) yields a finite series involving $S(-n+1,k)$.

Continuity and differentiability of the Stirling functions $S(\alpha,k)$ are given by the following result.

Proposition 4.9. *Let $k \in \mathbb{N}$ be arbitrary. There hold the following assertions.*

(a) *$S(\alpha,k)$ as a function of α is continuous for $\alpha \in \mathbb{C}$ ($\alpha \neq 0$).*
(b) *If $\alpha \in \mathbb{C}$ ($\alpha \neq 0$), then $S(\alpha,k)$ is arbitrarily often continuously differentiable, and for any $m \in \mathbb{N}$*

$$\left(\frac{d}{d\alpha}\right)^{m} S(\alpha,k) = \frac{1}{k!}\sum_{j=1}^{k}(-1)^{k-j}\binom{k}{j} j^{\alpha}(\log j)^{m}. \tag{4.4.68}$$

Proof. The continuity of $S(\alpha, k)$ for $\alpha \in \mathbb{C}$ ($\alpha \neq 0$) follows from (4.4.31).

When $\alpha \in \mathbb{C}$ and $\alpha \neq 0$, the function j^α as a function of α is arbitrarily often continuously differentiable and (4.4.31) gives the same for $S(\alpha, k)$. The relation (4.4.68) follows from (4.4.31) by differentiation. This completes the proof of (ii) and the proposition. □

4.4.6 Applications of Stirling functions $S(\alpha, k)$ ($\alpha \in \mathbb{C}$) to Hadamard-type fractional operatos

Now we apply the Stirling functions $S(\alpha, k)$ to express Hadamard fractional integration (1.3.123), (1.3.124) and differentiation (1.3.129), (1.3.130) in terms of $S(\alpha, k)$ and usual differentiation.

First we prove such results for the more general Hadamard-type fractional integration (1.3.125), (1.3.126) and differentiation (1.3.131), (1.3.132) in terms of more general Stirling functions generalizing $S_c(n, k)$ defined via (4.1.18). Such Stirling functions $S_c(\alpha, k)$, defined in [124], have the explicit representation in the form (4.4.31)

$$S_c(\alpha, k) = \frac{1}{k!}\sum_{j=0}^{k}(-1)^{k-j}\binom{k}{j}(c+j)^\alpha \quad (\alpha \in \mathbb{C};\ k \in \mathbb{N}_0;\ c \in \mathbb{R}), \tag{4.4.69}$$

established in [126], for which

$$\lim_{c\to 0} S_c(\alpha, k) = S(\alpha, k). \tag{4.4.70}$$

Alternatively, the $S_c(\alpha, k)$ can be defined by

$$S_c(\alpha, 0) = c^\alpha,\ \ S_c(\alpha+1, k) = S_c(\alpha, k-1) + (c+k)S_c(\alpha, k).$$

The properties of $S_c(\alpha, k)$ are generally similar to those of the $S(\alpha, k)$.

First we show that Hadamard-type fractional integration (1.3.125) and differentiation (1.3.131) can be expressed in terms of infinite series involving the $S_c(\alpha, k)$ and classical differentiation by the following statement.

Theorem 4.25. *Let $f(x)$, defined for $x > 0$, be an arbitrarily often differentiable function such that its Taylor series converges, and let $\alpha \in \mathbb{C}$, $c > 0$.*

(a) *When $\mathfrak{Re}(\alpha) \geq 0$, the Hadamard type fractional derivative ${}^{\mathrm{H}}D^\alpha_{0+,c}f$ is given by* (1.3.131) *if and only if there holds for $x > 0$ the relation*

$$\left({}^{\mathrm{H}}D^\alpha_{0+,c}f\right)(x) = \sum_{k=0}^{\infty} S_c(\alpha, k)x^k f^{(k)}(x). \tag{4.4.71}$$

(b) *When* $\Re e(\alpha) > 0$, *the Hadamard type fractional integral* ${}^{\mathrm{H}}I^{\alpha}_{0+,c}f$ *is given by* (1.3.125) *if and only if for* $x > 0$

$$\left({}^{\mathrm{H}}I^{\alpha}_{0+,c}f\right)(x) = \sum_{k=0}^{\infty} S_c(-\alpha, k)x^k f^{(k)}(x). \tag{4.4.72}$$

Proof. When $\alpha = 0$, then in accordance with (4.4.69) and (1.3.90) for $c \in \mathbb{R}$

$$S_c(0,0) = 1, \;\; S_c(0,k) = 0 \;(k \in \mathbb{N}),$$

and (4.4.71) takes on the form ${}^{\mathrm{H}}D^{0}_{0+,c}f = f$ which coincides with the one in (1.3.133).

Let $\Re e(\alpha) \geq 0$ $(\alpha \neq 0)$, and let $\left({}^{\mathrm{H}}D^{\alpha}_{0+,c}f\right)(x)$ be given for $x > 0$ by (1.3.131) with $n = [\Re e(\alpha)] + 1$

$$\left({}^{\mathrm{H}}D^{\alpha}_{0+,c}f\right)(x) = x^{-c}\left(x\frac{d}{dx}\right)^n x^c \frac{1}{\Gamma(n-\alpha)}\int_0^x \left(\frac{u}{x}\right)^c \left(\log\frac{x}{u}\right)^{n-\alpha-1} f(u)\frac{du}{u}. \tag{4.4.73}$$

Fix $x > 0$. By conditions of the theorem, for any $u \in [0, x]$ and any $y > 0$ we have

$$f(u) = \sum_{k=0}^{\infty} \frac{f^{(k)}(y)}{k!}(u-y)^k.$$

By the binomial formula this relation can be rewritten as

$$f(u) = \sum_{k=0}^{\infty} \frac{f^{(k)}(y)}{k!}\sum_{j=0}^{k}(-1)^{k-j}\binom{k}{j} y^{k-j}u^j, \tag{4.4.74}$$

or

$$f(u) = \sum_{j=0}^{\infty} a_j(y)u^j, \;\; a_j(y) = \sum_{k=0}^{\infty}\frac{f^{(k)}(y)}{k!}(-1)^{k-j}\binom{k}{j} y^{k-j}. \tag{4.4.75}$$

For any fixed $y > 0$, (4.4.75) is a convergent power series because it coincides with the Taylor series (4.4.74), being convergent by the condition of the theorem. Then we can apply Proposition 1.2(ii). Substituting relation (4.4.74) into (4.4.73) and using the formula (1.3.163), we have

$$\begin{aligned}\left({}^{\mathrm{H}}D^{\alpha}_{0+,c}f\right)(x) &= \sum_{k=0}^{\infty}\frac{f^{(k)}(y)}{k!}\sum_{j=0}^{k}(-1)^{k-j}\binom{k}{j} y^{k-j}({}^{\mathrm{H}}D^{\alpha}_{0+,c}u^j)(x)\\ &= \sum_{k=0}^{\infty} f^{(k)}(y)\frac{1}{k!}\sum_{j=0}^{k}(-1)^{k-j}\binom{k}{j} y^{k-j}(c+j)^{\alpha}x^j.\end{aligned}$$

Setting $y = x$ in this relation, we obtain

$$\left({}^{\mathrm{H}}D^{\alpha}_{0+,c}f\right)(x) = \sum_{k=0}^{\infty} x^k f^{(k)}(x) \left(\frac{1}{k!}\sum_{j=0}^{k}(-1)^{k-j}\binom{k}{j}(c+j)^{\alpha}\right),$$

which, combined with (4.4.69), yields (4.4.71).

Conversely, let ${}^{\mathrm{H}}D^{\alpha}_{0+,c}f$ have representation (4.4.71) for $x > 0$. Fixing this $x > 0$, noting (4.4.69) and taking any $y > 0$ we rewrite (4.4.71) in the form

$$\left({}^{\mathrm{H}}D^{\alpha}_{0+,c}f\right)(x) = \sum_{k=0}^{\infty} x^k f^{(k)}(x) y^{-j} \left(\frac{1}{k!}\sum_{j=0}^{k}(-1)^{k-j}\binom{k}{j} y^j (c+j)^{\alpha}\right).$$

Applying relation (4.4.75) (with x being replaced by y) and interchanging orders of summation and integration, being admissible again by Proposition 1.2(ii), we find for $\delta = x(d/dx)$,

$$\begin{aligned}\left({}^{\mathrm{H}}D^{\alpha}_{0+,c}f\right)(x) &= \sum_{k=0}^{\infty} \frac{f^{(k)}(x)}{k!} x^k y^{-j} \sum_{j=0}^{k}(-1)^{k-j}\binom{k}{j}({}^{\mathrm{H}}D^{\alpha}_{0+,c}u^j)(x) \\ &= x^{-c}\delta^n x^c \frac{1}{\Gamma(n-\alpha)}\int_0^x \left(\frac{u}{x}\right)^c \left(\log\frac{x}{u}\right)^{n-\alpha-1} \\ &\quad\times \left(\sum_{k=0}^{\infty}\frac{f^{(k)}(x)}{k!}x^k y^{-j}\sum_{j=0}^{k}(-1)^{k-j}\binom{k}{j}u^j\right)\frac{du}{u}.\end{aligned}$$

Setting $y = x$, we have

$$\begin{aligned}\left({}^{\mathrm{H}}D^{\alpha}_{0+,c}f\right)(x) &= x^{-c}\delta^n x^c \frac{1}{\Gamma(n-\alpha)}\int_0^x \left(\frac{u}{x}\right)^c \left(\log\frac{x}{u}\right)^{n-\alpha-1} \\ &\quad\times \left(\sum_{k=0}^{\infty}\frac{f^{(k)}(x)}{k!}\sum_{j=0}^{k}(-1)^{k-j}\binom{k}{j}x^{k-j}u^j\right)\frac{du}{u},\end{aligned}$$

and (1.3.131) follows by noting (4.4.74) with y being replaced by x.

The relation (4.4.72) is proved similarly by using Proposition 1.2(i). Thus the proof is complete. □

Observe that the fractional differentiation and integration operators ${}^{\mathrm{H}}D^{\alpha}_{0+,c}$ and ${}^{\mathrm{H}}I^{\alpha}_{0+,c}$ of Theorem 4.25 are both expressed as infinite series involving the classical powers $(xd/dx)^k$, with coefficients $S_c(\alpha, k)$ and

$S_c(-\alpha, k)$, respectively. Therefore, the relations (4.4.71) and (4.4.72) express in a clear way the natural transition from ${}^{\mathrm{H}}D^{\alpha}_{0+,c}$ to ${}^{\mathrm{H}}I^{\alpha}_{0+,c}$, and conversely.

Let Q be the "reflection" operator (see Eq. (2.5) of [126]) defined by

$$(Qg)(x) = g\left(\frac{1}{x}\right). \tag{4.4.76}$$

It is directly verified that the Hadamard-type fractional integration operators ${}^{\mathrm{H}}I^{\alpha}_{0+,c}$ and ${}^{\mathrm{H}}I^{\alpha}_{-,c}$ given by (1.3.125) and (1.3.126) are connected via the operator Q by

$${}^{\mathrm{H}}I^{\alpha}_{0+,c}Qf = Q^{\mathrm{H}}I^{\alpha}_{-,c}Qf,$$

and similarly for the Hadamard-type fractional differentiation operators ${}^{\mathrm{H}}D^{\alpha}_{0+,c}$ and ${}^{\mathrm{H}}D^{\alpha}_{-,c}$ defined by (1.3.131) and (1.3.132)

$${}^{\mathrm{H}}D^{\alpha}_{0+,c}Qf = Q^{\mathrm{H}}D^{\alpha}_{-,c}Qf.$$

Since the operator $Q^{-1} = Q$, we deduce from Theorem 4.25 the representation of the Hadamard-type fractional integration (1.3.126), ${}^{\mathrm{H}}I^{\alpha}_{-,c}$, and differentiation (1.3.132), ${}^{\mathrm{H}}D^{\alpha}_{-,c}$, in terms of Stirling functions $S_c(\alpha, k)$ and usual differentiation.

Theorem 4.26. *Let $f(x)$, defined for $x > 0$, be arbitrarily often differentiable such that its Taylor series converges. Let $\alpha \in \mathbb{C}$, $c > 0$, and let Q be the "reflection" operator* (4.4.76).

(a) *When $\mathfrak{Re}(\alpha) \geq 0$, the Hadamard-type fractional derivative ${}^{\mathrm{H}}D^{\alpha}_{-,c}f$ is given by* (1.3.132) *if and only if there holds for $x > 0$ the relation*

$$\left({}^{\mathrm{H}}D^{\alpha}_{-,c}f\right)(x) = \sum_{k=0}^{\infty} S_c(\alpha, k)x^{-k}(Qf)^{(k)}\left(\frac{1}{x}\right). \tag{4.4.77}$$

(b) *When $\mathfrak{Re}(\alpha) > 0$, the Hadamard-type fractional integral ${}^{\mathrm{H}}I^{\alpha}_{-,c}f$ is given by* (1.3.126) *if and only if for $x > 0$*

$$\left({}^{\mathrm{H}}I^{\alpha}_{-,c}f\right)(x) = \sum_{k=0}^{\infty} S_c(-\alpha, k)x^{-k}(Qf)^{(k)}\left(\frac{1}{x}\right). \tag{4.4.78}$$

Both $({}^{\mathrm{H}}D^{\alpha}_{0+,c}f)(x)$ and $({}^{\mathrm{H}}D^{\alpha}_{-,c}f)(x)$ are continuous functions as functions of c at $c = 0$ for fixed $\alpha \in \mathbb{C}$ ($\mathfrak{Re}(\alpha) \geq 0$), and likewise for $({}^{\mathrm{H}}I^{\alpha}_{0+,c}f)(x)$ and $({}^{\mathrm{H}}I^{\alpha}_{-,c}f)(x)$ for fixed $\alpha \in \mathbb{C}$ ($\mathrm{Re}(\alpha) > 0$). The functions $S_c(\alpha, k)$ are also continuous functions of $c \in \mathbb{R}$ at $c = 0$ for fixed $\alpha \in \mathbb{C}$ and $k \in \mathbb{N}_0$. Therefore letting $c \to 0$ in Theorem 4.25 and taking (4.4.70) into account we deduce the corresponding representations of Hadamard fractional differentiation (1.3.129), ${}^{\mathtt{H}}D^{\alpha}_{0+}$, and integration (1.3.123), ${}^{\mathtt{H}}I^{\alpha}_{0+}$, in terms of the Stirling functions $S(\alpha, k)$ and usual differentiation.

Theorem 4.27. *Let $f(x)$, defined for $x > 0$, be an arbitrarily often differentiable function such that its Taylor series converges, and let $\alpha \in \mathbb{C}$.*

(a) *When $\Re e(\alpha) \geq 0$, the Hadamard fractional derivative ${}^{\mathtt{H}}D^{\alpha}_{0+}f$ is given by (1.3.129) if and only if there holds for $x > 0$ the relation*

$$\left({}^{\mathtt{H}}D^{\alpha}_{0+}f\right)(x) = \sum_{k=0}^{\infty} S(\alpha, k) x^k f^{(k)}(x). \tag{4.4.79}$$

(b) *When $\Re e(\alpha) > 0$, the Hadamard-type fractional integral ${}^{\mathtt{H}}I^{\alpha}_{0+}f$ is given by (1.3.123) if and only if for $x > 0$*

$$\left({}^{\mathtt{H}}I^{\alpha}_{0+}f\right)(x) = \sum_{k=0}^{\infty} S(-\alpha, k) x^k f^{(k)}(x). \tag{4.4.80}$$

Corollary 4.7. *Let $f(x)$ defined for $x > 0$ be an arbitrarily often differentiable function such that its Taylor series converges, and let $n \in \mathbb{N}$. Then*

$$({}^{\mathtt{H}}I^{n}_{0+}f)(x) = \int_0^x \frac{du_1}{u_1} \int_0^{u_1} \frac{du_2}{u_2} \cdots \int_0^{u_{n-1}} \frac{du_n}{u_n} = \sum_{k=0}^{\infty} S(-n, k) x^k f^{(k)}(x). \tag{4.4.81}$$

Corollary 4.8. *Let $n \in \mathbb{N}$ and let $f(x)$ be a function defined for $x > 0$ and differentiable up to order n and let $x \in \mathbb{R}$. Then*

$$\delta^n f(x) = \sum_{k=0}^{n} S(n, k) x^k f^{(k)}(x), \quad \delta = x\frac{d}{dx}. \tag{4.4.82}$$

Similarly Theorem 4.26 yields the representation of Hadamard fractional differentiation (1.3.130), ${}^{\mathtt{H}}D^{\alpha}_{-}$, and integration (1.3.124), ${}^{\mathtt{H}}I^{\alpha}_{-}$, in terms of the Stirling functions $S(\alpha, k)$ and usual differentiation. Indeed, we have

Theorem 4.28. *Let $f(x)$ be an arbitrarily often differentiable function for $x > 0$ such that its Taylor series converges. Let $\alpha \in \mathbb{C}$ ($\Re e(\alpha) \geq 0$), and let Q be the "reflection" operator (4.4.76).*

(a) *When $\mathfrak{Re}(\alpha) \geq 0$, the Hadamard fractional derivative ${}^{\mathrm{H}}D_{-}^{\alpha} f$ is given by* (1.3.130) *if and only if for $x > 0$ there holds the relation*

$$({}^{\mathrm{H}}D_{-}^{\alpha} f)(x) = \sum_{k=0}^{\infty} S(\alpha, k) x^{-k} (Qf)^{(k)} \left(\frac{1}{x}\right). \tag{4.4.83}$$

(b) *When $\mathfrak{Re}(\alpha) > 0$, the Hadamard fractional integral ${}^{\mathrm{H}}I_{-}^{\alpha} f$ is given by* (1.3.124) *if and only if for $x > 0$*

$$\left({}^{\mathrm{H}}I_{-}^{\alpha} f\right)(x) = \sum_{k=0}^{\infty} S(-\alpha, k) x^{-k} (Qf)^{(k)} \left(\frac{1}{x}\right). \tag{4.4.84}$$

4.5 Generalized Stirling Functions of the Second Kind $S(n, \beta)$ with Complex Second Argument β

In this section we give extensions of the classical Stirling numbers of the second kind $S(n, k)$ to functions $S(n, \beta)$, whereby the second parameter k becomes any complex β, as well as to functions $S(\alpha, \beta)$, whereby also the first parameter n becomes any complex α. Also we give the representations of $S(\alpha, \beta)$ by Liouville fractional operators.

Firstly the second parameter k is extended from a nonnegative integer k to any complex β. We recall from Eq. (4.1.6) that the classical Stirling numbers of the second kind, $S(n, k)$ are defined by

$$S(n, k) = \frac{1}{k!} \lim_{x \to 0} \Delta^k (x^n), \tag{4.5.1}$$

where $x \in \mathbb{R}$, $n, k \in \mathbb{N}_0$, and Δ^k is the difference of order k given by

$$\begin{aligned}
\Delta^k f(x) &= \Delta \left(\Delta^{k-1} f\right)(x) = \sum_{j=0}^{k} (-1)^{k-j} \binom{k}{j} f(x + j) \\
&= D^m \left[\sum_{j=0}^{\infty} \left(\sum_{k=0}^{j} \frac{S(\alpha, k+1)}{(j-k)!} \right) x^{j+1} \right] \Bigg|_{x=0} \\
&= \sum_{j=0}^{k} (-1)^j \binom{k}{j} f(x + k - j)
\end{aligned}$$

with $\Delta^0 f(x) = f(x)$, $\Delta^1 f(x) = f(x+1) - f(x)$. This suggests that one may define $S(n, \beta)$ for $\beta \in \mathbb{C}$ by

$$S(n, \beta) = \frac{1}{\Gamma(\beta + 1)} \lim_{x \to 0} \Delta^\beta (x^n) \quad (x \in \mathbb{R};\ n \in \mathbb{N}_0), \tag{4.5.2}$$

where Δ^β is a suitable generalization of Δ^k from integer k to complex β.

4.5.1 *Definition and some basic properties*

In order to cover as large a class of functions as possible, an exponential factor is introduced in the corresponding definition.

Definition 4.2. The generalized fractional difference operator $\Delta^{\beta,\epsilon}$ for $\beta \in \mathbb{C}$, $\epsilon \geq 0$ is defined for "sufficiently good" functions f by

$$\Delta^{\beta,\epsilon} f(x) = \sum_{j=0}^{\infty} (-1)^j \binom{\beta}{j} e^{(\beta-j)\epsilon} f(x+\beta-j) \quad (x \in \mathbb{R}), \tag{4.5.3}$$

where $\binom{\beta}{j}$ are the general binomial coefficients given by

$$\binom{\beta}{j} = \frac{[\beta]_j}{j!} = \frac{\beta(\beta-1)\cdots(\beta-j+1)}{j!} \quad (j \in \mathbb{N}), \tag{4.5.4}$$

with $[\beta]_0 = 1$ and $\mathbb{N}$ being the set of all positive integers.

Since $\binom{k}{j} = 0$ for $k \in \mathbb{N}_0$, $0 \leq k < j$, the difference (4.1.6) for $\beta = k \in \mathbb{N}_0$ takes on the form

$$\begin{aligned} \Delta^{k,\epsilon} f(x) &= \sum_{j=0}^{k} (-1)^j \binom{k}{j} e^{(k-j)\epsilon} f(x+k-j) \\ &= \sum_{j=0}^{k} (-1)^{k-j} \binom{k}{j} e^{j\epsilon} f(x+j) \quad (x \in \mathbb{R}) \end{aligned} \tag{4.5.5}$$

which is precisely $\Delta^k f(x)$ of (4.1.4) for $\epsilon = 0$, so that the new fractional difference $\Delta^{k,0}$ reduce to the classical one, namely

$$\Delta^{k,0} f(x) = \Delta^k f(x) \quad (x \in \mathbb{R}).$$

Here $\Delta^{0,\epsilon} f(x)$ with $\beta = 0$, is the identity operator

$$\Delta^{0,\epsilon} f(x) = f(x). \tag{4.5.6}$$

It is important to remark here that there are several alternative forms to introduce a generalized difference operator Δ^β from our definition of $\Delta^{\beta,\epsilon}$. The most natural way is the $\Delta^\beta \equiv \Delta^{\beta,0}$, but another interesting possible definition could be $\overline{\Delta}^\beta \equiv \Delta^{\beta,\{\beta\}}$, where $\{\beta\} = \beta - [\beta]$ and $[\beta]$ is the integral part of β. The latter definition will be used below during the development of this chapter.

Definition 4.3. The generalized Stirling functions of the second kind, $S(n,\beta;\epsilon)$, are given for $\beta \in \mathbb{C}$ by

$$S(n,\beta;\epsilon) = \frac{1}{\Gamma(\beta+1)} \lim_{x\to 0} \Delta^{\beta,\epsilon}(x^n) \quad (x \in \mathbb{R};\ n \in \mathbb{N}_0;\ \epsilon \geq 0), \tag{4.5.7}$$

provided this limit exists.

In particular, we will write

$$S(n,\beta) = S(n,\beta;0). \tag{4.5.8}$$

Theorem 4.29. *The following assertions hold:*

(a) *If $n \in \mathbb{N}_0$ and either of the conditions $\beta \in \mathbb{C}$ $(\beta \notin \mathbb{Z})$, $\epsilon > 0$, or $\beta \in \mathbb{C}$ $(\mathfrak{Re}(\beta) > n,\ \beta \neq n+1, n+2, \ldots)$, $\epsilon = 0$ hold, then the generalized Stirling functions $S(n,\beta;\epsilon)$ can be represented in the form*

$$S(n,\beta;\epsilon) = \frac{1}{\Gamma(\beta+1)} \sum_{j=0}^{\infty} (-1)^j \binom{\beta}{j} e^{(\beta-j)\epsilon} (\beta-j)^n. \tag{4.5.9}$$

(b) *If $n \in \mathbb{N}_0$, $\beta = k \in \mathbb{N}$ and $\epsilon \geq 0$, then the generalized Stirling functions $S(n,k;\epsilon)$ have the representation*

$$\begin{aligned} S(n,k;\epsilon) &= \frac{1}{k!} \sum_{j=0}^{k} (-1)^j \binom{k}{j} e^{(k-j)\epsilon} (k-j)^n \\ &= \frac{1}{k!} \sum_{j=0}^{k} (-1)^{k-j} \binom{k}{j} e^{j\epsilon} j^n. \end{aligned} \tag{4.5.10}$$

(c) *If $n \subset \mathbb{N}_0$, $\beta = 0$ and $\epsilon \geq 0$, then*

$$S(0,0;\epsilon) = 1 \quad (\epsilon \geq 0), \tag{4.5.11}$$

$$S(n,0;\epsilon) = 0 \quad (n \in \mathbb{N},\ \epsilon \geq 0). \tag{4.5.12}$$

Proof. When $n \in \mathbb{N}_0$, $\beta \in \mathbb{C}$ $(\beta \notin \mathbb{Z})$ and $\epsilon \geq 0$, the $S(n,\beta;\epsilon)$ have the representation (4.5.9). This relation is valid provided that the series on the right-hand side of (4.5.9) is absolutely convergent. Rewrite this series in the form

$$\sum_{j=0}^{\infty} c_j \quad \text{with } c_j = (-1)^j \binom{\beta}{j} e^{(\beta-j)\epsilon} (\beta-j)^n, \tag{4.5.13}$$

and note the estimate

$$\left|\binom{\beta}{j}\right| \leq \frac{A}{j^{\Re e(\beta)+1}} \quad (\beta \in \mathbb{C};\ A > 0) \tag{4.5.14}$$

for any $\beta \in \mathbb{C}$ and sufficiently large $j \in \mathbb{N}$, $A > 0$ being a certain positive constant. For $\beta \neq -1, -2, \ldots$ this estimate follows from relation (1.51) of [542]. If $\beta = -m$ ($m \in \mathbb{N}$), then by (4.5.4)

$$\binom{\beta}{j} = \frac{(-1)^j}{(m-1)!}\frac{\Gamma(m+j)}{\Gamma(j+1)} \sim \frac{(-1)^j}{(m-1)!}\frac{1}{j^{1-m}} \quad (j \to \infty)$$

in accordance with Eq. (1.66) of [542]; this yields (4.5.14) with $\beta = m = -1, -2, \ldots$ for sufficiently large $j \in \mathbb{N}$. Then (4.5.13) and (4.5.14) lead to the estimate

$$|c_j| \leq B\frac{e^{-\epsilon j}}{j^{\mathrm{Re}(\beta)-n+1}}, \quad B = Ae^{\Re e(\beta)\epsilon}, \tag{4.5.15}$$

for sufficiently large $j \in \mathbb{N}$. This implies that the series in (4.5.9) is convergent when $\epsilon > 0$ or when $\epsilon = 0$ and $\Re e(\beta) > n$. Thus assertion (a) of the theorem follows.

When $n \in \mathbb{N}_0$, $\beta = k \in \mathbb{N}$ and $\epsilon \geq 0$, $S(n, k; \epsilon)$ has the representation (4.5.10). Replacing the index of summation j by $k - j$ and taking into account the well known formula

$$\binom{k}{k-j} = \binom{k}{j} \quad (k, j \in \mathbb{N}_0,\ 0 \leq j \leq k)$$

(see, *e. g.*, p. 822 of [4]), we obtain the second formula in (4.5.10), proving (b). This completes the proof of the theorem. □

Corollary 4.9. *When $\epsilon = 0$ and either of the conditions $n \in \mathbb{N}_0$, $\beta \in \mathbb{C}$ ($\Re e(\beta) > n$, $\beta \neq n+1, n+2, \ldots$), or $n \in \mathbb{N}_0$, $\beta = k \in \mathbb{N}$, or $n \in \mathbb{N}$, $\beta = 0$ are satisfied, then there holds the representation*

$$S(n, \beta; 0) = \frac{1}{\Gamma(\beta+1)}\sum_{j=0}^{\infty}(-1)^j\binom{\beta}{j}(\beta - j)^n$$

for the generalized Stirling functions $S(n, \beta) \equiv S(n, \beta; 0)$. In particular, for the classical Stirling numbers of the second kind $S(n, k)$ there hold the *relations*

$$S(n,k) \equiv S(n,k;0) = \frac{1}{k!}\sum_{j=0}^{k}(-1)^j \binom{k}{j} (k-j)^n$$
$$= \frac{1}{k!}\sum_{j=0}^{k}(-1)^{k-j} \binom{k}{j} j^n \qquad (4.5.16)$$

with $n \in \mathbb{N}_0$, $k \in \mathbb{N}$; *and*

$$S(n,0) \equiv S(n,0;0) = 0 \quad (n \in \mathbb{N}). \qquad (4.5.17)$$

Since $\binom{k}{j} = 0$ for $0 \leq k < j$, relation (4.5.12) can be considered as a particular case of (4.5.10) for $k = 0$. But such an observation is not true for (4.5.11). The next theorem yields a further representation for the $S(n,k;\epsilon)$, different to that in (4.5.10)–(4.5.12).

Theorem 4.30. *For* $n \in \mathbb{N}_0$, $\beta = k \in \mathbb{N}_0$ *and* $\epsilon \geq 0$, *the generalized Stirling functions* $S(n,k;\epsilon)$ *have the representation*

$$S(n,k;\epsilon) = \frac{(-1)^k}{k!}\lim_{x\to 1}\left[\left(x\frac{d}{dx}\right)^n (1-xe^{\epsilon})^k\right]. \qquad (4.5.18)$$

Proof. When $n = k = 0$ and $n \in \mathbb{N}$, $k = 0$, (4.5.18) coincides with (4.5.11) and (4.5.12), respectively. Let $n \in \mathbb{N}_0$ and $k \in \mathbb{N}$. Using the binomial formula

$$(a+b)^k = \sum_{j=0}^{k}\binom{k}{j} a^j b^{k-j} \quad (a,b \in \mathbb{C};\ k \in \mathbb{N}_0) \qquad (4.5.19)$$

with $a = 1$ and $b = -xe^{\epsilon}$, and the relation

$$\left(x\frac{d}{dx}\right)^n x^{\gamma} = \gamma^n x^{\gamma} \quad (\gamma > 0)$$

with $\gamma = j \in \mathbb{N}$, we have

$$\frac{(-1)^k}{k!}\left[\left(x\frac{d}{dx}\right)^n (1-xe^{\epsilon})^k\right] = \frac{(-1)^k}{k!}\left(x\frac{d}{dx}\right)^n \left[\sum_{j=0}^{k}(-1)^j\binom{k}{j} e^{\epsilon j}x^j\right]$$
$$= \frac{(-1)^k}{k!}\sum_{j=0}^{k}(-1)^j\binom{k}{j} e^{\epsilon j} j^n x^j.$$

Taking the limit $x \to 1$ and noting (4.5.10) we obtain (4.5.18). □

Corollary 4.10. *For $n \in \mathbb{N}_0$ and $\beta = k \in \mathbb{N}_0$, the classical Stirling numbers $S(n,k)$ have the representation*

$$S(n,k) = \frac{(-1)^k}{k!} \lim_{x \to 1} \left[\left(x \frac{d}{dx} \right)^n (1-x)^k \right]. \tag{4.5.20}$$

In particular, if $1 \leq n < k$ then

$$S(n,k) = 0 \quad (n,k \in \mathbb{N};\ 1 \leq n \leq k-1). \tag{4.5.21}$$

Next we consider particular cases of the generalized Stirling functions $S(n,\beta;\epsilon)$. For $n = 0$ there holds the following assertion.

Property 4.1. *Let either of the conditions $\beta \in \mathbb{C}$ $(\beta \notin \mathbb{Z})$, $\epsilon > 0$, or $\beta \in \mathbb{C}$ $(\Re e(\beta) > 0,\ \beta \notin \mathbb{N})$, $\epsilon = 0$ hold. Then*

$$S(0,\beta;\epsilon) = \frac{(e^\epsilon - 1)^\beta}{\Gamma(\beta+1)}. \tag{4.5.22}$$

Proof. Let $\beta \in \mathbb{C}$ $(\beta \notin \mathbb{N}_0)$, $\epsilon > 0$, or $\beta \in \mathbb{C}$ $(\Re e(\beta) > 0,\ \beta \notin \mathbb{N})$, $\epsilon = 0$. Then by Theorem 4.29(a) there holds the representation (4.5.9) with $n = 0$

$$\begin{aligned} S(0,\beta;\epsilon) &= \frac{1}{\Gamma(\beta+1)} \sum_{j=0}^{\infty} (-1)^j \binom{\beta}{j} e^{(\beta-j)\epsilon} \\ &= e^{\beta\epsilon} \frac{1}{\Gamma(\beta+1)} \sum_{j=0}^{\infty} (-1)^j \binom{\beta}{j} e^{-\epsilon j}. \end{aligned} \tag{4.5.23}$$

Applying the expansion

$$(1+w)^\beta = \sum_{j=0}^{\infty} \binom{\beta}{j} w^j \quad (w \in \mathbb{C},\ |w| < 1; \quad \beta \in \mathbb{C}) \tag{4.5.24}$$

with $w = -e^{-\epsilon}$ to (4.5.23), we obtain

$$S(0,\beta;\epsilon) = e^{\beta\epsilon} \frac{1}{\Gamma(\beta+1)} (1-e^{-\epsilon})^\beta = \frac{(e^\epsilon - 1)^\beta}{\Gamma(\beta+1)},$$

which proves (4.5.22). □

Property 4.2. *For $k \in \mathbb{N}_0$ and $\epsilon > 0$, one has*

$$S(0,k;\epsilon) = \frac{(e^\epsilon - 1)^k}{k!} \quad (k \in \mathbb{N}_0;\ \epsilon \geq 0). \tag{4.5.25}$$

In particular,

$$S(0,1;\epsilon) = e^\epsilon - 1 \quad (\epsilon \geq 0). \tag{4.5.26}$$

Proof. When $k = 0$, (4.5.25) is clear in view of (4.5.11) and since $0! = \Gamma(1) = 1$. When $k \in \mathbb{N}$, then by Theorem 4.29(b),

$$S(0,k;\epsilon) = \frac{1}{k!}\sum_{j=0}^{k}(-1)^{k-j}\binom{k}{j}e^{j\epsilon} \quad (k \in \mathbb{N},\ \epsilon \geq 0);$$

the binomial formula (4.5.19) applied with $a = e^{\epsilon}$ and $b = -1$ then yields the result in (4.5.25). From (4.5.25), when $k = 1$, (4.5.26) follows. □

Setting $\epsilon = 0$ in Properties 4.1 and 4.2 we obtain the following result.

Property 4.3. *When $\beta \in \mathbb{C}$ ($\Re e(\beta) > 0$), then*

$$S(0,\beta) \equiv S(0,\beta;0) = 0. \tag{4.5.27}$$

Theorem 4.29(b) also yields the following result.

Property 4.4. *There hold for $\epsilon > 0$ the relations*

$$S(n,1;\epsilon) = e^{\epsilon} \quad (n \in \mathbb{N}), \tag{4.5.28}$$

$$S(1,2;\epsilon) = e^{\epsilon}\left(e^{\epsilon} - 1\right), \quad S(2,2;\epsilon) = e^{\epsilon}\left(2e^{\epsilon} - 1\right). \tag{4.5.29}$$

The next property follows from Theorem 4.29(a) if we take into account that the series on the right hand side of (4.5.9) is convergent and the Gamma-function $\Gamma(z)$ has poles of first order at the points $z = 0, -1. -2, \ldots$ (see Section 1.1 of [226]).

Property 4.5. *When $n \in \mathbb{N}_0$, $m \in \mathbb{N}$ and $\epsilon > 0$, then*

$$S(n,-m;\epsilon) = 0. \tag{4.5.30}$$

We note that if $\epsilon > 0$, then representation (4.5.9) holds for any complex $\beta \in \mathbb{C}$ except when $\beta \in \mathbb{Z}$. Property 4.5 shows that the generalized Stirling functions $S(n,\beta;\epsilon)$ with $\beta = -1, -2, \ldots$ and $\epsilon > 0$ are equal to zero. Now we evaluate the value of $S(n,\beta;\epsilon)$ for $\beta = -1/2$. To formulate the result we need the generalized hypergeometric function ${}_pF_q[a_1, \cdots, a_p; b_1, \cdots, b_q; z]$, defined for $p, q \in \mathbb{N}_0$ and complex $a_i \in \mathbb{C}$ $(1 \leq i \leq p)$ and $b_j \in \mathbb{C}$ $(1 \leq j \leq q)$ by (see formula 4.1(1) of [226])

$${}_pF_q[a_1, \cdots, a_p; b_1, \cdots, b_q; z] = \sum_{j=0}^{\infty} \frac{(a)_1 \ldots (a)_p}{(b)_1 \ldots (b)_q}\frac{z^j}{j!}, \tag{4.5.31}$$

where $(a)_j$ $(j \in \mathbb{N}_0)$ is the Pochhammer symbol

$$(a)_0 = 1,\ (a)_j = a(a+1)\cdots(a+j-1) \quad (j = 1, 2, \cdots). \tag{4.5.32}$$

Property 4.6. *If $n \in \mathbb{N}_0$ and $\epsilon > 0$, then*

$$S\left(n, -\frac{1}{2}; \epsilon\right) = \frac{e^{\epsilon/2}}{2^n\sqrt{\pi}}\ {}_{n+1}F_n\left[\frac{1}{2}, \cdots, \frac{1}{2}; -\frac{1}{2}, \cdots, -\frac{1}{2}; e^{-\epsilon}\right]. \tag{4.5.33}$$

Proof. By (4.5.9)

$$S\left(n, -\frac{1}{2}; \epsilon\right) = \frac{1}{\Gamma(1/2)}\sum_{j=0}^{\infty}(-1)^j \binom{-1/2}{j} e^{(1/2-j)\epsilon}(1/2-j)^n. \tag{4.5.34}$$

Observing the relation

$$\binom{\beta}{j} = \frac{(-1)^j(-\beta)_j}{j!} \quad (\beta \in \mathbb{C};\ j \in \mathbb{N}_0)$$

and $\Gamma(1/2) = \sqrt{\pi}$ (see Eq. 1.2(10) of [226]), we rewrite (4.5.34) as

$$S\left(n, -\frac{1}{2}; \epsilon\right) = \frac{(-1)^n e^{\epsilon/2}}{\sqrt{\pi}}\sum_{j=0}^{\infty}(1/2)_j\left(j - \frac{1}{2}\right)^n \frac{e^{-j\epsilon}}{j!}. \tag{4.5.35}$$

But it is clear that

$$j - \frac{1}{2} = -\frac{(1/2)_j}{2(-1/2)_j} \quad (j \in \mathbb{N}_0).$$

Substituting the latter into (4.5.35) yields (4.5.33) in accordance with (4.5.31). □

The next result deals with the asymptotic behavior of $S(n, k; \epsilon)$ when $n \to \infty$.

Property 4.7. *For $k \in \mathbb{N}$ and $\epsilon \geq 0$ there holds the asymptotic estimate*

$$\lim_{n\to\infty}[k^{-n}S(n, k; \epsilon)] = \frac{1}{k!}e^{\epsilon k}. \tag{4.5.36}$$

Proof. By (4.5.10) we have

$$k^{-n}S(n, k, \epsilon) = \frac{1}{k!}e^{\epsilon k} + \sum_{j=0}^{k-1}(-1)^{k-j}\binom{k}{j}e^{j\epsilon}\left(\frac{j}{k}\right)^n;$$

taking the limit $n \to \infty$, we obtain (4.5.36). □

Observe that for $\epsilon = 0$ the estimate (4.5.36) takes the form

$$\lim_{n\to\infty}[k^{-n}S(n, k)] = \frac{1}{k!},$$

which coincides with the known asymptotic estimate for $S(\alpha, k)$ when $\alpha = n \to +\infty$ given in Eq. (8.3) of [129].

4.5.2 *Main properties*

In this section we present the main properties of the generalized Stirling functions $S(n, \beta; \epsilon)$. First we establish the recurrence relation.

Theorem 4.31. *Let* $n \in \mathbb{N}$, $\beta \in \mathbb{C}$ ($\beta \notin \mathbb{Z}$) *and either* $\epsilon > 0$ *or* $\epsilon = 0$, $\Re e(\beta) > n$. *The generalized Stirling functions* $S(n, \beta; \epsilon)$ *satisfy the recurrence relation*

$$S(n, \beta; \epsilon) = \beta S(n-1, \beta; \epsilon) + S(n-1, \beta-1; \epsilon). \tag{4.5.37}$$

Proof. In accordance with Theorem 4.29(a), all terms on the both sides of (4.5.37) are defined for the given range of parameters n, β and ϵ. Noting (4.5.9),

$$\begin{aligned}
&\beta S(n-1, \beta; \epsilon) + S(n-1, \beta-1; \epsilon)\\
&= \frac{\beta}{\Gamma(\beta+1)} \sum_{j=0}^{\infty} (-1)^j \binom{\beta}{j} e^{(\beta-j)\epsilon} (\beta-j)^{n-1}\\
&\quad + \frac{1}{\Gamma(\beta)} \sum_{j=0}^{\infty} (-1)^j \binom{\beta-1}{j} e^{(\beta-1-j)\epsilon} (\beta-1-j)^{n-1}\\
&= \frac{\beta}{\Gamma(\beta+1)} e^{\beta\epsilon} \beta^{n-1} + \frac{\beta}{\Gamma(\beta+1)} \sum_{j=1}^{\infty} (-1)^j \binom{\beta}{j} e^{(\beta-j)\epsilon} (\beta-j)^{n-1}\\
&\quad + \frac{\beta}{\Gamma(\beta+1)} \sum_{j=1}^{\infty} (-1)^{j+1} \binom{\beta-1}{j-1} e^{(\beta-j)\epsilon} (\beta-j)^{n-1}\\
&= \frac{\beta}{\Gamma(\beta+1)} e^{\beta\epsilon} \beta^{n-1}\\
&\quad + \frac{\beta}{\Gamma(\beta+1)} \sum_{j=1}^{\infty} (-1)^j \left[\binom{\beta}{j} - \binom{\beta-1}{j-1}\right] e^{(\beta-j)\epsilon} (\beta-j)^{n-1}.
\end{aligned}$$

Applying the binomial relation

$$\binom{\beta}{j} - \binom{\beta-1}{j-1} = \binom{\beta-1}{j} \qquad (\beta \in \mathbb{C};\ j \in \mathbb{N})$$

we can rewrite the last relation in the form

$$\begin{aligned}
&\beta S(n-1, \beta; \epsilon) + S(n-1, \beta-1; \epsilon) \qquad (4.5.38)\\
&= \frac{\beta^n}{\Gamma(\beta+1)} e^{\beta\epsilon} + \sum_{j=1}^{\infty} (-1)^j \left[\beta \binom{\beta-1}{j} \frac{1}{\beta-j}\right] e^{(\beta-j)\epsilon} (\beta-j)^n.
\end{aligned}$$

It is directly verified that

$$\beta \binom{\beta-1}{j} \frac{1}{\beta-j} = \binom{\beta}{j},$$

so that (4.5.38) takes the form

$$\begin{aligned} &\beta S(n-1,\beta;\epsilon) + S(n-1,\beta-1;\epsilon) \\ &= \frac{\beta^n}{\Gamma(\beta+1)} e^{\beta\epsilon} + \sum_{j=1}^{\infty} (-1)^j \binom{\beta}{j} e^{(\beta-j)\epsilon} (\beta-j)^n, \end{aligned}$$

which yields (4.5.37) if we take (4.5.9) into account. This completes the proof of the theorem. □

Corollary 4.11. *Let $n \in \mathbb{N}$ and $\beta \in \mathbb{C}$ ($\beta \notin \mathbb{Z}$) be such that* $\mathrm{Re}(\beta) > n$. *Then the generalized Stirling functions $S(n,\beta) \equiv S(n,\beta;0)$ satisfy the recurrence relation*

$$S(n,\beta) = \beta S(n-1,\beta) + S(n-1,\beta-1). \tag{4.5.39}$$

Theorem 4.31 is also true for $\beta = k \in \mathbb{N}$.

Corollary 4.12. *Let $n \in \mathbb{N}$, $k \in \mathbb{N}$ and $\epsilon \geq 0$. The generalized Stirling functions $S(n,k;\epsilon)$ satisfy the recurrence relation*

$$S(n,k;\epsilon) = kS(n-1,k;\epsilon) + S(n-1,k-1;\epsilon). \tag{4.5.40}$$

Proof. When $k > 1$, relation (4.5.40) is proved on the basis of the representation (4.5.10) similarly to the proof of formula (4.5.37) in Theorem 4.29. When $k = 1$, (4.5.40) takes the form

$$S(n,1;\epsilon) = S(n-1,1;\epsilon) + S(n-1,0;\epsilon). \tag{4.5.41}$$

If $n = 1$, then according to (4.5.28), (4.5.26) and (4.5.11),

$$S(1,1;\epsilon) = e^{\epsilon}, \quad S(0,1;\epsilon) = e^{\epsilon} - 1, \quad S(0,0;\epsilon) = 1$$

and so (4.5.41) holds for $n = 1$. When $n > 1$, (4.5.41) is also valid since in accordance with (4.5.28) and (4.5.12),

$$S(n,1;\epsilon) = S(n-1,1;\epsilon) = e^{\epsilon}, \; S(n-1,0;\epsilon) = 0.$$

Thus the proof is complete. □

Corollary 4.13. *For $n, k \in \mathbb{N}$, the classical Stirling numbers $S(n,k)$ satisfy the recurrence relation*

$$S(n,k) = kS(n-1,k) + S(n-1,k-1). \tag{4.5.42}$$

Now we construct the exponential generating function for $S(n,\beta;\epsilon)$.

Theorem 4.32. *Let $z \in \mathbb{C}$, $\beta \in \mathbb{C}$ and $\epsilon \geq 0$. The generating function for the generalized Stirling functions $S(n,\beta;\epsilon)$ is given by*

$$\left(e^{z+\epsilon} - 1\right)^{\beta} = \Gamma(\beta+1) \sum_{n=0}^{\infty} S(n,\beta;\epsilon) \frac{z^n}{n!} \tag{4.5.43}$$

for $\beta \notin \mathbb{Z}$ and $\epsilon > 0$, and by

$$\left(e^{z+\epsilon} - 1\right)^{k} = k! \sum_{n=0}^{\infty} S(n,k;\epsilon) \frac{z^n}{n!}. \tag{4.5.44}$$

for $\beta = k \in \mathbb{N}_0$ and $\epsilon \geq 0$.

Proof. If $\beta \in \mathbb{C}$ ($\beta \notin \mathbb{Z}$) and $\epsilon > 0$, the generalized Stirling functions $S(n,\beta;\epsilon)$ are given for $n \in \mathbb{N}_0$ by the representation (4.5.9) noting Theorem 4.29(a). Substitution of this representation into the right-hand side of (4.5.43) yields

$$\Gamma(\beta+1) \sum_{n=0}^{\infty} S(n,\beta;\epsilon) \frac{z^n}{n!} = \sum_{n=0}^{\infty} \left[\sum_{j=0}^{\infty} (-1)^j \binom{\beta}{j} e^{(\beta-j)\epsilon} (\beta-j)^n \right] \frac{z^n}{n!}. \tag{4.5.45}$$

On the other hand, using (4.5.24) with $w = -e^{-(z+\epsilon)}$ and the known expansion for the exponential function, we have

$$\begin{aligned}
\left(e^{z+\epsilon} - 1\right)^{\beta} &= e^{\beta(z+\epsilon)} \left[1 - e^{-(z+\epsilon)}\right]^{\beta} \\
&= e^{\beta(z+\epsilon)} \sum_{j=0}^{\infty} (-1)^j \binom{\beta}{j} e^{-j(z+\epsilon)} \\
&= \sum_{j=0}^{\infty} (-1)^j \binom{\beta}{j} e^{(\beta-j)\epsilon} e^{z(\beta-j)} \\
&= \sum_{j=0}^{\infty} (-1)^j \binom{\beta}{j} e^{(\beta-j)\epsilon} \left[\sum_{n=0}^{\infty} \frac{[z(\beta-j)]^n}{n!} \right].
\end{aligned} \tag{4.5.46}$$

But the series

$$e^{\beta(z+\epsilon)}\sum_{j=0}^{\infty}(-1)^j\binom{\beta}{j}e^{-j(z+\epsilon)}$$

is absolutely convergent for any fixed $z \in \mathbb{C}$ because, according to (4.5.14), it is majorized by the convergent series

$$1+Ae^{\beta[\mathfrak{Re}(z)+\epsilon]}\sum_{j=1}^{\infty}\frac{1}{j^{\mathfrak{Re}(\beta)+1}}e^{-j[\mathrm{Re}(z)+\epsilon]}.$$

Hence the double series

$$\sum_{j=0}^{\infty}(-1)^j\binom{\beta}{j}e^{(\beta-j)\epsilon}\left[\sum_{n=0}^{\infty}\frac{[z(\beta-j)]^n}{n!}\right]$$

is also absolutely convergent, and, by an analogue of Fubini's theorem (see, *e. g.*, Theorem 12-42 of [37]), we can interchange the order of summation in this double series. Performing such an interchange and taking (4.5.46) and (4.5.45) into account, we deduce (4.5.43).

If $\beta = k \in \mathbb{N}_0$ and $\epsilon \geq 0$, then $S(n,k,\epsilon) = O(k^n)$ $(n \to \infty)$ by (4.5.36), and hence in accordance with the Cauchy-Hadamard relation the series on the right hand side of (4.5.44) has radius of convergence $R = +\infty$, noting $\sqrt[n]{[|k|^n/n!]} \to 0$ $(n \to \infty)$. Thus relation (4.5.44) is proved similarly to (4.5.43) by applying (4.5.10) and (4.5.19) and using Theorem 12-42 in [37]:

$$\begin{aligned}\left(e^{z+\epsilon}-1\right)^k &= \sum_{j=0}^{k}(-1)^j\binom{k}{j}e^{j(z+\epsilon)} = \sum_{j=0}^{k}(-1)^j\binom{k}{j}e^{j\epsilon}\left[\sum_{n=0}^{\infty}\frac{(zj)^n}{n!}\right]\\ &= \sum_{n=0}^{\infty}\left[\sum_{j=0}^{k}(-1)^j\binom{k}{j}e^{j\epsilon}j^n\right]\frac{z^n}{n!} = k!\sum_{n=0}^{\infty}S(n,k;\epsilon)\frac{z^n}{n!}.\end{aligned}$$

□

Corollary 4.14. *For $z \in \mathbb{C}$ and $k \in \mathbb{N}_0$,*

$$(e^z-1)^k = k!\sum_{n=k}^{\infty}S(n,k)\frac{z^n}{n!}. \tag{4.5.47}$$

Proof. Let $\epsilon = 0$. Then from (4.5.44) and taking (4.5.21) into account, we obtain (4.5.47). □

From Theorem 4.32 we deduce a further property of the $S(n,\beta;\epsilon)$.

Property 4.8. *Let $n \in \mathbb{N}_0$, $\beta \in \mathbb{C}$ and $\epsilon \geq 0$.*

(a) *For $\beta \notin \mathbb{Z}$ and $\epsilon > 0$, the generalized Stirling functions $S(n, \beta; \epsilon)$ can be represented in the form (4.5.9) if and only if they are given by*

$$S(n, \beta; \epsilon) = \frac{1}{\Gamma(\beta+1)} \lim_{x \to +0} \left[\left(\frac{d}{dx} \right)^n (e^{x+\epsilon} - 1)^\beta \right]. \tag{4.5.48}$$

(b) *For $\beta = k \in \mathbb{N}$ and $\epsilon \geq 0$, the $S(n, k; \epsilon)$ are representable by (4.5.10) if and only if*

$$S(n, k; \epsilon) = \frac{1}{k!} \lim_{x \to 0} \left[\left(\frac{d}{dx} \right)^n (e^{x+\epsilon} - 1)^k \right].$$

In particular, $S(n, k)$ are representable by (4.5.16) if and only if

$$S(n, k) = \frac{1}{k!} \lim_{x \to 0} \left[\left(\frac{d}{dx} \right)^n (e^x - 1)^k \right] \quad (n \in \mathbb{N}_0;\ k \in \mathbb{N}). \tag{4.5.49}$$

Proof. Let $n \in \mathbb{N}_0$, $\beta \notin \mathbb{Z}$ and $\epsilon > 0$ and let $S(n, \beta; \epsilon)$ be representable by (4.5.9). Then by Theorem 4.14 there holds the representation (4.5.43) for $z = x$ near zero, and hence by Taylor's formula we obtain (4.5.48). Conversely, let (4.5.48) hold. Substituting (4.5.24) with $w = e^{-(x+\epsilon)}$ into (4.5.48) we have

$$\begin{aligned} S(n, \beta; \epsilon) &= \frac{1}{\Gamma(\beta+1)} \lim_{x \to +0} \left[\left(\frac{d}{dx} \right)^n \sum_{j=0}^{\infty} (-1)^j \binom{\beta}{j} e^{(x+\epsilon)(\beta-j)} \right] \\ &= \frac{1}{\Gamma(\beta+1)} \lim_{x \to +0} \left[\sum_{j=0}^{\infty} (-1)^j \binom{\beta}{j} e^{(x+\epsilon)(\beta-j)} (\beta-j)^n \right] \\ &= \frac{1}{\Gamma(\beta+1)} \sum_{j=0}^{\infty} (-1)^j \binom{\beta}{j} e^{\epsilon(\beta-j)} (\beta-j)^n, \end{aligned}$$

which yields (4.5.9). The assertion (b) is proved similarly. □

Remark 4.4. The results presented in this section are true for the generalized Stirling functions $S(n, \beta; \epsilon)$ with $\beta \in \mathbb{C}$ ($\beta \notin \mathbb{Z}$) and for $S(n, k; \epsilon)$ with $k \in \mathbb{N}_0$, and hold for any $\epsilon > 0$ or any $\epsilon \geq 0$. In particular, we can take $\epsilon = \{\beta\}$, where $\{\beta\}$ is the fractional part of β, *i. e.*,

$$\{\beta\} = \beta - [\beta],$$

$[\beta]$ being the integral part of β. For such ϵ we can introduce the generalized Stirling functions

$$S^*(n, \beta) \equiv S(n, \beta; \{\beta\}). \tag{4.5.50}$$

All results in Sections 4.5.1 and 4.5.2 including Theorems 4.29–4.32 can be reformulated for such generalized Stirling functions.

Remark 4.5. In Sections 4.5.1 and 4.5.2 we have considered properties $S(n,\beta)$ with $n \in \mathbb{N}_0$ and $\beta \in \mathbb{C}$ ($\Re e(\beta) > n$) or $\beta = k \in \mathbb{N}_0$. Relation (4.5.42) can be used to define $S(n,\beta)$ for negative integers $\beta \in \mathbb{Z}_- = \{-1,-2,\ldots\}$. We replace n by $n+1$ in (4.5.42) and rewrite it in the form

$$S(n,k-1) = S(n+1,k) - kS(n,k) \quad (n \in \mathbb{N};\ k \in \mathbb{N}). \tag{4.5.51}$$

The right hand side of (4.5.51) is defined for $k = 0$ and therefore we can define $S(n,-1)$, which in accordance with (4.5.17) yields

$$S(n,-1) = 0 \quad (n \in \mathbb{N}_0). \tag{4.5.52}$$

Now by (4.5.52) the right hand side of (4.5.51) is defined for $k = -1$, and hence we can define $S(n,-2)$ by (4.5.51) and obtain

$$S(n,-2) = 0 \quad (n \in \mathbb{N}_0).$$

Continuation of this process yields zero values for $S(n,\beta)$ when $\beta = -1,-2,\ldots$, *i. e.*

$$S(n,-m) = 0 \quad (n \in \mathbb{N}_0;\ m \in \mathbb{N}). \tag{4.5.53}$$

This result could be formally obtained directly from definition (4.5.7) if we take into account that the Gamma function $\Gamma(z)$ has poles of first order at points $z = 0,-1,-2,\ldots$. In this connection see also Property 4.5.

Remark 4.6. If $\epsilon > 0$, then relation(4.5.37) remains true for $n \in \mathbb{N}$ and $\beta = -m$ ($m \in \mathbb{N}$) by (4.5.30), because the left and right hand sides of (4.5.37) for such values of n, β and ϵ are equal to zero. Similarly, when $\epsilon = 0$, then in accordance with (4.5.17) and (4.5.53) the recurrence relation (4.5.39) remains true for $n \in \mathbb{N}$ and $\beta = -m$ ($m \in \mathbb{N}$).

4.6 Generalized Stirling Functions of the Second Kind $S(\alpha,\beta)$ with Complex Arguments

Observing Definition 4.3, one can extend it to the instance when the first parameter $n \in \mathbb{N}_0$ is replaced by any $\alpha \in \mathbb{C}$. Indeed,

Definition 4.4. The generalized Stirling functions of the second kind, $S(\alpha, \beta; \epsilon)$, for $\alpha \in \mathbb{C}$, $\beta \in \mathbb{C}$ ($\beta \notin \mathbb{N}_0$) and $\alpha \in \mathbb{C}$ ($\Re e(\alpha) > 0$), $\beta = k \in \mathbb{N}$ are given by

$$S(\alpha, \beta; \epsilon) = \frac{1}{\Gamma(\beta+1)} \lim_{x \to 0} \Delta^{\beta,\epsilon}(x^\alpha) \quad (\epsilon \geq 0). \tag{4.6.1}$$

When $\alpha \in \mathbb{C}$ ($\Re e(\alpha) \leq 0;\ \alpha \neq 0$) and $\beta = k \in \mathbb{N}_0$ then

$$S(\alpha, k; \epsilon) = \frac{1}{k!} \lim_{x \to 0} \Delta_*^{k,\epsilon}(x^\alpha) \quad (\epsilon \geq 0) \tag{4.6.2}$$

with the "cut" finite difference defined (the term $j = 0$ is missing) by

$$\Delta_*^{k,\epsilon} f(x) = \sum_{j=1}^{k} (-1)^{k-j} \binom{k}{j} e^{j\epsilon} f(x+j) \quad (x \in \mathbb{R};\ k \in \mathbb{N};\ \epsilon \geq 0), \tag{4.6.3}$$

$$\Delta_*^{0,\epsilon} f(x) \equiv 0 \quad (x \in \mathbb{R};\ \epsilon \geq 0). \tag{4.6.4}$$

4.6.1 *Basic properties*

In this section we present properties of the new generalized Stirling functions $S(\alpha, \beta; \epsilon)$ with complex $\alpha, \beta \in \mathbb{C}$ and $\epsilon \geq 0$.

When $\alpha \in \mathbb{C}$ ($\Re e(\alpha) > 0$), $\beta = k \in \mathbb{N}$ (4.6.1) takes the form

$$S(\alpha, k; \epsilon) = \frac{1}{k!} \lim_{x \to 0} \Delta^{k,\epsilon}(x^\alpha), (\epsilon \geq 0), \tag{4.6.5}$$

where $\Delta^{k,\epsilon}$ is given by (4.5.5). Explicit representations of $S(\alpha, \beta; \epsilon)$ are now given by the following result

Theorem 4.33. *The following three assertions hold:*

(a) *If $\alpha \in \mathbb{C}$ and either of the conditions $\beta \in \mathbb{C}$ ($\beta \notin \mathbb{Z}$), $\epsilon > 0$, or $\beta \in \mathbb{C}$ ($\beta \notin \mathbb{Z}$, $\mathrm{Re}(\beta) > \Re e(\alpha)$), $\epsilon = 0$ hold, then the generalized Stirling functions $S(\alpha, \beta; \epsilon)$ can be represented in the form*

$$S(\alpha, \beta; \epsilon) = \frac{1}{\Gamma(\beta+1)} \sum_{j=0}^{\infty} (-1)^j \binom{\beta}{j} e^{(\beta-j)\epsilon} (\beta - j)^\alpha. \tag{4.6.6}$$

(b) *If $\alpha \in \mathbb{C}$ ($\alpha \neq 0$), $\beta = k \in \mathbb{N}$ and $\epsilon \geq 0$, then functions $S(\alpha, k; \epsilon)$ have the representation*

$$S(\alpha, k; \epsilon) = \frac{1}{k!} \sum_{j=1}^{k} (-1)^{k-j} \binom{k}{j} e^{j\epsilon} j^\alpha. \tag{4.6.7}$$

(c) *If* $\alpha \in \mathbb{C}$ $(\alpha \neq 0)$ *and* $\epsilon \geq 0$, *then*

$$S(\alpha, 0; \epsilon) = 0. \tag{4.6.8}$$

Proof. The proof of (4.6.6) is the same as in Theorem 4.29, replacing n by α and using the estimate of the form (4.5.15)

$$|c_j| \leq B \frac{e^{-\epsilon j}}{j^{\Re e(\beta-\alpha)+1}}, \quad B = Ae^{\beta},$$

according to which the series on the right-hand side of (4.6.6) is convergent when either $\epsilon > 0$ or $\epsilon = 0$, $\Re e(\beta) > \Re e(\alpha)$.

Formula (4.6.7) is proved by using (4.6.5), (4.5.5) and (4.6.2),(4.6.3) in the cases $\Re e(\alpha) > 0$ and $\Re e(\alpha) \leq 0$ $(\alpha \neq 0)$, respectively.

When $\Re e(\alpha) > 0$, (4.6.8) follows from (4.6.5) and (4.5.6), while (4.6.2) with $k = 0$ and (4.6.4) yield (4.6.8) for $\Re e(\alpha) \leq 0$ $(\alpha \neq 0)$. Thus the theorem is proved. □

Corollary 4.15. *In particular, there hold the following assertions for the function* $S(\alpha, \beta) \equiv S(\alpha, \beta, 0)$:

(a) *Let* $\alpha \in \mathbb{C}$ *and* $\beta \in \mathbb{C}$ *be such that* $\beta \notin \mathbb{Z}$ *and* $\mathrm{Re}(\beta) > \Re e(\alpha)$. *The generalized Stirling functions* $S(\alpha, \beta)$ *have the representation*

$$S(\alpha, \beta) = \frac{1}{\Gamma(\beta+1)} \sum_{j=0}^{\infty} (-1)^j \binom{\beta}{j} (\beta - j)^{\alpha}. \tag{4.6.9}$$

(b) *For* $\alpha \in \mathbb{C}$ $(\alpha \neq 0)$ *and* $k \in \mathbb{N}$,

$$S(\alpha, k) = \frac{1}{k!} \sum_{j=1}^{k} (-1)^{k-j} \binom{k}{j} j^{\alpha}. \tag{4.6.10}$$

(c) *For* $\alpha \in \mathbb{C}$ $(\alpha \neq 0)$,

$$S(\alpha, 0) = 0. \tag{4.6.11}$$

The main properties obtained in Section 4.4.1 for $S(\alpha, k)$ can be extended to the generalized Stirling functions $S(\alpha, k; \epsilon)$ with any nonnegative $k \in \mathbb{N}_0$ and $\epsilon \geq 0$.

It should be noted that (4.6.6) gives the explicit representation for the generalized Stirling functions $S(\alpha, \beta; \epsilon)$ with any complex $\alpha \in \mathbb{C}$ and $\beta \in \mathbb{C}$ except for the case $\beta \in \mathbb{Z}$. In particular, α and β can be purely complex numbers. With $\alpha = n + i\phi$ ($n \in \mathbb{Z},\ \phi \in \mathbb{R},\ \phi \neq 0$) and $\beta = k + i\theta$ ($k \in \mathbb{Z},\ \theta \in \mathbb{R},\ \theta \neq 0$), Theorem 4.33 yields the following properties.

Property 4.9. *Let $n, k \in \mathbb{Z}$ and $\phi, \theta \in \mathbb{R}$ ($\phi \neq 0,\ \theta \neq 0$).*

(a) *For $\epsilon > 0$, the generalized Stirling functions $S(n + i\phi, k + i\theta; \epsilon)$ have the representation*

$$S(n + i\phi, k + i\theta; \epsilon) = \frac{1}{\Gamma(k + 1 + i\theta)} \sum_{j=0}^{\infty} (-1)^j \binom{k + i\theta}{j} e^{(k-j+i\theta)\epsilon} (k - j + i\theta)^{n+i\phi}. \tag{4.6.12}$$

(b) *For $n < k$, the generalized Stirling functions $S(n + i\phi, k + i\theta)$ are given by*

$$S(n + i\phi, k + i\theta) = \frac{1}{\Gamma(k + 1 + i\theta)} \sum_{j=0}^{\infty} (-1)^j \binom{k + i\theta}{j} (k - j + i\theta)^{n+i\phi}. \tag{4.6.13}$$

Property 4.10. *Let $n \in \mathbb{Z}$ and $\phi \in \mathbb{R}$ ($\phi \neq 0$). There hold the following three results.*

(a) *For $\beta \in \mathbb{C}$ and $\epsilon > 0$, the generalized Stirling functions $S(n + i\phi, \beta; \epsilon)$ have the representation*

$$S(n + i\phi, \beta; \epsilon) = \frac{1}{\Gamma(\beta + 1)} \sum_{j=0}^{\infty} (-1)^j \binom{\beta}{j} e^{(\beta-j)\epsilon} (\beta - j)^{n+i\phi}. \tag{4.6.14}$$

In particular, when $n = 0$

$$S(i\phi, \beta; \epsilon) = \frac{1}{\Gamma(\beta + 1)} \sum_{j=0}^{\infty} (-1)^j \binom{\beta}{j} e^{(\beta-j)\epsilon} (\beta - j)^{i\phi}. \tag{4.6.15}$$

(b) *Let $\beta \in \mathbb{C}$ be such that $\mathfrak{Re}(\beta) > n$. The generalized Stirling functions $S(n + i\phi, \beta)$ are given by*

$$S(n + i\phi, \beta) = \frac{1}{\Gamma(\beta + 1)} \sum_{j=0}^{\infty} (-1)^j \binom{\beta}{j} (\beta - j)^{n+i\phi}. \tag{4.6.16}$$

In particular, when $n = 0$, with $\mathfrak{Re}(\beta) > 0$,

$$S(i\phi, \beta) = \frac{1}{\Gamma(\beta + 1)} \sum_{j=0}^{\infty} (-1)^j \binom{\beta}{j} (\beta - j)^{i\phi}. \tag{4.6.17}$$

(c) *For $k \in \mathbb{N}$, the generalized Stirling functions $S(n+i\phi, k)$ have the representation*

$$S(n+i\phi, k) = \frac{1}{k!}\sum_{j=1}^{k}(-1)^{k-j}\binom{k}{j} j^{n+i\phi}. \tag{4.6.18}$$

In particular, when $n = 0$,

$$S(i\phi, k) = \frac{1}{k!}\sum_{j=0}^{k}(-1)^{k-j}\binom{k}{j} j^{i\phi}. \tag{4.6.19}$$

Property 4.11. *Let $k \in \mathbb{Z}$ and $\theta \in \mathbb{R}$ $(\theta \neq 0)$.*

(a) *For $\alpha \in \mathbb{C}$ and $\epsilon > 0$, the generalized Stirling functions $S(\alpha, k+i\theta; \epsilon)$ are given by*

$$S(\alpha, k+i\theta; \epsilon) = \frac{1}{\Gamma(i\theta+1)}\sum_{j=0}^{\infty}(-1)^{j}\binom{k+i\theta}{j} e^{(k-j+i\theta)\epsilon}(k-j+i\theta)^{\alpha}. \tag{4.6.20}$$

In particular, when $k = 0$,

$$S(\alpha, i\theta; \epsilon) = \frac{1}{\Gamma(i\theta+1)}\sum_{j=0}^{\infty}(-1)^{j}\binom{i\theta}{j} e^{(i\theta-j)\epsilon}(i\theta-j)^{\alpha}. \tag{4.6.21}$$

(b) *Let $\alpha \in \mathbb{C}$ be such that $\Re e(\alpha) < k$. Then the generalized Stirling functions $S(\alpha, k+i\theta)$ are given by*

$$S(\alpha, k+i\theta) = \frac{1}{\Gamma(i\theta+1)}\sum_{j=0}^{\infty}(-1)^{j}\binom{k+i\theta}{j}(k-j+i\theta)^{\alpha}. \tag{4.6.22}$$

In particular, when $k = 0$, with $\Re e(\alpha) < 0$,

$$S(\alpha, i\theta) = \frac{1}{\Gamma(i\theta+1)}\sum_{j=0}^{\infty}(-1)^{j}\binom{i\theta}{j}(i\theta-j)^{\alpha}. \tag{4.6.23}$$

According to Property 4.3, formulas (4.6.20)–(4.6.21) take on a simple form in the case $\alpha = 0$.

Property 4.12. *Let $k \in \mathbb{Z}$ and $\theta \in \mathbb{R}$ $(\theta \neq 0)$. For $\epsilon > 0$ there holds the relation*

$$S(0, k+i\theta; \epsilon) = \frac{(e^{\epsilon}-1)^{k+i\theta}}{\Gamma(k+1+i\theta)}. \tag{4.6.24}$$

The next formulas follow from Theorem 4.33(b).

Property 4.13. *If $\alpha \in \mathbb{C}$ $(\alpha \neq 0)$ and $\epsilon \geq 0$, then*

$$S(\alpha, 1; \epsilon) = e^{\epsilon}; \quad S(\alpha, 2; \epsilon) = e^{\epsilon}(2^{\alpha-1}e^{\epsilon} - 1), \tag{4.6.25}$$

$$S(\alpha, 3; \epsilon) = \frac{1}{6}\left[3e^{\epsilon} - 3e^{2\epsilon}2^{\alpha} + e^{3\epsilon}3^{\alpha}\right].$$

In particular,

$$S(\alpha, 1) = 1, \ S(\alpha, 2) = 2^{\alpha-1} - 1; \quad S(\alpha, 3) = \frac{1}{2}\left[1 - 2^{\alpha} + 3^{\alpha-1}\right].$$

The next property follows from Theorem 4.33 if we take into account that the series on the right hand side of (4.6.6) is convergent and the Gamma-function $\Gamma(z)$ has poles of the first order at $z = 0, -1. -2, \ldots$:

Property 4.14. *When $\alpha \in \mathbb{C}$, $\beta = -1, -2, \cdots$ and $\epsilon > 0$,*

$$S(\alpha, -m; \epsilon) = 0 \quad (m \in \mathbb{N}).$$

The result to follow presents recurrence relations for the generalized Stirling functions $S(\alpha, \beta; \epsilon)$.

Theorem 4.34. *There hold the following three results.*

(a) *For $\alpha \in \mathbb{C}$, $\beta \in \mathbb{C}$ $(\beta \notin \mathbb{Z})$ and $\epsilon > 0$, the generalized Stirling functions $S(\alpha, \beta; \epsilon)$ satisfy the recurrence relation*

$$S(\alpha, \beta; \epsilon) = \beta S(\alpha - 1, \beta; \epsilon) + S(\alpha - 1, \beta - 1; \epsilon). \tag{4.6.26}$$

(b) *Let $\alpha \in \mathbb{C}$, $\beta \in \mathbb{C}$ be such that $\beta \notin \mathbb{Z}$ and $\Re e(\beta) > \Re e(\alpha)$. The generalized Stirling functions $S(\alpha, \beta) \equiv S(\alpha, \beta; 0)$ satisfy the recurrence relation*

$$S(\alpha, \beta) = \beta S(\alpha - 1, \beta) + S(\alpha - 1, \beta - 1). \tag{4.6.27}$$

(c) *For $\alpha \in \mathbb{C}$, $k \in \mathbb{N}$ and $\epsilon \geq 0$, the generalized Stirling functions $S(\alpha, k; \epsilon)$ satisfy the recurrence relation*

$$S(\alpha, k; \epsilon) = kS(\alpha - 1, k; \epsilon) + S(\alpha - 1, k - 1; \epsilon). \tag{4.6.28}$$

In particular,

$$S(\alpha, k) = kS(\alpha - 1, k) + S(\alpha - 1, k - 1). \tag{4.6.29}$$

Proof. When $\alpha \in \mathbb{C}$, $\beta \in \mathbb{C}$ $(\beta \notin \mathbb{Z})$, $\epsilon > 0$ and $\alpha \in \mathbb{C}$, $\beta \in \mathbb{C}$ $(\beta \notin \mathbb{Z}$, $\mathfrak{Re}(\beta) > \mathfrak{Re}(\alpha))$, $\epsilon = 0$, then the left and right hand sides of (4.6.26) and (4.6.27) are defined in accordance with Theorem 4.33(a). These relations are proved similarly to the proof of formula (4.5.37) in Theorem 4.31 with $n \in \mathbb{N}$ being replaced by $\alpha \in \mathbb{C}$.

When $\alpha \in \mathbb{C}$, $k \in \mathbb{N}_2 = \{2, 3, \ldots\}$ and $\epsilon \geq 0$, formula (4.6.28) is proved similarly to the proof of this relation with $\epsilon = 0$ in 1.3.137. When $k = 1$, (4.6.28) takes the form

$$S(\alpha, 1; \epsilon) = S(\alpha - 1, 1; \epsilon) + S(\alpha - 1, 0; \epsilon), \tag{4.6.30}$$

When $\alpha \neq 0, 1$, (2.4.33) is true since $S(\alpha, 1; \epsilon) = S(\alpha - 1, 1; \epsilon) = e^{\epsilon}$ by (4.6.25), while $S(\alpha - 1, 0; \epsilon) = 0$ in accordance with (4.6.8). When $\alpha = 0$ and $\alpha = 1$, (4.6.30) has the form

$$S(0, 1; \epsilon) = S(-1, 1; \epsilon) + S(-1, 0; \epsilon) \tag{4.6.31}$$

and

$$S(1, 1; \epsilon) = S(0, 1; \epsilon) + S(0, 0; \epsilon), \tag{4.6.32}$$

respectively. Equation (4.6.31) is valid since $S(0, 1; \epsilon) = S(-1, 1; \epsilon) = e^{\epsilon}$ by (4.6.25) and $S(-1, 0; \epsilon) = 0$ in accordance with (4.6.8). (4.6.32) is also valid because $S(1, 1; \epsilon) = e^{\epsilon}$, $S(0, 1; \epsilon) = e^{\epsilon} - 1$ and $S(0, 0; \epsilon) = 1$ according to (4.6.25), (4.5.26) and (4.5.11), respectively. Equation (4.6.28) with $\epsilon = 0$ yields (4.6.29). This completes the proof of the theorem. □

Remark 4.7. In this section we have considered properties of the generalized Stirling functions $S(\alpha, \beta)$ with $\alpha \in \mathbb{C}$ and $\beta \in \mathbb{C}$ $(\beta \notin \mathbb{Z}$, $\mathfrak{Re}(\beta) > \mathfrak{Re}(\alpha))$ and $\beta = k \in \mathbb{N}_0$. Relation (4.6.29) can be used for the definition of $S(\alpha, \beta)$ for $\beta \in \mathbb{Z}_- = \{-1, -2, \ldots\}$. We replace α by $\alpha + 1$ in (4.6.29) and rewrite it in the form

$$S(\alpha, k - 1) = S(\alpha + 1, k) - kS(\alpha, k) \quad (\alpha \in \mathbb{C};\ k \in \mathbb{N}). \tag{4.6.33}$$

The right-hand side of (4.6.33) is defined for $k = 0$ and therefore we can define $S(\alpha, -1)$ by (4.6.33) with $k = 0$ by

$$S(\alpha, -1) = S(\alpha + 1, 0), \tag{4.6.34}$$

and so

$$S(\alpha, -1) = 0 \quad (\alpha \neq -1),\ S(-1, -1) = 1. \tag{4.6.35}$$

By (4.6.35) the right hand side of (4.6.33) is defined for $k = -1$, and so we can define $S(\alpha, -2)$ by (4.6.33) with $k = -1$ by

$$S(\alpha, -2) = S(\alpha + 1, -1) + S(\alpha, -1),$$

and obtain

$$S(\alpha, -2) = 0 \quad (\alpha \neq -2, -1), \;\; S(-2, -2) = S(-1, -2) = 1.$$

This process can be continued to obtain the values of the generalized Stirling functions $S(\alpha, \beta)$ for $\beta = -m$ $(m \in \mathbb{N})$ by the recurrence relation

$$S(\alpha, -m) = S(\alpha + 1, 1 - m) + (m - 1)S(\alpha, 1 - m) \quad (\alpha \in \mathbb{C};\; m \in \mathbb{N}).$$

In this way we obtain the relations

$$S(\alpha, -m) = 0 \quad (\alpha \in \mathbb{C};\; \alpha \neq -m, -(m-1), \cdots, -1;\; m \in \mathbb{N}) \qquad (4.6.36)$$

and

$$S(\alpha, -m) \neq 0 \quad (\alpha \in \mathbb{C};\; \alpha = -m, -(m-1), \cdots, -1;\; m \in \mathbb{N}). \qquad (4.6.37)$$

In particular, the induction yields the formula

$$S(-m, -m) = 1 \quad (m \in \mathbb{N}),$$

and (4.5.53) follows from (4.6.36) when $\alpha = n \in \mathbb{N}$.

4.6.2 *Representations by Liouville fractional operators*

Representations of the generalized Stirling functions $S(\alpha, k)$ in terms of the Liouville fractional differentiation operators ${}^{\mathrm{L}}D_{+}^{\alpha}$, ${}^{\mathrm{L}}D_{-}^{\alpha}$ and fractional integration operators ${}^{\mathrm{L}}I_{+}^{-\alpha}$, $I_{-}^{-\alpha}$ in the cases $\mathrm{Re}(\alpha) > 0$ and $\Re e(\alpha) < 0$, respectively, were established in Sections 4.4.1 and 4.4.2. In this section we give representations of the generalized Stirling functions $S(\alpha, \beta; \epsilon)$ in terms of such Liouville fractional operators.

First we consider the case $\Re e(\alpha) > 0$ and establish representations of $S(\alpha, \beta; \epsilon)$ in terms of the Liouville fractional differentiation operators ${}^{\mathrm{L}}D_{+}^{\alpha}$ and ${}^{\mathrm{L}}D_{-}^{\alpha}$.

Theorem 4.35. *Let $\alpha \in \mathbb{C}$ and $\beta \in \mathbb{C}$ $(\beta \notin \mathbb{Z})$ be such that $\Re e(\alpha) > 0$ and either $\epsilon > 0$ or $\Re e(\beta) > \Re e(\alpha)$, $\epsilon = 0$, and let ${}^{\mathrm{L}}D_{+}^{\alpha}$ and ${}^{\mathrm{L}}D_{-}^{\alpha}$ be the*

Liouvillle fractional differentiation operators (1.3.63) *and* (1.3.64). *Then the generalized Stirling functions* $S(\alpha, \beta; \epsilon)$ *have the representation*

$$S(\alpha,\beta;\epsilon) = \frac{1}{\Gamma(\beta+1)}\left({}^{\mathrm{L}}D_{+}^{\alpha}\sum_{j=0}^{m}(-1)^j\binom{\beta}{j}e^{(\beta-j)t}\right)(\epsilon)$$
$$+\frac{e^{\alpha\pi i}}{\Gamma(\beta+1)}\left({}^{\mathrm{L}}D_{-}^{\alpha}\left[(e^t-1)^{\beta}-\sum_{j=0}^{m}(-1)^j\binom{\beta}{j}e^{(\beta-j)t}\right]\right)(\epsilon), \tag{4.6.38}$$

where $m = [\Re e(\beta)]$, *and* $e^{\alpha\pi i}$ *is the principal branch of* $(-1)^{\alpha}$.

Proof. By the conditions of the theorem, $m < \Re e(\beta) < m+1$, and hence $\Re e(\beta - j) > 0$ for $j \leq m$, and applying (1.3.71), the first term in the right-hand side of (4.6.38) equals

$$\frac{1}{\Gamma(\beta+1)}\sum_{j=0}^{m}(-1)^j\binom{\beta}{j}\left({}^{\mathrm{L}}D_{+}^{\alpha}e^{(\beta-j)t}\right)(\epsilon)$$
$$=\frac{1}{\Gamma(\beta+1)}\sum_{j=0}^{m}(-1)^j\binom{\beta}{j}e^{(\beta-j)\epsilon}(\beta-j)^{\alpha}. \tag{4.6.39}$$

Further, for $t > 0$

$$(e^t-1)^{\beta} = e^{t\beta}(1-e^{-t})^{\beta} = \sum_{j=0}^{m}(-1)^j\binom{\beta}{j}e^{(\beta-j)t}$$

and hence

$$(e^t-1)^{\beta}-\sum_{j=0}^{m}(-1)^j\binom{\beta}{j}e^{(\beta-j)t} = \sum_{j=m+1}^{\infty}(-1)^j\binom{\beta}{j}e^{(\beta-j)t}. \tag{4.6.40}$$

According to the estimate (4.5.14), the last series is convergent when either $\epsilon > 0$ or $\Re e(\beta) > 0$, $\epsilon = 0$, at least one of which is true by the conditions of the theorem. Since $\Re e(\beta - j) < 0$ for $j \geq m+1$ and noting (4.6.40) and

(1.3.92), the second term on the right-hand side of (4.6.38) equals

$$\frac{e^{\alpha\pi i}}{\Gamma(\beta+1)}\left({}^{\mathrm{L}}D^{\alpha}_{-}\left[\sum_{j=m+1}^{\infty}(-1)^j\binom{\beta}{j}e^{(\beta-j)t}\right]\right)(\epsilon)$$

$$=\frac{e^{\alpha\pi i}}{\Gamma(\beta+1)}\sum_{j=m+1}^{\infty}(-1)^j\binom{\beta}{j}\left({}^{\mathrm{L}}D^{\alpha}_{-}e^{-(j-\beta)t}\right)(\epsilon)$$

$$=\frac{1}{\Gamma(\beta+1)}\sum_{j=m+1}^{\infty}(-1)^j\binom{\beta}{j}e^{-(j-\beta)\epsilon}[e^{\pi i}(j-\beta)]^{\alpha}$$

$$=\frac{1}{\Gamma(\beta+1)}\sum_{j=m+1}^{\infty}(-1)^j\binom{\beta}{j}e^{(\beta-j)\epsilon}(\beta-j)^{\alpha}. \tag{4.6.41}$$

Taking the sum of (4.6.39) and (4.6.41), we deduce

$$\frac{1}{\Gamma(\beta+1)}\left({}^{\mathrm{L}}D^{\alpha}_{+}\sum_{j=0}^{m}(-1)^j\binom{\beta}{j}e^{(\beta-j)t}\right)(\epsilon)$$

$$+\frac{e^{\alpha\pi i}}{\Gamma(\beta+1)}\left({}^{\mathrm{L}}D^{\alpha}_{-}\left[(e^t-1)^{\beta}-\sum_{j=0}^{m}(-1)^j\binom{\beta}{j}e^{(\beta-j)t}\right]\right)(\epsilon)$$

$$=\frac{1}{\Gamma(\beta+1)}\sum_{j=0}^{\infty}(-1)^j\binom{\beta}{j}e^{(\beta-j)\epsilon}(\beta-j)^{\alpha}$$

which, combined with (4.6.6), yields (4.6.38). This completes the proof. □

When $\mathfrak{Re}(\alpha)<0$, there hold representations of the generalized Stirling functions $S(\alpha,\beta;\epsilon)$ in terms of the Liouville fractional integration operators ${}^{\mathrm{L}}I^{\alpha}_{+}$ and ${}^{\mathrm{L}}I^{\alpha}_{-}$.

Theorem 4.36. *Let $\alpha\in\mathbb{C}$ and $\beta\in\mathbb{C}$ ($\beta\notin\mathbb{Z}$) be such that $\mathfrak{Re}(\alpha)<0$ and either $\epsilon>0$ or $\mathfrak{Re}(\beta)>\mathfrak{Re}(\alpha)$, $\epsilon=0$, and let ${}^{\mathrm{L}}I^{-\alpha}_{+}$ and ${}^{\mathrm{L}}I^{-\alpha}_{-}$ be the Liouvillle fractional integration operators* (1.3.61) *and* (1.3.62). *Then the generalized Stirling functions $S(\alpha,\beta;\epsilon)$ have the representation*

$$S(\alpha,\beta;\epsilon)=\frac{1}{\Gamma(\beta+1)}\left({}^{\mathrm{L}}I^{-\alpha}_{+}\sum_{j=0}^{m}(-1)^j\binom{\beta}{j}e^{(\beta-j)t}\right)(\epsilon) \tag{4.6.42}$$

$$+\frac{e^{\alpha\pi i}}{\Gamma(\beta+1)}\left({}^{\mathrm{L}}I^{-\alpha}_{-}\left[(e^t-1)^{\beta}-\sum_{j=0}^{m}(-1)^j\binom{\beta}{j}e^{(\beta-j)t}\right]\right)(\epsilon),$$

for $m=[\mathfrak{Re}(\beta)]$, $e^{-\alpha\pi i}$ being the principal branch of $(-1)^{-\alpha}$.

Theorem 4.36 is proved similarly to Theorem 4.35 by evaluating the two terms on the right-hand side of (4.6.42) and using relations (1.3.69) and (1.3.70).

4.6.3 *First application*

In this section we give an application of generalized Stirling functions $S(n, \beta; \epsilon)$ to represent fractional order differences in terms of such functions and usual differentiation. First we consider the fractional difference $\Delta^{\beta,\epsilon}$ defined by (4.5.3).

Theorem 4.37. *Let $f(x)$, defined for $x \in \mathbb{R}$, be an arbitrarily often differentiable function. There hold the following three statements.*

(a) *If $\beta \in \mathbb{C}$ ($\beta \notin \mathbb{Z}$) and $\epsilon > 0$, then*

$$\Delta^{\beta,\epsilon} f(x) = \Gamma(\beta+1) \sum_{n=0}^{\infty} S(n, \beta; \epsilon) \frac{f^{(n)}(x)}{n!}, \tag{4.6.43}$$

provided that the series on the right-hand side of (4.6.43) *is absolutely convergent.*

(b) *If $k \in \mathbb{N}_0$ and $\epsilon \geq 0$, then*

$$\Delta^{k,\epsilon} f(x) = k! \sum_{n=0}^{\infty} S(n, k; \epsilon) \frac{f^{(n)}(x)}{n!}, \tag{4.6.44}$$

provided that the series on the right-hand side of (4.6.44) *is absolutely convergent.*
In particular,

$$\Delta^k f(x) = k! \sum_{n=k}^{\infty} \frac{S(n,k)}{n!} f^{(n)}(x) \quad (x \in \mathbb{R}) \tag{4.6.45}$$

holds provided that the series on the right side of (4.6.45) *also has this property.*

Proof. By the conditions of theorem for any $j \in \mathbb{N}_0$ we have by Taylor's expansion that

$$f(x+\beta-j) = \sum_{n=0}^{\infty} \frac{f^{(n)}(x)}{n!} (\beta - j)^n. \tag{4.6.46}$$

Substituting this relation into (4.5.3) and interchanging the order of summation, which is possible by an analogue of Fubini's theorem for the series (see Theorem 12-42 of [37]), we obtain

$$\begin{aligned}\Delta^{\beta,\epsilon} f(x) &= \sum_{j=0}^{\infty}(-1)^j \binom{\beta}{j} e^{\epsilon(\beta-j)} \sum_{n=0}^{\infty} \frac{f^{(n)}(x)}{n!}(\beta-j)^n \\ &= \sum_{n=0}^{\infty}\left[\sum_{j=0}^{\infty}(-1)^j \binom{\beta}{j} e^{\epsilon(\beta-j)}(\beta-j)^n\right]\frac{f^{(n)}(x)}{n!}.\end{aligned}$$

Recalling (4.5.9), this yields (4.6.43). Relation (4.6.44) is proved similarly by using (4.5.5). This completes the proof of the theorem. □

Corollary 4.16. *Let $f(x)$ be defined for $x \in \mathbb{R}$ and be arbitrarily often differentiable on $\mathbb{R}$. Let $\beta > 0$ ($\beta \notin \mathbb{N}$), $\{\beta\}$ being the fractional part of β, and $S^*(n,\beta)$ be given by (4.5.50). There holds the relation*

$$\Delta^{\beta,\{\beta\}} f(x) = \Gamma(\beta+1) \sum_{n=0}^{\infty} S^*(n,\beta) \frac{f^{(n)}(x)}{n!} \tag{4.6.47}$$

provided that the series on the right-hand side of (4.6.47) is absolutely convergent.

Corollary 4.17. *Let $f(x)$, defined for $x \in \mathbb{R}$, be an arbitrarily often differentiable function, and let $\epsilon \geq 0$. Then*

$$\Delta^{1,\epsilon} f(x) = (e^{\epsilon}-1) f(x) + e^{\epsilon} \sum_{n=1}^{\infty} \frac{f^{(n)}(x)}{n!}. \tag{4.6.48}$$

In particular,

$$\Delta^{1} f(x) = \sum_{n=1}^{\infty} \frac{f^{(n)}(x)}{n!}. \tag{4.6.49}$$

Proof. Corollary 4.17 follows from Theorem 4.43(a) if we put $\epsilon = \{\beta\}$ and take (4.5.50) into account. Equation (4.6.44) with $k = 1$ yields (4.6.48) noting (4.5.26) and (4.5.28). The relation (4.6.49) follows from (2.6.5) when $\epsilon = 0$. □

Next we consider the previous result with respect to the classical fractional difference Δ^{β} with $\epsilon = 0$, namely

$$\Delta^{\beta} f(x) \equiv \Delta^{\beta,0} f(x) = \sum_{j=0}^{\infty}(-1)^j \binom{\beta}{j} f(x+\beta-j). \tag{4.6.50}$$

Theorem 4.38. *Let $f(x)$, defined for $x \in \mathbb{R}$, be an arbitrarily often differentiable function. There hold the following assertions.*

(a) *If $\beta \in \mathbb{C}$ $(\beta \notin \mathbb{Z})$ and $\epsilon > 0$, then*

$$\Delta^{\beta,\epsilon} f(x) = \Gamma(\beta+1) \sum_{n=0}^{\infty} \frac{S(n,\beta;\epsilon)}{n!} \sum_{m=0}^{n} \binom{n}{m} (-\epsilon)^{n-m} f^{(m)}(x) \tag{4.6.51}$$

provided that the series on the right-hand side of (4.6.51) is absolutely convergent.

(b) *If $k \in \mathbb{N}_0$ and $\epsilon \geq 0$, then*

$$\Delta^{k,\epsilon} f(x) = k! \sum_{n=0}^{\infty} \frac{S(n,k;\epsilon)}{n!} \sum_{m=0}^{n} \binom{n}{m} (-\epsilon)^{n-m} f^{(m)}(x) \tag{4.6.52}$$

provided that the series on the right-hand side of (4.6.52) is absolutely convergent. In particular, relation (4.6.45) holds provided again that the series on the right hand side of (4.6.45) is absolutely convergent.

Proof. Let $\beta \in \mathbb{C}$ $(\beta \notin \mathbb{Z})$ and $\epsilon > 0$. By the condition of the theorem, the right hand side of (4.6.51) is well-defined, and in accordance with Theorem 4.29(a) the generalized $S(n,\beta;\epsilon)$ have the representation (4.5.9) for any $n \in \mathbb{N}_0$. Substituting (4.5.9) into the right hand side of (4.6.50) we have

$$\begin{aligned}
&\Gamma(\beta+1) \sum_{n=0}^{\infty} \frac{S(n,\beta;\epsilon)}{n!} \sum_{m=0}^{n} \binom{n}{m} (-\epsilon)^{n-m} f^{(m)}(x) \\
&- \sum_{n=0}^{\infty} \frac{1}{n!} \left[\sum_{j=0}^{\infty} (-1)^j \binom{\beta}{j} e^{\epsilon(\beta-j)} (\beta-j)^n \right] \sum_{m=0}^{n} \binom{n}{m} (-\epsilon)^{n-m} f^{(m)}(x).
\end{aligned}$$

Since the series on the right side is absolutely convergent we can interchange orders of summation to deduce

$$\sum_{j=0}^{\infty}(-1)^j\binom{\beta}{j}e^{\epsilon(\beta-j)}\left[\sum_{n=0}^{\infty}\frac{1}{n!}(\beta-j)^n\sum_{m=0}^{n}\binom{n}{m}(-\epsilon)^{n-m}f^{(m)}(x)\right]$$
$$=\sum_{j=0}^{\infty}(-1)^j\binom{\beta}{j}e^{\epsilon(\beta-j)}\left[\sum_{m=0}^{\infty}f^{(m)}(x)\sum_{n=m}^{\infty}\binom{n}{m}(-\epsilon)^{n-m}\frac{(\beta-j)^n}{n!}\right]$$
$$=\sum_{j=0}^{\infty}(-1)^j\binom{\beta}{j}\sum_{m=0}^{\infty}\frac{f^{(m)}(x)}{m!}e^{\epsilon(\beta-j)}\left[\sum_{n=m}^{\infty}(-\epsilon)^{n-m}\frac{(\beta-j)^n}{(n-m)!}\right]$$
$$=\sum_{j=0}^{\infty}(-1)^j\binom{\beta}{j}\sum_{m=0}^{\infty}\frac{f^{(m)}(x)}{m!}(b-j)^m\left[e^{\epsilon(\beta-j)}\sum_{k=0}^{\infty}\frac{(-\epsilon(\beta-j))^k}{k!}\right]$$
$$=\sum_{j=0}^{\infty}(-1)^j\binom{\beta}{j}\left[\sum_{m=0}^{\infty}\frac{f^{(m)}(x)}{m!}(\beta-j)^m\right].$$

Now, applying (4.6.46) we have

$$\Gamma(\beta+1)\sum_{n=0}^{\infty}\frac{S(n,\beta;\epsilon)}{n!}\sum_{m=0}^{n}\binom{n}{m}(-\epsilon)^{n-m}f^{(m)}(x)$$
$$=\sum_{j=0}^{\infty}(-1)^j\binom{\beta}{j}f(x+\beta-j),$$

which, combined with (4.6.50), yields (4.6.51).

As to part (b), when $k=0$, then according to (4.5.11)–(4.5.12), Eq. (4.6.52) takes the form $\Delta^{0,\epsilon}f(x)=f(x)$ which is true by (4.5.6). When $k\in\mathbb{N}$ and $\epsilon\geq 0$, then substituting (4.5.10) into the right hand side of (4.6.52) and following the same arguments as above we deduce (4.6.52). Thus the theorem is proved. □

Corollary 4.18. *Let $f(x)$, defined for $x\in\mathbb{R}$, be an arbitrarily often differentiable function. Let $\beta>0$ ($\beta\notin\mathbb{N}$), $\{\beta\}$ being the fractional part of β, and $S^*(n,\beta)$ be given by (4.5.50). Then*

$$\Delta^{\beta,\{\beta\}}f(x)=\Gamma(\beta+1)\sum_{n=0}^{\infty}\frac{S^*(n,\beta)}{n!}\sum_{m=0}^{n}\binom{n}{m}(-\{\beta\})^{n-m}f^{(m)}(x) \tag{4.6.53}$$

provided that the series in the right hand side of (4.6.53) is absolutely convergent.

4.6.4 *Special examples*

In this section we apply the relations established in Theorems 4.37 and 4.38 to particular functions $f(x)$. This is a first attempt to see their range of applicability.

Example 4.1. Let $f(x) = \log(x+a)$, with $a \in \mathbb{R}$ and $x > -a$. Then for any $n \in \mathbb{N}$

$$D^n(\log(x+a)) = (-1)^{n-1}(n-1)!(x+a)^{-n}, \qquad D = \frac{d}{dx}, \tag{4.6.54}$$

hence (4.6.43) and (4.6.44) and (4.6.47) yield, noting (4.5.22), the following Corollary.

Corollary 4.19. *Let $a \in \mathbb{R}$, $x \geq a$.*

(a) If $\beta \in \mathbb{C}$, $\beta \notin \mathbb{Z}$ and $\epsilon > 0$, then

$$\begin{aligned}&\Delta^{\beta,\epsilon}(\log(x+a)) \\ &= \log(e)\left[(e^\epsilon-1)^\beta \log(x+a) + \Gamma(\beta+1)\sum_{n=1}^{\infty} S(n,\beta;\epsilon)\frac{(-1)^n}{n}(x+a)^{-n}\right].\end{aligned} \tag{4.6.55}$$

(b) If $k \in \mathbb{N}$ and $\epsilon \geq 0$, then

$$\begin{aligned}&\Delta^{k,\epsilon}(\log(x+a)) \\ &= \log(e)\left[(e^\epsilon-1)^k \log(x+a) + k!\sum_{n=1}^{\infty} S(n,k;\epsilon)\frac{(-1)^n}{n}(x+a)^{-n}\right].\end{aligned} \tag{4.6.56}$$

(c) If $\beta > 0$ and $\beta \notin \mathbb{N}$, then

$$\begin{aligned}&\Delta^{\beta,\{\beta\}}(\log(x+a)) \\ &= \log(e)\left[(e^{\{\beta\}}-1)^\beta \log(x+a) + \Gamma(\beta+1)\sum_{n=1}^{\infty} S^*(n,\beta)\frac{(-1)^n}{n}(x+a)^{-n}\right].\end{aligned} \tag{4.6.57}$$

In particular, for $\epsilon = 0$, (4.6.56) takes the form

$$\Delta^k(\log(x+a)) = \log(e)k!\sum_{n=k}^{\infty} S(n,k)\frac{(-1)^n}{n}(x+a)^{-n} \quad (k \in \mathbb{N}). \tag{4.6.58}$$

Further, if $\beta = 1$ and $\epsilon \geq 0$, then, in accordance with (4.5.28), Eq. (4.6.56) takes the form

$$\begin{aligned}&\Delta^{1,\epsilon}(\log(x+a)) \\ &= \log(e)\left[(e^\epsilon-1)\log(x+a) + e^\epsilon\sum_{n=1}^{\infty} \frac{(-1)^n}{n}(x+a)^{-n}\right].\end{aligned} \tag{4.6.59}$$

Finally, (4.6.51) gives

Corollary 4.20. *Let $\beta > 0$, $\beta \notin \mathbb{N}$, $a \in \mathbb{R}$, $x \geq -a$ and $\epsilon > 0$. Then*

$$\Delta^{\beta}(\log(x+a)) \tag{4.6.60}$$
$$= \log(e)(e^{\epsilon}-1)^{\beta}\log(x+a)$$
$$+ \log(e)\Gamma(\beta+1)\sum_{n=0}^{\infty}(-1)^{n-1}\frac{S(n,\beta;\epsilon)}{n!}\sum_{m=0}^{n}\binom{n}{m}\epsilon^{n-m}\frac{(x+a)^{-m}}{m!}.$$

Example 4.2. If $f(x) = (x+a)^{\gamma}$, with $a \in \mathbb{R}$, $x > -a$ and $\gamma \in \mathbb{R}$, then

$$D^{n}((x+a)^{\gamma}) = (-1)^{n}(-\gamma)_{n}(x+a)^{\gamma-n} \quad (n \in \mathbb{N}),$$

where $(\gamma)_n$ is given by (4.5.32).

So, again noting (4.5.22), then (4.6.43), (4.6.44) and (4.6.47) yield

Corollary 4.21. *Let $a \in \mathbb{R}$, $x \geq -a$ and $\gamma \in \mathbb{R}$.*

(a) *If $\beta \in \mathbb{C}$, $\beta \notin \mathbb{Z}$ and $\epsilon > 0$, then*

$$\Delta^{\beta,\epsilon}((x+a)^{\gamma}) \tag{4.6.61}$$
$$= (e^{\epsilon}-1)^{\beta}(x+a)^{\gamma} + \Gamma(\beta+1)\sum_{n=1}^{\infty}(-1)^{n}S(n,\beta;\epsilon)\frac{(-\gamma)_n}{n!}(x+a)^{\gamma-n}.$$

(b) *If $k \in \mathbb{N}$ and $\epsilon \geq 0$, then*

$$\Delta^{k,\epsilon}((x+a)^{\gamma}) \tag{4.6.62}$$
$$= (e^{\epsilon}-1)^{k}(x+a)^{\gamma} + k!\sum_{n=1}^{\infty}(-1)^{n}S(n,k;\epsilon)\frac{(-\gamma)_n}{n!}(x+a)^{\gamma-n}.$$

(c) *If $\beta > 0$ and $\beta \notin \mathbb{N}$, then*

$$\Delta^{\beta,\{\beta\}}((x+a)^{\gamma}) \tag{4.6.63}$$
$$= (e^{\{\beta\}}-1)^{\beta}(x+a)^{\gamma} + \Gamma(\beta+1)\sum_{n=1}^{\infty}(-1)^{n}S^{*}(n,\beta)\frac{(-\gamma)_n}{n!}(x+a)^{\gamma-n}.$$

In particular, for $\epsilon = 0$, Eq. (4.6.62) has the form

$$\Delta^{k}((x+a)^{\gamma}) = k!\sum_{n=k}^{\infty}(-1)^{n}S(n,k)\frac{(-\gamma)_n}{n!}(x+a)^{\gamma-n} \ (k \in \mathbb{N}).$$

Further, if $\beta = 1$ and $\epsilon \geq 0$, then (4.6.62) has the form,

$$\Delta^{1,\epsilon}((x+a)^{\gamma}) = (e^{\epsilon}-1)(x+a)^{\gamma} + e^{\epsilon}\sum_{n=1}^{\infty}(-1)^{n}\frac{(-\gamma)_n}{n!}(x+a)^{\gamma-n}.$$

Finally, (4.6.51) gives

Corollary 4.22. *Let $\beta > 0$, $\beta \notin \mathbb{N}$, $a \in \mathbb{R}$, $x \geq -a$ and $\epsilon > 0$. Then*

$$\Delta^{\beta,\epsilon}((x+a)^{\gamma}) \tag{4.6.64}$$
$$= (e^{\epsilon}-1)^{\beta}(x+a)^{\gamma}$$
$$+\Gamma(\beta+1)\sum_{n=0}^{\infty}(-1)^{n}\frac{S(n,\beta;\epsilon)}{n!}\sum_{m=0}^{n}\binom{n}{m}\epsilon^{n-m}(-\gamma)_{m}(x+a)^{\gamma-m}.$$

Example 4.3. Let $f(x) = \psi(x+a)$, with $a \in \mathbb{R}$ and $x > -a$, where $\psi(z)$ is the Euler psi-function given by

$$\psi(z) = \frac{d}{dz}\log\Gamma(z) = \frac{\Gamma'(z)}{\Gamma(z)}.$$

It is known [14, p. 326] that for $z \neq 0, -1, -2, \ldots$

$$D^{n}\psi(z) = (-1)^{n+1}n!\zeta(n+1,z) \quad (n \in \mathbb{N}),$$

where $\zeta(s,z)$ is the generalized Riemann zeta-function defined by

$$\zeta(s,z) = \sum_{m=0}^{\infty}\frac{1}{(z+m)^{s}} \quad (s, z \in \mathbb{C},\ \Re e(s) > 1;\ z \neq 0, -1, -2, \ldots).$$

So, again noting (4.5.22), then (4.6.43), (4.6.44) and (4.6.47) yield

Corollary 4.23. *Let $a \in \mathbb{R}$ and $x \geq -a$.*

(a) *If $\beta \in \mathbb{C}$, $\beta \notin \mathbb{Z}$ and $\epsilon > 0$, then*

$$\Delta^{\beta,\epsilon}\psi(x+a) \tag{4.6.65}$$
$$= (e^{\epsilon}-1)^{\beta}\psi(x+a) + \Gamma(\beta+1)\sum_{n=1}^{\infty}S(n,\beta;\epsilon)(-1)^{n+1}\zeta(n+1,x+a).$$

(b) *If $k \in \mathbb{N}$ and $\epsilon \geq 0$, then*

$$\Delta^{k,\epsilon}\psi(x+a) = (e^{\epsilon}-1)^{k}\psi(x+a) + k!\sum_{n=1}^{\infty}S(n,k;\epsilon)(-1)^{n+1}\zeta(n+1,x+a). \tag{4.6.66}$$

(c) *If $\beta > 0$ and $\beta \notin \mathbb{N}$, then*

$$\Delta^{\beta,\{\beta\}}\psi(x+a) \tag{4.6.67}$$
$$= (e^{\{\beta\}}-1)^{\beta}\psi(x+a) + \Gamma(\beta+1)\sum_{n=1}^{\infty}S^{*}(n,\beta)(-1)^{n+1}\zeta(n+1,x+a).$$

In particular, for $\epsilon = 0$, (4.6.66) gives

$$\Delta^k \psi(x+a) = k! \sum_{n=k}^{\infty} S(n,k)(-1)^{n+1}\zeta(n+1, x+a) \quad (k \in \mathbb{N}).$$

Further, if $\beta = 1$ and $\epsilon \geq 0$, then

$$\Delta^{1,\epsilon}\psi(x+a) = (e^{\epsilon}-1)^k \, \psi(x+a) + e^{\epsilon} \sum_{n=1}^{\infty} (-1)^{n+1}\zeta(n+1, x+a) \ (\epsilon \geq 0).$$

Furthermore, (4.6.51) delivers

Corollary 4.24. *If $\beta > 0$, $\beta \notin \mathbb{N}$, $a \in \mathbb{R}$, $x \geq -a$ and $\epsilon > 0$, then*

$$\begin{aligned} &\Delta^{\beta}\psi(x+a) \qquad (4.6.68) \\ &= (e^{\epsilon}-1)^{\beta}\psi(x+a) \\ &\quad + \Gamma(\beta+1) \sum_{n=0}^{\infty} (-1)^{n-1} \frac{S(n,\beta;\epsilon)}{n!} \sum_{m=0}^{n} \binom{n}{m} \epsilon^{n-m} m! \zeta(m+1, x). \end{aligned}$$

Finally the known relation

$$\Delta^k \psi(z+a) = (-1)^{k-1}(k-1)! \frac{\Gamma(z+a)}{\Gamma(z+a+k)},$$

with $k \in \mathbb{N}$; $z \in \mathbb{C}$, $z \neq -a, -a-1, \ldots, -a-k+1$, *cf.*, *e. g.* p. 328 of [320] or p. 19 of [226], yields

Property 4.15. *If $k \in \mathbb{N}$, $a \in \mathbb{R}$ and $x \in \mathbb{R}$ $(x \neq -a, -a-1, \ldots, -a-k+1)$, then there holds the summation formula*

$$\sum_{n=k}^{\infty} S(n,k)(-1)^n \zeta(n+1, x+a) = \frac{(-1)^k}{k} \frac{\Gamma(x+a)}{\Gamma(x+a+k)}.$$

4.7 Connections Between the Generalized Stirling Functions of the First and the Second Kind. Applications

In this section we prove connections between the Stirling functions of the first and second kind $s(\alpha, \beta)$ and $S(\alpha, k)$.

4.7.1 *Coincidence relations*

The Stirling functions of the first kind $s(-n,\beta)$, defined for $n \in \mathbb{N}_0$ and $\beta \in \mathbb{C}$ by (4.3.25), coincide, apart from a multiplicative factor, with the Stirling functions of the second kind $S(-\beta,n)$, defined in (4.4.3). The situation is similar for $s(-k,-\alpha)$ and $S(\alpha,k)$.

Theorem 4.39.

(a) *Let $\beta \in \mathbb{C}$ $(\beta \neq 0)$ and $n \in \mathbb{N}$. The Stirling functions $s(-n,\beta)$ coincide with $S(-\beta,n)$ apart from the constant multiplier $(-1)^n e^{-(\beta+1)\pi}$*
$$s(-n,\beta) = (-1)^n e^{-(\beta+1)\pi} S(-\beta,n). \tag{4.7.1}$$
In particular, for $m \in \mathbb{Z}$, $m \neq 0$,
$$s(-n,m) = (-1)^{n+m-1} S(-m,n).$$

(b) *Let $\alpha \in \mathbb{C}$ $(\alpha \neq 0)$ and $k \in \mathbb{N}$. The Stirling functions $S(\alpha,k)$ coincide with the $s(-k,-\alpha)$ apart from the constant multiplier $(-1)^k e^{(\alpha+1)\pi}$*
$$S(\alpha,k) = (-1)^k e^{(\alpha+1)\pi} s(-k,-\alpha). \tag{4.7.2}$$
In particular, for $m \in \mathbb{Z}$,
$$S(m,k) = (-1)^{m+k+1} s(-k,-m).$$

Proof. The result in (4.7.1) follows from Theorem 4.12 and Theorem 4.17, if we take into account the explicit representations for $s(-n,\beta)$ and $S(-\beta,n)$ given by (4.3.25) and (4.1.16). The relation (4.7.2) clearly follows from (4.7.1). □

The above theorem enables one to transfer several results we have established for the Stirling functions of second kind $S(\alpha,k)$ to such for the Stirling functions of first kind $s(-n,\beta)$. One such result is Theorem 4.18 that expresses $S(\alpha,k)$ in terms of Liouville derivatives in the form
$$S(\alpha,k) = \frac{(-1)^k}{k!} \lim_{x\to 0} \left({}^{\mathrm{L}}D_+^{\alpha} \left[\left(1-e^t\right)^k - 1\right]\right)(x) \tag{4.7.3}$$
for $\alpha \in \mathbb{C}$, $\mathfrak{Re}(\alpha) > 0$, and $k \in \mathbb{N}$. It can be transferred as follows.

Theorem 4.40. *Let $\beta \in \mathbb{C}$, $\mathfrak{Re}(\beta) < 0$, and $n \in \mathbb{N}$. Then the Stirling functions $s(-n,\beta)$ have the Liouville fractional derivative representation*
$$s(-n,\beta) = \frac{e^{(\beta+1)\pi i}}{n!} \lim_{x\to 0} \left({}^{\mathrm{L}}D_+^{-\beta} \left[\left(1-e^t\right)^k - 1\right]\right)(x). \tag{4.7.4}$$

Proof. The result in (4.7.4) follows directly from (4.7.3) and (4.7.1). □

Formula (4.7.4) can be used as an alternative definition of $s(-n,\beta)$.

4.7.2 *Results from sampling analysis*

For the counterpart of the classical orthogonality property for the Stirling functions we need the sampling theorem of signal analysis. Let $B^p_{\pi W}$, $W > 0$, $1 \leq p < \infty$, be the class of those functions $g \in L^p(\mathbb{R})$ having an extension to the complex plane $\mathbb{C}$ as an entire function of exponential type πW, namely

$$|g(z)| \leq \exp(\pi W|y|)\|g\|_C \qquad (z = x + iy; x, y \in \mathbb{R}).$$

The sampling theorem now states

Theorem 4.41 (Sampling theorem). *Any signal function $g \in B^p_{\pi W}$, $1 \leq p < \infty$, some $W > 0$, can be completely reconstructed from its sampled values $g(j/W)$ taken at the nodes j/W, $j \in \mathbb{Z}$, in terms of*

$$g(z) = \sum_{j=-\infty}^{\infty} g\left(\frac{j}{W}\right) \frac{\sin[\pi(Wz - j)]}{\pi(Wz - j)}, \tag{4.7.5}$$

the series converging absolutely and uniformly on compact subsets of $\mathbb{C}$.

For literature regarding sampling analysis see, *e. g.*, Butzer *et al.* [132, 133] and Higgins [302]. Details on the application of the special function $(\sin \pi x)/(\pi x)$, also known as the Sinc function or Whittaker's cardinal function, in this context may alo be found in the seminal work of Stenger [580] or the original publications by E. T. Whittaker [627] and his son, J. M. Whittaker [628].

A basic new application of this theorem to be needed below, which is definitely of independent interest, is given by Lemma 4.6, namely the sampling theorem for the power function $t^\alpha/\Gamma(\alpha + 1)$ in terms of $t^j/j!$ for $j \in \mathbb{N}_0 - \mathbb{N} \cup \{0\}$.

Lemma 4.6 (Sampling representation of the power function). *For $\alpha \in \mathbb{C}$ and $t > 0$ there holds the relation*

$$\frac{t^\alpha}{\Gamma(\alpha + 1)} = \sum_{j=0}^{\infty} \frac{t^j}{j!} \frac{\sin[\pi(\alpha - j)]}{\pi(\alpha - j)} \qquad (\alpha \in \mathbb{C}; \quad t > 0). \tag{4.7.6}$$

Proof. Set $g(z) = t^z/\Gamma(z + 1)$, where $z = x + iy = |z|e^{i \arg z}$ with $-\pi \leq \arg(z) < \pi$. $g(z)$ is clearly an entire function of $z \in \mathbb{C}$. Using the asymptotic relation for the Gamma function (formula 1.18(2) of [226]), *viz.*

$$\Gamma(z) \sim (2\pi)^{1/2} e^{-z} e^{(z-1/2)\log(z)} \qquad (z \to \infty), \tag{4.7.7}$$

we have

$$|g(z)| \sim (2\pi|z|)^{-1/2}\left(\frac{te}{|z|}\right)^{|z|} e^{y\arg z} \quad (|z| \to \infty).$$

If we choose $R > 0$ sufficiently large, this relation yields the estimate

$$|g(z)| \leq Ae^{\pi|y|} \qquad (A > 0; |z| \geq R). \tag{4.7.8}$$

If $|z| \leq R$ then, choosing $n \in \mathbb{N}$ such that $|z+n| \geq R$ and using the relation

$$(z)_k = \frac{\Gamma(z+k)}{z},$$

one has

$$|g(z)| \leq \frac{|t^{z+n}(z+1)\cdots(z+n)|}{t^n|\Gamma(z+n+1)|} \leq \frac{(R+1)\cdots(R+n)}{a^n}\frac{|t^{z+n}|}{|\Gamma(z+n+1)|},$$

and hence, according to (4.7.8),

$$|g(z)| \leq Be^{\pi|y|} \qquad (B = A(R+1)\cdots(R+n)a^n; \ \ |z| \geq R).$$

Using (4.7.7), it is directly verified that $g(x) \in L^p(\mathbb{R})$ for $1 \leq p < \infty$.

Thus, g is of exponential type π, $g \in B_\pi^p$, and an application of the sampling formula (4.7.5) (with $z = \alpha$ and $W = 1$) yields (4.7.6), noting that

$$g(j) = \frac{t^j}{\Gamma(j+1)} = 0 \quad \text{for} \quad j \in \mathbb{Z}^-. \tag{4.7.9}$$

□

Theorem 4.42 (Sampling theorem for Stirling functions $S(\alpha, k)$). *For $\alpha \in \mathbb{C}$ and $k \in \mathbb{N}_0$ there holds true*

$$\frac{S(\alpha,k)}{\Gamma(\alpha+1)} = \sum_{j=k}^{\infty} \frac{S(j,k)}{j!}\frac{\sin[\pi(\alpha-j)]}{\pi(\alpha-j)}. \tag{4.7.10}$$

Proof. If $\alpha = 0$, then (4.7.10) is clear. Indeed, $S(0,0) = 1$ and $S(k,m) = 0$ $(k, m \in \mathbb{N}_0;\ k < m)$, $S(0,k) = 0$ $(k \in \mathbb{N})$. Let $\alpha \neq 0$. By Proposition 4.5, for $z \in \mathbb{C}$ and $\alpha \in \mathbb{C}$ $(\alpha \neq 0)$ there holds the Newton series expansion

$$z^\alpha = \sum_{k=0}^{\infty} S(\alpha,k)[z]_k \tag{4.7.11}$$

(recall from Eq. (4.1.7) that $[z]_0 = 1$, and $[z]_k = z(z-1)\cdots(z-k+1)$ for $k \in \mathbb{N}$), the series being absolutely convergent for $|z| \geq \lambda + 1$, the abscissa of convergence being $\lambda \leq \max[0, \mathfrak{Re}(\alpha) - 1/2]$. Rewrite (4.7.11) with $z = t > 0$ in the form

$$\frac{t^\alpha}{\Gamma(\alpha+1)} = \sum_{k=0}^{\infty} \frac{S(\alpha,k)}{\Gamma(\alpha+1)}[t]_k \quad (t>0; \quad \alpha \in \mathbb{C}, \quad \alpha \neq 0). \tag{4.7.12}$$

Taking into account (4.7.12) and (4.7.6) and interchanging the order of summation yields

$$\frac{t^\alpha}{\Gamma(\alpha+1)} = \sum_{j=0}^{\infty} \left(\sum_{k=0}^{j} \frac{S(j,k)}{j!}[t]_k \right) \frac{\sin[\pi(\alpha-j)]}{\pi(\alpha-j)}$$

$$\sum_{k=0}^{\infty} \left(\sum_{j=k}^{\infty} \frac{S(j,k)}{j!} \frac{\sin[\pi(\alpha-j)]}{\pi(\alpha-j)} \right) [t]_k. \tag{4.7.13}$$

A comparison of the coefficients of the series (4.7.12) and (4.7.13) yields (4.7.10). □

4.7.3 *Generalized orthogonality properties*

It is well known that the classical Stirling numbers of the first and second kind $s(n,k)$ and $S(n,k)$ are connected by the basic orthogonality relation

$$\sum_{k=m}^{n} s(n,k)S(k,m) = \sum_{k=m}^{n} S(n,k)s(k,m) = \delta_{n,m} \quad (m,n \in \mathbb{N}_0), \tag{4.7.14}$$

where $\delta_{m,n} = 1$ for $m = n$ and $\delta_{m,n} = 0$ for $m \neq n$, the Kronecker delta. The counterpart for the Stirling functions reads

Theorem 4.43. *Let $\alpha \in \mathbb{C}$ and $m \in \mathbb{N}_0$. There holds for $\alpha \in \mathbb{C}$, $m \in \mathbb{N}_0$,*

$$\sum_{k=m}^{\infty} s(\alpha,k)S(k,m) = \sum_{k=m}^{\infty} S(\alpha,k)s(k,m) = \frac{\Gamma(\alpha+1)}{\Gamma(m+1)} \frac{\sin[\pi(\alpha-m)]}{\pi(\alpha-m)}. \tag{4.7.15}$$

Proof. Basic for the proof of the left-hand side of (4.7.15) is the sampling theorem for $s(\alpha,k)$, with $\alpha \in \mathbb{C}$ and $k \in \mathbb{N}_0$ (see Theorem 4.1 of [122])

$$\frac{s(\alpha,k)}{\Gamma(\alpha+1)} = \sum_{n=k}^{\infty} \frac{s(n,k)}{n!} \frac{\sin[(\alpha-n)\pi]}{(\alpha-n)\pi} \quad (\alpha \in \mathbb{C}; \quad k \in \mathbb{N}_0). \tag{4.7.16}$$

Taking into account the property $S(k,m) = 0$ $(k,m \in \mathbb{N}_0;\ k < m)$, applying the left-hand side of (4.7.14), changing the orders of summation and observing the known property $s(n,k) = 0$ ($n, k \in \mathbb{N}_0$ for $n < k$), there follow from (4.7.16)

$$\begin{aligned}
\sum_{k=m}^{\infty} s(\alpha,k)S(k,m) &= \sum_{k=0}^{\infty} s(\alpha,k)S(k,m) \\
&= \Gamma(\alpha+1)\sum_{k=0}^{\infty}\sum_{n=k}^{\infty} \frac{s(n,k)}{n!}\frac{\sin[\pi(\alpha-n)]}{\pi(\alpha-n)} S(k,m) \\
&= \Gamma(\alpha+1)\sum_{n=0}^{\infty}\sum_{k=0}^{n} s(n,k)S(k,m)\frac{\sin[\pi(\alpha-n)]}{n!\,\pi(\alpha-n)} \\
&= \Gamma(\alpha+1)\sum_{n=m}^{\infty}\left[\sum_{k=m}^{n} s(n,k)S(k,m)\right]\frac{\sin[\pi(\alpha-n)]}{n!\,\pi(\alpha-n)}.
\end{aligned}$$

Using (4.7.14), we deduce

$$\begin{aligned}
\sum_{k=0}^{\infty} s(\alpha,k)S(k,m) &= \Gamma(\alpha+1)\sum_{n=m}^{\infty} \delta_{n,m}\frac{\sin[\pi(\alpha-n)]}{n!\pi(\alpha-n)} \\
&= \frac{\Gamma(\alpha+1)}{m!}\frac{\sin[\pi(\alpha-m)]}{\pi(\alpha-m)},
\end{aligned}$$

which proves the left-hand side of (4.7.15).

The proof of the right-hand side of (4.7.15) is similar. In fact,

$$\begin{aligned}
\sum_{k=m}^{\infty} S(\alpha,k)s(k,m) &= \Gamma(\alpha+1)\sum_{k=0}^{\infty}\sum_{n=k}^{\infty} \frac{S(n,k)}{n!}\frac{\sin[\pi(\alpha-n)]}{\pi(\alpha-n)} s(k,m) \\
&= \Gamma(\alpha+1)\sum_{n=m}^{\infty}\left[\sum_{k=m}^{n} S(n,k)s(k,m)\right]\frac{\sin[\pi(\alpha-n)]}{n!\,\pi(\alpha-n)}.
\end{aligned}$$

This yields the right-hand side of (4.7.15) and the proof is complete. □

Observe that formula (4.7.15) reduces to the classical one (4.7.14) for $\alpha = n \in \mathbb{N}$.

4.7.4 *The $s(\alpha, k)$ connecting two types of fractional derivatives*

Here we obtain an expression for the classical Riemann-Liouville fractional derivative ${}^{\mathrm{RL}}D^{\alpha}_{0+}$ in terms of the Hadamard derivatives of integer order ${}^{\mathrm{H}}D^{k}_{0+}$. In some sense, such relation is the inverse to the sum formula (4.4.79).

Observe that if $n \in \mathbb{N}_0$, then (4.4.79) and (4.4.80) are the classical formulas, respectively,

$$\left({}^{\mathrm{H}}D_{0+}^{n} f\right)(x) \equiv (\delta^n f)(x) = \sum_{k=0}^{n} S(n,k) x^k f^{(k)}(x) \quad (n \in \mathbb{N}_0), \tag{4.7.17}$$

where $\delta = \left(x\frac{d}{dx}\right)$, $\delta^n = \left(x\frac{d}{dx}\right)^n$, and

$$x^n \left(\frac{d}{dx}\right)^n f(x) = \sum_{k=0}^{n} s(n,k) \delta^k f(x) \quad (n \in \mathbb{N}_0). \tag{4.7.18}$$

We shall now use a unified notation $\left({}^{\mathrm{RL}}\mathbf{D}_{0+}^{\alpha} f\right)(x)$ $(\alpha \in \mathbb{C})$ for the Riemann-Liouville fractional derivative (1.3.3) of order α and for the Riemann-Liouville fractional integral (1.3.1) of order $-\alpha$ in the cases $\Re e(\alpha) \geq 0$ and $\Re e(\alpha) < 0$

$${}^{\mathrm{RL}}\mathbf{D}_{0+}^{\alpha} f = {}^{\mathrm{RL}}D_{0+}^{\alpha} f \quad (\Re e(\alpha) \geq 0); \quad {}^{\mathrm{RL}}\mathbf{D}_{0+}^{\alpha} f = {}^{\mathrm{RL}}I_{0+}^{-\alpha} f \quad (\Re e(\alpha) < 0). \tag{4.7.19}$$

Theorem 4.44. *Let $f(x)$ be an arbitrarily often differentiable function on $x > 0$ such that $x^\alpha \left({}^{\mathrm{RL}}\mathbf{D}_{0+}^{\alpha} f\right)(x)/\Gamma(\alpha+1)$ as a function of $\alpha \in \mathbb{C}$ belongs to the class B_π^p for $1 \leq p < \infty$. Then there holds the expansion*

$$x^\alpha \left({}^{\mathrm{RL}}D_{0+}^{\alpha} f\right)(x) = \sum_{k=0}^{\infty} s(\alpha,k) \left({}^{\mathrm{H}}D_{0+}^{k} f\right)(x) \quad (\alpha \in \mathbb{C}, \quad \Re e(\alpha) \geq 0) \tag{4.7.20}$$

for the Riemann-Liuoville fractional derivative $\left({}^{\mathrm{RL}}D_{0+}^{\alpha} f\right)(x)$ of order $\alpha \in \mathbb{C}$, $\Re e(\alpha) \geq 0$, provided that the series in the right-hand side of (4.7.20) converges.

Proof. Inserting (4.7.17) (with $n = k$) into the right-hand side of (4.7.20) and changing the orders of summations and applying (4.7.15), we have

$$\begin{aligned}
&\sum_{k=0}^{\infty} s(\alpha,k) \left({}^{\mathrm{H}}D_{0+}^{k} f\right)(x) \\
&= \sum_{k=0}^{\infty} s(\alpha,k) \sum_{j=0}^{k} S(k,j) x^j f^{(j)}(x) \\
&= \sum_{j=0}^{\infty} \left(\sum_{k=j}^{\infty} s(\alpha,k) S(k,j) \right) x^j f^{(j)}(x) \\
&= \Gamma(\alpha+1) \sum_{j=0}^{\infty} \frac{\sin[(\alpha-j)\pi]}{(\alpha-j)\pi} \frac{x^j f^{(j)}(x)}{j!}.
\end{aligned} \tag{4.7.21}$$

By the assumption of the theorem, for $x > 0$, $x^{z\,\mathrm{RL}}\mathbf{D}_{0+}^{z} f/\Gamma(z+1)$ as a function of $z \in \mathbb{C}$ belongs to the class B_π^p. Applying the sampling formula (4.7.5) (with $z = \alpha$ and $W = 1$), and taking into account (4.7.9), noting $\left({}^{\mathrm{RL}}\mathbf{D}^j f\right)(x) \equiv ({}^{\mathrm{RL}}D_{0+}^j f)(x) = f^{(j)}(x)$ for $j \in \mathbb{N}_0$, we deduce for $\alpha \in \mathbb{C}$

$$\frac{x^\alpha \left({}^{\mathrm{RL}}\mathbf{D}_{0+}^\alpha f\right)(x)}{\Gamma(\alpha+1)} = \sum_{j=0}^{\infty} \frac{x^j f^{(j)}(x)}{j!} \frac{\sin[\pi(\alpha-j)]}{\pi(\alpha-j)}. \tag{4.7.22}$$

When $\alpha \in \mathbb{C}$, $\mathfrak{Re}(\alpha) \geq 0$, then (4.7.22) in combination with (4.7.19) yields

$$\frac{x^\alpha ({}^{\mathrm{RL}}D_{0+}^\alpha f)(x)}{\Gamma(\alpha+1)} = \sum_{j=0}^{\infty} \frac{x^j f^{(j)}(x)}{j!} \frac{\sin[\pi(\alpha-j)]}{\pi(\alpha-j)}. \tag{4.7.23}$$

Thus (4.7.20) follows from (4.7.21) and (4.7.23). □

Corollary 4.25. *If the conditions of Theorem 4.44 are satisfied, then for the integral* ${}^{\mathrm{RL}}I_{0+}^\alpha f$ *of order* $\alpha \in \mathbb{C}$, $\mathfrak{Re}(\alpha) > 0$, *there holds the sampling formula*

$$\frac{x^{-\alpha} \left({}^{\mathrm{RL}}I_{0+}^\alpha f\right)(x)}{\Gamma(1-\alpha)} = \sum_{j=0}^{\infty} \frac{x^j f^{(j)}(x)}{j!} \frac{\sin[\pi(\alpha+j)]}{\pi(\alpha+j)}.$$

To apply Theorem 4.44 we need the following auxiliary result

Lemma 4.7. *Let* $\lambda \in \mathbb{C}$, $\mathfrak{Re}(\lambda) < 1$, *and let* ${}^{\mathrm{RL}}\mathbf{D}_{0+}^\alpha f$ *be given by* (4.7.19). *Then* $x^{\alpha\,\mathrm{RL}}\mathbf{D}_{0+}^\alpha t^{-\lambda}/\Gamma(\alpha+1)$, $x^{\alpha\,\mathrm{RL}}\mathbf{D}_{0+}^\alpha t^{-\lambda}\log(t)/\Gamma(\alpha+1)$ *as functions of* $\alpha \in \mathbb{C}$ *are of exponential type* π,

$$\frac{x^\alpha ({}^{\mathrm{RL}}\mathbf{D}_{0+}^\alpha t^{-\lambda})(x)}{\Gamma(\alpha+1)} \in B_\pi^p, \qquad \frac{x^\alpha \left({}^{\mathrm{RL}}\mathbf{D}_{0+}^\alpha t^{-\lambda}\log(t)\right)(x)}{\Gamma(\alpha+1)} \in B_\pi^p.$$

In particular,

$$x^\alpha \left({}^{\mathrm{RL}}\mathbf{D}_{0+}^\alpha \log(t)\right)(x)/\Gamma(\alpha+1) \in B_\pi^p.$$

Proof. Let $\alpha \in \mathbb{C}$, $\lambda \in \mathbb{C}$, $\mathfrak{Re}(\lambda) < 1$. By (4.3.30) and the corresponding formula for the Riemann-Liouville fractional integral, we find that

$$\frac{x^\alpha \left({}^{\mathrm{RL}}\mathbf{D}_{0+}^\alpha t^{-\lambda}\right)(x)}{\Gamma(\alpha+1)} = \frac{\Gamma(1-\lambda)x^{-\lambda}}{\Gamma(1-\lambda-\alpha)\Gamma(\alpha+1)} \quad (x > 0;\ \ \mathfrak{Re}(\lambda) < 1). \tag{4.7.24}$$

It is directly verified, as in the proof of Lemma 4.6, that the right-hand side of (4.7.24) as a function of $\alpha \in \mathbb{C}$ is of exponential type π, and hence $x^\alpha \left({}^{\mathrm{RL}}\mathbf{D}_{0+}^\alpha t^{-\lambda}\right)(x)/\Gamma(\alpha+1) \in B_\pi^p$.

By the known formula for the Riemann-Liouville fractional integral (Eq. (2.50) in [542]), and an analoguous formula for the Riemann-Liouville fractional derivative, one has for $x > 0$ and $\Re e(\lambda) < 1$,

$$\frac{x^\alpha \left({}^{\mathrm{RL}}\mathbf{D}_{0+}^\alpha t^{-\lambda} \log(t)\right)(x)}{\Gamma(\alpha+1)} \tag{4.7.25}$$
$$= \frac{\Gamma(1-\lambda)x^{-\lambda}}{\Gamma(1-\lambda-\alpha)\Gamma(\alpha+1)}[\log(x) + \psi(1-\lambda) - \psi(1-\lambda-\alpha)].$$

The functions $\Gamma(z)$ and $\psi(z)$ have the same simple poles $z = -k$ ($k \in \mathbb{N}_0$). Therefore the right-hand side of (4.7.25) is an entire function of α. Using the asymptotic relation (4.7.7) for the Gamma function and the asymptotic estimate for $\psi(z)$ at infinity, given in the second formula of (4.3.42), it is established similarly as in the proof of Lemma 4.6, that the right-hand side of (4.7.25) as a function of $\alpha \in \mathbb{C}$ is of exponential type π, and hence $x^\alpha \left({}^{\mathrm{RL}}\mathbf{D}_{0+}^\alpha t^{-\lambda} \log(t)\right)(x)/\Gamma(\alpha+1) \in B_\pi^p$. In particular, when $\lambda = 0$, $x^\alpha \left({}^{\mathrm{RL}}\mathbf{D}_{0+}^\alpha \log(t)\right)(x)/\Gamma(\alpha+1) \in B_\pi^p$, and the lemma is proved. □

Example 4.4. Here we consider an application of Theorem 4.44, namely to the power function $f_1(x) = x^{-\lambda}$ ($x > 0$, $\lambda \in \mathbb{C}$). Then $\delta x^{-\lambda} = -\lambda x^{-\lambda}$ and $\delta^n x^{-\lambda} = (-\lambda)^n x^{-\lambda} = \left({}^{\mathrm{RL}}D_{0+}^n t^{-\lambda}\right)(x)$ for $n > 1$. But

$$\left(\frac{d}{dx}\right)^k (x^{-\lambda}) = [-\lambda]_k x^{-\lambda-k} \quad (\lambda \in \mathbb{C}, \quad k \in \mathbb{N}_0).$$

On the other hand, by (4.7.18), for $x > 0$ and $\lambda \in \mathbb{C}$, we obtain

$$x^n \left(\frac{d}{dx}\right)^n (x^{-\lambda}) = \sum_{k=0}^{n} s(n,k)(-\lambda)^k x^{-\lambda} \delta^k x^{-\lambda} = x^{-\lambda}[-\lambda]_n.$$

By Theorem 4.44 and Lemma 4.7, we can obtain for $f_1(x)$, with $x > 0$ and $\lambda \in \mathbb{C}$, $\Re e(\lambda) < 1$, $|\lambda| < 1$, the following relation

$$x^\alpha \left({}^{\mathrm{RL}}D_{0+}^\alpha t^{-\lambda}\right)(x) = \sum_{k=0}^{\infty} s(\alpha,k)\delta^k x^{-\lambda}$$
$$= x^{-\lambda} \sum_{k=0}^{\infty} s(\alpha,k)(-\lambda)^k = x^{-\lambda}[-\lambda]_\alpha. \tag{4.7.26}$$

The series above converge for $|\lambda| < 1$.

Example 4.5. As a second application of Theorem 4.44, take $f_2(x) = \log(x)$ $(x > 0)$. Then for $\delta = x\frac{d}{dx}$, $\delta \log(x) = 1$ and $\delta^m \log(x) = 0$ for $m > 1$. Then from (4.7.18) we have for $n \in \mathbb{N}$

$$x^n \left(\frac{d}{dx}\right)^n \log(x) = \sum_{k=0}^{n} s(n,k)\delta^k f(x) = s(n,0)\log(x) + s(n,1).$$

According to (4.2.10), (4.2.23) and (4.7.9) this yields for $x > 0$ and $n \in \mathbb{N}_0$,

$$x^n \left(\frac{d}{dx}\right)^n \log(x) = \frac{1}{\Gamma(1-n)}\log(x) + (-1)^{n-1}(n-1)! = (-1)^{n-1}\Gamma(n). \tag{4.7.27}$$

By Theorem 4.44 and Lemma 4.7 and (4.2.10), we obtain the following result.

Corollary 4.26. *For $\alpha \in \mathbb{C}$ and $x > 0$, there holds*

$$\begin{aligned} x^\alpha \left({}^{\mathrm{RL}}D_{0+}^{\alpha} \log(t)\right)(x) &= s(\alpha,0)\log(x) + s(\alpha,1) \\ &= \frac{\log(x) + \psi(1) - \psi(1-\alpha)}{\Gamma(1-\alpha)}. \end{aligned} \tag{4.7.28}$$

Observe that for $\alpha = n \in \mathbb{N}$ (4.7.28) coincides with (4.7.27),

$$\begin{aligned} x^n \left(\frac{d}{dx}\right)^n \log(x) &= \lim_{\alpha\to n} \left[x^\alpha \left({}^{\mathrm{RL}}D_{0+}^{\alpha} \log(t)\right)(x)\right] \\ &= \lim_{\alpha\to n} \frac{\log(x) + \psi(1) - \psi(1-\alpha)}{\Gamma(1-\alpha)} = -\lim_{\alpha\to n} \frac{\psi'(1-\alpha)}{\Gamma'(1-\alpha)} \\ &= (-1)^{n-1}\Gamma(n). \end{aligned}$$

4.7.5 *The representation of a general fractional difference operator via $s(\alpha, k)$*

Let Δ^k be the finite difference of order $k \in \mathbb{N}_0$ given by (4.1.8) and (4.1.9). A well-known operator in the calculus of finite differences, the operations of which are analogues to those of $\delta = xd/dx$, is the operator $\theta f(x) = x\Delta f(x)$, for which there holds the iterative formula

$$\theta^n f(x) = \sum_{k=1}^{n} [x+k-1]_k S(n,k)\Delta^k f(x) \quad (n \in \mathbb{N}), \tag{4.7.29}$$

where $S(n,k)$ $(n \in \mathbb{N},\ \ k = 1, 2, \cdots, n)$ are the Stirling numbers of the second kind for $\alpha = n$.

If we multiply equation (4.7.29) by the Stirling functions of the first kind $s(m,n)$, sum it from $n=1$ to $n=m\in\mathbb{N}$ and use the first orthogonality formula in (4.7.14), we obtain the *inversion* of the operator θ^n in the form (see, *e. g.*, p. 200 of [320])

$$\Delta^m f(x) = \frac{1}{[x+m-1]_m}\sum_{n=1}^{m} s(m,n)\theta^n f(x) \quad (m\in\mathbb{N}). \tag{4.7.30}$$

We now establish a generalization of this relation for a generalized "infinite" or fractional order difference $\boldsymbol{\Delta}^\alpha f$, with complex order $\alpha\in\mathbb{C}$, defined for suitable functions f by

$$\boldsymbol{\Delta}^\alpha f = \Delta^\alpha f \ \ (\Re e(\alpha)\geq 0); \quad \boldsymbol{\Delta}^\alpha f = \Delta^{-\alpha} f \ \ (\Re e(\alpha)<0), \tag{4.7.31}$$

where (it being a true infinite series)

$$\Delta^\alpha f(x) = e^{\alpha\pi i}\sum_{j=0}^{\infty}(-1)^j\binom{\alpha}{j} f(x+j) \quad (\alpha\in\mathbb{C},\ \ \Re e(\alpha)\geq 0). \tag{4.7.32}$$

Note that if $\alpha=k\in\mathbb{N}_0$, then, by (4.7.9), $\binom{k}{j}=0$ for $j=k+1,k+2,\ldots$, so that (4.7.32) turns out to be the classical finite difference $\Delta^k f$ given by (4.1.8).

Theorem 4.45. *Let $\alpha\in\mathbb{C}$ and $f(x)$, $x\in\mathbb{R}$, be a function such that*

$$g_\alpha(f) = \frac{[x+\alpha-1]_\alpha}{\Gamma(\alpha+1)}\boldsymbol{\Delta}^\alpha f(x)\in B^p_\pi \ \ \textit{for some} \ \ 1\leq p<\infty. \tag{4.7.33}$$

Then

$$\boldsymbol{\Delta}^\alpha f(x) = \frac{1}{[x+\alpha-1]_\alpha}\sum_{n=1}^{\infty} s(\alpha,n)\theta^n f(x), \tag{4.7.34}$$

if the series is convergent

Proof. Using (4.7.29) and interchanging the orders of summation, we have

$$\begin{aligned}\sum_{n=1}^{\infty} s(\alpha,n)\theta^n f(x) &= \sum_{n=1}^{\infty} s(\alpha,n)\sum_{k=1}^{n}[x+k-1]_k S(n,k)\Delta^k f(x)\\ &= \sum_{k=1}^{\infty}\left(\sum_{n=k}^{\infty} s(\alpha,n)S(n,k)\right)[x+k-1]_k\Delta^k f(x).\end{aligned}$$

Then an application of the orthogonality property (4.7.15) yields

$$\sum_{n=1}^{\infty} s(\alpha,n)\theta^n f(x) = \sum_{k=1}^{\infty} \frac{\Gamma(\alpha+1)}{\Gamma(k+1)} \frac{\sin[\pi(\alpha-k)]}{\pi(\alpha-k)} [x+k-1]_k \Delta^k f(x). \quad (4.7.35)$$

On the other hand, applying Theorem 4.41 to $(g_\alpha)(f)$, we have

$$\frac{[x+\alpha-1]_\alpha}{\Gamma(\alpha+1)} \boldsymbol{\Delta}^\alpha f(x) = \sum_{k=-\infty}^{\infty} \frac{[x+k-1]_k}{\Gamma(k+1)} \frac{\sin[\pi(\alpha-k)]}{\pi(\alpha-k)} \Delta^k f(x),$$

which, since $[x+k-1]_k/\Gamma(k+1) = 0$ for $k \in \mathbb{Z}^-$, yields

$$\frac{[x+\alpha-1]_\alpha}{\Gamma(\alpha+1)} \boldsymbol{\Delta}^\alpha f(x) = \sum_{k=0}^{\infty} \frac{[x+k-1]_k}{\Gamma(k+1)} \frac{\sin[\pi(\alpha-k)]}{\pi(\alpha-k)} \Delta^k f(x). \quad (4.7.36)$$

Therefore, (4.7.34) follows from (4.7.35) and (4.7.36). □

Corollary 4.27. *If the conditions of Theorem 4.45 are satisfied, then for the generalized fractional difference* $\Delta^\alpha f$ *of order* $\alpha \in \mathbb{C}$, $\mathfrak{Re}(\alpha) \geq 0$, *given by* (4.7.32), *there holds the series representation* (4.7.34), *provided the series converges.*

For the following application of the Theorem 4.45 we need the auxiliary result

Lemma 4.8. *Let* $\alpha \in \mathbb{C}$, *and let* $g_\alpha(f)$ *be given by* (4.7.33). *Then the specific functions* $g_\alpha\left(e^{\lambda x}\right)$ *with* $\lambda < 0$, $g_\alpha(\psi)$ *and* $g_\alpha\left(1/[x-1]_\lambda\right)$ *with* $\lambda \in \mathbb{C}$, $\mathfrak{Re}(\lambda) > -|\mathfrak{Re}(\alpha)|$, *as functions of* α, *are of exponential type* π,

$$g_\alpha\left(e^{\lambda x}\right) \in B_\pi^p \quad (\lambda < 0), \quad (4.7.37)$$

$$g_\alpha(\psi) \in B_\pi^p,$$

$$g_\alpha\left(\frac{1}{[x-1]_\lambda}\right) \in B_\pi^p \quad (\mathfrak{Re}(\lambda) > -|\mathfrak{Re}(\alpha)|). \quad (4.7.38)$$

Proof. First note that in accordance with (4.7.31) and (4.7.32) the infinite series $\boldsymbol{\Delta}^\alpha\left(e^{\lambda x}\right)$ converges for any $\alpha \in \mathbb{C}$ and $\lambda < 0$. Using (4.7.33) and noting (4.3.6) (with $z = e^\lambda$ and $\mu = \alpha$), we have for the first $g_\alpha\left(e^{\lambda x}\right)$

$$g_\alpha\left(e^{\lambda x}\right) = \frac{\Gamma(x+\alpha)}{\Gamma(\alpha+1)\Gamma(x)} e^{\alpha\pi i} e^{\lambda x}\left(1-e^\lambda\right)^\alpha \quad (\mathfrak{Re}(\alpha) \geq 0),$$

$$g_\alpha\left(e^{\lambda x}\right) = \frac{\Gamma(x+\alpha)}{\Gamma(\alpha+1)\Gamma(x)} e^{-\alpha\pi i} e^{\lambda x}\left(1-e^\lambda\right)^{-\alpha} \quad (\mathfrak{Re}(\alpha) < 0).$$

By the first relation in (4.3.42), with $t = \alpha$, $b = 1$, $a = x$, it is directly verified that

$$\lim_{\alpha\to\infty,\mathfrak{Re}(\alpha)>0} \frac{\Gamma(x+\alpha)}{\Gamma(\alpha+1)}\left(1-e^{\lambda}\right)^{\alpha} = 0; \tag{4.7.39}$$

$$\lim_{\alpha\to\infty,\mathfrak{Re}(\alpha)<0} \frac{\Gamma(x+\alpha)}{\Gamma(\alpha+1)}\left(1-e^{\lambda}\right)^{-\alpha} = 0. \tag{4.7.40}$$

If for any fixed $x > 0$ we choose $R > 0$ sufficiently large, then from (4.7.37) and (4.7.38) we deduce for any $\alpha \in \mathbb{C}$ the estimate

$$\left|g_{\alpha}\left(e^{\lambda x}\right)\right| \le Ae^{\pi|y|} \quad (z = x+iy;\ A > 0;\quad |z| \ge R).$$

If $|z| \le R$, then arguments similar to those in the proof of Lemma 4.6 gives

$$\left|g_{\alpha}\left(e^{\lambda x}\right)\right| \le Be^{\pi|y|} \quad (z = x+iy;\ B > 0;\quad |z| \le R).$$

If $\alpha \in \mathbb{R}$, then in accordance with (4.7.39), $g_{\alpha}\left(e^{\lambda x}\right) \in L^p(\mathbb{R})$ for any $1 < p < \infty$. This proves (4.7.37), namely that $g_{\alpha}\left(e^{\lambda x}\right) \in B^p_{\pi}$.

Using the second and first asymptotic estimates in (4.3.42), and taking (4.3.28) into account, we have for $\alpha \in \mathbb{C}$ the estimates

$$\left|(-1)^j\binom{\pm\alpha}{j}\psi(x+j)\right| \le A\frac{\log(j)}{(j)^{|\mathfrak{Re}(\alpha)|+1}} \qquad (A > 0;\quad j \in N),$$

$$\left|(-1)^j\binom{\pm\alpha}{j}\left(\frac{1}{[x+j-1]_{\lambda}}\right)\right| \le B\frac{1}{(j)^{|\mathfrak{Re}(\alpha)|+\mathfrak{Re}(\lambda)+1}} \quad (B > 0;\quad j \in N).$$

Thus the series $\boldsymbol{\Delta}^{\alpha}(\psi)$ and $\boldsymbol{\Delta}^{\alpha}\left(1/[x-1]_{\lambda}\right)$ converge when $|\mathfrak{Re}(\alpha)| > 0$ and $|\mathfrak{Re}(\alpha)| + \mathfrak{Re}(\lambda) > 0$, respectively. The result (4.7.37) and relation (4.7.38) are proved by using the asymptotic properties of the gamma and digamma functions, similarly to the above for (4.7.37). □

Example 4.6. Here we consider an application of Theorem 4.45, namely to the digamma function $f_3(x) = \psi(x) = \Gamma'(x)/\Gamma(x)$, $x > 0$. For this specific function, the operator $\theta = x\Delta$ satisfies

$$\theta\psi(x) = (x\Delta)\psi(x) = 1, \quad \theta^k\psi(x) = 0 \quad (k = \mathbb{N}\backslash\{1\}). \tag{4.7.41}$$

Thus the classical inversion formula (4.7.30) applied to the $f_3(x) = \psi(x)$ $(x > 0)$ takes on the form, also recalling (4.2.23),

$$\Delta^m\psi(x) = \frac{s(m,1)\ \theta\psi(x)}{[x+m-1]_m} = (-1)^{m-1}(m-1)!\frac{\Gamma(x)}{\Gamma(x+m)}. \tag{4.7.42}$$

Our new application is the extension of the finite difference result (4.7.42) to the generalized fractional order difference $\mathbf{\Delta}^{\alpha}\psi(x)$, given by

Corollary 4.28. *For the function $f(x) = \psi(x)$, $x > 0$, the series given by $\mathbf{\Delta}^{\alpha}\psi(x)$ converges for $\alpha \in \mathbb{C}$, and has the representation*

$$\mathbf{\Delta}^{\alpha}\psi(x) = \left[\frac{\psi(1) - \psi(1-\alpha)}{\Gamma(1-\alpha)}\right]\frac{\Gamma(x)}{\Gamma(x+\alpha)}. \tag{4.7.43}$$

Proof. By Lemma 4.8, $g_{\alpha}(\psi)$ as a function of α is of exponential type π, $g_{\alpha}(\psi) \in B^{p}_{\pi}$, and thus we can apply Theorem 4.45. Noting (4.7.41) and (4.2.16), (4.7.34) turns out to be for $x > 0$

$$\mathbf{\Delta}^{\alpha}\psi(x) = \frac{s(\alpha,1)\theta\psi(x)}{[x+\alpha-1]_{\alpha}} = \left[\frac{\psi(1) - \psi(1-\alpha)}{\Gamma(1-\alpha)}\right]\frac{\Gamma(x)}{\Gamma(x+\alpha)}. \qquad \square$$

Example 4.7. As another application of Theorem 4.45 let us take the function $f_4(x) = 1/[x-1]_{\lambda}$ with complex $x, \lambda \in \mathbb{C}$, for which

$$\theta^{k} f_4(x) = \frac{(-\lambda)^{k}}{[x-1]_{\lambda}} \quad (k \in \mathbb{N}), \tag{4.7.44}$$

see p. 200 of [320]).

Here (4.7.30) can readily be shown to take the form for $x, \lambda \in \mathbb{C}$,

$$\Delta^{m}\left(\frac{1}{[x-1]_{\lambda}}\right) = \frac{[-\lambda]_{m}}{[x+m-1]_{m}[x-1]_{\lambda}} = \frac{\Gamma(x-\lambda)}{\Gamma(x+m)}\frac{\Gamma(1-\lambda)}{\Gamma(1-\lambda-m)}.$$

Corollary 4.29. *Let $\alpha \in \mathbb{C}$ and $\lambda \in \mathbb{C}$ be such that $\Re e(\lambda) > -|\Re e(\alpha)|$ and $|\lambda| < 1$. For the function $f_4(x) = 1/[x-1]_{\lambda}$, $x \in \mathbb{R}$, the series given by $\mathbf{\Delta}^{\alpha}(1/[x-1]_{\lambda})$ converges for $\alpha \in \mathbb{C}$, and has the representation for $x \in \mathbb{R}$*

$$\begin{aligned}\mathbf{\Delta}^{\alpha}\left(\frac{1}{[x-1]_{\lambda}}\right) &= \frac{1}{[x+\alpha-1]_{\alpha}[x-1]_{\lambda}}\left([-\lambda]_{\alpha} - s(\alpha,0)\right)\\ &= \frac{\Gamma(x-\lambda)}{\Gamma(x+\alpha)}\left([-\lambda]_{\alpha} - \frac{1}{\Gamma(1-\alpha)}\right).\end{aligned} \tag{4.7.45}$$

Proof. By Lemma 4.8, we know that $g_{\alpha}(1/[x-1]_{\lambda})$ as a function of α is of exponential type π, $g_{\alpha}(1/[x-1]_{\lambda}) \in B^{p}_{\pi}$, and thus we can apply Theorem 4.45. Taking (4.7.44) and definition (4.2.3) into account, we have for $\alpha \in \mathbb{C}$

$$\begin{aligned}\mathbf{\Delta}^{\alpha}\left(\frac{1}{[x-1]_{\lambda}}\right) &= \frac{1}{[x+\alpha-1]_{\alpha}}\sum_{n=1}^{\infty} s(\alpha,n)\frac{(-\lambda)^{n}}{[x-1]_{\lambda}} =\\ &= \frac{1}{[x+\alpha-1]_{\alpha}[x-1]_{\lambda}}\left([-\lambda]_{\alpha} - s(\alpha,0)\right),\end{aligned}$$

which completes the proof. $\square$

Chapter 5

Fractional Variational Principles

The calculus of variations is dealing with the problem of extremising functionals. As it is known this problem represents a generalization of the problem of finding extremes of functions of several variables, or it is the problem of finding extremes of functions of an infinite number of variables. The roots of the calculus of variations can be seen in the works of the Greek thinkers.

The calculus of variations gives us analytical techniques to deal with important issues like construct the geodesic between two given points on a given surface, obtain the curve between two given points in the plane that yields a surface of revolution of minimum area when revolved around a given axis, deal with brachistochrone problem, and so on. The calculus of variations is used both to generate interesting differential equations, and to prove the existence of solutions, even in the case when these solutions cannot be obtained analytically. The Lagrangian and Hamiltonian formulation of dynamical systems represents one of the most important principles in physics. The classical Hamiltonian and Lagrangian mechanics are described in terms of derivatives of integer order. However, due to the presence of the friction the physical world is rather nonconservative. As it is known, the notion of dissipativity is a concept in system theory both from theoretical and practical point of view. Dissipativity is deeply connected to the notion of energy. When time evolves, a given dissipative system absorbs a fraction of its supplied energy and transforms it, for example, into heat. On the other hand the presence of frictional forces in physical models increases the complexity in the mathematical tools required to deal with them.

The fractional variational principles represent an emerging part of fractional calculus having an important impact in several areas of physics and engineering. In this context we refer to the works of Riewe [521, 522], Agrawal [15, 16, 18, 19], Klimek [343, 344], Atanacković *et al.* [43], Baleanu *et al.* [55, 56, 62, 63, 75, 445], Jumarie [322, 323], Cresson *et al.* [165, 166], Tarasov *et al.* [587, 589], Golmankhaneh [261], Herrmann [296], and Rabei *et al.* [500–502, 506, 508].

Particularly, the fractional variational principles are deeply connected to the fractional quantization procedure and to control theory, see Baleanu *et al.* [57–61, 64, 68, 69, 73, 76, 79, 84–86, 89, 154, 297, 316, 448, 505], Torres *et al.* [34, 218, 442], Cresson [164], Agrawal *et al.* [20–25, 444, 476], Laskin [358], Biswas and Sen [106], Jelicic and Petrovacki [319], Herrmann [295], Goldfain [260], and Tricaud and Chen [606].

The chapter is basically divided into five main sections. The first section contains an introductory part and the survey of results. After that, the fractional Euler-Lagrange equations for discrete and continuous systems are presented. The fractional Lagrangian formulation for field systems is discussed, followed by the fractional Euler-Lagrange equations in the presence of a delay. The results for both Riemann-Liouville and Caputo derivatives are listed. The recently established fractional discrete Euler-Lagrange equations are presented and some results about the fractional Lagrange-Finsler geometry are briefly illustrated. At the end of the first section of this chapter, some illustrative applications are shown. We mention the fractional variational principles with Riesz derivatives, the multi-order and the multi-term fractional variational approach with Hilfer derivatives, a fractional Lagrangian approach of the Schrödinger equation, the fractional generalization of two equivalent Lagrangians, the Euler-Lagrange equations in fractional space, the multi-time fractional Lagrangian equations, the generalization of the well know Faddeev-Jackiw formalism and the fractional variational calculus with generalized boundary conditions.

The second part of the chapter is devoted to fractional Hamiltonian dynamics. After a short introduction and an overview of the main results, we concentrate on the fractional Hamiltonian analysis of discrete and continuous systems. Here we present direct methods to obtain the fractional canonical equations with Riemann-Liouville, Caputo and Riesz-

Caputo derivatives. A special part is devoted to the fractional Hamiltonian formulation of the constrained systems. The fractional Hessian matrix is presented and the reduced phase space is discussed. After that, the fractional generalization of the Ostrogradski's construction is presented. In the application part we start by illustrating some examples for the discrete fractional constrained systems. After that,the fractional Hamiltonian formulation in fractional time is presented together with a fractional generalization of the Nambu mechanics. The fractional counterpart of the supersymmetric classical model and the fractional optimal control formulation with and without delay are also illustrated. Multi-time Hamiltonian equations, the Hamilton-Jacobi formulation with Caputo derivative and fractional dynamics on the extended phase space are described.

The third part deals with fractional variational problems within shifted orthonormal Legendre polynomials. The fourth part is devoted to a fractional Euler-Lagrange equations model for freely oscillating dynamical systems. The fifth part is dedicated to a study of the fractional Bateman-Feshbach-Tikochinsky oscillator.

5.1 Fractional Euler-Lagrange Equations

5.1.1 *Introduction and survey of results*

The Euler-Lagrange equation, developed by Leonhard Euler and Joseph-Louis Lagrange in the 1750s, represents the key formula of the calculus of variations. It gives a way to answer the question for functions which extremize a given cost functional. In the Lagrangian approach, the kinetic and the potential energies are expressed in terms of generalized coordinates and generalized velocities corresponding to each particle. The classical Lagrangian represents the difference between the kinetic and the potential energy. For example, the corresponding Lagrangian for dissipative systems depends explicitly on time, which implies that the Hamiltonian depends explicitly on time too. As a result, one of the good candidates to describe the behavior of the dissipative systems is the fractional calculus. The first application of fractional calculus reported by Niels Henrik Abel in 1823 was related to the formulation of the tautochrone problem [542]. It was proved that the fractional derivatives are the infinitesimal generators of a class of translation invariant convolution semigroups which appear universally as attractors [304]. However, the fractional Leibniz rule and the chain rule become more complicated than the forms encountered in the classical case and the integration by parts formula involves both the left and the right derivatives. As a result, during the last few years, all these properties of fractional calculus made the fractional variational principles an interesting area for many researchers from various fields. We notice that some physical interpretations of the fractional integrals and fractional derivatives have been proposed during the last years [301, 460, 578, 585, 586, 596, 597].

Non-local theories have been investigated and applied in several physical problems [95, 264–266, 374, 377]. It is well known that, due to the form of the fractional differential operators, the fractional Lagrangians and Hamiltonians are typical examples of non-local theories [16, 63, 506] .

Several methods to obtain the fractional Euler-Lagrange equations and the corresponding Hamiltonians have been proposed. These equations are new from the mathematical point of view because both the left and the right derivatives are involved.

The first attempt to formulate the fractional generalization of Lagrangian and Hamiltonian equations is due to Riewe [521, 522] who claimed that, by making use of the fractional derivatives of various orders, it is possible to choose Lagrangians that result in a wide range of dissipative Euler-Lagrange equations [521].

The symmetric fractional derivatives were investigated and the Euler-Lagrange equations for models depending on sequential derivatives of this type were obtained in [343]. These investigations were performed in both the Lagrangian and the Hamiltonian formalism.

In [16] an extension of the classical calculus of variations for systems containing Riemann-Liouville fractional derivatives was discussed.

In [63] the Lagrangians linear in velocities were investigated by using the fractional calculus with Riemann-Liouville derivatives and the corresponding Euler-Lagrange equations were obtained.

The Euler-Lagrange equations for fields were investigated within the fractional Lagrangian formulation in [75].

The Euler-Lagrange equations and the transversality conditions for fractional variational problems were discussed in [19].

A generalization of Noether's theorem leading to conservation laws for fractional Euler-Lagrangian equation was obtained in [43].

A version of Noether's theorem for fractional variational problems with Riesz-Caputo derivatives was reported in [245].

In [322] an extension of the Lagrange analytical mechanics was proposed to deal with dynamics of fractal nature. This approach used a slight modification of the Riemann-Liouville derivative definition.

The fractional generalization of nonholonomic constraints defined by equations with fractional derivatives was examined in [589] and the corresponding equations of motion were derived by using a variational principle.

In [34] the Euler-Lagrange fractional equations and the sufficient optimality conditions for problems of the calculus of variations with functionals containing both fractional derivatives and fractional integrals of Riemann-Liouville type were investigated.

The fractional Euler-Lagrange equations with Riemann-Liouville fractional derivatives in the presence of delay were derived in [73]; the fractional variational principles with Caputo derivatives in the presence of delay derivatives were examined in [316].

The optimal control problems in the presence of delay in the state variables as well as the presence of the Riemann-Liouville fractional derivatives of the state variables were investigated in [315].

Fractional embedding of differential operators and Lagrangian systems were explained in [164].

Fractional Euler-Lagrange equations were reported in the presence of the elements of Berezin algebra in [76].

A class of fractional differential equations which were obtained by using the fractional variational principles was illustrated in [84].

The fractional multi-time Lagrangian equations were derived for dynamical systems within Riemann-Liouville derivatives [66].

The fractional Faddeev-Jackiw formalism was constructed in [262].

Based on the conventions established in [612, 613], a nonholonomic deformation of Fedosov type quantization of fractional Lagrange-Finsler geometries was done in [87]. The constructions are obtained for a fractional almost Kähler model encoding equivalently all data for fractional Euler-

Lagrange equations with Caputo derivative [87].

The paper [299] presents the necessary and sufficient optimality conditions for the Euler-Lagrange fractional equations of fractional variational problems with determining in which spaces the functional must exist where the functional contains right and left fractional derivatives in the Riemann-Liouville sense and the upper bound of integration less than the upper bound of the interval of the fractional derivative

In [514] a new formulation for the constrained optimization problems in which the objective function was considered in the fractional integral form.

Some recent manuscripts describe the importance of the fractional Euler-Lagrange equations associated with systems of real world applications [72, 71, 65, 82, 70, 81].

5.1.2 *Fractional Euler-Lagrange equations for discrete and continuous systems*

5.1.2.1 *Fractional Euler-Lagrange equations for discrete systems*

Assume that α_j $(j = 1, \ldots, n_1)$ and β_k $(k = 1, \ldots, n_2)$ denote two sets of positive real numbers, $\alpha_{\max} = \max(\alpha_1, \ldots, \alpha_{n_1}, \beta_1, \ldots, \beta_{n_2})$, and M represents an integer fulfilling $M - 1 \leq \alpha_{\max} \leq M$. In addition we assume that $L_f\left(t, q^\rho, {}^{\mathrm{RL}}D_{a+}^{\alpha_1} q^\rho, \ldots, {}^{\mathrm{RL}}D_{a+}^{\alpha_{n1}} q^\rho, {}^{\mathrm{RL}}D_{b-}^{\beta_1} q^\rho, \ldots, {}^{\mathrm{RL}}D_{b-}^{\beta_{n2}} q^\rho\right)$ denotes a function having continuous first and second (partial) derivatives with respect to all its arguments.

Theorem 5.1 (see [16]). *Let $J[q^\rho]$ be a functional of the type*

$$J[q^\rho] = \int_a^b L_f\left(t, q^\rho, {}^{\mathrm{RL}}D_{a+}^{\alpha_1} q^\rho, \ldots, {}^{\mathrm{RL}}D_{a+}^{\alpha_{n1}} q^\rho, {}^{\mathrm{RL}}D_{b-}^{\beta_1} q^\rho, \ldots, {}^{\mathrm{RL}}D_{b-}^{\beta_{n2}} q^\rho\right) dt, \tag{5.1.1}$$

defined on the set of n functions q^ρ, $\rho = 1, \ldots, n$ which have continuous left Riemann-Liouville fractional derivatives of orders α_j, $j = 1, \ldots, n_1$, and right Riemann-Liouville fractional derivatives of orders β_j, $j = 1, \ldots, n_2$, in $[a, b]$ and obey the boundary conditions $(q^\rho)^{(j)}(a) = q^\rho_{aj}$ and $(q^\rho)^{(j)}(b) = q^\rho_{bj}$, $j = 1, \ldots, M - 1$. A necessary condition for $J[q^\rho]$ to admit an extremum for given functions $q^\rho(t)$, $\rho = 1, \ldots, n$, is that $q^\rho(t)$ satisfy Euler-Lagrange *equations*

$$\frac{\partial L}{\partial q^\rho} + \sum_{j=1}^{n_1} {}^{\mathrm{RL}}D_{b-}^{\alpha_j} \frac{\partial L_f}{\partial\, {}^{\mathrm{RL}}D_{a+}^{\alpha_j} q^\rho} + \sum_{j=1}^{n_2} {}^{\mathrm{RL}}D_{a+}^{\beta_j} \frac{\partial L_f}{\partial\, {}^{\mathrm{RL}}D_{b-\,t}^{\beta_j} q^\rho} = 0. \tag{5.1.2}$$

We mention that when α_j becomes an integer, ${}^{\mathrm{RL}}D_{a+}^{\alpha_j}$ and ${}^{\mathrm{RL}}D_{b-}^{\alpha_j}$ are replaced by the ordinary derivatives $(d/dt)^{\alpha_j}$ and $(-d/dt)^{\alpha_j}$, respectively. We notice that the method initiated by Agrawal [16] was applied by Baleanu and co-workers to several problems of physical interest [54, 55, 63, 75].

As an example, we consider a mechanical system described by the classical Lagrangian

$$L(x, y, z) = \dot{x}\dot{z} + yz^3. \tag{5.1.3}$$

Making use of (5.1.3), the classical solutions of the Euler-Lagrange equations can be written as

$$x(t) = at + b, \quad z(t) = 0. \tag{5.1.4}$$

We notice that $y(t)$ has an undetermined evolution and a and b are constants to be determined by making use of the initial conditions.

The next step is to generalize (5.1.3) to the fractional case. Among several possibilities we choose

$$L_f = ({}^{\mathrm{RL}}D_{a+}^{\alpha} x)\, {}^{\mathrm{RL}}D_{a+}^{\alpha} z + yz^3. \tag{5.1.5}$$

Thus, the fractional Euler-Lagrange equations of (5.1.5) become

$${}^{\mathrm{RL}}D_{b-}^{\alpha}({}^{\mathrm{RL}}D_{a+}^{\alpha} z) = 0, \quad z^3 = 0, \quad {}^{\mathrm{RL}}D_{b-}^{\alpha}({}^{\mathrm{RL}}D_{a+}^{\alpha} x) + 3yz^2 = 0. \tag{5.1.6}$$

From (5.1.6) we observe that $z = 0$, and we conclude that y is undetermined. We notice that the equation for $x(t)$ reads as follows

$${}^{\mathrm{RL}}D_{b-}^{\alpha}({}^{\mathrm{RL}}D_{a+}^{\alpha} x) = 0. \tag{5.1.7}$$

The solution of equation (5.1.7), when $1 < \alpha < 2$, is given by

$$\begin{aligned} x(t) &= A(t-a)^{\alpha-1} + B(t-a)^{\alpha-2} \\ &\quad + C(t-a)^{\alpha}\, {}_2F_1\left(1, 1-\alpha, 1+\alpha, \frac{t-a}{b-a}\right) \\ &\quad + D(t-a)^{\alpha}\, {}_2F_1\left(1, 2-\alpha, 1+\alpha, \frac{t-a}{b-a}\right), \end{aligned} \tag{5.1.8}$$

where ${}_2F_1$ is the Gauss hypergeometric function and A, B, C, D are real constants. When $\alpha \to 1^+$ and $a = 0$, the classical linear solution of one-dimensional space is recovered, namely

$$x(t) = A + Ct. \tag{5.1.9}$$

In conclusion we notice that the fractional dynamics includes, as a particular case, the classical one.

5.1.3 *Fractional Lagrangian formulation of field systems*

It is well known that the Euler-Lagrange equations are used to describe central force motion, scattering, perturbation theory, Noether's theorem and so on. Extensions to continuous and relativistic systems and classical electrodynamics were also investigated. Since the Euler-Lagrange equations are fundamental for the field theory, we briefly present their fractional generalization in the following.

We notice that a covariant form of the action would involve a classical Lagrangian density $\mathcal{L}$ via

$$S = \int \mathcal{L} d^4 x = \int \mathcal{L} d^3 x dt, \tag{5.1.10}$$

where

$$\mathcal{L} = \mathcal{L}(\phi, \partial_\mu \phi). \tag{5.1.11}$$

The classical covariant Euler-Lagrange equation corresponding to (5.1.10) is

$$\frac{\partial \mathcal{L}}{\partial \phi} - \partial_\mu \frac{\partial \mathcal{L}}{\partial(\partial_\mu \phi)} = 0, \tag{5.1.12}$$

where ϕ denotes the field variable.

The next step is to present the fractional generalization of (5.1.10) as it was reported in [75]. For these reasons, we consider the following action S, namely

$$S = \int \mathcal{L}\left[\phi(x), ({}^{\mathrm{RL}}D^{\alpha_k}_{a_k-}\phi(x)), ({}^{\mathrm{RL}}D^{\alpha_k}_{a_k+}\phi(x)), x\right] d^3 x dt. \tag{5.1.13}$$

Here $0 < \alpha_k \leq 1$, and the a_k correspond to x_1, x_2, x_3 and t respectively. In the following the limits of integration are $-\infty$ and $+\infty$, respectively.

By using the corresponding fractional integration by parts formula, the fractional Euler-Lagrange equation corresponding to (5.1.13) has the form

$$\frac{\partial \mathcal{L}}{\partial \phi} + \sum_{k=1}^{4} \left\{ ({}^{\mathrm{RL}}D^{\alpha_k}_{-\infty+}) \frac{\partial \mathcal{L}}{\partial({}^{\mathrm{RL}}D^{\alpha_k}_{\infty-})\phi} + ({}^{\mathrm{RL}}D^{\alpha_k}_{\infty-}) \frac{\partial \mathcal{L}}{\partial({}^{\mathrm{RL}}D^{\alpha_k}_{-\infty+})\phi} \right\} - 0, \tag{5.1.14}$$

see [75].

As it is expected for $\alpha_k \to 1$, Eq. (5.1.14) becomes the usual Euler-Lagrange equation for the classical fields [75].

5.1.4 *Fractional Euler-Lagrange equations with delay*

We believe that the inclusion of a delay (see, *e. g.*, [26, 27] and the references therein) in a fractional Lagrangian can give better results in many problems involving dynamics of complex systems. For example the fractional generalization of the Bloch equation [393, 395] that includes both fractional derivatives and time delays was investigated in [100]. The existence of the fractional derivative on the left side of the Bloch equation encodes a degree of system memory in the dynamic model for magnetization. The appearance of a time delay on the right side of the equation balances the equation by adding a degree of system memory [100]. Some other results on combined use of fractional derivatives and delay can be found in [83, 99, 153, 538] as well as in the references therein. In the following we present some results of fractional variational principles with both fractional derivatives and delay .

5.1.4.1 *Riemann-Liouville fractional Euler-Lagrange equations with delay*

First we state a lemma used later in the proofs, namely

Lemma 5.1. *Let $\alpha > 0$, $p, q \geq 1$, $r \in T = (t_1, t_2)$ and $\frac{1}{p} + \frac{1}{q} \leq 1 + \alpha$ ($p \neq 1$ and $q \neq 1$ in the case when $\frac{1}{p} + \frac{1}{q} = 1 + \alpha$).*

(a) *If $\varphi \in L_p(t_1, t_2)$ and $\psi \in L_q(t_1, t_2)$, then*

$$\int_{t_1}^{r} \varphi(t)({}^{\mathrm{RL}}I_{t_1+}^{\alpha}\psi)(t)dt = \int_{t_1}^{r} \psi(t)({}^{\mathrm{RL}}I_{r-}^{\alpha}\varphi)(t)dt \tag{5.1.15}$$

and hence, if $g \in {}^{\mathrm{RL}}I_{t_2-}^{\alpha}(L_p)$ and $f \in {}^{\mathrm{RL}}I_{t_1+}^{\alpha}(L_q)$, then

$$\int_{t_1}^{r} g(t)({}^{\mathrm{RL}}D_{t_1+}^{\alpha}f)(t)dt = \int_{t_1}^{r} f(t)({}^{\mathrm{RL}}D_{r-}^{\alpha}g)(t)dt \tag{5.1.16}$$

(b) *If $\varphi \in L_p(t_1, t_2)$ and $\psi \in L_q(t_1, t_2)$ then*

$$\int_{r}^{t_2} \varphi(t)({}^{\mathrm{RL}}I_{t_1+}^{\alpha}\psi)(t)dt = \int_{r}^{t_2} \psi(t)({}^{\mathrm{RL}}I_{t_2-}^{\alpha}\varphi)(t)dt \tag{5.1.17}$$
$$+ \frac{1}{\Gamma(\alpha)}\int_{t_1}^{r} \psi(t)\left(\int_{r}^{t_2} \phi(s)(s-t)^{\alpha-1}ds\right)dt$$

and hence, if $g \in {}^{\mathrm{RL}}I_{t_2-}^{\alpha}(L_p)$ *and* $f \in {}^{\mathrm{RL}}I_{t_1+}^{\alpha}(L_q)$, *then*

$$\begin{aligned}\int_r^{t_2} g(t)({}^{\mathrm{RL}}D_{t_1+}^{\alpha}f)(t)dt \\ = \int_r^{t_2} f(t)({}^{\mathrm{RL}}D_{t_2-}^{\alpha}g)(t)dt \\ - \frac{1}{\Gamma(\alpha)}\int_{t_1}^{r}({}^{\mathrm{RL}}D_{t_1+}^{\alpha}f)(t)\left(\int_r^{t_2}({}^{\mathrm{RL}}D_{t_2-}^{\alpha}g)(s)(s-t)^{\alpha-1}ds\right)dt\end{aligned} \tag{5.1.18}$$

which implies

$$\begin{aligned}\int_r^{t_2} g(t)({}^{\mathrm{RL}}D_{t_1+}^{\alpha}f)(t)dt \\ = \int_r^{t_2} f(t)({}^{\mathrm{RL}}D_{t_2-}^{\alpha}g)(t)dt \\ - \frac{1}{\Gamma(\alpha)}\int_{t_1}^{r} f(t)^{\mathrm{RL}}D_{r-}^{\alpha}\left(\int_r^{t_2}({}^{\mathrm{RL}}D_{t_2-}^{\alpha}g)(s)(s-t)^{\alpha-1}ds\right)dt.\end{aligned} \tag{5.1.19}$$

The proof can be seen in [73].

The next step is to analyze a modified problem when both the fractional derivatives and delay appear in the Lagrangian. The starting point is the one-dimensional problem [73]: Minimize

$$J(y) = \int_{t_1}^{t_2} F(t, y(t), {}^{\mathrm{RL}}D_{t_1+}^{\alpha}y(t), y(t-\tau), y'(t-\tau))dt, \tag{5.1.20}$$

in such a way that

$$y(t_2) = b, \qquad y(t) = \phi(t) \quad (t \in [t_1-\tau, t_1]), \tag{5.1.21}$$

where $t_1 < t_2$ and $0 < \tau < t_2 - t_1$.

By using the corresponding delay notations [208], namely

$$y_\tau = y(t-\tau), \quad y'_\tau = y'(t-\tau) \tag{5.1.22}$$

Eq. (5.1.20) is written as

$$J(y) = \int_{t_1}^{t_2} F(t, y(t), {}^{\mathrm{RL}}D_{t_1+}^{\alpha}y(t), y_\tau, y'_\tau)dt. \tag{5.1.23}$$

We are now in a position to formulate the following theorem.

Theorem 5.2. *Let* $J(y)$ *be a functional of the form* [73]

$$J(y) = \int_{t_1}^{t_2} F(t, y(t), {}^{\mathrm{RL}}D_{t_1+}^{\alpha}y(t), y(t-\tau), y'(t-\tau))dt, \tag{5.1.24}$$

defined on a set of continuous functions $y(t)$ which have continuous left Riemann-Liouville derivatives of order α in $[t_1, t_2]$ and satisfy the boundary conditions $y(t_1) = y(t_2) = b$, $y(t) = \phi(t)$, $t \in [t_1 - \tau, t_1]$ and $y(t_2) = a_2$. The necessary condition for $J(y)$ to possess an extremum for a given function $y(t)$ is that $y(t)$ fulfills the Euler-Lagrange equations

$$\begin{aligned} 0 = F_y(t) + F_{y_\tau}(t+\tau) + {}^{\mathrm{RL}}D^{\alpha}_{t_2-\tau,-} \frac{\partial F(t)}{\partial({}^{\mathrm{RL}}D^{\alpha}_{t_1+}y(t))} \\ - \frac{dF_{y'_\tau}(t+\tau)}{dt} \\ - \frac{1}{\Gamma(\alpha)} {}^{\mathrm{RL}}D^{\alpha}_{t_2-\tau,-} \int_{t_2-\tau}^{t_2} \left[{}^{\mathrm{RL}}D^{\alpha}_{t_2-} \frac{\partial F(t)}{\partial({}^{\mathrm{RL}}D^{\alpha}_{t_1+}y(t))}) \right] (z)(z-t)^{\alpha-1} dz \end{aligned} \tag{5.1.25}$$

for $t_1 \le t \le t_2 - \tau$,

$$F_y(t) + {}^{\mathrm{RL}}D^{\alpha}_{t_2-} \left(\frac{\partial F(t)}{\partial({}^{\mathrm{RL}}D^{\alpha}_{t_1+}y(t))} \right) = 0, \tag{5.1.26}$$

for $t_2 - \tau \le t \le t_2$, as well as the boundary condition

$$F_{y'_\tau}(t+\tau)\eta(t) \mid_{t_1}^{(t_2-\tau)^-} = 0. \tag{5.1.27}$$

From (5.1.25), (5.1.26) and (5.1.27) we conclude that when the delay terms are absent and in the limit $\alpha \to 1$ the classical results are reobtained.

We briefly mention the generalization of Theorem 5.2 with fixed end points and several functions. To this end, let us assume that the functional $J(y_1, y_2, \ldots, y_n)$ has the form [73]

$$\begin{aligned} J(y_1, y_2, \ldots, y_n) \\ = \int_{t_1}^{t_2} F(t, y_1(t), \ldots y_n(t), {}^{\mathrm{RL}}D^{\alpha}_{t_1+}y_1(t), \ldots, {}^{\mathrm{RL}}D^{\alpha}_{t_1+}y_n(t), \\ y_1(t-\tau), \ldots, y_n(t-\tau), y'_1(t-\tau), \ldots, y'_n(t-\tau))dt, \end{aligned} \tag{5.1.28}$$

and that it fulfills the boundary conditions

$$y_i(t_2) = y_{it_2}, \quad y_i(b) = y_{ib}, \quad y_i(t) = \phi_i(t) \ (i = 1, \ldots, n; \, t \in [t_1 - \tau, t_1]) \tag{5.1.29}$$

where $t_1 < t_2$ and $0 < \tau < t_2 - t_1$. Besides we assume that F possesses continuous partial derivatives with respect to all of its parameters and that the functions ϕ_i, $i = 1, 2, \ldots n$, are smooth.

Theorem 5.3. *A necessary condition for the curve $y_i = y_i(t), i = 1, \cdots, n$, fulfilling the boundary conditions (5.1.29) to be extremal for the functional (5.1.28) is that*

$$0 = F_{y_i}(t) + F_{(y_i)_\tau}(t+\tau) \tag{5.1.30}$$
$$+ {}^{\mathrm{RL}}D^{\alpha}_{t_2-\tau,-}\frac{\partial F(t)}{\partial({}^{\mathrm{RL}}D^{\alpha}_{t_1+}y_i(t))} - \frac{dF_{(y_i)'_\tau}(t+\tau)}{dt}$$
$$- \frac{1}{\Gamma(\alpha)}{}^{\mathrm{RL}}D^{\alpha}_{t_2-\tau,-}\int_{t_2-\tau}^{t_2}\left[{}^{\mathrm{RL}}D^{\alpha}_{t_2-}\left(\frac{\partial F(t)}{\partial({}^{\mathrm{RL}}D^{\alpha}_{t_1+}y_i(t))}\right)\right](z)(z-t)^{\alpha-1}dz$$

for $t_1 \le t \le t_2 - \tau$,

$$F_{y_i}(t) + {}^{\mathrm{RL}}D^{\alpha}_{t_2-}\left(\frac{\partial F(t)}{\partial({}^{\mathrm{RL}}D^{\alpha}_{t_1+}y_i(t))}\right) = 0 \tag{5.1.31}$$

for $t_2 - \tau \le t \le t_2$, and that the boundary conditions

$$\left(F_{(y_i)'_\tau}\right)(t+\tau)\eta(t)\,\Big|_{t_1}^{(t_2-\tau)^-} = 0 \tag{5.1.32}$$

are fulfilled for $i = 1, \ldots, n$.

For more details about this topic see [73].

5.1.4.2 *Caputo fractional Euler-Lagrange equations with delay*

The case when the Riemann-Liouville derivative is replaced by a Caputo derivative is of interest in its own right from both the theoretical and the applied point of view. Therefore, we now discuss the following problem [316]: Minimize

$$J(y) = \int_a^b F(t, y(t), {}^{\mathrm{C}}D_{a+}y(t), {}^{\mathrm{C}}D^{\beta}_{b-}y(t), y(t-\tau), y'(t-\tau))dt \tag{5.1.33}$$

such that

$$y(b) = c, \qquad y(t) = \phi(t)(\quad t \in [a-\tau, a]), \tag{5.1.34}$$

where $a < b$, $0 < \tau < b - a$, $0 < \alpha < 1$, $0 < \beta < 1$, and where c is a constant and F is a function with continuous first and second partial derivatives with respect to all of its arguments. The corresponding results are contained in the following theorem.

Theorem 5.4. *Let $J(y)$ be a functional of the form*

$$J(y) = \int_a^b F(t, y(t), {}^{\mathrm{C}}D^{\alpha}_{a+}y(t), {}^{\mathrm{C}}D^{\beta}_{b-}y(t), y(t-\tau), y'(t-\tau))dt, \tag{5.1.35}$$

with $0 < \alpha, \beta < 1$*, defined on a set of continuous functions* $y(t)$ *which have continuous left Caputo derivatives of order* α *and right derivatives of order* β *in* $[a, b]$ *and satisfy the conditions* $y(t) = \phi(t)$ $(t \in [a-\tau, a])$ *and* $y(b) = c$. *Moreover, let* $F : [a, b] \times \mathbb{R}^5 \to \mathbb{R}$ *have continuous partial derivatives with respect to all its arguments. Then, the necessary condition for* $J(y)$ *to possess an extremum for a given function* $y(t)$ *is that* $y(t)$ *satisfies the Euler-Lagrange equations*

$$\begin{aligned}
0 = &\frac{\partial F}{\partial y(t)}(t) + {}^{\mathrm{RL}}D^{\alpha}_{b-\tau,-}\left(\frac{\partial F}{\partial\, {}^{\mathrm{C}}D^{\alpha}_{a+}y(t)}\right)(t) + {}^{\mathrm{RL}}D^{\beta}_{a+}\left(\frac{\partial F}{\partial\, {}^{\mathrm{C}}D^{\beta}_{b-}y(t)}\right)(t)\\
&+ \frac{\partial F}{\partial y(t-\tau)}(t+\tau) - \frac{d}{dt}\frac{\partial F}{\partial y'(t-\tau)}(t+\tau)\\
&- \frac{1}{\Gamma(\alpha)}\,{}^{\mathrm{RL}}D^{\alpha}_{b-\tau,-}\left(\int_{b-\tau}^{b}\left({}^{\mathrm{RL}}D^{\alpha}_{b-}\frac{\partial F}{\partial\, {}^{\mathrm{C}}D^{\alpha}_{a+}y(t)}\right)(s)(s-t)^{\alpha-1}ds\right)
\end{aligned} \tag{5.1.36}$$

for $a \le t \le b - \tau$,

$$\begin{aligned}
0 = &\frac{\partial F}{\partial y(t)}(t) + {}^{\mathrm{RL}}D^{\alpha}_{b-}\left(\frac{\partial F}{\partial\, {}^{\mathrm{C}}D^{\alpha}_{a+}y(t)}\right)(t)\\
&+ {}^{\mathrm{RL}}D^{\beta}_{b-\tau,+}\left(\frac{\partial F}{\partial\, {}^{\mathrm{C}}D^{\beta}_{b-}y(t)}\right)(t)\\
&- \frac{1}{\Gamma(\beta)}\,{}^{\mathrm{RL}}D^{\beta}_{b-\tau,+}\left(\int_{a}^{b-\tau}\left({}^{\mathrm{RL}}D^{\beta}_{a+}\frac{\partial F}{\partial\, {}^{\mathrm{C}}D^{\beta}_{b-}y(t)}\right)(s)(s-t)^{\beta-1}ds\right)
\end{aligned} \tag{5.1.37}$$

for $b - \tau \le t \le b$*, and the transversality condition*

$$\frac{\partial F}{\partial y'(t-\tau)}(t+\tau)\eta(t)\ \Big|_a^{b-\tau} = 0, \tag{5.1.38}$$

for any admissible function η *satisfying* $\eta(t) = 0$ $(t \in [a-\tau, a])$ *and* $\eta(b) = 0$.

The generalization of the above theorem reads as follows (see [316]).

Theorem 5.5. *Let* $J(y_1, y_2, ..., y_n)$ *be a functional of the form*

$$\begin{aligned}
&J(y_1, y_2, ..., y_n)\\
&= \int_a^b F(t, y_1(t), y_2(t), ..., y_n(t), {}^{\mathrm{C}}D^{\alpha}_{a+}y_1(t), {}^{\mathrm{C}}D^{\alpha}_{a+}y_2(t), \ldots, {}^{\mathrm{C}}D^{\alpha}_{a+}y_n(t),\\
&\qquad {}^{\mathrm{C}}D^{\beta}_{b-}y_1(t), {}^{\mathrm{C}}D^{\beta}_{b-}y_2(t), \ldots, {}^{\mathrm{C}}D^{\beta}_{b-}y_n(t),\\
&\qquad y_1(t-\tau), y_2(t-\tau), \ldots, y_n(t-\tau),\\
&\qquad y_1'(t-\tau), y_2'(t-\tau), \ldots, y_n'(t-\tau))dt,
\end{aligned} \tag{5.1.39}$$

defined on a set of continuous functions $y_i(t)$, $i = 1, 2, \ldots, n$ that have left Caputo fractional derivatives of order α and right Caputo fractional derivatives of order β in the interval $[a, b]$ and satisfy the conditions

$$y_i(b) = c_i, \quad y_i(t) = \phi_i(t) \qquad (i = 1, 2 \ldots, n; t \in [a - \tau, a]), \tag{5.1.40}$$

where $a < b$, $0 < \tau < b - a$, the c_i are constant and the $\phi_i(t)$ are smooth. Assume also that $F : \mathbb{R}^{4n+1} \to \mathbb{R}$ has continuous second partial derivatives with respect to all of its arguments. Then, the necessary conditions for the curves $y_i(t)$, $i = 1, 2, \ldots, n$, satisfying the conditions (5.1.40) to be extremal for the functional (5.1.39) are

$$\begin{aligned} 0 = {} & \frac{\partial F}{\partial y_i(t)}(t) + {}^{\mathrm{RL}}D^{\alpha}_{b-\tau,-}\left(\frac{\partial F}{\partial^{\mathrm{C}} D^{\alpha}_{a+} y_i(t)}\right)(t) \\ & + {}^{\mathrm{RL}}D^{\beta}_{a+}\left(\frac{\partial F}{\partial^{\mathrm{C}} D^{\beta}_{b-} y_i(t)}\right)(t) + \frac{\partial F}{\partial y_i(t-\tau)}(t+\tau) - \frac{d}{dt}\frac{\partial F}{\partial y_i'(t-\tau)}(t+\tau) \\ & - \frac{1}{\Gamma(\alpha)}{}^{\mathrm{RL}}D^{\alpha}_{b-\tau,-}\left(\int_{b-\tau}^{b}\left({}^{\mathrm{RL}}D^{\alpha}_{b-}\frac{\partial F}{\partial^{\mathrm{C}} D^{\alpha}_{a+} y_i(t)}\right)(s)(s-t)^{\alpha-1}ds\right) \end{aligned} \tag{5.1.41}$$

for $a \le t \le b - \tau$ and $i = 1, 2, \ldots, n$,

$$\begin{aligned} 0 = {} & \frac{\partial F}{\partial y_i(t)}(t) + {}^{\mathrm{RL}}D^{\alpha}_{b-}\left(\frac{\partial F}{\partial^{\mathrm{C}} D^{\alpha}_{a+} y_i(t)}\right)(t) \\ & + {}^{\mathrm{RL}}D^{\beta}_{b-\tau,+}\left(\frac{\partial F}{\partial^{\mathrm{C}} D^{\beta}_{b-} y_i(t)}\right)(t) \\ & - \frac{1}{\Gamma(\beta)}{}^{\mathrm{RL}}D^{\beta}_{b-\tau,+}\left(\int_{a}^{b-\tau}\left({}^{\mathrm{RL}}D^{\beta}_{a+}\frac{\partial F}{\partial^{\mathrm{C}} D^{\beta}_{b-} y_i(t)}\right)(s)(t-s)^{\beta-1}ds)\right) \end{aligned} \tag{5.1.42}$$

for $b - \tau \le t \le b$ and $i = 1, 2, \ldots, n$, and the transversality conditions

$$\frac{\partial F}{\partial y_i'(t-\tau)}(t+\tau)\eta_i(t) \mid_a^{b-\tau} = 0, \qquad i = 1, 2, \ldots, n, \tag{5.1.43}$$

for any admissible functions η_i satisfying $\eta_i(t) = 0$ for $t \in [a - \tau, a]$ and $\eta_i(b) = 0$.

An example and more details about this topic are given in [316].

5.1.5 *Fractional discrete Euler-Lagrange equations*

The aim of this part is to present the formula of integration by parts as well as the corresponding discrete fractional Euler-Lagrange equations with the help of the right fractional sum and difference following the time scale calculus. This kind of calculus represents a unification of the theory of difference and differential equations, it unifies the integral and differential calculus with the calculus of finite differences [1, 2, 10, 36, 90, 110, 111, 307, 328]. As a result, we obtain a formalism to investigate a discrete-continuous dynamical system.

Let us assume that $\alpha > 0$ and $\sigma(s) = s + 1$. Then, the αth (forward) fractional sum of f was defined in [429] and it was used in [44] and [45] as

$$\Delta^{-\alpha} f(t) = \frac{1}{\Gamma(\alpha)} \sum_{s=a}^{t-\alpha} (t - \sigma(s))^{(\alpha-1)} f(s). \tag{5.1.44}$$

If $\alpha > 0$ and $\rho(s) = s - 1$, then we define the αth (backward) fractional sum of f as

$$\nabla^{-\alpha} f(t) = \frac{1}{\Gamma(\alpha)} \sum_{s=t+\alpha}^{b} (\rho(s) - t)^{(\alpha-1)} f(s). \tag{5.1.45}$$

Using the notation

$$N_a = \{a, a+1, a+2, a+3, \ldots\} \quad \text{and} \quad {}_bN = \{b, b-1, b-2, b-3, \ldots\},$$

a by-part formula relating these two operators was obtained in [2].

Proposition 5.1. *Let $\alpha > 0$, and consider $a, b \subset \mathbb{R}$ such that $a < b$ and $b \equiv a + \alpha(\mathrm{mod} 1)$. If f is defined on N_a and g is defined on ${}_bN$, then we obtain*

$$\sum_{s=a+\alpha}^{b} (\Delta^{-\alpha} f)(s) g(s) = \sum_{s=a}^{b-\alpha} f(s) \nabla^{-\alpha} g(s). \tag{5.1.46}$$

A by-part formula for fractional differences was also obtained in [2]:

Proposition 5.2. *Let $\alpha > 0$ be non-integer and let us assume that $b \equiv a + n - \alpha(\mathrm{mod} 1)$. If f is defined on ${}_bN$ and g is defined on N_a, then we have*

$$\sum_{s=a+(n-\alpha)+1}^{b-n+1} f(s) \Delta^{\alpha} g(s) = \sum_{s=a+n-1}^{b-(n-\alpha)+1} g(s) \nabla^{\alpha} f(s). \tag{5.1.47}$$

We mention that when $\alpha = 1$ we obtain the classical counterpart of the formula, namely

$$\sum_{s=a}^{b-1} f(s)\Delta g(s) = f(s)g(s)|_a^b - \sum_{s=a+1}^{b} g(s)\nabla f(s) \tag{5.1.48}$$

$$= f(s)g(s)|_a^b - \sum_{s=a}^{b-1} g(s+1)\Delta f(s).$$

The key point in this subsection is to consider the functional [2] $J : S \to \mathbb{R}$ with

$$J(y) = \sum_{s=a-\alpha}^{b} L(s, y(s), \Delta^\alpha y(s)) \tag{5.1.49}$$

where

$$a, b \in \mathbb{R}, \quad 0 < \alpha < 1, \quad b \equiv a - \alpha \ (\text{mod}1),$$

$$L : (N_{a-\alpha} \cap \ {}_bN) \times (\mathbb{R}^n)^2 \to \mathbb{R},$$

and

$$S = \{y : N_{a-\alpha} \cap {}_bN \to \mathbb{R}^n : y(a) = y_0 \text{ and } y(b+\alpha) = y_1\}.$$

Moreover, we assume that the function y fits the discrete time scales N_a and $N_{a-\alpha}$. That is, $y(s) = y(s-\alpha)$ for all $s \in N_a$.

From (5.1.49) we obtain the corresponding Euler-Lagrange equations as (*cf.* [2])

$$\frac{\partial L(s)}{\partial y} + \nabla^\alpha \frac{\partial L(s)}{\partial \Delta^\alpha y} = 0. \tag{5.1.50}$$

5.1.6 *Fractional Lagrange-Finsler geometry*

Since we are dealing with fractional differential geometry, some adequate symbols should be used. As a result, the left and the right Caputo derivatives were denoted in [612, 613] as

$${}_{1x}\overset{\alpha}{\underline{\partial}}_x f(x) = \frac{1}{\Gamma(s-\alpha)} \int_{1x}^{x} (x - x')^{s-\alpha-1} \left(\frac{\partial}{\partial x'}\right)^s f(x')dx', \tag{5.1.51a}$$

$${}_{x}\overset{\alpha}{\underline{\partial}}_{2x} f(x) = \frac{1}{\Gamma(s-\alpha)} \int_{x}^{2x} (x' - x)^{s-\alpha-1} \left(-\frac{\partial}{\partial x'}\right)^s f(x')dx', \tag{5.1.51b}$$

respectively, where, as usual, $s = \lceil \alpha \rceil$ denotes the smallest integer not less than α. For example, we emphasize that the integral is considered from ${}_1x$ to x in the symbol of partial derivative ${}_{1x}\overset{\alpha}{\underline{\partial}}_x$. We put α over a symbol in order to emphasize that the constructions are considered within fractional calculus with $\alpha \in (0,1)$.

We recall (see, *e. g.*, [330]) that a Lagrange space $L^n = (M, L)$ of integer dimension n is defined by a Lagrange fundamental function $L(x,y)$, namely a regular real function $L : TM \to \mathbb{R}$, where TM is the tangent bundle over the manifold M, such that the Hessian

$$ {}_L g_{ij} = \frac{1}{2} \frac{\partial^2 L}{\partial y^i \partial y^j} \qquad (5.1.52) $$

is not degenerate.

We say that a Lagrange space L^n is a Finsler space F^n if and only if its fundamental function L is positive and two-homogeneous with respect to the variables y^i, *i. e.* $L = F^2$. In the following we work with Lagrange spaces and their fractional generalizations, considering the Finsler spaces to form a more particular, homogeneous subclass.

Definition 5.1. A (target) fractional Lagrange space $\overset{\alpha}{\underline{L^n}} = (\overset{\alpha}{\underline{M}}, \overset{\alpha}{L})$ of fractional dimension $\alpha \in (0,1)$, is defined by a regular real function $\overset{\alpha}{L} : \overset{\alpha}{\underline{T}}M \to \mathbb{R}$, such that the fractional Hessian is [88]

$$ {}_L\overset{\alpha}{g}_{ij} = \frac{1}{4}\left(\overset{\alpha}{\underline{\partial}}_i \overset{\alpha}{\underline{\partial}}_j + \overset{\alpha}{\underline{\partial}}_j \overset{\alpha}{\underline{\partial}}_i \right) \overset{\alpha}{L} \neq 0. \qquad (5.1.53) $$

We consider the coefficients ${}_L\overset{\alpha}{g}{}^{ij}$ as being inverse to the ${}_L\overset{\alpha}{g}_{ij}$ given in Eq. (5.1.53). Any $\overset{\alpha}{\underline{L^n}}$ can be associated to a prime integer Lagrange space L^n [88].

The notion of nonlinear connection (N-connection) on $\overset{\alpha}{\underline{L^n}}$ can be defined in analogy to that on nonholonomic fractional manifolds [612, 613] by considering the fractional tangent bundle $\overset{\alpha}{\underline{T}}M$.

Definition 5.2. An N-connection $\overset{\alpha}{\mathbf{N}}$ on $\overset{\alpha}{\underline{T}}M$ is defined by a nonholonomic distribution (Whitney sum) with conventional h- and v-subspaces, $\underline{h\overset{\alpha}{T}M}$ and $\underline{v\overset{\alpha}{T}M}$, when

$$ \underline{\overset{\alpha}{T}\overset{\alpha}{T}M} = \underline{h\overset{\alpha}{T}M} \oplus \underline{v\overset{\alpha}{T}M}. \qquad (5.1.54) $$

Locally, a fractional N-connection is defined by a set of coefficients, $\overset{\alpha}{\mathbf{N}} = \{{}^{\alpha}N_i^a\}$, where

$$\overset{\alpha}{\mathbf{N}} = {}^{\alpha}N_i^a(u)(dx^i)^{\alpha} \otimes \overset{\alpha}{\underline{\partial}}_a. \tag{5.1.55}$$

Proposition 5.3 (see [88]). *There is a class of N-adapted fractional (co-) frames linearly depending on* ${}^{\alpha}N_i^a$,

$${}^{\alpha}\mathbf{e}_{\beta} = \left[{}^{\alpha}\mathbf{e}_j = \overset{\alpha}{\underline{\partial}}_j - {}^{\alpha}N_j^a \overset{\alpha}{\underline{\partial}}_a, \ {}^{\alpha}e_b = \overset{\alpha}{\underline{\partial}}_b \right], \tag{5.1.56}$$

$${}^{\alpha}\mathbf{e}^{\beta} = [{}^{\alpha}e^j = (dx^j)^{\alpha}, \ {}^{\alpha}\mathbf{e}^b = (dy^b)^{\alpha} + {}^{\alpha}N_k^b(dx^k)^{\alpha}]. \tag{5.1.57}$$

The above mentioned bases are nonholonomical and they are characterized by

$$[{}^{\alpha}\mathbf{e}_{\alpha}, \ {}^{\alpha}\mathbf{e}_{\beta}] = {}^{\alpha}\mathbf{e}_{\alpha}{}^{\alpha}\mathbf{e}_{\beta} - {}^{\alpha}\mathbf{e}_{\beta}{}^{\alpha}\mathbf{e}_{\alpha} = {}^{\alpha}W_{\alpha\beta}^{\gamma}\,{}^{\alpha}\mathbf{e}_{\gamma},$$

where the nontrivial nonholonomy coefficients ${}^{\alpha}W_{\alpha\beta}^{\gamma}$ are calculated as ${}^{\alpha}W_{ib}^a = \overset{\alpha}{\underline{\partial}}_b\,{}^{\alpha}N_i^a$ and ${}^{\alpha}W_{ij}^a = {}^{\alpha}\Omega_{ji}^a = {}^{\alpha}\mathbf{e}_i\,{}^{\alpha}N_j^a - {}^{\alpha}\mathbf{e}_j\,{}^{\alpha}N_i^a$, respectively. The values ${}^{\alpha}\Omega_{ji}^a$ define the coefficients of the N-connection curvature.

We consider the values $y^k(\tau) = dx^k(\tau)/d\tau$, for $x(\tau)$ parametrizing the smooth curves on a given manifold M such that $\tau \in [0,1]$. The fractional counterparts of these configurations are determined by replacing $d/d\tau$ with the fractional Caputo derivative $\overset{\alpha}{\underline{\partial}}_{\tau} = {}_{1\tau}\overset{\alpha}{\underline{\partial}}_{\tau}$ when ${}^{\alpha}y^k(\tau) = \overset{\alpha}{\underline{\partial}}_{\tau}x^k(\tau)$.

Using the techniques developed in [432, 611, 614] but for the case of the fractional derivatives and integrals, we formulate the following theorem:

Theorem 5.6. *Any* $\overset{\alpha}{L}$ *defines the fundamental geometric objects determining canonically a nonholonomic fractional Riemann–Cartan geometry on* $\overset{\alpha}{\underline{T}}M$ *being satisfied the properties* [88]*:*

(1) *The fractional Euler-Lagrange equations*

$$\overset{\alpha}{\underline{\partial}}_{\tau}\left({}_{1y^i}\overset{\alpha}{\underline{\partial}}_i\overset{\alpha}{L}\right) \quad {}_{1x^i}\overset{\alpha}{\underline{\partial}}_i\overset{\alpha}{L} = 0$$

are equivalent to the fractional nonlinear geodesic (equivalently, semi–spray) equations

$$\left(\overset{\alpha}{\underline{\partial}}_{\tau}\right)^2 x^k + 2\overset{\alpha}{G^k}(x, {}^{\alpha}y) = 0,$$

where

$$\overset{\alpha}{G}{}^k = \frac{1}{4}\, {}_L\overset{\alpha}{g}{}^{kj}\left[y^j\, {}_{1y^j}\overset{\alpha}{\underline{\partial}}_j\left({}_{1x^i}\overset{\alpha}{\underline{\partial}}_i\overset{\alpha}{L}\right) - {}_{1x^i}\overset{\alpha}{\underline{\partial}}_i\overset{\alpha}{L}\right]$$

defines the canonical N-connection

$$ {}^{\alpha}_{L}N^a_j = {}_{1y^j}\overset{\alpha}{\underline{\partial}}_j\overset{\alpha}{G}{}^k(x, {}^{\alpha}y). \tag{5.1.58}$$

(2) *There is a canonical (Sasaki type) metric structure,*

$$ {}_L\overset{\alpha}{\mathbf{g}} = {}^{\alpha}_{L}g_{kj}(x,y)\ {}^{\alpha}e^k \otimes\ {}^{\alpha}e^j +\ {}^{\alpha}_{L}g_{cb}(x,y)\ {}^{\alpha}_{L}\mathbf{e}^c \otimes\ {}^{\alpha}_{L}\mathbf{e}^b, \tag{5.1.59}$$

where the preferred frame structure (defined linearly by ${}^{\alpha}_{L}N^a_j$*) is* ${}^{\alpha}_{L}\mathbf{e}_\nu = ({}^{\alpha}_{L}\mathbf{e}_i, e_a)$.

(3) *There is a canonical metrical distinguished connection*

$$ {}^{\alpha}_{c}\mathbf{D} = (h\ {}^{\alpha}_{c}D, v\ {}^{\alpha}_{c}D) = \{{}^{\alpha}_{c}\mathbf{\Gamma}^{\gamma}{}_{\alpha\beta} = ({}^{\alpha}\widehat{L}^i{}_{jk},\ {}^{\alpha}\widehat{C}^i_{jc})\}\ ,$$

which is a linear connection preserving under parallelism the splitting (5.1.54) and metric compatible, i. e. ${}^{\alpha}_{c}\mathbf{D}\left({}_L\overset{\alpha}{\mathbf{g}}\right) = 0$, *when*

$$ {}^{\alpha}_{c}\mathbf{\Gamma}^i{}_j = {}^{\alpha}_{c}\mathbf{\Gamma}^i{}_{j\gamma}\ {}^{\alpha}_{L}\mathbf{e}^\gamma = \widehat{L}^i{}_{jk}e^k + \widehat{C}^i_{jc}\ {}^{\alpha}_{L}\mathbf{e}^c,$$

for $\widehat{L}^i{}_{jk} = \widehat{L}^a{}_{bk}, \widehat{C}^i_{jc} = \widehat{C}^a_{bc}$ *in* ${}^{\alpha}_{c}\mathbf{\Gamma}^a{}_b = {}^{\alpha}_{c}\mathbf{\Gamma}^a{}_{b\gamma}\ {}^{\alpha}_{L}\mathbf{e}^\gamma = \widehat{L}^a{}_{bk}e^k + \widehat{C}^a_{bc}\ {}^{\alpha}_{L}\mathbf{e}^c$, *and*

$$\begin{aligned} {}^{\alpha}\widehat{L}^i_{jk} &= \frac{1}{2}\ {}^{\alpha}_{L}g^{ir}\left({}^{\alpha}_{L}\mathbf{e}_k\ {}^{\alpha}_{L}g_{jr} +\ {}^{\alpha}_{L}\mathbf{e}_j\ {}^{\alpha}_{L}g_{kr} -\ {}^{\alpha}_{L}\mathbf{e}_r\ {}^{\alpha}_{L}g_{jk}\right), \\ {}^{\alpha}\widehat{C}^a_{bc} &= \frac{1}{2}\ {}^{\alpha}_{L}g^{ad}\left({}^{\alpha}e_c\ {}^{\alpha}_{L}g_{bd} +\ {}^{\alpha}e_c\ {}^{\alpha}_{L}g_{cd} -\ {}^{\alpha}e_d\ {}^{\alpha}_{L}g_{bc}\right) \end{aligned} \tag{5.1.60}$$

are the generalized Christoffel indices.

5.1.7 *Applications*

5.1.7.1 *Fractional variational principles with Riesz derivatives*

For $0 < \alpha < 1$ we define the Riesz derivative

$$ {}^{\mathrm{R}}D^{\alpha}_{[a,b]}q(t) = \frac{1}{2}({}^{\mathrm{RL}}D^{\alpha}_{a+}q(t) - {}^{\mathrm{RL}}D^{\alpha}_{b-}q(t))$$

and the Riesz-Caputo derivative

$$ {}^{\mathrm{RC}}D^{\alpha}_{[a,b]}q(t) = \frac{1}{2}({}^{\mathrm{C}}D^{\alpha}_{a+}q(t) - {}^{\mathrm{C}}D^{\alpha}_{b-}q(t)),$$

respectively.

Let us assume that $L_f(t, q, {}^{\mathrm{RC}}D^{\alpha}_{[a,b]}q(t))$ is a function with continuous first and second (partial) derivatives with respect to all its arguments. Then, the fractional Euler-Lagrange equations with Riesz's derivatives are described by the following theorem [20].

Theorem 5.7. *Let $J[q]$ be a functional of the form*

$$J[q] = \int_a^b L_f(t, q, {}^{\mathrm{RC}}D^{\alpha}_{[a,b]}q(t))dt$$

defined on the set of functions which have continuous Riesz-Caputo fractional derivative of order α in $[a, b]$ and which satisfy the boundary conditions $q(a) = q_a$ and $q_b = q_b$. Then a necessary condition for $J[q]$ to have a maximum for a given function $q(t)$ is that $q(t)$ satisfies the Euler-Lagrange equation

$$\frac{\partial L_f}{\partial q} - {}^{\mathrm{R}}D^{\alpha}_{[a,b]}\left(\frac{\partial L_f}{\partial\, {}^{\mathrm{RC}}D^{\alpha}_{[a,b]}q}\right) = 0. \tag{5.1.61}$$

5.1.7.2 *Multi-order and multi-term fractional variational formulations with Hilfer derivatives*

Hilfer [304, 604] has introduced the left two-parameter fractional derivative

$${}^{\mathrm{Hi}}D^{\alpha,\beta}_{a+}y(t) = {}^{\mathrm{RL}}I^{(1-\beta)(n-\alpha)}_{a+}\, D^n\, {}^{\mathrm{RL}}I^{\beta(n-\alpha)}_{a+}y(t) \tag{5.1.62}$$

and the right two-parameter fractional derivative

$${}^{\mathrm{Hi}}D^{\alpha,\beta}_{b-}y(t) = {}^{\mathrm{RL}}I^{(1-\beta)(n-\alpha)}_{b-}\, (-D)^n\, {}^{\mathrm{RL}}I^{\beta(n-\alpha)}_{b-}y(t) \tag{5.1.63}$$

where $\alpha \notin \mathbb{N}$ denotes the order of the fractional derivative, $n = \lceil\alpha\rceil$, and $\beta \in [0, 1]$ represents a parameter. It is easily seen that these derivatives reduce to the Caputo operators for $\beta = 0$ and to the Riemann-Liouville operators for $\beta = 1$.

A functional in terms of multiple fractional Hilfer derivatives [304, 604] can be defined as [28]

$$J[y] = \int_a^b F(t, y, {}^{\mathrm{Hi}}D^{\alpha_{11},\beta_{11}}_{a+}y, \ldots, {}^{\mathrm{Hi}}D^{\alpha_{1n},\beta_{1n}}_{a+}y, {}^{\mathrm{Hi}}D^{\alpha_{21},\beta_{21}}_{b-}y, \ldots, {}^{\mathrm{Hi}}D^{\alpha_{2m},\beta_{2m}}_{b-}y)dt. \tag{5.1.64}$$

For this functional, the corresponding Euler-Lagrange equation and the terminal conditions have been derived in [28]; they take the forms

$$0 = \frac{\partial F}{\partial y} + \sum_{j=1}^{n} {}^{\mathrm{Hi}}D_{b-}^{\alpha_{1j},1-\beta_{1j}} \frac{\partial F}{\partial\, {}^{\mathrm{Hi}}D_{a+}^{\alpha_{1j},\beta_{1j}} y} + \sum_{k=1}^{m} {}^{\mathrm{Hi}}D_{a+}^{\alpha_{2k},1-\beta_{2k}} \frac{\partial F}{\partial\, {}^{\mathrm{Hi}}D_{b-}^{\alpha_{2k},\beta_{2k}} y}, \tag{5.1.65}$$

$$0 = \left[\sum_{j=1}^{n} {}^{\mathrm{RL}}I_{b-}^{(1-\beta_{1j})(1-\alpha_{1j})} \frac{\partial F}{\partial\, {}^{\mathrm{Hi}}D_{a+}^{\alpha_{1j},\beta_{1j}} y} {}^{\mathrm{RL}}I_{a+}^{\beta_{1j}(1-\alpha_{1j})} \eta(t)\right]_{t=a} - \left[\sum_{k=1}^{m} {}^{\mathrm{RL}}I_{a+}^{(1-\beta_{2k})(1-\alpha_{2k})} \frac{\partial F}{\partial\, {}^{\mathrm{Hi}}D_{b-}^{\alpha_{2k},\beta_{2k}} y} {}^{\mathrm{RL}}I_{b-}^{\beta_{2k}(1-\alpha_{2k})} \eta(t)\right]_{t=a}, \tag{5.1.66}$$

and

$$0 = \left[\sum_{j=1}^{n} {}^{\mathrm{RL}}I_{b-}^{(1-\beta_{1j})(1-\alpha_{1j})} \frac{\partial F}{\partial\, {}^{\mathrm{Hi}}D_{a+}^{\alpha_{1j},\beta_{1j}} y} {}^{\mathrm{RL}}I_{a+}^{\beta_{1j}(1-\alpha_{1j})} \eta(t)\right]_{t=b} - \left[\sum_{k=1}^{m} {}^{\mathrm{RL}}I_{a+}^{(1-\beta_{2k})(1-\alpha_{2k})} \frac{\partial F}{\partial\, {}^{\mathrm{Hi}}D_{b-}^{\alpha_{2k},\beta_{2k}} y} {}^{\mathrm{RL}}I_{b-}^{\beta_{2k}(1-\alpha_{2k})} \eta(t)\right]_{t=b}. \tag{5.1.67}$$

When the functional has multiple functions, then it can be written as

$$J[y] = \int_a^b F(t, \mathbf{y}, {}_aD_t^{\alpha_1,\beta_1}\mathbf{y}, {}_tD_b^{\alpha_2,\beta_2}\mathbf{y})dt, \tag{5.1.68}$$

where $\mathbf{y} = [y_1(t), \cdots, y_m(t)]^T$ represents an m-dimensional vector. For this functional the Euler-Lagrange equation and the terminal conditions are written as [28]

$$0 = \frac{\partial F}{\partial \mathbf{y}} + {}^{\mathrm{Hi}}D_{b-}^{\alpha_1,1-\beta_1} \frac{\partial F}{\partial\, {}^{\mathrm{Hi}}D_{a+}^{\alpha_1,\beta_1}\mathbf{y}} + {}^{\mathrm{Hi}}D_{a+}^{\alpha_2,1-\beta_2} \frac{\partial F}{\partial\, {}^{\mathrm{Hi}}D_{b-}^{\alpha_2,\beta_2}\mathbf{y}}, \tag{5.1.69}$$

$$0 = \left[{}^{\mathrm{RL}}I_{b-}^{(1-\beta_1)(1-\alpha_1)} \frac{\partial F}{\partial\, {}^{\mathrm{Hi}}D_{a+}^{\alpha_1,\beta_1} y_j} {}^{\mathrm{RL}}I_{a+}^{\beta_1(1-\alpha_1)} \eta_j(t)\right]_{t=a} - \left[{}^{\mathrm{RL}}I_{a+}^{(1-\beta_2)(1-\alpha_2)} \frac{\partial F}{\partial\, {}^{\mathrm{Hi}}D_{b-}^{\alpha_2,\beta_2} y_j} {}^{\mathrm{RL}}I_{b-}^{\beta_2(1-\alpha_2)} \eta_j(t)\right]_{t=a} \tag{5.1.70}$$

for $j = 1, 2 \ldots, m$, and

$$0 = \left[{}^{\mathrm{RL}}I_{b-}^{(1-\beta_1)(1-\alpha_1)} \frac{\partial F}{\partial\, {}^{\mathrm{Hi}}D_{a+}^{\alpha_1,\beta_1} y_j} {}^{\mathrm{RL}}I_{a+}^{\beta_1(1-\alpha_1)} \eta_j(t) \right]_{t=b} - \left[{}^{\mathrm{RL}}I_{a+}^{(1-\beta_2)(1-\alpha_2)} \frac{\partial F}{\partial\, {}^{\mathrm{Hi}}D_{b-}^{\alpha_2,\beta_2} y_j} {}^{\mathrm{RL}}I_{b-}^{\beta_2(1-\alpha_2)} \eta_j(t) \right]_{t=b} \quad (5.1.71)$$

for $j = 1, \ldots, m$.

We notice that when a functional contains both, multiple fractional derivatives and multiple functions, the necessary condition for this function can be obtained by making use of the same procedure used here.

5.1.7.3 *A fractional Lagrangian approach of Schrödinger equations*

The variant of the Schrödinger equation containing fractional derivatives was proposed by several researchers. In particular, in [207] the authors solved the fractional Schrödinger equation by making use of the quantum Riesz fractional operator suggested in [358]. A Caputo version of the time-fractional Schrödinger equation was suggested in [449], and a generalized fractional Schrödinger equation with space-time fractional derivatives was developed in [623]. Moreover, in [444] the fractional Schrödinger equation was obtained with a fractional variational principle and a fractional Klein-Gordon equation.

The classical Schrödinger equation in 1+1 dimensions is given by

$$i\hbar \frac{d\psi}{dt} + \frac{\hbar^2}{2m} \triangle\psi - V(x)\psi = 0. \quad (5.1.72)$$

The expression of the classical Lagrangian corresponding to (5.1.72) is given by

$$\mathcal{L} = \frac{i\hbar}{2}(\psi^\dagger \frac{d\psi}{dt} - \psi \frac{d\psi^\dagger}{dt}) - \frac{\hbar^2}{2m}(\nabla\psi\nabla\psi^\dagger) - V(x)\psi\psi^\dagger. \quad (5.1.73)$$

Based on (5.1.73), the Lagrangian density for the fractional Schrödinger equation can be written in the form

$$\mathcal{L}' = \frac{i\hbar^\alpha}{2}[\psi^\dagger({}^{\mathrm{RL}}D_{a+,t}^{\alpha}\psi) + \psi({}^{\mathrm{RL}}D_{b-,t}^{\alpha}\psi^\dagger)] + \frac{\hbar^{2\alpha}}{2m^\alpha}[{}^{\mathrm{RL}}D_{a+,x}^{\alpha}\psi({}^{\mathrm{RL}}D_{b-,x}^{\alpha}\psi^\dagger)] - V(x)\psi\psi^\dagger. \quad (5.1.74)$$

The corresponding equations for ψ and $\psi^\dagger$ are

$$i\hbar^\alpha({}^{\mathrm{RL}}D_{a+,t}^{\alpha}\psi) + \frac{\hbar^{2\alpha}}{2m^\alpha} {}^{\mathrm{RL}}D_{a+,x}^{\alpha}({}^{\mathrm{RL}}D_{a+,x}^{\alpha}\psi) - V(x)\psi = 0, \quad (5.1.75)$$

$$i\hbar^{\alpha}({}^{\mathrm{RL}}D^{\alpha}_{b-,t}\psi^{\dagger}) + \frac{\hbar^{2\alpha}}{2m^{\alpha}}{}^{\mathrm{RL}}D^{\alpha}_{b-,x}({}^{\mathrm{RL}}D^{\alpha}_{b-,x}\psi^{\dagger}) - V(x)\psi = 0, \tag{5.1.76}$$

respectively.

5.1.7.4 *Fractional Lagrangians which differ by a fractional Riesz derivative*

Following [59], let us consider the classical Lagrangian $L(q^{\sigma}(t), \dot{q}^{\sigma}(t))$, $\sigma \in \{1, \ldots, n\}$, together with its fractional counterpart

$$L_f(q^{\sigma}(t), {}^{\mathrm{RC}}D^{\alpha}_{[a,b]}q^{\sigma}(t)). \tag{5.1.77}$$

with the aim of presenting the generalization of the notion of the equivalent Lagrangians in the fractional case. We add ${}^{\mathrm{RC}}D^{\alpha}_{[a,b]}q(t)$ to (5.1.77) and we get a new fractional Lagrangian,

$$L = L_f(q^{\sigma}(t), {}^{\mathrm{RC}}D^{\alpha}_{[a,b]}q^{\sigma}(t)) + C({}^{\mathrm{RC}}D^{\alpha}_{[a,b]}q^{m}(t)). \tag{5.1.78}$$

Here m represents the given coordinate while C denotes a real constant. The new fractional Euler-Lagrange equations arising from (5.1.78) have been shown in [59] to take the form

$$\frac{\partial L_f}{\partial q^{\sigma}(t)} + {}^{\mathrm{R}}D^{\alpha}_{[a,b]}\frac{\partial L_f}{\partial\, {}^{\mathrm{RC}}D^{\alpha}_{[a,b]}q^{\sigma}(t)} = 0, \tag{5.1.79}$$

$$\frac{\partial L_f}{\partial q^{m}(t)} + {}^{\mathrm{R}}D^{\alpha}_{[a,b]}\frac{\partial L_f}{\partial\, {}^{\mathrm{RC}}D^{\alpha}_{[a,b]}q^{m}(t)} + \frac{C}{2}\left(\frac{(t-a)^{-\alpha}}{\Gamma(1-\alpha)} - \frac{(b-t)^{-\alpha}}{\Gamma(1-\alpha)}\right) = 0, \tag{5.1.80}$$

$$\frac{\partial L_f}{\partial q^{\delta}(t)} + {}^{\mathrm{R}}D^{\alpha}_{[a,b]}\frac{\partial L_f}{\partial\, {}^{\mathrm{RC}}D^{\alpha}_{[a,b]}q^{\delta}(t)} = 0, \tag{5.1.81}$$

for $\sigma = 1, \ldots, m-1$ and $\delta = m+1, \ldots, n$. We notice that the last term of (5.1.80) characterizes the fractional dynamics [59]. However, when $\alpha = 1$ we recover the classical case.

5.1.7.5 *Euler-Lagrange equations in fractional space*

Let us now consider the action function of the form

$$S = \frac{1}{\Gamma(\alpha)}\int_a^b L_f(t, {}^{\mathrm{RL}}D^{\beta}_{a+}q, {}^{\mathrm{RL}}D^{\gamma}_{b-}q)(t-\tau)^{\alpha-1}d\tau, \tag{5.1.82}$$

where $0 \leq \alpha \leq 1$, $0 \leq \gamma \leq 1$, $0 \leq \beta \leq 1$. From [446] we conclude that the Euler-Lagrange equations corresponding to (5.1.82) are given by

$$0 = \frac{\partial L_f}{\partial q} + \frac{1}{(t-\tau)^{\alpha-1}} {}^{\mathrm{RL}}D_{b-}^{\beta} \left(\frac{\partial L_f}{\partial\, {}^{\mathrm{RL}}D_{a+}^{\beta}} (t-\tau)^{\alpha-1} \right) \tag{5.1.83}$$
$$+ \frac{1}{(t-\tau)^{\alpha-1}} + {}^{\mathrm{RL}}D_{a+}^{\gamma} \left(\frac{\partial L_f}{\partial\, {}^{\mathrm{RL}}D_{b-}^{\gamma}} (t-\tau)^{\alpha-1} \right).$$

When $\beta = 1$ and $\gamma = 1$, the classical results are reobtained, namely

$$\frac{\partial L}{\partial q} - \frac{d}{dt}\left(\frac{\partial L}{\partial \dot{q}}\right) - \frac{\alpha-1}{t-\tau}\left(\frac{\partial L}{\partial \dot{q}}\right) = 0. \tag{5.1.84}$$

5.1.7.6 *Multi time fractional Lagrangian equations*

We consider in the following the multi time variable defined as

$$(t^{\nu}) = (t^1, t^2, \ldots, t^p) \in \mathbb{R}^p, \qquad \nu = 1, \ldots, p. \tag{5.1.85}$$

Let us denote the limits of the range of integration by $a = (a^{\nu})$ and $b = (b^{\nu})$, and define the functions as

$$q^i : \mathbb{R}^p \to \mathbb{R}, \quad (t^1, t^2, \ldots, t^p) \mapsto q^i(t^1, t^2, \ldots, t^p), \quad i = 1, \ldots, n. \tag{5.1.86}$$

Following [66], the multi time Lagrangian is written as

$$L_f : \mathbb{R}^{p+n+np} \to \mathbb{R}, \quad (t^{\nu}, q^i, ({}^{\mathrm{RL}}D_{a+,t^{\nu}}^{\alpha})q^i) \mapsto L_f(t^{\nu}, q^i, \ ({}^{\mathrm{RL}}D_{a+,t^{\nu}}^{\alpha})q^i) \tag{5.1.87}$$

and as a result, the corresponding fractional Euler-Lagrangian equations of (5.1.87) have the form

$$\frac{\partial L_f}{\partial q^i} + {}^{\mathrm{RL}}D_{b-,t^{\nu}}^{\alpha} \frac{\partial L_f}{\partial ({}^{\mathrm{RL}}D_{a+,t^{\nu}}^{\alpha} q^i)} = 0. \tag{5.1.88}$$

5.1.7.7 *Fractional Faddeev-Jackiw formalism*

The Faddeev-Jackiw formalism [230] is one of the most frequently used approaches to deal with the reduction of constrained and unconstrained systems. In the Faddeev-Jackiw procedure we start from a first order Lagrangian in time derivative and after that we write down the Lagrange density

$$L(q^i, \dot{q}^i, p_i, \dot{p}_i) = \frac{1}{2}\begin{pmatrix} p_i & q^i \end{pmatrix}\begin{pmatrix} 0 & 1 \\ -1 & 0 \end{pmatrix}\begin{pmatrix} \dot{p}^i \\ \dot{q}_i \end{pmatrix} - H(p_i, q^i). \tag{5.1.89}$$

It is possible to construct a first order in time derivatives Lagrangian in such a way that the configuration space coincides with the Hamiltonian phase space. This property can be achieved by enlarging the given n-dimensional configuration space to a $2n$-dimensional configuration space. In the fractional mechanics approach let us consider, as in [262], the fractional Lagrangian

$$L_f = \frac{1}{2}(q^i \,{}^{\mathrm{RL}}D^{\alpha}_{b-} p_i + p_i \,{}^{\mathrm{RL}}D^{\alpha}_{a+} q^i) - H(q^i, p_i) \tag{5.1.90}$$

that can be written in matrix notation as

$$L_f = \frac{1}{2}\begin{pmatrix} \xi_n \, \xi_m \end{pmatrix}\begin{pmatrix} 0 & 1 \\ 1 & 0 \end{pmatrix}\begin{pmatrix} {}^{\mathrm{RL}}D^{\alpha}_{a+}\xi_n \\ {}^{\mathrm{RL}}D^{\alpha}_{b-}\xi_m \end{pmatrix} - H(q,p), \tag{5.1.91}$$

where $\xi_i = p_i$ for $i = 1, \ldots, n$ and $\xi_i = q^i$ for $i = n+1, \ldots, 2n$. The corresponding Euler-Lagrange equations of (5.1.91) have the form [262]

$$\frac{\partial L_f}{\partial q^i} + {}^{\mathrm{RL}}D^{\alpha}_{b-}\frac{\partial L_f}{\partial \,{}^{\mathrm{RL}}D^{\alpha}_{a+} q^i} = 0, \tag{5.1.92}$$

$$\frac{\partial L_f}{\partial p_i} + {}^{\mathrm{RL}}D^{\alpha}_{a+}\frac{\partial L_f}{\partial \,{}^{\mathrm{RL}}D^{\alpha}_{b-} p_i} = 0. \tag{5.1.93}$$

The equations that we have obtained are equivalent with the fractional Hamilton equations, namely

$$\begin{pmatrix} {}^{\mathrm{RL}}D^{\alpha}_{a+}\xi^n \\ {}^{\mathrm{RL}}D^{\alpha}_{b-}\xi^m \end{pmatrix} = \begin{pmatrix} 0 & 1 \\ 1 & 0 \end{pmatrix}\begin{pmatrix} \partial H/\partial \xi^n \\ \partial H/\partial \xi^m \end{pmatrix}. \tag{5.1.94}$$

As a result, the basic Faddeev-Jackiw brackets become

$$\{q^i, p_j\}_{FJ} = \delta_{ij}, \qquad \{q^i, q^j\}_{FJ} = \{p_i, p_j\}_{FJ} = 0. \tag{5.1.95}$$

Finally, in bracket notation we obtain the following equations [262]

$$\begin{pmatrix} {}^{\mathrm{RL}}D^{\alpha}_{a+}\xi^n \\ {}^{\mathrm{RL}}D^{\alpha}_{b-}\xi^m \end{pmatrix} = \{\xi^m, \xi^n\}\begin{pmatrix} \partial H/\partial \xi^n \\ \partial H/\partial \xi^m \end{pmatrix}, \tag{5.1.96}$$

where

$$\{\xi^m, \xi^n\} = \begin{pmatrix} 0 & 1 \\ 1 & 0 \end{pmatrix}. \tag{5.1.97}$$

We notice that this matrix is symmetric while in the classical case it is anti-symmetric [262].

5.1.7.8 *Fractional variational calculus with generalized boundary condition*

We consider the type of fractional variational calculus induced by functionals of the form

$$J(y) = {}^{\mathrm{RL}}I_{a+}^{\gamma} L_f(x, y(x), {}^{\mathrm{RL}}D_{a+}^{\alpha} y, y(a)). \tag{5.1.98}$$

According to [298], the corresponding Euler-Lagrange equation becomes

$$\frac{(x-t)^{\gamma-1}}{\Gamma(\gamma)}\frac{\partial L_f}{\partial y} + {}^{\mathrm{C}}D_{x-}^{\alpha}\left(\frac{(x-t)^{\gamma-1}}{\Gamma(\gamma)}\frac{\partial L_f}{\partial^{\mathrm{RL}} D_{a+}^{\alpha} y}\right) = 0 \tag{5.1.99}$$

with the natural boundary condition (transversality conditions)

$$\left(\frac{(x-t)^{\gamma-1}}{\Gamma(\gamma)}\frac{\partial L_f}{\partial\, {}^{\mathrm{C}}D_{a+}^{\alpha} y}\right)\bigg|_{t=x} = 0. \tag{5.1.100}$$

If $y(a)$ is given then we have $\eta(a) = 0$, but in the opposite case we obtain the boundary condition

$$\int_a^x \frac{(x-t)^{\gamma-1}}{\Gamma(\gamma)}\frac{\partial L_f}{\partial y(a)} dt = 0. \tag{5.1.101}$$

We notice that these conditions are only necessary for an extremum [298].

As a special case, we consider the following problem. If y is a local extremizer to

$$J(y) = \int_a^b L_f(t, y(t), {}^{\mathrm{RL}}D_{a+}^{\alpha} y) dt, \tag{5.1.102}$$

and if we consider $\gamma = 1$ and $x = b$ in (5.1.99), (5.1.100) and (5.1.101), respectively, then, according to [298], we obtain the fractional Euler-Lagrange equation as

$$\frac{\partial L_f}{\partial y} + {}^{\mathrm{C}}D_{b-}^{\alpha}\left(\frac{\partial L_f}{\partial\, {}^{\mathrm{RL}}D_{a+}^{\alpha} y}\right) = 0 \tag{5.1.103}$$

for all $t \in [a, b]$, with the corresponding boundary condition

$$\left(\frac{\partial L_f}{\partial^{\mathrm{RL}} D_{a+}^{\alpha} y}\right)\bigg|_{t=b} = 0. \tag{5.1.104}$$

The sufficient conditions for the existence of an extremum were discussed elsewhere [298]. More details about this topic can be found in [19, 21, 297, 403].

5.2 Fractional Hamiltonian Dynamics

5.2.1 *Introduction and overview of results*

The classical dynamics of Hamiltonian systems is characterized by conservation of phase-space volume under time evolution. This aspect is a cornerstone of conventional statistical mechanics of Hamiltonian systems. The main characteristic of the fractional Hamiltonian is its nonlocality. For these reasons, several techniques were applied to discuss this issue in the fractional case.

A Hamiltonian formalism was developed for non-local field theories in d space-time dimensions by considering auxiliary $(d + 1)$-dimensional field theories which are local with respect to the evolution time [264]. The physical reduced space of non-local theories around the fixed points of these systems were analyzed in [265]. The space-time non-commutative field theories are acausal and the unitarity is lost as it was shown in [254, 562].

The higher-derivatives theories [258, 453] appear naturally as corrections to general relativity and cosmic strings.

The Hamiltonian treatment of non-local theories and Ostrogradski's formalism [258, 293] was discussed in [95] such that we recast the second class constraints into first class constraints by invoking the boundary Poisson bracket.

The passage from the Lagrangian containing fractional derivatives to the Hamiltonian was achieved in [506]. Also, the fractional Hamilton's equations of motion were obtained in a similar manner to the classical mechanics [506].

The definition of the fractional Hessian matrix within Riemann-Liouville fractional derivatives was introduced in [54].

Fractional Hamiltonian equations in terms of Riesz fractional derivatives were obtained in [20].

Using the fact that an extremum of variation of generalized action can lead to the fractional dynamics in the case of systems with long-range interaction and long-term memory function, the generalized Noether's theorem and Hamiltonian type equation were investigated in [590].

A generalized classical mechanics with fractional derivatives based on the time-clock randomization of momenta and coordinates taken from the conventional phase space was considered in [579]. The fractional equations of motion were derived using the Hamiltonian formalism.

It was proved in [262] that for the fractional constraint Hamiltonian formulation, by using Dirac brackets, we obtain the same equations as those obtained from fractional Euler-Lagrange equations.

The fractional Hamilton-Jacobi formulations for systems containing Riesz fractional derivatives was derived in [507]. The classical Nambu mechanics was generalized to involve fractional derivatives using two approaches, namely by using the fractional exterior derivative as well as by extending the standard velocities to the fractional ones [263].

5.2.2 *Fractional Hamiltonian analysis for discrete and continuous systems*

5.2.2.1 *A direct method within Riemann-Liouville fractional derivatives*

In the following, we introduce the meaning of the fractional Hamiltonian. As a starting point, for simplicity, below we consider the following form of the fractional Euler-Lagrange equations

$$\frac{\partial L_f}{\partial q^\rho(t)} + {}^{\mathrm{RL}}D^{\alpha}_{b-} \frac{\partial L_f}{\partial\, {}^{\mathrm{RL}}D^{\alpha}_{a+} q^\rho(t)} = 0, \quad 0<\alpha<1, \quad \rho = 1,\ldots,N, \tag{5.2.1}$$

where L_f denotes the corresponding fractional Lagrangian given as

$$L_f(q^\rho, {}^{\mathrm{RL}}D^{\alpha}_{a+} q^\rho, t), \quad 0<\alpha<1. \tag{5.2.2}$$

In the following we use (5.2.1) to define the generalized momenta as in [506], *viz.*

$$p_{\alpha_\rho} = \frac{\partial L_f}{\partial\, {}^{\mathrm{RL}}D^{\alpha}_{a+} q^\rho}, \quad \rho = 1,\ldots,N. \tag{5.2.3}$$

As a consequence of (5.2.1) and (5.2.3), a fractional Hamiltonian function is defined as

$$H_f = p_{\alpha_\rho} {}^{\mathrm{RL}}D^{\alpha}_{a+} q^\rho - L_f. \tag{5.2.4}$$

The canonical equations corresponding to (5.2.4) are

$$\frac{\partial H_f}{\partial t} = -\frac{\partial L_f}{\partial t}, \tag{5.2.5}$$

$$\frac{\partial H_f}{\partial p_{\alpha_\rho}} = {}^{\mathrm{RL}}D_{a+}^{\alpha} q^{\rho}, \tag{5.2.6}$$

$$\frac{\partial H_f}{\partial q^{\rho}} = {}^{\mathrm{RL}}D_{b-}^{\alpha} p_{\alpha_\rho}, \quad 0 < \alpha < 1, \quad \rho = 1, \ldots, N. \tag{5.2.7}$$

For the derivation of these equations and more results on this topic see [506].

5.2.2.2 *A direct method with Caputo fractional derivatives*

In the following, we present briefly the Hamiltonian formulation with Caputo's fractional derivatives as developed in [62]. Let us consider the fractional Lagrangian

$$L_f(q, {}^{\mathrm{C}}D_{a+}^{\alpha} q, t), \quad 0 < \alpha < 1. \tag{5.2.8}$$

By making use of (5.2.8), the canonical momenta p_α become

$$p_\alpha = \frac{\partial L_f}{\partial\, {}^{\mathrm{C}}D_{a+}^{\alpha} q}. \tag{5.2.9}$$

As a result we define the fractional canonical Hamiltonian as

$$H_f = p_\alpha \, ({}^{\mathrm{C}}D_{a+}^{\alpha} q) - L_f. \tag{5.2.10}$$

The expressions of the fractional Hamilton equations are

$$\frac{\partial H_f}{\partial t} = -\frac{\partial L_f}{\partial t}, \tag{5.2.11a}$$

$$\frac{\partial H_f}{\partial p_\alpha} = {}^{\mathrm{C}}D_{a+}^{\alpha} q, \tag{5.2.11b}$$

$$\frac{\partial H_f}{\partial q} = {}^{\mathrm{RL}}D_{b-}^{\alpha} p_\alpha. \tag{5.2.11c}$$

In [62] a detailed description of this formulation is provided.

We mention that the fractional order Hamiltonian equations corresponding to the functional with fractional integral and fractional derivative in the Caputo sense, namely

$$J(y) = {}^{\mathrm{RL}}I_{a+}^{\alpha} L_f(t, q, {}^{\mathrm{C}}D_{a+}^{\beta} q), \qquad t \in [a, b], \quad q(a) = q_o, \tag{5.2.12}$$

and are given by [297]

$$\frac{\partial H_f}{\partial t} = -\frac{\partial L_f}{\partial t}, \tag{5.2.13a}$$

$$\frac{\partial H_f}{\partial p_\beta} = {}^{\mathrm{C}}D^{\beta}_{a+}q, \tag{5.2.13b}$$

$$\frac{\partial H_f}{\partial q} = (x-t)^{1-\alpha}\,{}^{\mathrm{RL}}D^{\beta}_{x-}\left((x-t)^{\alpha-1}\frac{\partial Lf}{\partial\,{}^{\mathrm{C}}D^{\beta}_{a+}q}\right). \tag{5.2.13c}$$

5.2.2.3 *A direct method within Riesz-Caputo fractional derivatives*

The fractional canonical momenta of the fractional Lagrangian

$$L = L_f(q, {}^{\mathrm{RC}}D^{\alpha}_{[a,b]}q) \tag{5.2.14}$$

are defined as

$$p_\alpha = \frac{\partial L_f}{\partial\,{}^{\mathrm{RC}}D^{\alpha}_{[a,b]}q}. \tag{5.2.15}$$

As a result, the corresponding fractional Hamiltonian becomes

$$H_f(q, p_\alpha) = p_\alpha({}^{\mathrm{RC}}D^{\alpha}_{[a,b]}q) - L_f. \tag{5.2.16}$$

By using (5.2.14), (5.2.15) and (5.2.16) we obtain the fractional Hamilton's equations (see [20])

$${}^{\mathrm{R}}D^{\alpha}_{[a,b]}p_\alpha = -\frac{\partial H_f}{\partial q}, \qquad {}^{\mathrm{RC}}D^{\alpha}_{[a,b]}q = \frac{\partial H_f}{\partial p_\alpha}. \tag{5.2.17}$$

5.2.3 *Fractional Hamiltonian formulation for constrained systems*

5.2.3.1 *Fractional Hessian matrix*

The fractional canonical equations described in the previous paragraphs are valid for the case when no primary constraints exist. As it is well known, several dynamical systems of physical interest have constraints [54, 55, 57].

In the classical case, Dirac's formalism provides a way to manage the constraints, namely they are divided into first and second class type, the Dirac's bracket is introduced and the second class constraints are eliminated [258].

Let us consider the equations of motion which can be obtainable using the variational principle that states the extremum of the action S,

$$S = \int L dt \tag{5.2.18}$$

under certain given boundary conditions. Lagrangian theories are divided into two parts. If the determinant of the Hessian matrix

$$H_{ij} = \frac{\partial^2 L}{\partial \dot{q}^i \partial \dot{q}^j} \tag{5.2.19}$$

is nonzero the theory is called nonsingular. In the opposite case the theory is called singular.

The main aim is to define properly the fractional Hessian of a given fractional singular Lagrangian. We have to have in mind the restriction to recover the same rank as in the classical case while $\alpha \to 1$.

In the following we consider the case where the fractional Lagrangian to start with is

$$L_f \left(t, q^1, \cdots, q^n, {}^{\mathrm{RL}}D_{a+}^{\alpha} q^1, \cdots, {}^{\mathrm{RL}}D_{a+}^{\alpha} q^n\right). \tag{5.2.20}$$

Using (5.2.19) and (5.2.20) a natural definition of the fractional Hessian matrix is

$$H_{ij} = \frac{\partial^2 L_f}{\partial\, {}^{\mathrm{RL}}D_{a+}^{\alpha} q^i \partial\, {}^{\mathrm{RL}}D_{a+}^{\alpha} q^j}. \tag{5.2.21}$$

We notice that the main advantage of using (5.2.21) is that the classical result of the Hessian matrix is recovered in the limit process.

Nevertheless, if ${}^{\mathrm{RL}}D_{a+}^{\alpha}$ is replaced by ${}^{\mathrm{RL}}D_{b-}^{\alpha}$ in (5.2.20) the obtained fractional Lagrangian

$$L_f \left(t, q^1, \cdots, q^n, {}^{\mathrm{RL}}D_{b-}^{\alpha} q^1, \cdots, {}^{\mathrm{RL}}D_{b-}^{\alpha} q^n\right) \tag{5.2.22}$$

leads us to define the fractional Hessian matrix as

$$H_{ij} = \frac{\partial^2 L_f}{\partial\, {}^{\mathrm{RL}}D_{b-}^{\alpha} q^i \partial\, {}^{\mathrm{RL}}D_{b-}^{\alpha} q^j}. \tag{5.2.23}$$

From (5.2.21) and (5.2.23) we conclude that the representation of the fractional Hessian matrix depends on the fractional derivatives involved in the given Lagrangian.

5.2.3.2 *The reduced phase space*

Finding the reduced phase space in the fractional case is a fundamental step in identification of the true degrees of freedom for a given physical system. For a given fractional Lagrangian $L_f = L_f(q^i, {}^{\mathrm{RC}}D^{\alpha}_{[a,b]}q^i)$, $i = 1, \ldots, n$, involving Riesz derivatives we suppose that $\Phi_m(p^i_\alpha, q^i) = 0$, $m = 1, \ldots, M$, with $M < n$, *i. e.* we have M primary constraints. Therefore, the fractional Hamiltonian equations are obtained from the fractional variational principle as (see [59])

$$\delta \int_{t1}^{t2} \left(p^i_\alpha({}^{\mathrm{RC}}D^{\alpha}_{[a,b]}q^i) - H_f - u^m\Phi_m\right) dt = 0. \tag{5.2.24}$$

The new variables u^m appear as Lagrange multipliers enforcing the primary constraints $\Phi_m = 0$. By using (5.2.24) we notice that the theory is invariant under $H_f \to H_f + c^m\Phi_m$, because this change merely results in changing $u^m \to u^m + c^m$. Finally, we obtain

$${}^{\mathrm{R}}D^{\alpha}_{[a,b]}p_{\alpha j} = -\frac{\partial H_f}{\partial q^j} - u^m\frac{\partial \Phi_m}{\partial q^j}, \quad {}^{\mathrm{RC}}D^{\alpha}_{[a,b]}q^j = -\frac{\partial H_f}{\partial p_{\alpha j}} + u^m\frac{\partial \Phi_m}{\partial p_{\alpha j}}. \tag{5.2.25}$$

5.2.3.3 *Fractional Ostrogradski's approach*

In the classical calculus the k-th order derivative of $\phi(t)$ is given by the Faà di Bruno formula [4]

$$\frac{d^k}{dt^k}F(h(t)) = k! \sum_{m=1}^{k} F^{(m)}(h(t)) \sum \prod_{r-1}^{k} \frac{1}{a_r!}\left(\frac{h^{(r)}(t)}{r!}\right)^{a_r}, \tag{5.2.26}$$

where the sum $\sum$ extends over all combinations of non-negative integer values of $a_1, \ldots, a_k$ such that the constraints

$$\sum_{r=1}^{k} r a_r = k \tag{5.2.27}$$

and

$$\sum_{r=1}^{k} a_r = m \tag{5.2.28}$$

are satisfied. As a result, the fractional derivative of a composition function is given by [60]

$$
\begin{aligned}
&{}^{C}D^{\alpha}_{a+}F(h(t)) \qquad (5.2.29)\\
&= \frac{(t-a)^{-p}}{\Gamma(1-p)}(F(h(t)-F(h(a)))\\
&+\sum_{k=1}^{\infty}\binom{\alpha}{k}k!\frac{(t-a)^{k-\alpha}}{\Gamma(k-\alpha+1)}\sum_{m=1}^{k}F^{(m)}(h(t))\sum\prod_{r=1}^{k}\frac{1}{a_r!}\left(\frac{h^{(r)}(t)}{r!}\right)^{a_r},
\end{aligned}
$$

where $\sum$ and the coefficients a_r have the same meaning as above. This form of the fractional generalization of the Faà di Bruno formula is appropriate for the so called fractional Ostrogradski Hamiltonian formulation [60].

In this line of thought we consider the dynamical variable $q(t)$ to be a 1+1 dimensional field $Q(x,t)$ such that the chirality condition [95]

$$
\frac{dQ(x,t)}{dt} = \partial_x Q(x,t), \qquad (5.2.30)
$$

is fulfilled. We notice that $Q(x,t) = q(x+t)$ assures the one-to-one correspondence between $q(t)$ and $Q(x,t)$ [95, 377].

Ostrogradski's coordinates are defined by

$$
Q^{(n)}(t) = (\partial_x)^n Q(x,t)\mid_{x=x_0}, \qquad (5.2.31)
$$

where the discontinuity curve $x_0(t) = x_0$ is a constant [95], which implies that

$$
Q(x,t) = \sum_{n=0}^{\infty}\frac{(x-x_0)^n}{n!}Q^{(n)}(t). \qquad (5.2.32)
$$

A corresponding boundary Poisson bracket was introduced in [95] as

$$
\{F(t),G(t)\} = \sum_{k,l=0}^{\infty} c_{k,l}\int_{-\infty}^{\infty}dx(\partial_x)^{k+l}\left[\frac{\delta F(t)}{\delta Q^{(k)}}(x,t)\frac{\delta G(t)}{\delta P^{(l)}}(x,t)\right] - (F\leftrightarrow G), \qquad (5.2.33)
$$

where the coefficients $c_{k,l}$ are constants and normalized in such a manner to satisfy the Jacobi identity.

The classical canonical relation becomes

$$
\{Q(x,t),P(x',t)\} = \delta_R(x-x'), \qquad (5.2.34)
$$

and Ostrogradski's momenta $P_{(n)}(t)$ are defined as

$$P_{(n)}(t) = \int_{-\infty}^{\infty} dx \frac{(x-x_0)^n}{n!} P(x,t), \tag{5.2.35}$$

therefore we conclude that [60, 95, 377]

$$P(x,t) = \sum_{n=0}^{\infty} P_{(n)}(t)(-\partial_x)^n \delta_R(x-x_0). \tag{5.2.36}$$

By using (5.2.36) the expression for $P_{(n)}(t)$ is

$$P_{(n)}(t) = \sum_{m=n}^{\infty} (-\partial_t)^{m-n} \frac{\partial L_f[Q](t)}{\partial Q^{(m+1)}(t)}, \tag{5.2.37}$$

where $L_f[Q](t)$ denotes the fractional counterpart of a given classical Lagrangian [60, 77].

As a result the Hamilton equations are written as

$$\dot{P}_{(n)}(t) + P_{(n-1)}(t) = \frac{\partial L'_f[Q](t)}{\partial Q^{(n)}(t)}, \qquad n \in \mathbb{N}, \tag{5.2.38}$$

$$\dot{P}_{(0)}(t) = \frac{\partial L'_f[Q](t)}{\partial Q^{(0)}(t)}. \tag{5.2.39}$$

By inspection we conclude that (5.2.39) is nothing but the Euler-Lagrange equation corresponding to the Lagrangian $L_f[Q](t)$. Finally, the expression of the corresponding Hamiltonian becomes

$$H = \sum_{n=0}^{\infty} P_{(n)}(t) Q^{(n+1)}(t) - L'_f[Q](t), \tag{5.2.40}$$

see [60, 77] where additional details on this topic may also be found.

5.2.4 *Applications*

5.2.4.1 *Discrete fractional constrained systems*

In the following we give some illustrative examples of fractional constrained dynamics as well as the exact solutions of their fractional Euler-Lagrange and fractional Hamiltonian equations.

Example 5.1. Let us consider the classical Lagrangian

$$L = \frac{1}{2}(\dot{q}_1(t) + \dot{q}_2(t))^2. \tag{5.2.41}$$

By making use of (5.2.41) we calculate canonical momenta as

$$p_{1\alpha} = p_{2\alpha} = \dot{q}_1(t) + \dot{q}_2(t). \tag{5.2.42}$$

The corresponding fractional Lagrangian is written as

$$L_f = \frac{1}{2}({}^{\mathrm{C}}D^{\alpha}_{a+}q_1(t) + {}^{\mathrm{C}}D^{\alpha}_{a+}q_2(t))^2, \tag{5.2.43}$$

where $0 < \alpha < 1$. Therefore, the fractional Hamiltonian corresponding to (5.2.43) becomes

$$H_f = \frac{(p_{1\alpha})^2}{2} + \lambda(p_{2\alpha} - p_{1\alpha}), \tag{5.2.44}$$

where $\lambda = {}^{\mathrm{C}}D^{\alpha}_{a+}q_2(t)$. As a result, the fractional canonical equations are

$${}^{\mathrm{C}}D^{\alpha}_{a+}q_2 = \lambda,\ {}^{\mathrm{C}}D^{\alpha}_{a+}q_1 = p_{1\alpha} - \lambda,\ {}^{\mathrm{RL}}D^{\alpha}_{b-}p_{1\alpha} = 0,\ {}^{\mathrm{RL}}D^{\alpha}_{b-}p_{2\alpha} = 0. \tag{5.2.45}$$

The Hamilton's equations in the extended phase-space are equivalent with the fractional Euler-Lagrange equations on the surface of constraint $p_{2\alpha} - p_{1\alpha} = 0$. The solutions of (5.2.45) are

$$p_{1\alpha} = p_{2\alpha} = (b-t)^{\alpha-1}, \qquad p_{3\alpha} = 0, \tag{5.2.46}$$

$$q_1(t) + q_2(t) = b_0 + \frac{1}{\Gamma(\alpha)}\int_0^t \frac{(b-t)^{\alpha-1}}{(t-\tau)^{1-\alpha}}d\tau, \quad \lambda = {}^{\mathrm{C}}D^{\alpha}_{a+}q_2(t). \tag{5.2.47}$$

Example 5.2. The classical Lagrangian is given by

$$L = \dot{q}_1\dot{q}_2 - q_3(q_1 + q_2). \tag{5.2.48}$$

and its fractional counterpart becomes

$$L_f = ({}^{\mathrm{C}}D^{\alpha}_{a+}q_1)({}^{\mathrm{C}}D^{\alpha}_{a+}q_2) - q_3(q_1 + q_2). \tag{5.2.49}$$

The Euler-Lagrange equations are given by

$${}^{\mathrm{RL}}D^{\alpha}_{b-}({}^{\mathrm{C}}D^{\alpha}_{a+}q_2) - q_3 = 0,\ \ {}^{\mathrm{RL}}D^{\alpha}_{b-}({}^{\mathrm{C}}D^{\alpha}_{a+}q_1) - q_3 = 0,\ \ q_1 + q_2 = 0. \tag{5.2.50}$$

By using (5.2.50), we obtain

$${}^{\mathrm{RL}}D^{\alpha}_{b-}({}^{\mathrm{C}}D^{\alpha}_{a+}q_2) = 0,\ \ q_1 = -q_2,\ \ q_3 = 0. \tag{5.2.51}$$

The canonical momenta associated to q_1, q_2 and q_3 have the forms

$$p_{\alpha 1} = {}^{\mathrm{C}}D^{\alpha}_{a+}q_2, \qquad p_{\alpha 2} = {}^{\mathrm{C}}D^{\alpha}_{a+}q_1, \qquad p_{\alpha 3} = 0. \tag{5.2.52}$$

The canonical Hamiltonian is given by

$$H_f = p_{\alpha 1}p_{\alpha 2} + q_3(q_1 + q_2) + \lambda p_{\alpha 3}, \tag{5.2.53}$$

where $\lambda = {}^{\mathrm{C}}D^{\alpha}_{a+}q_3$. The canonical equations in the extended phase space are

$$p_{\alpha 1} = {}^{\mathrm{C}}D^{\alpha}_{a+}q_2, \qquad p_{\alpha 2} = {}^{\mathrm{C}}D^{\alpha}_{a+}q_1, \qquad \lambda = {}^{\mathrm{C}}D^{\alpha}_{a+}q_3, \qquad {}^{\mathrm{RL}}D^{\alpha}_{b-}p_{\alpha 1} = q_3,$$

$${}^{\mathrm{RL}}D^{\alpha}_{b-}p_{\alpha 2} = q_3, \qquad {}^{\mathrm{RL}}D^{\alpha}_{b-}p_{\alpha 3} = q_1 + q_2, \qquad p_{\alpha 3} = 0, \qquad p_{\lambda} = 0.$$

Finally, the solution of (5.2.54) is given as

$$\begin{aligned} &q_3(t) = 0, \qquad q_1(t) + q_2(t) = 0, \qquad p_{\alpha 3} = 0, \\ &p_{\lambda} = 0, \qquad p_{\alpha 1} = (b-t)^{\alpha-1}, \qquad p_{\alpha 2} = -(b-t)^{\alpha-1}, \\ &q_2(t) = q_2(a) + {}^{\mathrm{RL}}I^{\alpha}_{a+}(b-t)^{\alpha-1}. \end{aligned} \tag{5.2.54}$$

5.2.4.2 *Fractional Hamiltonian formulation in fractional time*

We define the Lagrangian L_f as

$$L = L_f(t, {}^{\mathrm{RL}}D^{\beta}_{a+}q, {}^{\mathrm{RL}}D^{\gamma}_{b-}q)(t-\tau)^{\alpha-1}. \tag{5.2.55}$$

By using (5.2.55), we define the canonical momenta as

$$p^{\alpha}_{\mu} = \frac{\partial L_f}{\partial\, {}^{\mathrm{RL}}D^{\mu}_{a+}q}, \tag{5.2.56}$$

$$p^{\alpha}_{\nu} = \frac{\partial L_f}{\partial\, {}^{\mathrm{RL}}D^{\nu}_{b-}q}. \tag{5.2.57}$$

The canonical Hamiltonian has the form

$$H_f = p^{\alpha}_{\mu}\, {}^{\mathrm{RL}}D^{\mu}_{a+}q + p^{\alpha}_{\nu}\, {}^{\mathrm{RL}}D^{\nu}_{a+}q - L_f \tag{5.2.58}$$

and the canonical equations become

$$\begin{aligned} &\frac{\partial H_f}{\partial p^{\alpha}_{\mu}} = {}^{\mathrm{RL}}D^{\mu}_{a+}q, \qquad \frac{\partial H_f}{\partial p^{\alpha}_{\nu}} = {}^{\mathrm{RL}}D^{\nu}_{b-}q, \\ &\frac{\partial H_f}{\partial q} = {}^{\mathrm{RL}}D^{\mu}_{b-}p^{\alpha}_{\mu} + {}^{\mathrm{RL}}D^{\nu}_{a+}p^{\alpha}_{\nu}, \qquad \frac{\partial H_f}{\partial \tau} = -\frac{\partial L_f}{\partial \tau}, \end{aligned} \tag{5.2.59}$$

see [447].

5.2.4.3 *Fractional Nambu mechanics*

In 1973, Nambu proposed a generalization of the classical Hamiltonian formalism and he introduced so called the classical Nambu bracket [451]. This bracket, involving a dynamical quantity and two or more Hamiltonians, describes the time-evolution of a quantity in a generalization of Hamilton's equations of motion for special physical systems. Nambu brackets in phase space describe the generic classical evolution of systems with many independent integrals of motion beyond those required to complete integrability of a given system. The superintegrable systems are described within Nambu's mechanics. The conditions when a Killing-Yano tensor becomes a Nambu tensor were presented in [53].

In the following we introduce the fractional generalization of the Nambu mechanics as presented in [67]. We introduce firstly the basic classical notions and after that we present the fractional generalization.

The Poincaré-Cartan 1-form $\Omega^{(1)}$ is given by

$$\Omega^{(1)} = p\, dq - H(p,q)dt.$$

For the three-dimensional case we define $\Omega^{(2)}$ as

$$\Omega^{(2)} = qdp \wedge dr - H_1 dH_2 \wedge dt.$$

The fractional generalization of the Poincaré-Cartan 1-form can be defined as

$$\Omega^{(1)}_\alpha = p(dq)^\alpha - H(p,q)(dt)^\alpha. \tag{5.2.60}$$

Note that $\Omega^{(1)}_\alpha$ is a fractional 1-form that can be called a fractional Poincaré-Cartan 1-form. We notice that we have

$$d\Omega^{(1)}_\alpha = \{(dp)^\alpha + {}^{\mathrm{RL}}D^\alpha_{a+,q} H(dt)^\alpha\} \wedge \{(dq)^\alpha - {}^{\mathrm{RL}}D^\alpha_{a+,p} H(dt)^\alpha\}.$$

Thus, by generalized Pfaffian equations, we obtain

$$\frac{(dq)^\alpha}{(dt)^\alpha} = {}^{\mathrm{RL}}D^\alpha_{a+,p} H, \tag{5.2.61a}$$

$$\frac{(dp)^\alpha}{(dt)^\alpha} = -{}^{\mathrm{RL}}D^\alpha_{a+,q} H, \tag{5.2.61b}$$

which are called the fractional Hamiltonian equations.

Besides, the fractional generalization of the 2-form is defined as

$$\Omega^{(2)}_\alpha = q(dp)^\alpha \wedge (dr)^\alpha - H_1 d^\alpha H_2 \wedge (dt)^\alpha, \tag{5.2.62}$$

and as a result the fractional exterior derivative of the fractional 2-form becomes

$$d^\alpha \Omega^{(2)}_\alpha = d^\alpha (q(dp)^\alpha \wedge (dr)^\alpha) - d^\alpha (H_1 d^\alpha H_2 \wedge (dt)^\alpha). \tag{5.2.63}$$

After some direct calculations we obtain

$$
\begin{aligned}
&d^{\alpha}\Omega_{\alpha}^{(2)} \\
&= ({}^{\mathrm{RL}}D^{\alpha}_{a+,q}H_1\,{}^{\mathrm{RL}}D^{\alpha}_{a+,p}H_2 - {}^{\mathrm{RL}}D^{\alpha}_{a+,p}H_1\,{}^{\mathrm{RL}}D^{\alpha}_{a+,q}H_2)(dq)^{\alpha}\wedge(dp)^{\alpha}\wedge(dt)^{\alpha} \\
&\quad + ({}^{\mathrm{RL}}\mathrm{D}^{\alpha}_{\mathrm{a+,q}}\mathrm{H}_1\,{}^{\mathrm{RL}}\mathrm{D}^{\alpha}_{\mathrm{a+,r}}\mathrm{H}_2 - {}^{\mathrm{RL}}\mathrm{D}^{\alpha}_{\mathrm{a+,r}}\mathrm{H}_1\,{}^{\mathrm{RL}}\mathrm{D}^{\alpha}_{\mathrm{a+,q}}\mathrm{H}_2)(\mathrm{dq})^{\alpha}\wedge(\mathrm{dr})^{\alpha}\wedge(\mathrm{dt})^{\alpha} \\
&\quad + ({}^{\mathrm{RL}}D^{\alpha}_{a+,p}H_1\,{}^{\mathrm{RL}}D^{\alpha}_{a+,r}H_2 - {}^{\mathrm{RL}}D^{\alpha}_{a+,r}H_1\,{}^{\mathrm{RL}}D^{\alpha}_{a+,p}H_2)(dp)^{\alpha}\wedge(dr)^{\alpha}\wedge(dt)^{\alpha}.
\end{aligned}
$$

By using the definitions

$$
\begin{aligned}
\frac{\partial^{\alpha}(H_1,H_2)}{\partial^{\alpha}(p,r)} &= ({}^{\mathrm{RL}}D^{\alpha}_{a+,p}H_1\,{}^{\mathrm{RL}}D^{\alpha}_{a+,r}H_2 - {}^{\mathrm{RL}}D^{\alpha}_{a+,r}H_1\,{}^{\mathrm{RL}}D^{\alpha}_{a+,p}H_2), \\
\frac{\partial^{\alpha}(H_1,H_2)}{\partial^{\alpha}(q,r)} &= ({}^{\mathrm{RL}}D^{\alpha}_{a+,q}H_1\,{}^{\mathrm{RL}}D^{\alpha}_{a+,r}H_2 - {}^{\mathrm{RL}}D^{\alpha}_{a+,r}H_1\,{}^{\mathrm{RL}}D^{\alpha}_{a+,q}H_2), \quad (5.2.64) \\
\frac{\partial^{\alpha}(H_1,H_2)}{\partial^{\alpha}(q,p)} &= ({}^{\mathrm{RL}}D^{\alpha}_{a+,q}H_1\,{}^{\mathrm{RL}}D^{\alpha}_{a+,p}H_2 - {}^{\mathrm{RL}}D^{\alpha}_{a+,p}H_1\,{}^{\mathrm{RL}}D^{\alpha}_{a+,q}H_2),
\end{aligned}
$$

we obtain the generalized Pfaffian equations

$$(dq)^{\alpha} - \frac{\partial^{\alpha}(H_1,H_2)}{\partial^{\alpha}(p,r)}(dt)^{\alpha} = 0, \quad (5.2.65)$$

$$(dp)^{\alpha} - \frac{\partial^{\alpha}(H_1,H_2)}{\partial^{\alpha}(q,r)}(dt)^{\alpha} = 0, \quad (5.2.66)$$

$$(dr)^{\alpha} - \frac{\partial^{\alpha}(H_1,H_2)}{\partial^{\alpha}(q,p)}(dt)^{\alpha} = 0. \quad (5.2.67)$$

By using the analogy to the integer order situation, these equations are called fractional Nambu equations [67].

5.2.4.4 *A fractional supersymmetric model*

Now let us recall the development of [76] and consider the Lagrangian

$$L = \frac{1}{2}\dot{x}^2 - \frac{1}{2}\Phi^2(x) + \frac{i}{2}(\bar{\psi}\dot{\psi} - \dot{\bar{\psi}}\psi) - \Phi'(x)\bar{\psi}\psi. \quad (5.2.68)$$

As it is known, this Lagrangian characterizes pseudoclassical systems being supersymmetric [74, 76].

By inspection we conclude that the fractional counterpart of (5.2.68) is

$$L_f = \frac{1}{2}({}^{\mathrm{RL}}D^{\alpha}_{a+}x)^2 - \frac{1}{2}\Phi(x)^2 + \frac{i}{2}(\bar{\psi}\,{}^{\mathrm{RL}}D^{\alpha}_{a+}\psi - ({}^{\mathrm{RL}}D^{\alpha}_{a+}\bar{\psi})\psi) - \Phi'(x)\bar{\psi}\psi. \quad (5.2.69)$$

The fractional Hessian matrix corresponding to (5.2.69) has rank one, therefore we have two primary constraints. The expressions of the canonical momenta are written as

$$p = {}^{\mathrm{RL}}D^{\alpha}_{a+}x, \quad \pi = \frac{\partial_r L}{\partial\, {}^{\mathrm{RL}}D^{\alpha}_{a+}\psi} = -\frac{i}{2}\bar{\psi}, \quad \bar{\pi} = \frac{\partial_r L}{\partial\, {}^{\mathrm{RL}}D^{\alpha}_{a+}\bar{\psi}} = -\frac{i}{2}\psi \tag{5.2.70}$$

By making use of (5.2.70) we obtain two second-class constraints as

$$\chi_1 = \pi + \frac{i}{2}\bar{\psi} = 0, \qquad \chi_2 = \bar{\pi} + \frac{i}{2}\psi = 0. \tag{5.2.71}$$

Finally, the total fractional Hamiltonian is given by

$$H_T = \frac{p^2}{2} + \frac{\Phi^2(x)}{2} + \Phi'(x)\bar{\psi}\psi + (\pi + \frac{i}{2}\bar{\psi})\lambda_1 + (\bar{\pi} + \frac{i}{2}\psi)\lambda_2. \tag{5.2.72}$$

The Hamiltonian equations are then written as

$${}^{\mathrm{RL}}D^{\alpha}_{a+}x = \frac{\partial H_f}{\partial p}, \quad {}^{\mathrm{RL}}D^{\alpha}_{a+}\psi = \frac{\partial H_f}{\partial \pi}, \quad {}^{\mathrm{RL}}D^{\alpha}_{a+}\bar{\psi} = \frac{\partial H_f}{\partial \bar{\pi}}, \tag{5.2.73}$$

$${}^{\mathrm{RL}}D^{\alpha}_{b-}p = \frac{\partial H_f}{\partial x}, \quad {}^{\mathrm{RL}}D^{\alpha}_{b-}\pi = \frac{\partial H_f}{\partial \psi}, \quad {}^{\mathrm{RL}}D^{\alpha}_{b-}\bar{\pi} = \frac{\partial H_f}{\partial \bar{\psi}}. \tag{5.2.74}$$

Finally, by using (5.2.73) we conclude that

$${}^{\mathrm{RL}}D^{\alpha}_{a+}x = {}^{\mathrm{RL}}D^{\alpha}_{a+}x, \quad {}^{\mathrm{RL}}D^{\alpha}_{a+}\psi = \lambda_1, \quad {}^{\mathrm{RL}}D^{\alpha}_{a+}\psi = \lambda_2. \tag{5.2.75}$$

5.2.4.5 *Fractional optimal control formulation*

The main problem investigated here is to minimize the performance index

$$J(u) = \int_a^b F(x, u, t)dt \tag{5.2.76}$$

such that

$${}^{\mathrm{RL}}D^{\alpha}_{a+}x = G(x, u, t), \tag{5.2.77}$$

where the terminal conditions $x(a) = c$ and $x(b) = d$ are given.

The corresponding fractional order counterpart formulation of this problem for the case of scalar variables and functions was proposed in [18]. More details on this topic can be found in [18, 24, 25].

A modified performance index can be defined as

$$\bar{J}(u) = \int_a^b [H(x,u,t) - \lambda^{\mathrm{T}}\, {}^{\mathrm{RL}}D_{a+}^{\alpha} x] dt, \tag{5.2.78}$$

where $H(x,u,\lambda,t)$ represents the Hamiltonian

$$H(x,u,\lambda,t) = F(x,u,t) + \lambda^{\mathrm{T}} G(x,u,t), \tag{5.2.79}$$

and λ denotes the vector of Lagrange multipliers. By making use of (5.2.78), (5.2.79) and the fractional integration by parts, the necessary conditions for the fractional control problem become [18]

$${}^{\mathrm{RL}}D_{b-}^{\alpha}\lambda = \frac{\partial H}{\partial x}, \tag{5.2.80}$$

$$\frac{\partial H}{\partial u} = 0, \tag{5.2.81}$$

$${}^{\mathrm{RL}}D_{a+}^{\alpha} x = \frac{\partial H}{\partial \lambda}. \tag{5.2.82}$$

As an example, we present the following problem [25]:

Minimize

$$J = \frac{1}{2}\int_0^2 [{}^{\mathrm{RL}}D_{0+}^{\alpha}\, {}^{\mathrm{RL}}D_{0+}^{\alpha}\theta]^2 dt \tag{5.2.83}$$

with respect to the dynamic constraint ${}^{\mathrm{RL}}D_{0+}^{\alpha}\, {}^{\mathrm{RL}}D_{0+}^{\alpha}\theta(t) = u(t)$. Introducing the change of variables $\theta(t) = x_1(t)$ and ${}^{\mathrm{RL}}D_{0+}^{\alpha}\theta(t) = x_2(t)$, the modified performance index in (5.2.78) becomes

$$J = \int_0^2 [H(\mathbf{x},u,\lambda) - \lambda^{\mathrm{T}}\, {}^{\mathrm{RL}}D_{0+}^{\alpha}\mathbf{x}(t)] dt \tag{5.2.84}$$

with the Hamiltonian

$$H(\mathbf{x},u,\lambda) = \frac{1}{2}u^2(t) + \lambda^T(A\mathbf{x}(t) + \mathbf{b}u(t)). \tag{5.2.85}$$

In this system,

$$\begin{aligned} \mathbf{x}(t) &= \begin{pmatrix} x_1(t) \\ x_2(t) \end{pmatrix}, & \lambda(t) &= \begin{pmatrix} \lambda_1(t) \\ \lambda_2(t) \end{pmatrix}, \\ \mathbf{b} &= \begin{pmatrix} 0 \\ 1 \end{pmatrix}, & A &= \begin{pmatrix} 0 & 1 \\ 0 & 0 \end{pmatrix}. \end{aligned} \tag{5.2.86}$$

By making use of (5.2.80)–(5.2.82), we finally obtain the system of equations

$${}^{\mathrm{RL}}D_{2-}^{\alpha}\lambda_1 = 0, \qquad {}^{\mathrm{RL}}D_{2-}^{\alpha}\lambda_2 - \lambda_1 = 0, \qquad u + \lambda_2 = 0, \tag{5.2.87a}$$

$${}^{\mathrm{RL}}D_{0+}^{\alpha}x_1 - x_2 = 0, \qquad {}^{\mathrm{RL}}D_{0+}^{\alpha}x_2 - u = 0. \tag{5.2.87b}$$

The terminal conditions $x_1(0) = x_2(0) = 0$ and $x_1(2) = x_2(2) = 1$ are considered in [25] where, in particular, a method based on a Grünwald-Letnikov approximation scheme is presented to solve (5.2.87) numerically, and where the exact classical solution of (5.2.87) is shown to have the form

$$x_1(t) = \frac{t^2}{4}, \qquad x_2(t) = \frac{t}{2}, \tag{5.2.88}$$

$$\lambda_1(t) = 0, \qquad \lambda_2(t) = -u(t) = -\frac{1}{2}. \tag{5.2.89}$$

For the numerical solution of this problem, the numerical results of [25] indicated that the variables $x_1(t)$ and $x_2(t)$ and the control variable $u(t)$ converges for several values of α as the number of grid points N is increased. On the other hand, it was concluded that near the end point $t = 2$, the value of $u(t)$ grows rapidly. The results show that as α approaches to 1, the analytical solution is recovered, as expected. Finally, it was reported that this result was found to be consistent with the analytical solution and with other results presented in [24] and [25].

5.2.4.6 *The fractional optimal control approach with delay*

Next, as in [317], we find the optimal control variable $u(t)$ which minimizes the performance index

$$\begin{aligned} J(y,u) = \int_a^b & F[t, u(t), {}^{\mathrm{C}}D_{a+}^{\alpha_1} y(t), {}^{\mathrm{C}}D_{a+}^{\alpha_2} y(t), \ldots, {}^{\mathrm{C}}D_{a+}^{\alpha_n} y(t), \\ & {}^{\mathrm{C}}D_{a+}^{\beta_1} y(t), {}^{\mathrm{C}}D_{a+}^{\beta_2} y(t), \ldots, {}^{\mathrm{C}}D_{a+}^{\beta_m} y(t), y(t), y'(t), \ldots, y^{(k)}(t), \\ & y(t-\tau), y'(t-\tau), \ldots, y^{(k)}(t-\tau)]dt, \end{aligned} \tag{5.2.90}$$

subject to the constraint

$$\begin{aligned} 0 = G[& t, u(t), {}^{\mathrm{C}}D_{a+}^{\alpha_1} y(t), {}^{\mathrm{C}}D_{a+}^{\alpha_2} y(t), \ldots, {}^{\mathrm{C}}D_{a+}^{\alpha_n} y(t), \\ & {}^{\mathrm{C}}D_{b-}^{\beta_1} y(t), {}^{\mathrm{C}}D_{b-}^{\beta_2} y(t), \ldots, {}^{\mathrm{C}}D_{b-}^{\beta_m} y(t), y(t), y'(t), \ldots, y^{(k)}(t), \\ & y(t-\tau), y'(t-\tau), \ldots, y^{(k)}(t-\tau)], \end{aligned} \tag{5.2.91}$$

in such a way that

$$y^{(l)}(b) = c_l \quad (l = 0, 1, \ldots, k-1) \text{ and } y(t) = \phi(t) \quad (t \in [a-\tau, a]), \tag{5.2.92}$$

where $a < b$, $0 < \tau < b - a$, $\alpha_i \in \mathbb{R}$ $(i = 1, 2, ..., n)$, $\beta_j \in \mathbb{R}$ $(j = 1, 2, ..., m)$. Here the c_l are constants and F and G are functions from

$[a-\tau, b] \times \mathbb{R}^{n+m+2k+3}$ to $\mathbb{R}$ possessing continuous first and second partial derivatives with respect to all of their arguments.

We define a modified performance index as

$$\hat{J}(y,u) = \int_a^b (F + \lambda(t)G)\, dt, \tag{5.2.93}$$

where λ represents a Lagrange multiplier or an adjoint variable.

According to [315], the Euler-Lagrange equations are

$$\begin{aligned}
0 &= \sum_{i=1}^{n} {}^{\mathrm{RL}}D_{b-\tau,-}^{\alpha_i}\left(\frac{\partial F}{\partial\, {}^{\mathrm{C}}D_{a+}^{\alpha_i}y(t)}\right)(t) + \sum_{i=1}^{n} {}^{\mathrm{RL}}D_{b-\tau,-}^{\alpha_i}\left(\lambda\frac{\partial G}{\partial\, {}^{\mathrm{C}}D_{a+}^{\alpha_i}y(t)}\right)(t)\\
&+ \sum_{j=1}^{m} {}^{\mathrm{RL}}D_{a+}^{\beta_j}\left(\frac{\partial F}{\partial\, {}^{\mathrm{C}}D_{b-}^{\beta_j}y(t)}\right)(t) + \sum_{j=1}^{m} {}^{\mathrm{RL}}D_{a+}^{\beta_j}\left(\lambda\frac{\partial G}{\partial\, {}^{\mathrm{C}}D_{b-}^{\beta_j}y(t)}\right)(t)\\
&+ \sum_{p=0}^{k}(-1)^p\frac{d^p}{dt^p}\left(\frac{\partial F}{\partial y^{(p)}(t)}\right)(t) + \sum_{p=0}^{k}(-1)^p\frac{d^p}{dt^p}\left(\lambda\frac{\partial G}{\partial y^{(p)}(t)}\right)(t)\\
&+ \sum_{p=0}^{k}(-1)^p\frac{d^p}{dt^p}\left(\frac{\partial F}{\partial y^{(p)}(t-\tau)}\right)(t+\tau)\\
&+ \sum_{p=0}^{k}(-1)^p\frac{d^p}{dt^p}\left(\lambda\frac{\partial G}{\partial y^{(p)}(t-\tau)}\right)(t+\tau)\\
&- \sum_{i=1}^{n}\frac{1}{\Gamma(\alpha_i)}{}^{\mathrm{RL}}D_{b-\tau,-}^{\alpha_i}\left(\int_{b-\tau}^{b}({}^{\mathrm{RL}}D_{b-}^{\alpha_i}\left(\frac{\partial F}{\partial\, {}^{\mathrm{C}}D_{a+}^{\alpha_i}y(t)}\right)(s)(s-t)^{\alpha_i-1}ds\right)\\
&- \sum_{i=1}^{n}\frac{1}{\Gamma(\alpha_i)}{}^{\mathrm{RL}}D_{b-\tau,-}^{\alpha_i}\left(\int_{b-\tau}^{b}({}^{\mathrm{RL}}D_{b-}^{\alpha_i}\left(\lambda\frac{\partial G}{\partial\, {}^{\mathrm{C}}D_{a+}^{\alpha_i}y(t)}\right)(s)(s-t)^{\alpha_i-1}ds\right)\\
&+ \frac{\partial F}{\partial u(t)}(t) + \lambda(t)\frac{\partial G}{\partial u(t)}(t)
\end{aligned} \tag{5.2.94}$$

for $a \le t \le b-\tau$ and

$$\begin{aligned}
0 &= \sum_{i=1}^{n} {}^{\mathrm{RL}}D_{b-}^{\alpha_i}\left(\frac{\partial F}{\partial\, {}^{\mathrm{C}}D_{a+}^{\alpha_i}y(t)}\right)(t) + \sum_{i=1}^{n} {}^{\mathrm{RL}}D_{b-}^{\alpha_i}\left(\lambda\frac{\partial G}{\partial\, {}^{\mathrm{C}}D_{a+}^{\alpha_i}y(t)}\right)(t)\\
&+ \sum_{j=1}^{m} {}^{\mathrm{RL}}D_{b-\tau,-}^{\beta_j}\left(\frac{\partial F}{\partial\, {}^{\mathrm{C}}D_{b-}^{\beta_j}y(t)}\right)(t) + \sum_{j=1}^{m} {}^{\mathrm{RL}}D_{b-\tau,-}^{\beta_j}\left(\lambda\frac{\partial G}{\partial\, {}^{\mathrm{C}}D_{b-}^{\beta_j}y(t)}\right)(t)
\end{aligned}$$

$$
\begin{aligned}
&+\sum_{p=0}^{k}(-1)^p\frac{d^p}{dt^p}\left(\frac{\partial F}{\partial y^{(p)}(t)}\right)(t)+\sum_{p=0}^{k}(-1)^p\frac{d^p}{dt^p}\left(\lambda\frac{\partial G}{\partial y^{(p)}(t)}\right)(t)\\
&+\sum_{j=1}^{m}\frac{1}{\Gamma(\beta_j)}{}^{\mathrm{RL}}D_{b-\tau,+}^{\beta_j}\left(\int_a^{b-\tau}{}^{\mathrm{RL}}D_{a+}^{\beta_j}\left(\frac{\partial F}{\partial\,{}^{\mathrm{C}}D_{b-}^{\beta_j}y(t)}\right)(s)(t-s)^{\beta_j-1}ds\right)\\
&+\sum_{j=1}^{m}\frac{1}{\Gamma(\beta_j)}{}^{\mathrm{RL}}D_{b-\tau,+}^{\beta_j}\left(\int_a^{b-\tau}{}^{\mathrm{RL}}D_{a+}^{\beta_j}\left(\lambda\frac{\partial L}{\partial\,{}^{\mathrm{C}}D_{b-}^{\beta_j}y(t)}\right)(s)(t-s)^{\beta_j-1}ds\right)\\
&+\frac{\partial F}{\partial u(t)}(t)+\lambda(t)\frac{\partial G}{\partial u(t)}(t)
\end{aligned}
\tag{5.2.95}
$$

for $b-\tau\le t\le b$, whereas the transversality conditions are

$$
\begin{aligned}
0=&\sum_{p=1}^{k}\sum_{q=0}^{p-1}(-1)^q\frac{d^q}{dt^q}\left(\frac{\partial F}{\partial y^{(p)}(t)}\right)(t)\eta^{p-q-1}(t)\Big|_a^{b-\tau}\\
&+\sum_{p=1}^{k}\sum_{q=0}^{p-1}(-1)^q\frac{d^q}{dt^q}\left(\lambda\frac{\partial G}{\partial y^{(p)}(t)}\right)(t)\eta^{p-q-1}(t)\Big|_a^{b-\tau}\\
&+\sum_{p=1}^{k}\sum_{q=0}^{p-1}(-1)^q\frac{d^q}{dt^q}\left(\frac{\partial F}{\partial y^{(p)}(t-\tau)}\right)(t+\tau)\eta^{p-q-1}(t+\tau)\Big|_a^{b-\tau}\\
&+\sum_{p=1}^{k}\sum_{q=0}^{p-1}(-1)^q\frac{d^q}{dt^q}\left(\lambda\frac{\partial G}{\partial y^{(p)}(t-\tau)}\right)(t+\tau)\eta^{p-q-1}(t+\tau)\Big|_a^{b-\tau}\\
&+\sum_{p=1}^{k}\sum_{q=0}^{p-1}(-1)^q\frac{d^q}{dt^q}\left(\frac{\partial F}{\partial y^{(p)}(t)}\right)(t)\eta^{p-q-1}(t)\Big|_{b-\tau}^{b}\\
&+\sum_{p=1}^{k}\sum_{q=0}^{p-1}(-1)^q\frac{d^q}{dt^q}\left(\lambda\frac{\partial G}{\partial y^{(p)}(t)}\right)(t)\eta^{p-q-1}(t)\Big|_{b-\tau}^{b},
\end{aligned}
\tag{5.2.96}
$$

where η denotes any admissible function fulfilling $\eta(t)\equiv 0$ for $t\in[a-\tau,a]$ and $\eta^{(l)}(b)=0$ for $l=0,1,\ldots,k-1$.

5.2.4.7 *Fractional multi time Hamiltonian equations*

In 1935, the multi time Hamilton equations were introduced into classical mechanics by de Donder and Weyl [176, 626]. A quantization of field theory based on the de Donder-Weyl covariant Hamiltonian formalism was investigated in [326]. Following [66], we now present its fractional version by using the generalized canonical momenta

$$p_i^\nu = \frac{\partial L_f}{\partial\, {}^{\mathrm{RL}}D^\alpha_{a+,t^\nu} q^i} \tag{5.2.97}$$

and the Hamiltonian defined as

$$H = {}^{\mathrm{RL}}D^\alpha_{a+t^\nu} q^i \frac{\partial L}{\partial\, {}^{\mathrm{RL}}D^\alpha_{a+,t^\nu} q^i} - L_f(t^\nu, q^i, {}^{\mathrm{RL}}D^\alpha_{a+,t^\nu} q^i). \tag{5.2.98}$$

The transformations $D^\alpha_{a+,t^\nu} q^i \to p_i^\nu$ defines the corresponding fractional Legendre transform. Therefore, by using the fractional Legendre transformation we get the fractional multi time Hamiltonian equations as

$$\begin{aligned} \frac{\partial H}{\partial q^i} &= {}^{\mathrm{RL}}D^\alpha_{b-,t^\nu} p_i^\nu, \qquad i = 1, \ldots, n, \quad \nu = 1, \ldots, p, \\ \frac{\partial H}{\partial p_i^\nu} &= {}^{\mathrm{RL}}D^\alpha_{a+,t^\nu} q^i, \qquad i = 1, \ldots, n, \quad \nu = 1, \ldots, p, \\ \frac{\partial H}{\partial t^\nu} &= -\frac{\partial L_f}{\partial t^\nu}, \qquad \nu = 1, \ldots, p. \end{aligned} \tag{5.2.99}$$

By inspection we conclude that Eqs. (5.2.99) are $np + n$ equations and they are equivalent to the fractional Euler-Lagrange equation on $\mathbb{R}^n$.

5.2.4.8 *Hamilton-Jacobi formulation with Caputo fractional derivative*

In order to determine the Hamilton-Jacobi partial differential equation with fractional Caputo derivative, we start with the fractional Hamiltonian written as

$$H = p_\beta\, {}^{\mathrm{C}}D^\beta_{a+} q - L_f(t, q, {}^{\mathrm{C}}D^\beta_{a+} q). \tag{5.2.100}$$

The next step is to consider the canonical transform with a generating function $S(I^{1-\beta}_{a+} q, P_\beta, t)$, therefore the new Hamiltonian becomes

$$K = P_\beta\, {}^{\mathrm{C}}D^\beta_{a+} Q - L'_f(t, Q, {}^{\mathrm{C}}D^\beta_{a+} Q), \tag{5.2.101}$$

where Q, P_β denote the new canonical coordinates and L' is the new Lagrangian. As shown in [300], these coordinates fulfill the fractional Hamilton's principle

$$\delta(I^\alpha(p_\beta\, {}^{\mathrm{C}}D^\beta_{a+} q - H)) = 0 \quad \text{and} \quad \delta(I^\alpha(P_\beta\, {}^{\mathrm{C}}D^\beta_{a+} Q - K)) = 0. \tag{5.2.102}$$

We can say that this is satisfied if we have

$$\frac{(x-t)^{\alpha-1}}{\Gamma(\alpha)}(p_\beta\,{}^{\mathrm{C}}D^{\beta}_{a+}q-H)=\frac{(x-t)^{\alpha-1}}{\Gamma(\alpha)}(P_\beta\,{}^{\mathrm{C}}D^{\beta}_{a+}Q-K)+\frac{dF}{dt} \quad (5.2.103)$$

where the function F has the form

$$F=S({}^{\mathrm{RL}}I^{1-\beta}_{a+}q,P_\beta,t)-P_\beta\,{}^{\mathrm{RL}}I^{1-\beta}_{a+}Q. \quad (5.2.104)$$

By direct calculations, see [300], we obtain the Hamilton-Jacobi equations of motion as

$$p_\beta=(x-t)^{1-\alpha}\Gamma(\alpha)\frac{\partial S}{\partial\,{}^{\mathrm{RL}}I^{1-\beta}_{a+}q}, \quad (5.2.105)$$

$${}^{\mathrm{RL}}I^{1-\beta}_{a+}Q=\frac{\partial S}{\partial P_\beta}, \quad (5.2.106)$$

and

$$\begin{aligned}K+(\Gamma(\alpha)(x-t)^{1-\alpha}-1)P_\beta\,{}^{\mathrm{C}}D^{\beta}_{a+}Q+\frac{(t-a)^{-\beta}}{\Gamma(1-\beta)}Q(a)P_\beta \quad (5.2.107)\\ =H+\Gamma(\alpha)(x-t)^{1-\alpha}\frac{\partial S}{\partial t}+\frac{(x-a)^{\alpha-1}}{\Gamma(\alpha)}\frac{(t-a)^{-\beta}}{\Gamma(1-\beta)}q(a)p_\beta.\end{aligned}$$

When the transformed Hamiltonian K is identically zero we obtain from the Hamiltonian equations that $Q=E_1$ and $P_\beta=E_2$ where E_1,E_2 are two constants. As a result, the generating function satisfies the partial differential equation

$$H+\Gamma(\alpha)(x-t)^{1-\alpha}\frac{\partial S}{\partial t}=\frac{(t-a)^{-\beta}}{\Gamma(1-\beta)}E_1E_2-\frac{(x-a)^{\alpha-1}}{\Gamma(\alpha)}\frac{(t-a)^{-\beta}}{\Gamma(1-\beta)}q(a)p_\beta. \quad (5.2.108)$$

Therefore, this is the fractional Hamilton-Jacobi equation for the fractional variational problem given by the Caputo derivative. Finally, we notice, see also [300], that the generating function is

$$\begin{aligned}S=\int_{t_1}^{t_2}\Bigg(&\frac{(x-t)^{\alpha-1}}{\Gamma(\alpha)}L(t,q,\,{}^{\mathrm{C}}D^{\beta}_{a+}q)-\frac{(t-a)^{1-\beta}(x-t)^{\alpha-1}}{\Gamma(\alpha)\Gamma(1-\beta)}E_1E_2\\ &-\frac{(t-a)^{-\beta}(x-a)^{\alpha-1}(x-t)^{\alpha-1}}{(\Gamma(\alpha))^2\Gamma(1-\beta)}q(a)p_\beta\Bigg)\,dt \quad (5.2.109)\end{aligned}$$

For more details about fractional Hamilton–Jacobi equation see Refs. [499, 503, 504] and the references therein.

5.2.4.9 *Fractional dynamics on extended phase space*

The main idea is to start with a given Hamiltonian $H(q,p)$ and to find $L^q(q,\dot{q})$ as a solution of the differential equation [452]

$$H(q,\frac{\partial L^q}{\partial \dot{q}})-\dot{q}\frac{\partial L^q}{\partial \dot{q}}+L^q=0. \tag{5.2.110}$$

It is known [452] that the extended Hamiltonian has the form

$$H(q,\pi_q,p,\pi_p)=\dot{q}\pi_q+\dot{p}\pi_p-L(q,\dot{q},p,\dot{p}) \tag{5.2.111}$$

and that the final Hamiltonian reads

$$H(q,\pi_q,p,\pi_p)=\sum_{n=0}^{N}\frac{1}{n!}\left[\frac{\partial^n H}{\partial p^n}\pi_q^n-\frac{\partial^n H}{\partial q^n}\pi_p^n\right]. \tag{5.2.112}$$

From here on we follow [78] and first denote by $H_f(q,p)=p\,{}^{\mathrm{C}}D_{a+}^{\alpha}q-L_f^q$ the fractional Hamiltonian where L_f^q represents the fractional Lagrangian. By using a similar technique as in the classical case we then obtain

$$H_f(q,\frac{\partial L^q}{\partial\,{}^{\mathrm{C}}D_{a+}^{\alpha}q})-{}^{\mathrm{C}}D_{a+}^{\alpha}q\frac{\partial L_f^q}{\partial\,{}^{C}D_{a+}^{\alpha}q}+L_f^q=0, \tag{5.2.113}$$

and

$$H_f(p,\frac{\partial L_f^p}{\partial\,{}^{\mathrm{C}}D_{a+}^{\alpha}p})-{}^{\mathrm{C}}D_{a+}^{\alpha}p\frac{\partial L_f^p}{\partial\,{}^{\mathrm{C}}D_{a+}^{\alpha}p}+L_f^p=0. \tag{5.2.114}$$

Therefore, the fractional extended Lagrangian becomes

$$\begin{aligned}&L(q,{}^{\mathrm{C}}D_{a+}^{\alpha}q,p,{}^{\mathrm{C}}D_{a+}^{\alpha}p)\\&=-p\,{}^{\mathrm{C}}D_{a+}^{\alpha}q-q\,{}^{\mathrm{C}}D_{a+}^{\alpha}p+L_f^q(q,{}^{\mathrm{C}}D_{a+}^{\alpha}q)+L_f^p(p,{}^{\mathrm{C}}D_{a+}^{\alpha}p),\end{aligned} \tag{5.2.115}$$

and this implies the fractional Euler-Lagrange equations as

$${}^{\mathrm{RL}}D_{b-}^{\alpha}\frac{\partial L_f^q}{\partial\,{}^{\mathrm{C}}D_{a+}^{\alpha}q}-\frac{\partial L_f^q}{\partial q}-{}^{\mathrm{RL}}D_{b-}^{\alpha}p-{}^{\mathrm{C}}D_{a+}^{\alpha}p=0 \tag{5.2.116}$$

and

$${}^{\mathrm{RL}}D_{b-}^{\alpha}\frac{\partial L_f^p}{\partial\,{}^{\mathrm{C}}D_{a+}^{\alpha}p}-\frac{\partial L_f^p}{\partial p}-{}^{\mathrm{RL}}D_{b-}^{\alpha}q-{}^{\mathrm{C}}D_{a+}^{\alpha}q=0. \tag{5.2.117}$$

Having in mind that the fractional extended momenta are

$$\pi_q^f=\frac{\partial L_f^q}{\partial\,{}^{\mathrm{C}}D_{a+}^{\alpha}q}-p,\qquad \pi_p^f=\frac{\partial L_f^q}{\partial\,{}^{\mathrm{C}}D_{a+}^{\alpha}p}-q, \tag{5.2.118}$$

the fractional extended Hamiltonian was reported in [78] as

$$H_f(q, \pi_q^f, p, \pi_p^f) = \sum_{n=0}^{N} \frac{1}{n!} \left[\frac{\partial^n H_f}{\partial p^n} (\pi_q^f)^n - \frac{\partial^n H_f}{\partial q^n} (\pi_p^f)^n \right]. \tag{5.2.119}$$

We notice [78] that the corresponding fractional generalization of the Poisson brackets can be written as

$$\{F, G\}_f = \frac{\partial F}{\partial q}\frac{\partial G}{\partial \pi_q^f} - \frac{\partial F}{\partial \pi_q^f}\frac{\partial G}{\partial q} + \{F, G\} + \frac{\partial F}{\partial p}\frac{\partial G}{\partial \pi_p^f} - \frac{\partial F}{\partial \pi_p^f}\frac{\partial G}{\partial p}. \tag{5.2.120}$$

5.3 Fractional Variational Problems Within Shifted Orthonormal Legendre Polynomials

We recall that the classical Legendre polynomial $P_k(z)$ are orthogonal on the interval $[-1, 1]$ and can be found utilizing the recurrence relation, namely [229]

$$\begin{aligned} P_{k+1}(z) &= \tfrac{2k+1}{k+1} z P_k(z) - \tfrac{k}{k+1} P_{k-1}(z), \qquad 1 \leq k, \\ P_0(z) &= 1, \qquad P_1(z) = z. \end{aligned}$$

Having in mind to utilize the Legendre polynomials on $[0, t_f]$, we define the variable t; $z = \frac{2t}{t_f} - 1$. As a result, the shifted Legendre polynomials $P^*_{k,t_f}(t)$ become orthogonal on the interval $[0, t_f]$ and may be found with the help of the recurrence relation [229]

$$\begin{aligned} P^*_{k+1,t_f}(t) &= \tfrac{2k+1}{k+1}\left(\tfrac{2t}{t_f} - 1\right) P^*_{k,t_f}(t) - \tfrac{k}{k+1} P^*_{k-1,t_f}(t), \qquad 1 \leq k, \\ P^*_{0,t_f}(t) &= 1, \qquad P^*_{1,t_f}(t) = \tfrac{2t}{t_f} - 1. \end{aligned}$$

Thus, the orthogonality relation becomes [229]

$$\int_0^{t_f} P^*_{j,t_f}(t) P^*_{k,t_f}(t)\,dt = \begin{cases} \frac{t_f}{2k+1}, & \text{for } j = k, \\ 0, & \text{for } j \neq k. \end{cases} \tag{5.3.1}$$

In the next step we define the shifted orthonormal Legendre polynomials $P^\star_{k,t_f}(t)$; $P^\star_{k,t_f}(t) \equiv \sqrt{\frac{2k+1}{t_f}} P^*_{k,t_f}(t)$, then we conclude that [229]

$$\int_0^{t_f} P^\star_{j,t_f}(t) P^\star_{k,t_f}(t)\,dt = \begin{cases} 1, & \text{for } j = k, \\ 0, & \text{for } j \neq k. \end{cases} \tag{5.3.2}$$

We recall that the shifted orthonormal Legendre polynomials can be determined by utilizing the analytical form [229]

$$P^{\star}_{k,t_f}(t) = \sqrt{\frac{2k+1}{t_f}} \sum_{i=0}^{k} (-1)^{k+i} \frac{(k+i)!}{(k-i)!\,(i!)^2\,(t_f)^i}\, t^i. \tag{5.3.3}$$

Any function $y(t)$, square integrable on the interval $[0, t_f]$, can be expressed in terms of $P^{\star}_{k,t_f}(t)$, as below [229]

$$y(t) = \sum_{k=0}^{\infty} y_k P^{\star}_{k,t_f}(t),$$

where

$$y_k = \int_0^{t_f} y(t) P^{\star}_{k,t_f}(t) dt, \qquad 0 \le k. \tag{5.3.4}$$

By making use of the first $(N+1)$ shifted orthonormal Legendre polynomials terms, $y(t)$ can be approximated, namely [229]

$$y_N(t) = \sum_{k=0}^{N} y_k P^{\star}_{k,t_f}(t), \tag{5.3.5}$$

or in a matrix form

$$y_N(t) \simeq \mathbf{Y}^T \Delta_{N,t_f}(t), \tag{5.3.6}$$

such that

$$\mathbf{Y} = \begin{pmatrix} y_0 \\ y_1 \\ \vdots \\ y_N \end{pmatrix}, \qquad \Delta_{N,t_f}(t) = \begin{pmatrix} P^{\star}_{0,t_f}(t) \\ P^{\star}_{1,t_f}(t) \\ \vdots \\ P^{\star}_{N,t_f}(t) \end{pmatrix}. \tag{5.3.7}$$

Theorem 5.8. *According to Riemann-Liouville, the integral of order ν (fractional) of the vector $\Delta_{N,t_f}(t)$ may be determined as* [229]

$${}^{\mathrm{RL}}I^{\nu} \Delta_{N,t_f}(t) = \boldsymbol{I}^{(\nu)} \Delta_{N,t_f}(t), \tag{5.3.8}$$

where $\boldsymbol{I}^{(\nu)}$ is the $(N+1)\times(N+1)$ operational matrix of fractional integral of order ν and is defined by

$$\boldsymbol{I}^{(\nu)} = \begin{pmatrix} \Theta_\nu(0,0) & \Theta_\nu(0,1) & \cdots & \Theta_\nu(0,j) & \cdots & \Theta_\nu(0,N) \\ \Theta_\nu(1,0) & \Theta_\nu(1,1) & \cdots & \Theta_\nu(1,j) & \cdots & \Theta_\nu(1,N) \\ \vdots & \vdots & \ddots & \vdots & \ddots & \vdots \\ \Theta_\nu(i,0) & \Theta_\nu(i,1) & \cdots & \Theta_\nu(i,j) & \cdots & \Theta_\nu(i,N) \\ \vdots & \vdots & \ddots & \vdots & \ddots & \vdots \\ \Theta_\nu(N,0) & \Theta_\nu(N,1) & \cdots & \Theta_\nu(N,j) & \cdots & \Theta_\nu(N,N) \end{pmatrix},$$

where

$$\Theta_\nu(i,j) = \sum_{k=0}^{i} \delta_\nu(i,j,k) \tag{5.3.9}$$

and

$$\delta_\nu(i.j,k) = \sum_{l=0}^{j} \frac{(-1)^{i+j+k+l}\sqrt{2j+1}\sqrt{2i+1}(i+k)!(l+j)!(t_f)^\nu}{(i-k)!k!\Gamma(k+\nu+1)(j-l)!(l!)^2(k+l+\nu+1)}.$$

Proof. The proof can be seen in [101]. □

Let us consider $f(t)$ and $g(t)$ be two square integrable functions defined on $[0,t_f]$. Then we obtain the following decomposition [229]

$$\begin{aligned} f_N(t) &\simeq \textstyle\sum_{k=0}^{N} f_k P^\star_{k,t_f}(t) = \mathbf{F}^T \Delta_{N,t_f}(t), \\ g_N(t) &\simeq \textstyle\sum_{k=0}^{N} g_k P^\star_{k,t_f}(t) = \mathbf{G}^T \Delta_{N,t_f}(t), \end{aligned} \tag{5.3.10}$$

where f_k and g_k; $j=0,1,\ldots,N$, are given by [229]

$$\begin{aligned} f_k &= \textstyle\int_0^{t_f} f(t)P^\star_{k,t_f}(t)dt, \\ g_k &- \textstyle\int_0^{t_f} g(t)P^\star_{k,t_f}(t)dt, \end{aligned} \qquad 0 \le k.$$

Then, we conclude that

$$\begin{aligned} f(t)g(t) &\simeq \left(\mathbf{F}^T\Delta_{N,t_f}(t)\right)\left(\mathbf{G}^T\Delta_{N,t_f}(t)\right) \\ &= \mathbf{F}^T\Delta_{N,t_f}(t)\Delta^T_{N,t_f}(t)\mathbf{G}. \end{aligned} \tag{5.3.11}$$

Assuming that

$$\Delta_{N,t_f}(t)\Delta^T_{N,t_f}(t)\mathbf{G} = \mathbf{H}^T\Delta_{N,t_f}(t), \tag{5.3.12}$$

such that $\mathbf{H}$ is a $(N+1)\times(N+1)$ matrix, we conclude [229]

$$\sum_{k=0}^{N} g_k P^\star_{j,t_f}(t)P^\star_{k,t_f}(t) = \sum_{k=0}^{N} H_{kj}P^\star_{k,t_f}(t), \qquad j=0,1,\ldots,N.$$

Now, we multiply both sides by $P^\star_{i,t_f}(t)$, $i=0,1,\ldots,N$, and integrate the result from 0 to t_f, we obtain

$$\sum_{k=0}^{N} g_k \int_0^{t_f} P^\star_{i,t_f}(t)P^\star_{j,t_f}(t)P^\star_{k,t_f}(t)dt = \sum_{k=0}^{N} H_{ij}\int_0^{t_f} P^\star_{k,t_f}(t)P^\star_{i,t_f}(t)dt,$$
$$i,j=0,1,\ldots,N.$$

Thus, $\mathbf{H}$ becomes [229]

$$\mathbf{H} = \left(H_{ij}\right)_{0\le i,j\le N},$$

where

$$H_{ij} = \sum_{k=0}^{N} g_k \int_0^{t_f} P^\star_{i,t_f}(t)P^\star_{j,t_f}(t)P^\star_{k,t_f}(t)dt, \qquad i,j=0,1,\ldots,N.$$

From (5.3.10)-(5.3.12), the product of $f(t)$ and $g(t)$ becomes

$$f(t)g(t) \simeq \mathbf{F}^T\mathbf{H}^T\Delta_{N,t_f}(t). \tag{5.3.13}$$

Example Below, the following fractional variational problem is investigated [229]:

$$\textit{Minimum} \quad J = \int_0^{t_f} \left({}^{\mathrm{C}}D^{\nu}x(t) - g(t)\right)^2 dt, \qquad 0 < \nu \leq 1, \tag{5.3.14}$$

such that,

$$x(0) = a, \quad x(t_f) = b. \tag{5.3.15}$$

In the first step we approximate ${}^{\mathrm{C}}D^{\nu}x(t)$ by $P^{\star}_{k,t_f}(t)$ as given below [229]

$${}^{\mathrm{C}}D^{\nu}x(t) \simeq \mathbf{X}^T \Delta_{N,t_f}(t), \tag{5.3.16}$$

where $\mathbf{X}$ denotes an unknown coefficient matrix having the form [229]

$$\mathbf{X} = \begin{pmatrix} x_0 \\ x_1 \\ \vdots \\ x_N \end{pmatrix}. \tag{5.3.17}$$

We know that

$${}^{\mathrm{RL}}I^{\nu\,\mathrm{C}}D^{\nu}x(t) = x(t) - x(0), \tag{5.3.18}$$

and also adopting (5.3.8) together with (5.3.16), we end up with [229]

$${}^{\mathrm{RL}}I^{\nu\,\mathrm{C}}D^{\nu}x(t) \simeq \mathbf{X}^T \mathbf{I}^{(\nu)} \Delta_{N,t_f}(t). \tag{5.3.19}$$

Using the two previous equations, we conclude [229]

$$x(t) \simeq \mathbf{X}^T \mathbf{I}^{(\nu)} \Delta_{N,t_f}(t) + x(0). \tag{5.3.20}$$

By approximating $x(0)$ in terms of $P^{\star}_{k,t_f}(t)$ as

$$x(0) \simeq \mathbf{A}^T \Delta_{N,t_f}(t), \tag{5.3.21}$$

such that

$$\mathbf{A} = \begin{pmatrix} a \\ 0 \\ \vdots \\ 0 \end{pmatrix}, \tag{5.3.22}$$

the function $x(t)$ is approximated as

$$x(t) \simeq \left(\mathbf{X}^T \mathbf{I}^{(\nu)} + \mathbf{A}^T\right) \Delta_{N,t_f}(t), \tag{5.3.23}$$

In order to expand the performance index (5.3.14) in terms of shifted orthonormal Legendre polynomials, we consider the performance index (5.3.14) as

$$J = \int_0^{t_f} \left(\left({}^{\mathrm{C}}D^{\nu}x(t)\right)^2 - 2g(t){}^{\mathrm{C}}D^{\nu}x(t) + \left(g(t)\right)^2 \right) dt, \qquad 0 < \nu \leq 1. \quad (5.3.24)$$

Assuming that the $g(t)$ may be expanded as [229]

$$g_N(t) \simeq \sum_{k=0}^{N} g_k P^{\star}_{k,t_f}(t) = \mathbf{G}^T \Delta_{N,t_f}(t), \qquad (5.3.25)$$

where $g_k;\ j = 0, 1, \ldots, N$, are

$$g_k = \int_0^{t_f} g(t) P^{\star}_{k,t_f}(t) dt, \qquad 0 \leq k. \qquad (5.3.26)$$

Utilizing (5.3.16) and (5.3.25), the performance index (5.3.24) may be approximated as

$$\begin{aligned} J &= \textstyle\int_0^{t_f} \Big(\big(\mathbf{X}^T \Delta_{N,t_f}(t)\big)\big(\mathbf{X}^T \Delta_{N,t_f}(t)\big) - 2\big(\mathbf{G}^T \Delta_{N,t_f}(t)\big)\big(\mathbf{X}^T \Delta_{N,t_f}(t)\big) \\ &\qquad + \big(\mathbf{G}^T \Delta_{N,t_f}(t)\big)\big(\mathbf{G}^T \Delta_{N,t_f}(t)\big)\Big) dt \\ &= \textstyle\int_0^{t_f} \Big(\mathbf{X}^T \Delta_{N,t_f}(t)\Delta^T_{N,t_f}(t)\mathbf{X} - 2\mathbf{G}^T \Delta_{N,t_f}(t)\Delta^T_{N,t_f}(t)\mathbf{X} \\ &\qquad + \mathbf{G}^T \Delta_{N,t_f}(t)\Delta^T_{N,t_f}(t)\mathbf{G}\Big) dt \\ &= \textstyle\int_0^{t_f} \left(\mathbf{X}^T\mathbf{H}_1^T - 2\mathbf{G}^T\mathbf{H}_1^T + \mathbf{G}^T\mathbf{H}_2^T\right)\Delta_{N,t_f}(t) dt, \end{aligned} \qquad (5.3.27)$$

where $\mathbf{H}_1$ and $\mathbf{H}_2$ are $N \times N$ matrices, namely [229]

$$\mathbf{H}_1 = \left(H_{1,ij}\right)_{0 \leq i,j \leq N}, \qquad \mathbf{H}_2 = \left(H_{2,ij}\right)_{0 \leq i,j \leq N},$$

and

$$\begin{aligned} H_{1,ij} &= \textstyle\sum_{k=0}^{N} x_k \int_0^{t_f} P^{\star}_{i,t_f}(t) P^{\star}_{j,t_f}(t) P^{\star}_{k,t_f}(t) dt, \\ H_{2,ij} &= \textstyle\sum_{k=0}^{N} g_k \int_0^{t_f} P^{\star}_{i,t_f}(t) P^{\star}_{j,t_f}(t) P^{\star}_{k,t_f}(t) dt, \end{aligned} \qquad i, j = 0, 1, \ldots, N.$$

Applying the above mentioned Theorem, we conclude

$$\begin{aligned} J \equiv J[\mathbf{X}] &= \left(\mathbf{X}^T\mathbf{H}_1^T - 2\mathbf{G}^T\mathbf{H}_1^T + \mathbf{G}^T\mathbf{H}_2^T\right)\left(I^1 \Delta_{N,t_f}(t)\right)_{t=0}^{t=t_f} \qquad (5.3.28) \\ &= \left(\mathbf{X}^T\mathbf{H}_1^T - 2\mathbf{G}^T\mathbf{H}_1^T + \mathbf{G}^T\mathbf{H}_2^T\right)\mathbf{I}^{(1)}\left(\Delta_{N,t_f}(t)\right)_{t=0}^{t=t_f} \\ &= \left(\mathbf{X}^T\mathbf{H}_1^T - 2\mathbf{G}^T\mathbf{H}_1^T + \mathbf{G}^T\mathbf{H}_2^T\right)\mathbf{I}^{(1)}\Big(\Delta_{N,t_f}(t_f) - \Delta_{N,t_f}(0)\Big). \end{aligned}$$

Using Eq. (5.3.23), the boundary condition (5.3.15) becomes

$$\left(\mathbf{X}^T\mathbf{I}^{(\nu)} + \mathbf{A}^T\right)\Delta_{N,t_f}(t_f) - b = 0. \qquad (5.3.29)$$

Now, we use the Lagrange multiplier method in order to merge the two previous equations as

$$J^{\star}[\mathbf{X},\lambda] = J[\mathbf{X}] + \mathbf{B}^T\lambda, \tag{5.3.30}$$

where

$$\lambda = \begin{pmatrix} \lambda_0 \\ \lambda_1 \\ \vdots \\ \lambda_N \end{pmatrix}, \tag{5.3.31}$$

is the unknown Lagrange multiplier, and

$$\mathbf{B} = \begin{pmatrix} \left(\mathbf{X}^T\mathbf{I}^{(\nu)} + \mathbf{A}^T\right)\Delta_{N,t_f}(t_f) - b \\ 0 \\ \vdots \\ 0 \end{pmatrix}. \tag{5.3.32}$$

The necessary conditions for the extremum of the fractional variational problem (5.3.14)-(5.3.15) are

$$\frac{\partial J^{\star}[\mathbf{X},\lambda]}{\partial x_k} = 0, \qquad k = 0, 1, \cdots, N, \tag{5.3.33}$$

$$\frac{\partial J^{\star}[\mathbf{X},\lambda]}{\partial \lambda_k} = 0, \qquad k = 0, 1, \cdots, N. \tag{5.3.34}$$

By making use of any standard iteration technique, the system of $2(N+1)$ algebraic equations given from the two previous equations can be solved. Thus, the unknown coefficients x_k and $\lambda_k,\ k = 0, 1, \cdots, N$, may be obtained.

5.4 Fractional Euler-Lagrange Equations Model for Freely Oscillating Dynamical Systems

In this part we recall a discussion from [14] regarding a freely oscillating single pendulum with length l and uniform mass distribution, such that the damping force is proportional to the fractional derivative of θ with time varying order, *i. e.* with the Coimbra derivative as defined in Eq. (1.3.200), namely

$$\ddot{\theta}(t) + \frac{\alpha - 1}{t - \tau}\, {}^{\mathrm{Co}}D_0^{\beta(t)}\theta(t) + \omega_n^2 \sin\theta(t) = 0. \tag{5.4.1}$$

Here $\beta(t)$ represents the variable order of fractional derivative of the angular displacement θ, $\omega_n = \sqrt{2g/l}$ denotes the natural frequency of the system, τ represents an intrinsic time, and t is the observing time. For more details regarding this equation as well as its connection with variational principles we recommend the readers [14].

In the following we show the numerical solution of the linearized system (5.4.1) [14]. For this purpose, the classical centered finite difference method is utilized to approximate the first term, *i. e.* the second order derivative of θ, and the second term (the Coimbra fractional derivative) is approximated with the help of an appropriately adapted finite difference technique. The third term is linearized in the traditional manner which is acceptable if we assume that the oscillation angle is small [14].

Applying the finite difference method to $\ddot{\theta}(t)$ we conclude [14]

$$\theta_{n+1} \approx \theta(t_n+h) = \theta(t_n) + h\dot{\theta}(t_n) + \frac{h^2}{2!}\ddot{\theta}(t_n) + \frac{h^3}{3!}\dddot{\theta}(t_n) + \ldots, \quad (5.4.2a)$$

$$\theta_{n-1} \approx \theta(t_n-h) = \theta(t_n) - h\dot{\theta}(t_n) + \frac{h^2}{2!}\ddot{\theta}(t_n) - \frac{h^3}{3!}\dddot{\theta}(t_n) + \ldots, \quad (5.4.2b)$$

$$\theta_n \approx \theta(t_n), \quad (5.4.2c)$$

where h is the time increment. Subsequently, $\dot{\theta}_n(t) \approx \dot{\theta}(t_n)$, and so on.

By direct calculations we show that [14]

$$\dot{\theta}(t) = \frac{\theta_{n+1} - \theta_{n-1}}{2h} + O\left(h^2\right) \quad (5.4.3)$$

and

$$\ddot{\theta}(t) = \frac{\theta_{n+1} - 2\theta_n + \theta_{n-1}}{h^2} + O\left(h^2\right), \quad (5.4.4)$$

respectively.

We recall that the Coimbra fractional derivative was defined in Eq. (1.3.200) (see also [180]). The discretization of the integral provided by (1.3.200) can be done as follows for $\theta(0^+) = \theta(0^-)$ [563]: We start with the immediately obvious representation

$${}^{\mathrm{Co}}D_0^{\beta(t_n)}\theta(t_n) = \frac{1}{\Gamma[1-\beta(t_n)]}\sum_{j=0}^{n-1}\int_{t_j}^{t_{j+1}}(t_n-\zeta)^{-\beta(t_n)}D\theta(\zeta)d\zeta. \quad (5.4.5)$$

By applying the forward difference approximation to $D\theta(\zeta)$ we conclude that [14]

$$
\begin{aligned}
&{}^{\mathrm{Co}}D_0^{\beta(t_n)}\theta(t_n)\\
&\approx \frac{1}{\Gamma\left[1-\beta\left(t_n\right)\right]}\sum_{j=0}^{n-1}\int_{t_j}^{t_{j+1}}\left(t_n-\zeta\right)^{-\beta(t_n)}\left(\frac{\theta\left(t_{j+1}\right)-\theta\left(t_j\right)}{h}+h\cdot E_1\right)d\zeta\\
&=\frac{1}{\Gamma\left[1-\beta\left(t_n\right)\right]}\sum_{j=0}^{n-1}\frac{\theta\left(t_{j+1}\right)-\theta\left(t_j\right)}{h}\int_{t_j}^{t_{j+1}}\left(t_n-\zeta\right)^{-\beta(t_n)}d\zeta+E_{1n}. \quad (5.4.6)
\end{aligned}
$$

Here the error E_{1n} has the form

$$
E_{1n}=\frac{1}{\Gamma\left[1-\beta\left(t_n\right)\right]}\sum_{j=0}^{n-1}\int_{t_j}^{t_{j+1}}\left(t_n-\zeta\right)^{-\beta(t_n)}h\cdot E_1 d\zeta\,. \quad (5.4.7)
$$

Now, by solving the first term of the right hand side of (5.4.6) we obtain [14]

$$
\begin{aligned}
{}^{\mathrm{Co}}D_0^{\beta(t_n)}\theta(t_n)\approx\frac{1}{\Gamma\left[2-\beta(t_n)\right]}\sum_{j=0}^{n-1}\frac{\theta(t_{j+1})-\theta(t_j)}{h} \quad (5.4.8)\\
\times\left[\left(t_n-t_j\right)^{1-\beta(t_n)}-\left(t_n-t_{j+1}\right)^{1-\beta(t_n)}\right].
\end{aligned}
$$

When $j=0,1,2,...,n-1$ we may write

$$
-\left[\left(t_n-t_{j+1}\right)^a-\left(t_n-t_j\right)^a\right]=h^a\left[\left(n-j\right)^a-\left(n-j-1\right)^a\right]\ \text{where } a\in\mathbb{R}.
$$

Then, Eq. (5.4.8) becomes

$$
\begin{aligned}
&{}^{\mathrm{Co}}D_0^{\beta(t_n)}\theta(t_n) \quad (5.4.9)\\
&\approx\frac{h^{1-\theta(t_n)}}{\Gamma\left[2-\beta\left(t_n\right)\right]}\sum_{j=0}^{n-1}\frac{\theta\left(t_{j+1}\right)-\theta\left(t_j\right)}{h}\left[\left(n-j\right)^{1-\beta(t_n)}-\left(n-j-1\right)^{1-\beta(t_n)}\right]\\
&=\frac{h^{-\theta(t_n)}}{\Gamma\left[2-\beta\left(t_n\right)\right]}\sum_{j=0}^{n-1}\left[\left(n-j\right)^{1-\beta(t_n)}-\left(n-j-1\right)^{1-\beta(t_n)}\right]\left[\theta\left(t_{j+1}\right)-\theta\left(t_j\right)\right].
\end{aligned}
$$

By utilizing Eqs. (5.4.4), (5.4.9) and (5.4.1) and considering the linearization we obtain

$$
\begin{aligned}
0=&\sum_{j=0}^{n-1}\left\{\left[\left(n-j\right)^{1-\beta(t_n)}-\left(n-j-1\right)^{1-\beta(t_n)}\right]\left[\theta\left(t_{j+1}\right)-\theta\left(t_j\right)\right]\right\}+\omega_n^2\theta\left(t_n\right)\\
&+\frac{\theta\left(t_{n+1}\right)-2\theta\left(t_n\right)+\theta\left(t_{n-1}\right)}{h^2}+\frac{\alpha-1}{t_n-\tau}\,\frac{h^{-\theta(t_n)}}{\Gamma\left[2-\beta\left(t_n\right)\right]}. \quad (5.4.10)
\end{aligned}
$$

If we open the summation for $j = n-1$ and rewrite (5.4.10) we conclude that

$$0 = \omega_n^2 \theta(t_n) + \frac{\theta(t_{n+1}) - 2\theta(t_n) + \theta(t_{n-1})}{h^2} \tag{5.4.11}$$

$$+ \frac{\alpha - 1}{t_n - \tau} \frac{h^{-\beta(t_n)}}{\Gamma[2 - \beta(t_n)]}$$

$$\times \sum_{j=0}^{n-2} \left[(n-j)^{1-\beta(t_n)} - (n-j-1)^{1-\beta(t_n)} \right] [\theta(t_{j+1}) - \theta(t_j)]$$

$$+ \frac{\alpha - 1}{t_n - \tau} \frac{h^{-\theta(t_n)}}{\Gamma[2 - \beta(t_n)]} [\theta(t_n) - \theta(t_{n-1})]$$

$$\times \left[(n-(n-1))^{1-\beta(t_n)} - (n-(n-1)-1)^{1-\beta(t_n)} \right].$$

Multiply both sides of (5.4.11) by h^2, consider the initial conditions, and utilizing $(n-(n-1))^{1-\beta(t_n)} - (n-(n-1)-1)^{1-\beta(t_n)} = 1$ we get

$$\theta(t_{n+1}) - 2\theta(t_n) + \theta(t_{n-1}) + \frac{h^{2-\beta(t_n)}(\alpha - 1)}{(t_n - \tau)\{\Gamma[2 - \beta(t_n)]\}}$$

$$\times \sum_{j=0}^{n-2} \left\{ \left[(n-j)^{1-\beta(t_n)} - (n-j-1)^{1-\beta(t_n)} \right] [\theta(t_{j+1}) - \theta(t_j)] \right\}$$

$$+ \frac{h^{2-\beta(t_n)}(\alpha - 1)}{(t_n - \tau)\{\Gamma[2 - \beta(t_n)]\}} [\theta(t_n) - \theta(t_{n-1})] + h^2 \omega_n^2 \theta(t_n) = 0\,. \tag{5.4.12}$$

Below, the second order homogenous variable coefficients differential equation is considered as a particular case of (5.4.1).

The effect of α is investigated in Figure 5.1 in which the following can be concluded. For $\alpha < 1$, as α decreases the divergence of the system increases, while for $\alpha > 1$, as α increases the convergence of the system increases for the same values of β. For $\alpha = 1$ there is no dissipation force acting on the system.

The system is stable if $\alpha > 1$ and this is due to the positive damping force effect. The analysis of the system stability due to variation of the β values is depicted in Figure 5.2. We conclude that for various values of $\alpha > 1$ the system is stable, if $0 < \beta < 2$.

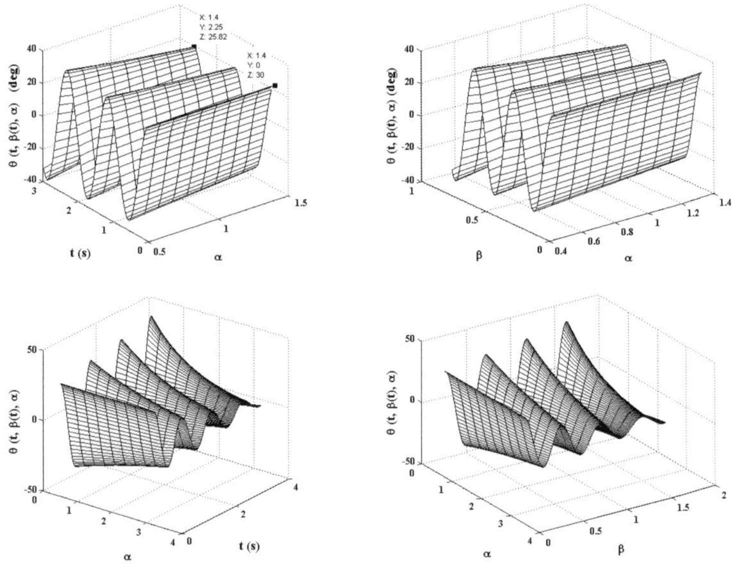

Fig. 5.1 The effect of α value on the system responses.

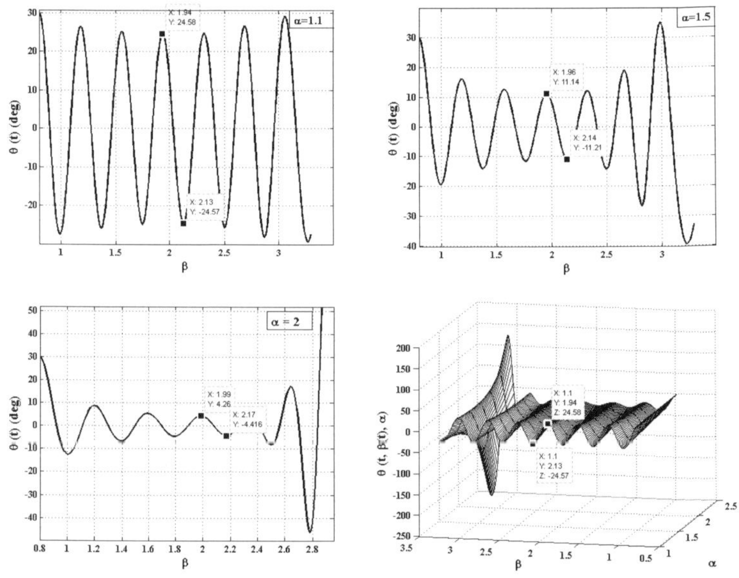

Fig. 5.2 The effect of β value on the system stability.

5.5 Bateman-Feshbach-Tikochinsky Oscillator With Fractional Derivative

Below we discuss the Lagrangian associated to the classical Bateman-Feshbach-Tikochinsky oscillator [71], namely

$$L = m\dot{q}\dot{y} + \frac{\gamma}{2}(q\dot{y} - \dot{q}y) - Kqy, \tag{5.5.1}$$

where q represents the damped harmonic oscillator coordinate, y is the time- reversed counterpart and m, K and γ are time independent.

The next step is to consider the corresponding the fractional Lagrangian as [71]

$$L^F = m({}^{\mathrm{RL}}D^{\alpha}_{a+}q)({}^{\mathrm{RL}}D^{\alpha}_{a+}y) + \frac{\gamma}{2}\left[q({}^{\mathrm{RL}}D^{\alpha}_{a+}y) - ({}^{\mathrm{RL}}D^{\alpha}_{a+}q)y\right] = Kqy. \tag{5.5.2}$$

The canonical momenta are given below [71]

$$\begin{aligned}
P_{\alpha,q} &= \frac{\partial L}{\partial {}^{\mathrm{RL}}D^{\alpha}_{a+}q} = m{}^{\mathrm{RL}}D^{\alpha}_{a+}y - \frac{\gamma}{2}y,\\
P_{\beta,q} &= \frac{\partial L}{\partial {}^{\mathrm{RL}}D^{\beta}_{b-}q} = 0,\\
P_{\alpha,y} &= \frac{\partial L}{\partial {}^{\mathrm{RL}}D^{\alpha}_{a+}y} = m{}^{\mathrm{RL}}D^{\alpha}_{a+}q + \frac{\gamma}{2}q,\\
P_{\beta,y} &= \frac{\partial L}{\partial {}^{\mathrm{RL}}D^{\beta}_{b-}y} = 0.
\end{aligned} \tag{5.5.3}$$

The fractional Hamiltonian is written as

$$H = P_{\alpha,q}\,{}^{\mathrm{RL}}D^{\alpha}_{a+}q + P_{\beta,q}\,{}^{\mathrm{RL}}D^{\beta}_{b-}q + P_{\alpha,y}\,{}^{\mathrm{RL}}D^{\alpha}_{a+}y + P_{\beta,y}\,{}^{\mathrm{RL}}D^{\beta}_{b-}y - L, \tag{5.5.4}$$

or

$$H = m({}^{\mathrm{RL}}D^{\alpha}_{a+}q){}^{\mathrm{RL}}D^{\alpha}_{a+}y + Kqy. \tag{5.5.5}$$

The first Hamiltonian equation of motion reads as [71]

$$m{}^{\mathrm{RL}}D^{\alpha}_{b-}\,{}^{\mathrm{RL}}D^{\alpha}_{a+}y - {}^{\mathrm{RL}}D^{\alpha}_{b-}\frac{\gamma}{2}y = Ky \tag{5.5.6}$$

and the second one has the following form [71]

$$m{}^{\mathrm{RL}}D^{\alpha}_{b-}\,{}^{\mathrm{RL}}D^{\alpha}_{a+}q + {}^{\mathrm{RL}}D^{\alpha}_{b-}\frac{\gamma}{2}q = Kq. \tag{5.5.7}$$

Below we illustrate the solution of (5.5.6) and (5.5.7).

For the numerical solution of these equations we utilize the decomposition to its canonical form with the substitutions of $y \equiv x_1$, and $q \equiv x_2$. Thus, we obtain [71]

$$\begin{aligned} {}^{\mathrm{RL}}D^{\alpha}_{a+}x_1 &= x_3, \\ {}^{\mathrm{RL}}D^{\alpha}_{b-}\left[mx_3 - \frac{\gamma}{2}x_1\right] &= Kx_1, \end{aligned} \tag{5.5.8}$$

$$\begin{aligned} {}^{\mathrm{RL}}D^{\alpha}_{a+}x_2 &= x_4, \\ {}^{\mathrm{RL}}D^{\alpha}_{b-}\left[mx_4 + \frac{\gamma}{2}x_2\right] &= Kx_2. \end{aligned} \tag{5.5.9}$$

Below we utilized four initial conditions, namely $x_1(0) \equiv y(0)$, $x_2(0) \equiv q(0)$ and $x_3(0) \equiv {}^{\mathrm{RL}}D^{\alpha}_{a+}y(0)$, $x_2(0) \equiv {}^{\mathrm{RL}}D^{\alpha}_{a+}q(0)$ together with the left and right Grünwald-Letnikov derivatives can be used. One can use directly the formula derived from the Grünwald-Letnikov definitions, backward and forward, respectively, corresponding to the discrete time step $kh, k = 1, 2, 3, \ldots$. The time interval $[a, b]$ is discretized by $(N + 1)$ equal grid points, such that $N = (b - a)/h$. Thus, we report the following results:

$${}^{\mathrm{GL}}D^{\alpha}_{a+}x_k = h^{-\alpha}\sum_{i=0}^{k} c_i x_{k-i}, \quad k = 0, \ldots, N, \tag{5.5.10}$$

$${}^{\mathrm{GL}}D^{\alpha}_{b-}x_k = h^{-\alpha}\sum_{i=0}^{Nk} c_i x_{k+i}, \quad k = N, \ldots, 0, \tag{5.5.11}$$

respectively, where $x_k \approx x(t_k)$ and $t_k = kh$. We recall that c_i, $i = 1, 2, 3, \ldots$, is computing as

$$c_i = \left(1 - \frac{1+\alpha}{i}\right) c_{i-1} \tag{5.5.12}$$

for $c_0 = 1$. Thus, we have [71]

$${}^{\mathrm{GL}}D^{\alpha}_{a+}x(t) = f(x(t), t) \tag{5.5.13}$$

under initial conditions: $y^{(k)}(0) = y_0^{(k)}$, $k = 0, 1, \ldots, n-1$ and $n - 1 < \alpha < n$, can be expressed for discrete time $t_k = kh$ as [71]

$$x(t_k) = f(x(t_k), t_k)h^{\alpha} - \sum_{i=m}^{k} c_i x(t_{k-i}), \tag{5.5.14}$$

where $m = 0$ if we do not use a short memory principle, otherwise it can be related to memory length.

Chapter 6

Continuous-Time Random Walks and Fractional Diffusion Models

6.1 Introduction

This chapter contains an introduction to continuous-time random walks (CTRWs) and their link to space-time fractional diffusion. The material covered below is standard and rather elementary in the analysis of stochastic processes, but applied scientists may have not been previously exposed to it, at least in this form. The basic idea is to derive a theorem relating the continuous or hydrodynamic limit of CTRWs to the space-time fractional diffusion as quickly as possible. The cost to be paid is generality, even if we give pointers to the literature helping in filling gaps. We assume that our readers are more interested in basic ideas rather than in mathematical details. Incidentally, Theorem 6.17 shows how the Caputo time derivative naturally emerges when some known results on normal diffusion are generalized to fractional diffusion. One can go on and use an equivalent formulation in terms of Riemann-Liouville fractional derivatives, but there is no point in doing that. A similar presentation of this material, but with more emphasis on compound Poisson processes and a discussion of the so-called Montroll-Weiss equation [440] can be found in a chapter of a collective book [549].

CTRWs are used in physics to model single particle (tracer) diffusion when the tracer time of residence in a site is much larger than the time needed to jump to another site [425, 426, 439, 440, 554, 555, 564]. CTRWs are phenomenological models, they can approximate microscopic theories for tracer motion, but they have an unphysical character. Indeed, if the stochastic process represents the position of a tracer particle, the sudden jumps of CTRWs occur with infinite speed. Moreover, the reader is warned that the processes called CTRWs in the literature on physics and chemical physics are known as generalized compound Poisson processes or compound renewal processes in the mathematical literature and they have a long history. Compound Poisson processes can be used to approximate Lévy processes. Indeed, as discussed by Feller in Chapter 17 of the second volume of his book [232] every infinitely divisible distribution can be represented as the limit of a sequence of compound Poisson distributions. The importance of these processes prompted de Finetti to devote part of his second volume in probability theory to compound Poisson processes as well [177]. In Theorem 6.13, we will derive a distribution given by Equation (6.2.84). When $P(n,t)$ is the distribution of the Poisson process, Equation (6.2.84) is also known as generalized Poisson distribution and it was discussed by Feller in his 1943 paper [231]. From the modelling viewpoint, this distribution is quite versatile. In the words of Feller [231]:

> Consider independent random events for which the simple Poisson distribution may be assumed, such as: telephone calls, the occurrence of claims in an insurance company, fire accidents, sickness and the like. With each event there may be associated a random variable X. Thus, in the above examples, X may represent the length of the ensuing conversation, the sum under risk, the damage, the cost (or length) of hospitalization, respectively.

The applications to insurance problems were indeed available at the beginning of the XXth Century [163, 389]. Already in 1943, Feller also wrote [231]:

> In view of the above examples, it is not surprising that the law, or special cases of it, have been discovered, by various means and sometimes under disguised forms, by many authors.

This process of rediscovery went on also after Feller's paper; as outlined above, it is the case of physics, where X is interpreted as tracer's position. More recently, for financial applications, X is seen as the log-return for a stock [402, 408, 548, 550]. More on that will be presented in the next chapter.

6.2 The Definition of Continuous-Time Random Walks

In this section, we shall formally define CTWRs as random walks with discrete time modified by a counting renewal process. Essentially, we need two basic ingredients:

(1) the random walk X_n;
(2) the counting process $N(t)$.

Let us begin with the random walk. This is a stochastic process given by a sum of independent and identically distributed (i.i.d.) random variables.

Definition 6.1. Let $\{Y_i\}_{i=1}^{\infty}$ be a sequence of i.i.d. random variables with cumulative distribution function given by $F_Y(y) = \mathbb{P}(Y \leq y)$, then the process X_n defined as

$$X_0 = 0 \tag{6.2.1}$$

$$X_n = \sum_{i=1}^{n} Y_i, \; n \geq 1 \tag{6.2.2}$$

is a random walk.

In this book, we do not consider proofs of existence of stochastic processes, however it is useful to see that the definition is not void. Let us consider the following example.

Example 6.1. Let $\{Y_i\}_{i=1}^{\infty}$ be a sequence of Bernoullian random variables with

$$F_Y(y) = \frac{1}{2}\theta(y) + \frac{1}{2}\theta(y-1) \tag{6.2.3}$$

where $\theta(x)$ is the càdlàg (*continue à droite, limite à gauche*) version of Heaviside function. This means that, for all $i \geq 1$, one has $Y_i = 0$ with probability $1/2$ or $Y_i = 1$ with probability $1/2$. In this case, the random walk X_n is just the number of *successes* up to time step n. Its one-point distribution is given by the binomial distribution of parameters $1/2$ and n. In other words, one can write $X_n \sim \text{Bin}(1/2, n)$, or, more explicitly

$$P(k,n) = \mathbb{P}(X_n = k) = \binom{n}{k}\frac{1}{2^n}. \tag{6.2.4}$$

In Example 6.1, we were able to derive the one-point distribution function $P(k,n)$ of the random walk X_n using a well-known result of elementary probability theory. Elementary probability theory is helpful in deriving a general formula for the cumulative distribution function $F_{X_n}(x) = \mathbb{P}(X_n \leq x)$ and for a generic random walk.

Theorem 6.1. *Let $\{Y_i\}_{i=1}^{\infty}$ be a sequence of i.i.d. random variables with cumulative distribution function given by $F_Y(y)$. Then the cumulative distribution function of the corresponding random walk X_n is given by the n-fold convolution of $F_Y(y)$, in symbols one gets*

$$F_{X_n}(x) = F_Y^{*n}(x). \tag{6.2.5}$$

Proof. Let us use induction on n for $n \geq 2$. To see that the formula is true for $n = 2$, let us consider the random variable $X_2 = Y_1 + Y_2$. One has the following chain of equalities

$$\begin{aligned} F_{X_2}(x) &= \mathbb{P}(X_2 \leq x) = \mathbb{P}(Y_1 + Y_2 \leq x) = \mathbb{E}(I_{Y_1+Y_2 \leq x}) \\ &= \mathbb{E}\left(I_{\{Y_1 \in \mathbb{R}\} \cap \{Y_2 \leq x - Y_1\}}\right) = \mathbb{E}\left(I_{Y_1 \in \mathbb{R}} I_{Y_2 \leq x - Y_1}\right) \\ &= \int_{-\infty}^{+\infty} dF_Y(y_1) \int_{-\infty}^{x-y_1} dF_Y(y_2) \\ &= \int_{-\infty}^{+\infty} dF_Y(y_1) F_Y(x - y_1) = F_Y^{*2}(x). \end{aligned} \tag{6.2.6}$$

Now suppose that the formula is true for $n-1$ and let us prove that it holds also for n. The inductive hypothesis is $F_{X_{n-1}}(x) = F_Y^{*(n-1)}(x)$. Taking into account definition 6.1, one has $X_n = X_{n-1} + Y_n$ and X_{n-1} is independent of Y_n, therefore, as before, we can write

$$\begin{aligned} F_{X_n}(x) &= \mathbb{P}(X_n \leq x) = \mathbb{P}(X_{n-1} + Y_n \leq x) = \mathbb{E}(I_{X_{n-1}+Y_n \leq x}) \\ &= \mathbb{E}\left(I_{\{X_{n-1} \in \mathbb{R}\} \cap \{Y_n \leq x - X_{n-1}\}}\right) = \mathbb{E}\left(I_{X_{n-1} \in \mathbb{R}} I_{Y_n \leq x - X_{n-1}}\right) \\ &= \int_{-\infty}^{+\infty} dF_Y^{*(n-1)}(u) \int_{-\infty}^{x-u} dF_Y(w) \\ &= \int_{-\infty}^{+\infty} dF_Y^{*(n-1)}(u) F_Y(x - u) = F_Y^{*n}(x). \end{aligned} \tag{6.2.7}$$

The latter chain of equalities completes the inductive proof. □

Remark 6.1. In the above derivation, I_A denotes the indicator function of event A. Moreover, we have used the fact that $\mathbb{P}(A) = \mathbb{E}(I_A)$ and that $I_{A \cap B} = I_A I_B$ so that $\mathbb{P}(A \cap B) = \mathbb{E}(I_A I_B)$.

Remark 6.2. Note that one has $F_Y^{*1}(x) = F_Y(x)$. The meaning of $F_Y^{*0}(x)$ is more interesting. Indeed, it is possible to show that $F_Y^{*0}(x) = \theta(x)$ where $\theta(x)$ denotes the càdlàg version of the Heaviside function.

Remark 6.3. The convolution in Equation (6.2.5) is known as Lebesgue-Stieltjes convolution (or convolution of measures) [102]. This is connected to the Lebesgue convolution (or convolution of functions) discussed in Section 1.1. The following corollary establishes the connection. Note that some authors, such as Feller [232], use a different notation for the two convolutions. The convolution of measures is denoted by $\star$ and the convolution of functions by $*$.

Corollary 6.1. *Let $\{Y_i\}_{i=1}^{\infty}$ be a sequence of i.i.d. random variables with cumulative distribution function given by $F_Y(y)$. Assume that the probability density function $f_Y(y) = dF_Y(y)/dy$ exists. Then the probability density function of the corresponding random walk X_n is given by the n-fold convolution of $f_Y(y)$, in symbols one can write*

$$f_{X_n}(x) = f_Y^{*n}(x) \tag{6.2.8}$$

Proof. It is sufficient to derive Equation (6.2.5) in order to get this result. Recall that if $f_Y(x)$ exists, one can write $dF_Y(x) = f_Y(x)dx$. □

Remark 6.4. Note that the derivative of Heaviside function, $\theta(x)$, is a generalized function [561] known as Dirac delta, $\delta(x)$.

Theorem 6.1 and Corollary 6.1 show that the one-point measure of random walk is the n-fold convolution of the *jump* measure. For this reason, it is useful to use Fourier transforms when dealing with sums of independent (and identically distributed) random variables. Given a random variable X, one can define its characteristic function as follows.

Definition 6.2. Let X be a random variable, its characteristic function $\widehat{f}_X(\kappa)$ is given by

$$\widehat{f}_X(\kappa) = \mathbb{E}\left(\mathrm{e}^{i\kappa X}\right). \tag{6.2.9}$$

Theorem 6.2. *If the random variable X has a probability density function $f_X(x)$, then its characteristic function is just the Fourier transform of the* *probability density function. In symbols, one has*

$$\widehat{f}_X(\kappa) = \mathcal{F}[f_X(x)](\kappa) = \int_{-\infty}^{+\infty} dx\, e^{i\kappa x} f_X(x). \tag{6.2.10}$$

Proof. The proof immediately follows from the definition

$$\widehat{f}_X(\kappa) = \mathbb{E}\left(e^{i\kappa X}\right) = \int_{-\infty}^{+\infty} dx\, e^{i\kappa x} f_X(x), \tag{6.2.11}$$

an elementary result which will be very useful. □

Remark 6.5. The conditions on $f_X(x)$ for the existence of the Fourier transform are not too demanding. For instance, the Fourier transform may also exist for generalized functions. In the case of Dirac delta, one has

$$\mathcal{F}[\delta(x)](\kappa) = \int_{-\infty}^{+\infty} dx\, \delta(x) e^{i\kappa x} = 1. \tag{6.2.12}$$

If Lebesgue-Stieltjes integrals are used, a generic probability density function $f_X(x)$ will be a non-negative generalized function satisfying the constraint

$$\int_{-\infty}^{+\infty} dx f_X(x) = 1, \tag{6.2.13}$$

and its Fourier transform will exist [149]. It is possible to prove that the characteristic function has the following properties:

(1) it is continuous for every $\kappa \in \mathbb{R}$;
(2) $\widehat{f}_X(0) = 1$;
(3) $\widehat{f}_X(\kappa)$ is a positive semi-definite function.

Of these three properties, only the third one needs further illustration. This is a rather technical condition. Take an arbitrary integer n and a set of real numbers $\kappa_1, \ldots, \kappa_n$, then build the matrix $a_{i,j} = \widetilde{f}(\kappa_i - \kappa_j)$. Then this matrix is positive semi-definite. A theorem due to Bochner shows that the converse is true, that is any function with the three properties above is the characteristic function of a random variable [109].

Remark 6.6. The derivatives of the characteristic function in $\kappa = 0$ are related to the moments of the corresponding random variable if they exist. The reader can check that

$$\mathbb{E}[Y] = -i \left.\frac{d f_Y(\kappa)}{d\kappa}\right|_{k=0}, \tag{6.2.14}$$

and that

$$\mathbb{E}[Y^2] = -\left.\frac{d^2 f_Y(\kappa)}{d\kappa^2}\right|_{k=0}. \tag{6.2.15}$$

The convolution theorem for Fourier transforms, *i. e.* Eq. (1.1.9), is very important when studying the sum of independent random variables, as it transforms convolutions into algebraic equations. The proof of this theorem, with different levels of detail, can be found in any book on Fourier methods. See reference [149] for example. The convolution theorem for Fourier transform has an immediate consequence on the characteristic function of the random walk X_n.

Corollary 6.2. *Let $\{Y_i\}_{i=1}^{\infty}$ be a sequence of i.i.d. random variables with cumulative distribution function given by $F_Y(y)$ and generalized probability distribution function denoted by $f_Y(y)$. Let X_n be the corresponding random walk. Then the characteristic function of the random walk is given by*

$$\widehat{f}_{X_n}(\kappa) = \mathbb{E}\left(e^{ikX_n}\right) = [\widehat{f}_Y(\kappa)]^n. \tag{6.2.16}$$

Proof. This statement is a direct consequence of Corollary 6.1 and of the convolution theorem as stated in Eq. (1.1.9). □

The following example shows a simple way to introduce the concept of stable random variables.

Example 6.2. Let $\{Y_n\}_{n=1}^{\infty}$ be a sequence of i.i.d. random variables with probability density function given by

$$f_Y(y) = \frac{1}{\sqrt{4\pi}} e^{-y^2/4}. \tag{6.2.17}$$

This is the normal (Gaussian) probability density function with expectation $\mu = 0$ and variance $\sigma^2 = 2$; in other terms, we have $Y_i \sim N(0, 2)$. Which is the characteristic function of the corresponding random walk X_n? For the Y_is, the characteristic function is

$$\widehat{f}_Y(\kappa) = \frac{1}{\sqrt{4\pi}} \int_{-\infty}^{+\infty} dy\, e^{i\kappa y - y^2/4} = e^{-k^2}. \tag{6.2.18}$$

Therefore, from Equation (6.2.16), one gets for X_n

$$\widehat{f}_{X_n}(\kappa) = \left[e^{-k^2}\right]^n = e^{-nk^2}. \tag{6.2.19}$$

This Fourier transform can be easily inverted to get the probability density function $p(x, n) = f_{X_n}(x)$

$$p(x, n) = f_{X_n}(x) = \frac{1}{\sqrt{4n\pi}} \mathrm{e}^{-x^2/4n}. \quad (6.2.20)$$

The reader can prove that when $Y_i \sim N(\mu, \sigma^2)$, that is when one has

$$f_Y(x) = \frac{1}{\sqrt{2\pi n\sigma^2}} \mathrm{e}^{-(x-\mu)^2/2\sigma^2}, \quad (6.2.21)$$

the characteristic function for jumps is

$$\widehat{f}_Y(\kappa) = \mathrm{e}^{i\mu\kappa - \sigma^2\kappa^2/2}, \quad (6.2.22)$$

the characteristic function of the random walk is

$$\widehat{f}_{X_n}(\kappa) = \mathrm{e}^{in\mu\kappa - n\sigma^2\kappa^2/2}, \quad (6.2.23)$$

and its probability density function is

$$p(x, n) = f_{X_n}(x) = \frac{1}{\sqrt{2\pi n\sigma^2}} \mathrm{e}^{-(x-n\mu)^2/2n\sigma^2}. \quad (6.2.24)$$

By comparing Equations (6.2.21) and (6.2.24), one can see that the n-fold convolution of a normal distribution is still a normal distribution, but with parameters rescaled by n. Another way of expressing this result is through the concept of stable random variable.

Definition 6.3. Let Y_1 and Y_2 be two independent and identically distributed random variables that can be seen as copies of a random variable Y. Then, Y is said to be stable or stable in the broad sense if for any constants a and b, the sum $aY_1 + bY_2$ is distributed as $cY + d$ for some constants c and d. If $d = 0$, then the random variable Y is called strictly stable [462].

Remark 6.7. There are alternative and equivalent definitions of stable random variables. In the following, we will use the characterization in terms of characteristic functions [462]. The proof of equivalence can be found in [543].

Definition 6.4. A random variable Y is stable if and only if it can be written as $Y = aZ + b$ and the characteristic function of Z is given by

$$\widehat{f}_Z(\kappa) = \mathbb{E}(\mathrm{e}^{i\kappa Z}) = \mathrm{e}^{-|\kappa|^\alpha[1 - i\gamma\tan(\pi\alpha/2)\mathrm{sign}(\kappa)]} \quad (6.2.25)$$

for $0 < \alpha < 1$, $-1 \le \gamma \le 1$, $a > 0$ and $b \in \mathbb{R}$, and

$$\widehat{f}_Z(\kappa) = \mathbb{E}(\mathrm{e}^{i\kappa Z}) = \mathrm{e}^{-|\kappa|[1 + i(2/\pi)\mathrm{sign}(\kappa)\log(|u|)]} \quad (6.2.26)$$

for the special case $\alpha = 1$.

The reader might wish to check directly that normal random variables are stable, being invariant under convolution or by studying the characteristic function of the normal distribution. Another important property of stable random variables is that they are attractors: under suitable hypothesis, when convolutions are repeated infinitely many times, the limiting distribution is given by a stable random variable. This is essentially the content of the celebrated central limit theorem. We will now discuss it in its Lindenberg-Lévy version. For that, we need to define the convergence in distribution for a sequence of random variables. Let us first define the so-called weak convergence for sequences of functions.

Definition 6.5. Let $\{F_i(x)\}_{i=1}^{\infty}$ be a sequence of cumulative probability distribution functions defined on $\mathbb{R}$. The sequence converges weakly to the cumulative probability distribution $F(x)$ and we write

$$\lim_{i\to\infty} F_i(x) \stackrel{w}{=} F(x) \tag{6.2.27}$$

if one has

$$\lim_{i\to\infty} F_i(x) = F(x) \tag{6.2.28}$$

on all the $x \in \mathcal{C}_F$, where $\mathcal{C}_F$ is the continuity set of $F(x)$.

Once weak convergence is defined, we can use it to define convergence in distribution for random variables.

Definition 6.6. Let $\{X_i\}_{i=1}^{\infty}$ be a sequence of random variables. We say that the sequence converges in distribution to a random variable X, and we write

$$\lim_{i\to\infty} X_i \stackrel{d}{=} X \tag{6.2.29}$$

if one has

$$\lim_{i\to\infty} F_{X_i}(x) = \mathbb{P}(X_i \leq x) \stackrel{w}{=} F_X(x). \tag{6.2.30}$$

The Lévy continuity theorem provides a necessary condition for the convergence in distribution.

Theorem 6.3. *Let $\{X_i\}_{i=1}^{\infty}$ be a sequence of random variables with cumulative distribution functions $F_{X_i}(x)$. Let $\widehat{f}_{X_i}(\kappa) = \mathbb{E}(\mathrm{e}^{i\kappa X_i})$ be the corresponding characteristic functions. Assume that there is a function $\widehat{f}(\kappa)$ such that for any $\kappa \in \mathbb{R}$ (pointwise), one has*

$$\lim_{i\to\infty} \widehat{f}_{X_i}(\kappa) = \widehat{f}(\kappa), \tag{6.2.31}$$

with $\widehat{f}(\kappa)$ continuous for $\kappa = 0$, $\widehat{f}(0) = 1$ and positive semi-definite. Then, there is a random variable X and a corresponding cumulative distribution function $F_X(x)$ such that $\widehat{f}(\kappa)$ is the characteristic function of X and

$$\lim_{i \to \infty} F_{X_i}(x) \stackrel{w}{=} F_X(x). \tag{6.2.32}$$

Proof. A nice proof of this theorem is contained in a book by Williams [630], Chapter 18. □

Theorem 6.4. *Let $\{Y_i\}_{i=1}^{\infty}$ be a sequence of i.i.d. random variables, such that their common expected value is $\mu_Y = \mathbb{E}(Y) < \infty$ and their common variance is $\sigma_Y^2 = \mathbb{E}[(Y - \mu_Y)^2] < \infty$. Let X_n be the corresponding random walk. Define the random variable $Z_n = \sqrt{n}(X_n/n - \mu_Y)/\sigma_Y$, then for $n \to \infty$, Z_n converges in distribution to a normally distributed random variable $Z \sim N(0, 1)$.*

Proof. As a consequence of the convolution theorem, *i. e.* Eq. (1.1.9), and of Equation (6.2.15), one gets for a generic $\kappa \in \mathbb{R}$

$$\widehat{f}_{Z_n}(\kappa) = \left[\widehat{f}_Y\left(\frac{\kappa}{\sqrt{n}\sigma_Y}\right)\right]^n = \left[1 - \frac{1}{2}\frac{\kappa^2}{n} + \mathrm{o}\left(\frac{\kappa^2}{n\sigma_Y^2}\right)\right]^n, \tag{6.2.33}$$

where the first two non-vanishing terms of the MacLaurin expansion for the characteristic function are highlighted. Therefore, one has that

$$\lim_{n \to \infty} \widehat{f}_{Z_n}(\kappa) = \mathrm{e}^{-\kappa^2/2}. \tag{6.2.34}$$

Equation (6.2.22) and the Lévy continuity theorem immediately imply that $Z_n \stackrel{d}{\to} Z$ for $n \to \infty$ with $Z \sim N(0, 1)$. □

The central limit theorem can be generalized in different directions. In our opinion, the fastest path to understand the connection between fractional diffusion and continuous-time random walks is to use symmetric α-stable random variables [313, 462, 543, 651].

Definition 6.7. A symmetric α-stable random variable Y_α has the following characteristic function

$$\widehat{f}_{Y_\alpha}(\kappa) = \mathrm{e}^{-|\kappa|^\alpha}, \tag{6.2.35}$$

for $\alpha \in (0, 2]$.

Remark 6.8. Note that this definition is a simplified version of Definition 6.4.

Remark 6.9. In general, neither the cumulative distribution function $F_{Y_\alpha}(y)$ nor the probability density $f_{Y_\alpha}(y)$ can be written in terms of elementary functions. There are exceptions. As a consequence of Equation (6.2.22), one has that $Y_2 \sim N(0,2)$. Another remarkable case is $\alpha = 1$ which coincides with the Cauchy distribution. Note that these distributions have infinite second moment for $\alpha \in (0,2)$, whereas their first moment is finite for $\alpha \in (1,2]$. In the applied literature, a scale parameter $h > 0$ is often included in the definition, and one writes $\widehat{f}_{Y_\alpha}(\kappa|h) = \mathrm{e}^{-|h\kappa|^\alpha}$. Sometimes, the scale parameter has the form $c = h^\alpha$, so that one has $\widehat{f}_{Y_\alpha}(\kappa|c) = \mathrm{e}^{-c|\kappa|^\alpha}$. It is a useful exercise to check that Y_α defined in Definition 6.7 is indeed a stable random variable according to Definition 6.3.

Similarly to the Normal distribution, symmetric α-stable distributions are invariant as well as attractors for the convolution. Gnedenko and Kolmogorov and Lévy proved a generalization of the central limit theorem, involving sums of independent and identically distributed random variables with infinite second moment [259, 367].

Theorem 6.5. *Let $\{Y\}_{i=1}^\infty$ be a sequence of i.i.d. random variables such that their characteristic function has the following behavior in the neighborhood of $\kappa = 0$*

$$\widehat{f}_Y(\kappa) = 1 - |\kappa|^\alpha + \mathrm{o}(|\kappa|^\alpha). \tag{6.2.36}$$

Let X_n be the corresponding random walk and define $Z_n = X_n/n^{1/\alpha}$, then one has

$$\lim_{n\to\infty} Z_n \overset{d}{=} Z, \tag{6.2.37}$$

where Z is a symmetric α-stable distribution.

Proof. Given a $\kappa \in \mathbb{R}$, one has the following chain of equalities

$$\widehat{f}_{Z_n}(\kappa) = \left[\widehat{f}_Y\left(\frac{\kappa}{n^{1/\alpha}}\right)\right]^n = \left[1 - \frac{|\kappa|^\alpha}{n} + \mathrm{o}\left(\frac{|\kappa|^\alpha}{n}\right)\right]^n, \tag{6.2.38}$$

so that

$$\lim_{n\to\infty} \widehat{f}_{Z_n}(\kappa) = \mathrm{e}^{-|k|^\alpha}. \tag{6.2.39}$$

The Lévy continuity theorem yields the thesis. □

A natural question is whether condition (6.2.36) is satisfied by some random variable. Indeed, random variables whose cumulative distribution function has a power-law behavior for $|y| \to \infty$ do satisfy (6.2.36). This result can be presented with different levels of sophistication (see *e. g.* [543]). Here, we present a simplified version; for more details, the reader can consult a paper by Gorenflo and Abdel-Rehim [268].

Theorem 6.6. *Let Y be a symmetric random variable and assume that its cumulative distribution function $F_Y(y)$ has the following asymptotic behavior for large y and for $\alpha \in (0, 2)$*

$$\lim_{y \to \infty} \frac{b}{\alpha} \frac{F_Y(y)}{1/y^{\alpha}} = 1, \tag{6.2.40}$$

with

$$b = \frac{\Gamma(\alpha + 1) \sin(\alpha\pi/2)}{\pi}, \tag{6.2.41}$$

then the characteristic function of Y satisfies condition (6.2.36).

Proof. This theorem is proved in Chapter 8 of [102]. □

It is now possible to define the counting process $N(t)$, the second ingredient needed for the continuous-time random walk. Here, we shall only consider counting processes of renewal type.

Definition 6.8. Let $\{J\}_{i=1}^{\infty}$ be a sequence of i.i.d. positive random variables interpreted as sojourn times between subsequent events arriving at random times. They define a renewal process whose epochs of renewal (time instants at which the events occur) are the random times $\{T\}_{n=0}^{\infty}$ defined by

$$\begin{aligned} T_0 &= 0, \\ T_n &= \sum_{i=1}^{n} J_i. \end{aligned} \tag{6.2.42}$$

The name renewal process is due to the fact that at any epoch of renewal, the process starts again from the beginning.

Definition 6.9. Associated to any renewal process, there is the process $N(t)$ defined as

$$N(t) = \max\{n : T_n \leq t\} \tag{6.2.43}$$

counting the number of events up to time t.

Remark 6.10. The counting process $N(t)$ is the Poisson process if and only if $J \sim \exp(\lambda)$, that is if and only if sojourn times are i.i.d. exponentially distributed random variables with parameter λ. Incidentally, this is the only case of Lévy and Markov counting process related to a renewal process (see Çinlar's book [147] for a proof of this statement).

Remark 6.11. We shall always assume that the counting process has càdlàg (*continue à droite et limite à gauche, i. e.* right continuous with left limits) sample paths. This means that the realizations are represented by step functions. If t_k is the epoch of the k-th jump, we have $N(t_k^-) = k-1$ and $N(t_k^+) = k$.

Remark 6.12. In Equation (6.2.43), max is used instead of the more general sup as only processes with finite (but arbitrary) number of jumps in $(0, t]$ are considered here.

Given the cumulative probability distribution function $F_J(t)$ for the sojourn times, one immediately gets the distribution for the corresponding epochs.

Theorem 6.7. *Let $\{J\}_{i=1}^{\infty}$ be a sequence of i.i.d. sojourn times with cumulative distribution function $F_J(t)$, then one gets for the generic epoch T_n*

$$F_{T_n}(t) = F_J^{*n}(t). \tag{6.2.44}$$

Proof. This theorem is the same as Theorem 6.1 with T_n playing the role of X_n. The only difference is that the cumulative distribution function is non-vanishing only for positive support. □

Example 6.3. Assume that $J \sim \exp(\lambda)$, then it can be proved by direct calculation that the epochs T_n follow the so-called Erlang distribution

$$F_{T_n}(t) = 1 - \mathrm{e}^{-\lambda t} \sum_{i=0}^{n-1} \frac{(\lambda t)^i}{i!}. \tag{6.2.45}$$

The distribution of the random variable $N(t)$ can be derived from the knowledge of $F_J(t)$ as well.

Theorem 6.8. *Let $\{J\}_{i=1}^{\infty}$ be a sequence of i.i.d. sojourn times with cumulative distribution function $F_J(t)$, then one has*

$$P(n,t) = \mathbb{P}(N(t) = n) = (f_J^{*n} * \bar{F}_J)(t) = \int_0^t du f_J^{*n}(u)\bar{F}_J(t-u), \quad (6.2.46)$$

where $f_J(t)$ is the probability density function of sojourn times J and $\bar{F}_J(t) = 1 - F_J(t)$ is the complementary cumulative distribution function.

Proof. The event $\{N(t) = n\}$ is equivalent to the event $\{T_n \leq t\} \cap \{T_{n+1} > t\} = \{T_n \leq t, T_{n+1} > t\}$. Further observe that $T_{n+1} = T_n + J_{n+1}$ and T_n and J_{n+1} are independent random variables. Now, the following chain of equalities holds true

$$\begin{aligned}
\mathbb{P}(N(t) = n) &= \mathbb{P}(T_n \leq t, T_{n+1} > t) = \mathbb{P}(T_n \leq t, J_{n+1} > t - T_n) \\
&= \mathbb{E}\left(I_{\{T_n \leq t\}} I_{\{J_{n+1} > t - T_n\}}\right) \\
&= \int_{T_n \leq t} du\, f_J^{*n}(u) \int_{J_{n+1} > t - T_n} dw\, f_J(w) \\
&= \int_0^t du\, f_J^{*n}(u) \int_{t-u}^{\infty} dw\, f_j(w) = \int_0^t du\, f_J^{*n}(u)[1 - F_J(t-u)] \\
&= \int_0^t du f_J^{*n}(u)\bar{F}_J(t-u) = (f_J^{*n} * \bar{F}_J)(t). \qquad (6.2.47)
\end{aligned}$$

In the above proof, we have used the properties of indicator functions presented in Remark 6.1. Moreover, the independence of T_n and J_{n+1} implies that their joint probability density function is the product of the two marginals $f_{T_n,J_{n+1}}(u,w) = f_{T_n}(u) f_{J_{n+1}}(w) = f_J^{*n}(u) f_J(w)$. □

For positive random variables, the Laplace transform defined in Equation (1.1.15) plays the same role as the Fourier transform.

Definition 6.10. Let Y be a positive random variable, then its Laplace transform can be written as

$$\widetilde{f}_Y(s) = \mathbb{E}\left(e^{-sY}\right), \qquad (6.2.48)$$

with $s \in \mathbb{C}$.

Theorem 6.9. *Let $f_Y(y)$ (with $y > 0$) denote the (generalized) probability density function of a positive random variable Y, then, for $s \in \mathbb{C}$, the Laplace transform is given by*

$$\widetilde{f}_Y(s) = \mathbb{E}(e^{-sY}) = \mathcal{L}[f_Y(y)](s) = \int_0^{\infty} dy\, f_Y(y) e^{-st}. \qquad (6.2.49)$$

Proof. This is an immediate consequence of the definition (and indeed it could be incorporated in the definition itself). If $f_Y(y)$ is a generalized function, Equation (6.2.49) is often called the Laplace-Stieltjes transform of the random variable or the Laplace-Stieltjes transform of the cumulative distribution function $F_Y(y)$. □

Remark 6.13. Let $s = \mathrm{Re}(s) + i\mathrm{Im}(s)$, then one can write

$$\mathcal{L}[f_Y(y)](s) = \widetilde{f}_Y(s) = \int_0^\infty dy\, f_Y(y)\mathrm{e}^{-\mathrm{Re}(s)y}\mathrm{e}^{i(-\mathrm{Im}(s))y}. \tag{6.2.50}$$

In other words, the Laplace transform can be seen as the Fourier transform calculated for $\kappa = -\mathrm{Im}(s)$ for the function $g_Y(y)$ which is 0 for $y < 0$ and equals $f_Y(y)\mathrm{e}^{-\mathrm{Re}(s)y}$ for $y > 0$. For values of s in which $g_Y(y) \in L^1(\mathbb{R})$, the Laplace transform exists. A classical reference on Laplace transforms is the book by Widder [629].

As discussed in Section 1.1, a convolution theorem holds true also for Laplace transforms. This is given in Equation (1.1.22). This theorem is proved in any textbook on Laplace transforms, see *e. g.* the book by LePage [363]. Two important quantities for random variables and stochastic processes with discrete values such as $N(t)$ are the *probability-generating function* and the *moment-generating function.*

Definition 6.11. The probability-generating function for the counting process $N(t)$ is defined as

$$G_{N(t)}(z,t) = \mathbb{E}\left(z^{N(t)}\right) = \sum_{n=0}^{\infty} \mathbb{P}(N(t) = n)z^n = \sum_{n=0}^{\infty} P(n,t)z^n. \tag{6.2.51}$$

Remark 6.14. The power series in (6.2.51) has non-negative coeffcients and $\lim_{z\to 1} G(z,t) = 1$, therefore, as a consequence of Abel's convergence theorem, its radius of convergence is at least 1.

Remark 6.15. The probability distribution function of $N(t)$ can be obtained from the power series expansion of the probability-generating function around $z = 0$. In particular, one has $P(n,t) = G^{(n)}(0,t)/n!$.

Definition 6.12. The moment-generating function of $N(t)$ is defined as

$$M_{N(t)}(w,t) = \mathbb{E}\left(\mathrm{e}^{wN(t)}\right), \quad w \in \mathbb{R}. \tag{6.2.52}$$

Remark 6.16. One has that $M_{N(t)}(0,t) = 1$ always, but the moment-generating function may not exist. If it exists, one has $M_{N(t)}(w,t) = G_{N(t)}(\mathrm{e}^w, t)$ immediately from the above definitions.

Theorem 6.10. *If the moment-generating function exists in an open interval around $w = 0$, one has that the moments of the distribution are given by the n-th derivative of $M_{N(t)}(w,t)$ in $w = 0$:*

$$\mathbb{E}(N^k(t)) = M^{(n)}_{N(t)}(0,t). \tag{6.2.53}$$

Proof. This is an immediate consequence of Definition 6.12. □

Example 6.4. The probability-generating function for the Poisson process ($J \sim \exp(\lambda)$) is given by

$$G_{N(t)}(z,t) = \mathrm{e}^{\lambda t(z-1)}. \tag{6.2.54}$$

The moment-generating function for the Poisson process is

$$M_{N(t)}(w,t) = \mathrm{e}^{\lambda t(\mathrm{e}^w - 1)}. \tag{6.2.55}$$

It is now possible to define the fractional Poisson process [359, 400, 516, 551] as a counting renewal process.

Definition 6.13. The Mittag-Leffler renewal process is the sequence $\{J_{\beta,i}\}_{i=1}^{\infty}$ of positive independent and identically distributed random variables with complementary cumulative distribution function $\bar{F}_{J_\beta}(0,t)$ given by

$$\bar{F}_{J_\beta}(t) = E_\beta\left(-t^\beta\right), \tag{6.2.56}$$

where $E_\beta(z)$ is the one-parameter Mittag-Leffler function defined in Equation (1.2.30).

Remark 6.17. The one-parameter Mittag-Leffler function in (6.2.56) is a generalization of the exponential function. It coincides with the exponential function for $\beta = 1$. For $\beta \in (0,1]$, the function $E_\beta(-t^\beta)$ is completely monotone and it is 1 for $t = 0$. This means that it is a legitimate survival function. To be more specific, a $C^\infty[0,\infty)$ function $f(t)$ is completely monotone if $(-1)^n f^{(n)}(t) \geq 0$ for all non-negative integers n and for $t > 0$.

It was proved by Mainardi and Gorenflo [396] that $E_\beta(-t^\beta)$ can be written as a mixture of exponential distributions

$$E_\beta(-t^\beta) = \int_0^\infty dr\, \mathrm{e}^{-rt}\, \frac{1}{\pi} \frac{r^{\beta-1}\sin(\beta\pi)}{r^{2\beta}+2r^\beta\cos(\beta\pi)+1}, \tag{6.2.57}$$

with

$$E_\beta(-t^\beta)|_{t=0} = \int_0^\infty dr \frac{1}{\pi} \frac{r^{\beta-1}\sin(\beta\pi)}{r^{2\beta}+2r^\beta\cos(\beta\pi)+1} = 1. \tag{6.2.58}$$

For this reason, complete monotonicity is an immediate consequence of Bernstein's theorem. The interested reader can consult [96, 492, 557].

Remark 6.18. The function $E_\beta(-t^\beta)$ behaves as a stretched exponential for $t \to 0$:

$$E_\beta\left(-t^\beta\right) \simeq 1 - \frac{t^\beta}{\Gamma(\beta+1)} \simeq \mathrm{e}^{-t^\beta/\Gamma(\beta+1)}, \quad \text{for } 0 < t \ll 1, \tag{6.2.59}$$

and as a power-law for $t \to \infty$:

$$E_\beta(-t^\beta) \simeq \frac{\sin(\beta\pi)}{\pi}\frac{\Gamma(\beta)}{t^\beta}, \quad \text{for } t \gg 1. \tag{6.2.60}$$

Remark 6.19. For applications, it is often convenient to include a scale parameter in the definition (6.2.56), and one can write

$$\bar{F}_{J_\beta}(t) = E_\beta\left(-(t/\gamma)^\beta\right). \tag{6.2.61}$$

The scale factor can be introduced in different ways, and the reader is warned to pay attention to its definition. The assumption $\gamma = 1$ made in (6.2.56) is equivalent to a change of time unit.

Theorem 6.11. *The counting process $N_\beta(t)$ associated to the renewal pro cess defined by Equation (6.2.56) has the following distribution*

$$P_\beta(n,t) = \mathbb{P}(N_\beta(t) = n) = \frac{t^{\beta n}}{n!} E_\beta^{(n)}(-t^\beta), \tag{6.2.62}$$

where $E_\beta^{(n)}(-t^\beta)$ denotes the n-th derivative of $E_\beta(z)$ evaluated at the point $z = -t^\beta$.

Proof. The Laplace transform of $P_\beta(0,t) = \bar{F}_{J_\beta}(t) = E_\beta(-t^\beta)$ is given by [487]

$$\widetilde{P}_\beta(0,s) = \frac{s^{\beta-1}}{1+s^\beta}. \tag{6.2.63}$$

Therefore, the Laplace transform of the probability density function $f_{J_\beta}(t) = -dP_\beta(0,t)/dt$ is given by (see *e. g.* [363] for the Laplace transform of the derivative)

$$\widetilde{f}_{J_\beta}(s) = \frac{1}{1+s^\beta}; \tag{6.2.64}$$

recalling Equation (6.2.46) and the convolution theorem for Laplace transforms, *i. e.* Eq. (1.1.22), one immediately has

$$\widetilde{P}_\beta(n,s) = \frac{1}{(1+s^\beta)^n}\frac{s^{\beta-1}}{1+s^\beta}. \tag{6.2.65}$$

Using Equation (1.80) in Podlubny's book [487] for the inversion of the Laplace transform in (6.2.65), one gets the thesis (6.2.62). □

Remark 6.20. The preceding theorem was proved by Scalas *et al.* [400, 551]. Notice that $N_1(t)$ is the Poisson process with parameter $\lambda = 1$. Recently, Meerschaert *et al.* [416] proved that the fractional Poisson process $N_\beta(t)$ coincides with the process defined by $N_1(D_\beta(t))$ where $D_\beta(t)$ is the functional inverse of the standard β-stable subordinator. The latter process was also known as fractal time Poisson process. This result unifies different approaches to fractional calculus [91, 416].

Remark 6.21. For $0 < \beta < 1$, the fractional Poisson process is semi-Markov, but not Markovian and is not Lévy. The process $N_\beta(t)$ is not Markovian as the only Markovian counting process is the Poisson process [147]. It is not Lévy as its distribution is not infinitely divisible.

As a corollary of this result, one can get the probablity-generating function of the fractional Poisson process.

Corollary 6.3. *The probability-generating function of the fractional Poisson process is given by*

$$G_{N_\beta(t)}(z,t) = E_\beta(t^\beta(z-1)). \tag{6.2.66}$$

Proof. By direct calculation, the coefficients of the MacLaurin series expansion of $G_{N_\beta(t)}(z,t)$ given by (6.2.66) are

$$\frac{t^{\beta n}}{n!}E_\beta^{(n)}(-t^\beta),$$

and they do coincide with $P_\beta(n,t)$ given in (6.2.62). □

Corollary 6.4. *The moment-generating function of the fractional Poisson process is given by*

$$M_{N_\beta(t)}(w,t) = E_\beta(t^\beta(\mathrm{e}^w - 1)). \tag{6.2.67}$$

Remark 6.22. Given that the series defining the Mittag-Leffler function in Equation (6.2.67) converges in an open interval around $w = 0$, the moments of all order of $N_\beta(t)$ exist. In, particular, straightforward calculations lead to the following values for the mean and the variance of the fractional Poisson process (see [359] for a detailed derivation):

$$\mathbb{E}(N_\beta(t)) = \frac{t^\beta}{\Gamma(\beta+1)}; \tag{6.2.68}$$

$$\mathrm{Var}(N_\beta(t)) = \mathbb{E}(N_\beta(t)) + [\mathbb{E}(N_\beta(t))]^2 \left(\frac{\beta B(\beta, 1/2)}{2^{2\beta-1}} - 1\right), \tag{6.2.69}$$

where $B(x, y) - \Gamma(x)\Gamma(y)/\Gamma(x + y)$ is Euler beta function.

Corollary 6.5. *The probabilities $P_\beta(n,t)$ introduced in Theorem 6.11 obey the following governing equations:*

$${}^{\mathrm{C}}D_t^\beta P_\beta(0,t) = -P_\beta(0,t), \tag{6.2.70}$$

$${}^{\mathrm{C}}D_t^\beta P_\beta(n,t) = P_\beta(n-1,t) - P_\beta(n,t), \text{ if } n \geq 1, \tag{6.2.71}$$

with the initial conditions $P_\beta(0,0) = 1$ and $P_\beta(n,0) = 0$ for $n \geq 1$. In (6.2.70) *and* (6.2.71), ${}^{\mathrm{C}}D_t^\beta$ *is the Caputo derivative* (1.3.36), *whose Laplace symbol is given by*

$$(\mathcal{L}[{}^{\mathrm{C}}D_t^\beta g(t)])(s) = s^\beta \widetilde{g}(s) - s^{\beta-1} g(0^+), \tag{6.2.72}$$

where g is a function whose Laplace trasform exists. Equation (6.2.72) *is a particular case of* (1.3.56).

Proof. Also this corollary is a straightforward consequence of Theorem 6.11. To begin with, use the Laplace transform of (6.2.70) and the initial condition $P_\beta(0,0) = 1$ to get

$$\widetilde{P}_\beta(0,s) = \frac{s^{\beta-1}}{1+s^\beta}, \tag{6.2.73}$$

thus leading to (as expected, see the proof of Theorem 6.11)

$$P_\beta(0,t) = E_\beta(-t^\beta). \tag{6.2.74}$$

The same technique applied to the recursions (6.2.71) leads to the equation

$$\widetilde{P}_\beta(n,s) = \frac{1}{s^\beta + 1}\widetilde{P}_\beta(n-1,s), \tag{6.2.75}$$

which, by induction, together with (6.2.73) gives

$$\widetilde{P}_\beta(n,s) = \frac{1}{(s^\beta + 1)^n}\frac{s^{\beta-1}}{1+s^\beta}, \tag{6.2.76}$$

and, inverting the Laplace transform as in Theorem 6.11, one gets

$$P_\beta(n,t) = \frac{t^{\beta n}}{n!}E_\beta^{(n)}(-t^\beta).$$

□

Remark 6.23. The governing equations (6.2.70) and (6.2.71) with the appropriate initial conditions are not sufficient to define a stochastic process. They lead to the one-point probability distribution (6.2.62) which is the one-point distribution of the fractional Poisson process defined above. However, other counting processes share the the same one-point probability distribution as illustrated with specific examples in [91]. A full characterization of the fractional Poisson process defined above in terms of its finite-dimensional distributions can be found in [491].

It is now possible to define continuous-time random walks as compound renewal processes.

Definition 6.14. Let $\{Y_i\}_{i=1}^\infty$ be a sequence of i.i.d. random variables with cumulative probability distribution $F_Y(y)$ and let X_n be the corresponding random walk. We give to the Y_is the meaning of jump widths for a diffusing particle. Let $\{J\}_{i=1}^\infty$ be a sequence of i.i.d. random variables with cumulative probability distribution $F_J(t)$ and with the meaning of sojourn times. Let $N(t)$ be the corresponding counting process. We then define the following stochastic process

$$X(t) = X_{N(t)} = \sum_{i=1}^{N(t)} Y_i \tag{6.2.77}$$

and we call it compound renewal process or continuous-time random walk (CTRW).

Remark 6.24. The CTRW is a random walk X_n as a function of a counting process, that is a random sum of independent random variables. Note that we have not said if the couple (J_i, X_i) consists of independent random variables. If this is the case, we have an uncoupled (or decoupled) CTRW. This is the simplest case in which durations are independent of jumps. This remark leads us to consider a particular class of non-Markovian stochastic processes, the so-called semi-Markov processes [147, 236, 237, 314]. In the following, the reader will find a quick and dirty introduction to semi-Markov processes.

Definition 6.15. A Markov renewal process is a two-component Markov chain $\{X_n, T_n\}_{n=0}^{\infty}$, where X_n, $n \geq 0$ is a Markov chain and T_n, $n \geq 0$ is the n-th epoch of a renewal process, homogeneous with respect to the second component and with transition probability given by

$$\mathbb{P}(Y_{n+1} \in A, J_{n+1} \leq t | Y_0, \ldots Y_n, J_1, \ldots, J_n) = \mathbb{P}(Y_{n+1} \in A, J_{n+1} \leq t | Y_n), \tag{6.2.78}$$

where $A \subset \mathbb{R}$ is a Borel set and $J_{n+1} = T_{n+1} - T_n$.

Remark 6.25. We will also assume homogeneity with respect to the first component. In other words, if $X_n = x$, the probability on the right-hand side of Equation (6.2.78) does not explicitly depend on n.

Remark 6.26. The positive function

$$Q(x, A, t) = \mathbb{P}(X_{n+1} = y \in A, J_{n+1} \leq t | X_n = x), \tag{6.2.79}$$

is called semi-Markov kernel with $x \in \mathbb{R}$, $A \subset \mathbb{R}$ a Borel set, and $t \geq 0$.

Definition 6.16. Let $N(t)$ denote the counting process defined as in Equation (6.2.43), the stochastic process $X(t)$ defined as

$$X(t) = X_{N(t)} \tag{6.2.80}$$

is the semi-Markov process associated to the Markov renewal process X_n, T_n, $n \geq 0$.

Theorem 6.12. *Uncoupled CTRWs are semi-Markov processes with semi-Markov kernel given by*

$$Q(x, A, t) = P(x, A) F_J(t), \tag{6.2.81}$$

where $P(x, A)$ is the Markov kernel (a.k.a. Markov transition function or transition probability kernel) of the random walk

$$P(x, A) \stackrel{\text{def}}{=} \mathbb{P}(X_{n+1} = y \in A | X_n = x), \tag{6.2.82}$$

and $F_J(t)$ is the cumulative probability distribution function of sojourn times. Moreover, let $f_Y(y)$ denote the probability density function of jumps, one has

$$P(x, A) = \int_{A-x} f_Y(u)\, du, \tag{6.2.83}$$

where $A - x$ is the set of values in A translated of x towards left.

Proof. The compound renewal process is a semi-Markov process by construction, where the couple X_n, T_n, $n \geq 0$ defining the corresponding Markov renewal process is made up of a random walk X_n, $n \geq 0$ with $X_0 = 0$ and a renewal process with epochs given by T_n, $n \geq 0$ with $T_0 = 0$. Equation (6.2.81) is an immediate consequence of the independence between the random walk and the renewal process. Finally, Equation (6.2.83) is the standard Markov kernel of a random walk whose jumps are i.i.d. random variables with probability density function $f_Y(y)$. □

The cumulative distribution function of an uncoupled compound renewal process can be obtained by means of purely probabilistic considerations.

Theorem 6.13. *Let $\{Y\}_{i=1}^{\infty}$ be a sequence of i.i.d. real-valued random variables with cumulative distribution function $F_Y(y)$ and let $N(t)$, $t \geq 0$ denote a counting process independent of the previous sequence and such that the number of events in the interval $[0, t]$ is a finite but arbitrary integer $n = 0, 1, \ldots$. Let $X(t)$ denote the corresponding compound renewal process. Then if $P(n, t) = \mathbb{P}(N(t) = n)$, the cumulative distribution function of $X(t)$ is*

$$F_{X(t)}(x, t) = \sum_{n=0}^{\infty} P(n, t) F_Y^{*n}(x), \tag{6.2.84}$$

where $F_Y^{*n}(x)$ is the n-fold convolution of $F_Y(y)$.

Proof. Assume that $X(0) = 0$ and that, at time t, there have been $N(t)$ jumps, with $N(t)$ assuming integer values starting from 0 ($N(t) = 0$ means no jumps up to time t). Consider a realization of $N(t)$, that is suppose one has $N(t) = n$. This means that

$$X(t) = \sum_{i=1}^{N(t)} Y_i = \sum_{i=1}^{n} Y_i = X_n. \tag{6.2.85}$$

In this case, one finds

$$F_{X_n}(x) = \mathbb{P}(X_n \leq x) = \mathbb{P}\left(\sum_{i=1}^{n} Y_i \leq x\right) = F_Y^{*n}(x). \tag{6.2.86}$$

Given the independence between $N(t)$ and the Y_is, one further has that

$$\mathbb{P}(X_n \leq x, N(t) = n) = \mathbb{P}(N(t) = n)\mathbb{P}(X_n \leq x) = P(n,t)F_Y^{*n}(x). \tag{6.2.87}$$

The events $\{X_n \leq x, N(t) = n\}$ for $n \geq 0$ are mutually exclusive and exhaustive, and this yields

$$\{X(t) \leq x\} = \cup_{n=0}^{\infty}\{X_n \leq x, N(t) = n\}, \tag{6.2.88}$$

and, for any $m \neq n$,

$$\{X_m \leq x, N(t) = m\} \cap \{X_n \leq x, N(t) = n\} = \emptyset. \tag{6.2.89}$$

Calculating the probability of the two sides in Equation (6.2.88) and using Equation (6.2.87) and the axiom of infinite additivity leads to

$$\begin{aligned} F_{X(t)}(x,t) &= \mathbb{P}(X(t) \leq x) = \mathbb{P}\left(\cup_{n=0}^{\infty}\{X_n \leq x, N(t) = n\}\right) \\ &= \sum_{n=0}^{\infty} \mathbb{P}(X_n \leq x, N(t) = n) = \sum_{n=0}^{\infty} P(n,t)F_Y^{*n}(x) \end{aligned} \tag{6.2.90}$$

which is our thesis. □

Remark 6.27. For $n = 0$, one assumes $F_{Y_0}^{*0}(y) = \theta(y)$ where $\theta(y)$ is the Heaviside function. Note that $P(0,t)$ is nothing else but the survival function at $y = 0$ of the counting process. Therefore, Equation (6.2.84) can be equivalently written as

$$F_{X(t)}(x,t) = P(n,0)\,\theta(x) + \sum_{n=1}^{\infty} P(n,t)F_Y^{*n}(x), \tag{6.2.91}$$

where $\theta(x)$ is the càdlàg version of Heaviside step function.

Remark 6.28. The series (6.2.84) is uniformly convergent for $x \neq 0$ and for any value of $t \in (0,\infty)$. This statement can be proved using Weierstrass M test. For $x = 0$ there is a jump in the cumulative distribution function of amplitude $P(0,t)$.

Example 6.5. As an example of compound renewal process, consider the case in which $Y_i \sim \mathcal{N}(\mu, \sigma^2)$, so that their cumulative distribution function is

$$F_Y(y) = \Phi(y|\mu, \sigma^2) = \frac{1}{2}\left(1 + \mathrm{erf}\left(\frac{y-\mu}{\sqrt{2\sigma^2}}\right)\right), \tag{6.2.92}$$

where

$$\mathrm{erf}(y) = \frac{2}{\sqrt{\pi}} \int_0^y \mathrm{e}^{-u^2}\, du \tag{6.2.93}$$

is the error function. In this case, the convolution $F_Y^{*n}(x)$ is given by $\Phi(x|n\mu, n\sigma^2)$. The sojourn times are $J_i \sim \exp(\lambda)$ and one finds

$$P(n,t) = \mathrm{e}^{-\lambda t}\frac{(\lambda t)^n}{n!}. \tag{6.2.94}$$

As a consequence of Theorem 6.13 one gets

$$F_{X(t)}(x,t) = \mathrm{e}^{-\lambda t} \sum_{n=0}^{\infty} \frac{(\lambda t)^n}{n!} \Phi(x|n\mu, n\sigma^2). \tag{6.2.95}$$

Equation (6.2.95) can be directly used for numerical estimates of $F_{X(t)}(x,t)$.

Corollary 6.6. *In the same hypotheses as in Theorem 6.13, the probability density* $f_{X(t)}(y,t)$ *of the process* $X(t)$ *is given by*

$$f_{X(t)}(x,t) = P(0,t)\,\delta(x) + \sum_{n=1}^{\infty} P(n,t) f_Y^{*n}(x), \tag{6.2.96}$$

where $f_Y^{*n}(x)$ *is the n-fold convolution of the probability density function* $f_Y(y) = dF_Y(y)/dy$.

Proof. The sought probability density function is $f_{X(t)}(x,t) = dF_{X(t)}(x,t)/dy$; Equation (6.2.96) is the formal derivative of Equation (6.2.84). If $x \neq 0$, there is no singular term and the series converges uniformly ($f_Y^{*n}(x)$ is bounded and Weierstrass M test applies), therefore, for any x the series converges to the derivative of $F_{X(t)}(x,t)$. This is so also in the case $x = 0$ for $n \geq 1$ and the jump in $x = 0$ gives the singular term of weight $P(0,t)$ (see Equation (6.2.91)). □

Among all the compound renewal processes, we will now define the compound fractional Poisson process [359].

Definition 6.17. Let $\{Y_i\}_{i=1}^{\infty}$ be a sequence of i.i.d. random variables with cumulative distribution function given by $F_Y(y)$ and let $N_\beta(t)$ be the fractional Poisson process, then the process $X_\beta(t)$ defined as

$$X_\beta(t) = X_{N_\beta(t)} = \sum_{i=1}^{N_\beta(t)} Y_i \tag{6.2.97}$$

is called compound fractional Poisson process.

Remark 6.29. The process $X_1(t)$ coincides with the compound Poisson process of parameter $\lambda = 1$.

Theorem 6.14. *Let $X_\beta(t)$ be a compound fractional Poisson process, then*

(1) its cumulative distribution function $F_{X_\beta(t)}(x,t)$ is given by

$$F_{X_\beta(t)}(x,t) = E_\beta\left(-t^\beta\right)\theta(x) + \sum_{n=1}^{\infty}\frac{t^{\beta n}}{n!}E_\beta^{(n)}\left(-t^\beta\right)F_Y^{*n}(x); \tag{6.2.98}$$

(2) its probability density function $f_{X_\beta(t)}(x,t)$ is given by

$$f_{X_\beta(t)}(y,t) = E_\beta\left(-t^\beta\right)\delta(x) + \sum_{n=1}^{\infty}\frac{t^{\beta n}}{n!}E_\beta^{(n)}\left(-t^\beta\right)f_Y^{*n}(x); \tag{6.2.99}$$

(3) its characteristic function $\widehat{f}_{X_\beta(t)}(\kappa,t)$ is given by

$$\widehat{f}_{X_\beta(t)}(\kappa,t) = E_\beta\left[t^\beta(\widehat{f}_Y(\kappa)-1)\right]. \tag{6.2.100}$$

Proof. The first two Equations (6.2.98) and (6.2.99) are a straightforward consequence of Theorem 6.13, Corollary 6.6 and Theorem 6.11. Equation (6.2.100) is the Fourier transform of (6.2.99). This can be verified by expanding it into a Taylor series with centre $-t^\beta$. □

Remark 6.30. For $0 < \beta < 1$, the compound fractional Poisson process is neither Markovian nor Lévy (see Remark 6.21). However, it is a semi-Markov process as a consequence of Theorem 6.12.

It is now possible to discuss the relationship between CTRWs and the space-time fractional diffusion equation. This will be the subject of the next section.

6.3 Fractional Diffusion and Limit Theorems

In order to link the processes introduced in the previous section to fractional diffusion, let us first consider the following Cauchy problem.

Theorem 6.15. *Consider the Cauchy problem for the space-time fractional diffusion equation with* $0 < \alpha \le 2$ *and* $0 < \beta \le 1$

$$\begin{aligned} {}^{\mathrm{SR}}D_x^{\alpha} u_{\alpha,\beta}(x,t) &= {}^{\mathrm{C}}D_t^{\beta} u_{\alpha,\beta}(x,t) \\ u_{\alpha,\beta}(x,0^+) &= \delta(x), \end{aligned} \tag{6.3.1}$$

then the Green function

$$u_{\alpha,\beta}(x,t) = \frac{1}{t^{\beta/\alpha}} W_{\alpha,\beta}\left(\frac{x}{t^{\beta/\alpha}}\right), \tag{6.3.2}$$

where

$$W_{\alpha,\beta}(u) = \frac{1}{2\pi}\int_{-\infty}^{+\infty} d\kappa\, \mathrm{e}^{-i\kappa u} E_{\beta}(-|\kappa|^{\alpha}), \tag{6.3.3}$$

solves the Cauchy problem [401, 551].

Proof. In Equation (6.3.1), ${}^{\mathrm{SR}}D_x^{\alpha}$ denotes the symmetric Riesz operator [539, 542, 643] whose Fourier symbol is $-|\kappa|^{\alpha}$; more precisely, for a suitable function $f(x)$ one can write

$$(\mathcal{F}[{}^{\mathrm{SR}}D_x^{\alpha} f(x)])(\kappa) = -|\kappa|^{\alpha} \widehat{f}(\kappa). \tag{6.3.4}$$

Moreover, ${}^{\mathrm{C}}D_t^{\beta}$ is the Caputo derivative defined in (1.3.36), whose Laplace symbol is given by (6.2.72). The application of the Laplace-Fourier transform to Equation (6.3.1) implies that

$$-|\kappa|^{\alpha} \widehat{\widetilde{u}}_{\alpha,\beta}(\kappa,s) = s^{\beta} \widehat{\widetilde{u}}_{\alpha,\beta}(\kappa,s) - s^{\beta-1}, \tag{6.3.5}$$

so that the Laplace-Fourier transform of the sought Green function is

$$\widehat{\widetilde{u}}_{\alpha,\beta}(\kappa,s) = \frac{s^{\beta-1}}{|\kappa|^{\alpha} + s^{\beta}}. \tag{6.3.6}$$

A comparison between (6.2.63) and (6.3.6) immediately shows that the inversion of the Laplace transform gives the Fourier transform of $u_{\alpha,\beta}(x,t)$

$$\widehat{u}_{\alpha,\beta}(\kappa,t) = E_{\beta}\left(-|\kappa|^{\alpha} t^{\beta}\right). \tag{6.3.7}$$

A further inversion of the Fourier transform leads to the thesis. □

Remark 6.31. In the above derivation, the role of Fourier and Laplace transforms is interchangeable. One can first invert the Fourier transform and then the Laplace transform and get the same result.

Remark 6.32. The Green function $u_{\alpha,\beta}(x,t)$ is a probability density function for any $t > 0$, that is

$$\int_{-\infty}^{+\infty} dx\, u_{\alpha,\beta}(x,t) = 1. \tag{6.3.8}$$

When $\alpha = 2$ and $\beta = 1$, the symmetric Riesz derivative coincides with the second derivative with respect to x: ${}^{\mathrm{SR}}D_x^2 = \partial^2/\partial x^2$ and the Caputo derivative becomes the first derivative with respect to time: ${}^{\mathrm{C}}D_t^1 = \partial/\partial t$. Then, Equation (6.3.1) defines the Cauchy problem for ordinary diffusion, with diffusion coefficient equal to 1; the corresponding Green function is

$$u_{2,1}(x,t) = \frac{1}{\sqrt{4\pi t}}\, \mathrm{e}^{-x^2/4t}. \tag{6.3.9}$$

Definition 6.18. A random variable X is called infinitely divisible if, for any n, it can be written as the sum of n independent and identically distributed random variables. Equivalently, the corresponding cumulative distribution function is called infinitely divisible if, for any n, it can be written as $F_X(x) = F_{Y_n}^{*n}(x)$, that is there exists a random variable Y_n whose n-fold convolution gives $F_X(x)$.

Remark 6.33. Stable distributions are infinitely divisible, but the converse is not true, not every infinitely divisible distribution is stable.

Remark 6.34. The Poisson distribution and the compound Poisson distribution are infinitely divisible and it can be shown that any infinitely divisible distribution is the limit of compound Poisson distributions.

Theorem 6.16. *For $0 < \alpha \leq 2$, and $\beta = 1$, the characteristic function of the Green function is*

$$\widehat{u}_{\alpha,1}(\kappa,t) = \mathrm{e}^{-|\kappa|^\alpha t} \tag{6.3.10}$$

and it is infinitely divisible.

Proof. For $\beta = 1$, the Mittag-Leffler function coincides with the exponential function and Equation (6.3.7) yields Equation (6.3.10). In order to show that the random variable $U_{\alpha,1}(t)$ is infinitely divisible for every t, one has to show that its cumulative distribution function is the n-fold convolution of n identical distribution functions. But then, it is enough to choose a random variable whose characteristic function is given by

$$[\mathrm{e}^{-|\kappa|^\alpha t}]^{1/n} = \mathrm{e}^{-|\kappa|^\alpha t/n}, \tag{6.3.11}$$

to prove infinite divisibility of $U_{\alpha,1}(t)$. □

Remark 6.35. It can be proved that to every infinitely divisible distribution there corresponds a unique càdlàg extension of a Lévy process. In the case under scrutiny, $u_{2,1}(x,t)$ corresponds to the Wiener process whose increments follow the $N(0,2t)$ distribution. The density $u_{\alpha,1}(x,t)$ for $0 < \alpha < 2$ corresponds to processes (called Lévy flights in the applied literature) that generalize the Wiener process and whose increments follow the symmetric α-stable distribution. For more information on infinitely divisible distribution, Lévy processes and related pseudo-differential operators the reader can consult the following references [38, 97, 232, 311, 348, 547, 559].

A simple way to understand the connection between CTRWs and space-time fractional diffusion is to consider the following stochastic process.

Definition 6.19. Let $\{Y_{\alpha,i}\}_{i=1}^{\infty}$ be a sequence of i.i.d. symmetric α-stable distributions whose characteristic function is given by

$$\widehat{f}_{Y_\alpha}(\kappa) = \mathrm{e}^{-|k|^\alpha} \tag{6.3.12}$$

and let $X_{\alpha,n}$ be the corresponding random walk. The following compound fractional Poisson process

$$X_{\alpha,\beta}(t) = X_{\alpha,N_\beta(t)} = \sum_{i=1}^{N_\beta(t)} Y_{\alpha,i} \tag{6.3.13}$$

is the fractional compound Poisson process with symmetric α-stable jumps.

Corollary 6.7. *The characteristic function of the fractional compound Poisson process with symmetric α-stable jumps is given by*

$$\widehat{f}_{X_{\alpha,\beta}(t)}(\kappa) = E_\beta\left[t^\beta\left(\mathrm{e}^{-|\kappa|^\alpha} - 1\right)\right]. \tag{6.3.14}$$

Proof. This result is an immediate consequence of Equations (6.2.100) and (6.3.12). □

If properly rescaled, the random variable $X_{\alpha,\beta}(t)$ can be made to converge weakly to $U_{\alpha,\beta}(t)$, the random variable whose distribution is characterized by the probability density function (6.3.2) that solves the Cauchy problem for the space-time fractional diffusion Equation (6.3.1). The trick is to build a sequence of random variables whose characteristic function converges to (6.3.7). Indeed, we can prove the following theorem.

Theorem 6.17. *Let $X_{\alpha,\beta}(t)$ be a compound fractional Poisson process with symmetric α-stable jumps and let h and r be two scaling factors such that*

$$X_{\alpha,n}(h) = hY_{\alpha,1} + \ldots + hY_{\alpha,n} \tag{6.3.15}$$

$$T_{\beta,n}(r) = rJ_{\beta,1} + \ldots + rJ_{\beta,n}, \tag{6.3.16}$$

and

$$\lim_{h,r\to 0} \frac{h^\alpha}{r^\beta} = 1, \tag{6.3.17}$$

with $0 < \alpha \leq 2$ and $0 < \beta \leq 1$. Given the assumption on the jumps $Y_{\alpha,i}$, for $h \to 0$, one has

$$\widehat{f}_{Y_\alpha}(h\kappa) = 1 - h^\alpha|\kappa|^\alpha + \mathrm{o}(h^\alpha|\kappa|^\alpha), \tag{6.3.18}$$

then, for $h, r \to 0$ with $h^\alpha/r^\beta \to 1$, $f_{hX_{\alpha,\beta}(rt)}(x,t)$ weakly converges to $u_{\alpha,\beta}(x,t)$, the Green function of the fractional diffusion equation.

Proof. In order to prove weak convergence, it suffices to show the convergence of the characteristic function (6.2.100) as a consequence of the Lévy continuity Theorem 6.3. Indeed, one can write

$$\widehat{f}_{hX_{\alpha,\beta}(rt)}(\kappa,t) = E_\beta\left[-\frac{t^\beta}{r^\beta}\left(\mathrm{e}^{-h^\alpha|\kappa|^\alpha} - 1\right)\right] \overset{h,r\to 0}{\to} E_\beta(-t^\beta|\kappa|^\alpha), \tag{6.3.19}$$

which completes the proof and establishes the connection between CTRWs and the space-time fractional diffusion equation. □

Remark 6.36. This result can be generalized to compound fractional Poisson processes with heavy tails both in the jump and in the sojourn time distributions. A more general proof can be found in [551]. The relationship between fractional diffusion and CTRWs is discussed in several physics papers with different levels of detail [159, 238, 529, 539, 643]. Hilfer and Anton realized the important role played by the Mittag-Leffler function in this derivation and rigorously discussed the link between fractional diffusion and CTRWs [306].

Remark 6.37. Up to now, we have focused on the (weak) convergence of random variables and not of stochastic processes. The convergence of processes is delicate as one must use techniques related to functional central limit theorems in appropriate functions spaces [312]. Let $L_\alpha(t)$ denote the symmetric α-stable Lévy process. Equation (6.3.7) is the characteristic function of the process $U_{\alpha,\beta}(t) = L_\alpha(D_\beta(t))$, that is of the symmetric α-stable Lévy process subordinated to *the inverse β-stable subordinator, $D_\beta(t)$, the functional inverse of the β-stable subordinator.* This remark leads to conjecture that $U_{\alpha,\beta}(t)$ is the functional limit of $X_{\alpha,\beta}(t)$, the compound fractional Poisson process with α-stable jumps. This conjecture can be found in a paper by Magdziarz and Weron [391] and is proved in Meerschaert *et al.* [416] using the methods discussed by Meerschaert and Scheffler [418].

Chapter 7

Applications of Continuous-Time Random Walks

7.1 Introduction to Financial Markets

In financial markets, when one considers tick-by-tick trades, not only price fluctuations, but also waiting times between consecutive trades vary at random. This fact is pictorially represented in Figure 7.1. In this figure, the value of the FTSE MIB Index is plotted for trades occurring on February 3rd, 2011. The FTSE MIB Index is a weighted average of the prices of the thirty more liquid stocks in the Italian Stock Exchange and it is updated every time a trade occurs. This is a consequence of trading rules.

In many regulated financial markets trading is performed by means of the so-called continuous-double auction. Here, we just present the basic idea of this microstructural market mechanism for an order driven market; details may vary from stock exchange to stock exchange. For every stock traded in the exchange there is a *book* where orders are registered. Traders can either place buy orders (*bids*) or sell orders (*asks*) and this explains why the auction is called double. Moreover, orders can be placed at any time, and, for this reason, the auction is called continuous. There are many different kinds of orders, but the typical order is the *limit order*. A bid limit order is an order to buy $q_b^{(T)}$ units of the share at a price not larger than a limit price $p_b^{(T)}$ selected by the trader, where T is a label identifying the trader. An ask limit order is an order to sell $q_a^{(T)}$ units of the share at a price not smaller than a limit price $p_a^{(T)}$ selected by the trader. The couples $(p_b^{(T)}, q_b^{(T)})$ are stored in the book and ordered from the best bid to the worst bid, the best bid being $\widehat{p}_b = \max_{T \in I_b}(p_b^{(T)})$, where I_b is the set of traders placing bids. The couples $(p_a^{(T)}, q_a^{(T)})$ are also stored in the

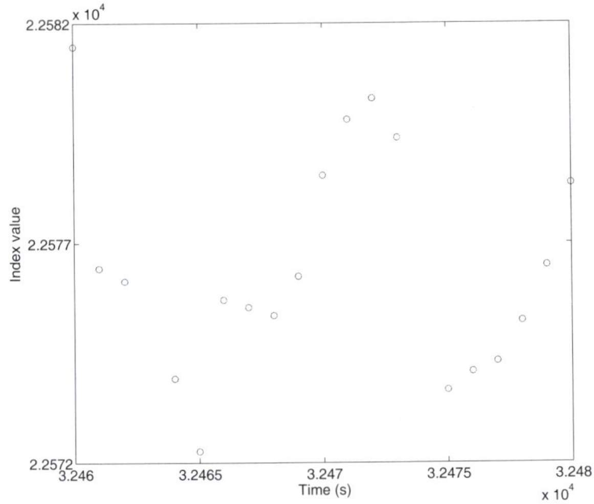

Fig. 7.1 Tick-by-tick price fluctuations. As explained in the text, this is the FTSE MIB Index recorded on February 3rd, 2011. Time is given in seconds since midnight.

book and ordered from the best ask to the worst ask, the best ask being $\widehat{p}_a = \min_{T \in I_a}(p_a^{(T)})$, where I_a is the set of traders placing asks. At a generic time t, one has that $\widehat{p}_a(t) > \widehat{p}_b(t)$. The difference

$$s(t) = \widehat{p}_a(t) - \widehat{p}_b(t) \tag{7.1.1}$$

is called *bid-ask spread*. Occasionally, a trader may accept an existing best bid or best ask, and the i-th trade takes place at the epoch t_i. This is called a *market order*. Market rules specify which are the priorities for limit orders placed at the same price and what to do when the quantity requested in a market order is not fully available at the best price. Several authors use the mid-price defined as

$$p_m(t) = \frac{\widehat{p}_b(t) + \widehat{p}_a(t)}{2} \tag{7.1.2}$$

in order to summarize and study the above process. Both the bid-ask spread and the mid-price can be represented as step functions varying at random times. Jumps in spread and midprice may occur when a better limit order enters the book or when a trade takes place.

Another important process is the one of realized trades represented in Figures 7.1 and 7.2. In Figure 7.2, a càdlàg representation of the process is given, the so-called *previous tick interpolation*, where it is assumed that the price remains constant between two consecutive trades. With these definitions in mind, we can now show how CTRWs can be used in financial modelling [408, 548]. Figures 7.3 to 7.5 represent consecutive magnifications of the same price process and they show how scaling breaks down. In particular, visual inspection shows that scaling is no longer valid already at the time scale of Figure 7.5. In other words, the behaviour of high-frequency price fluctuations cannot be described by self-similar, or self-affine, or even multifractal processes.

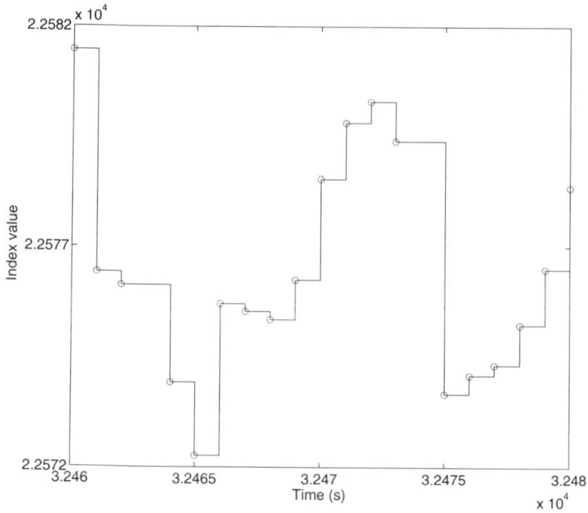

Fig. 7.2 Tick-by-tick price fluctuations represented as a càdlàg step function. FTSE MIB Index, February 3rd, 2011. These are the same data as in Figure 7.1.

7.2 Models of Price Fluctuations in Financial Markets

Let $p(t)$ represent the price of an asset at time t. Define $p_0 = p(0)$ the initial price. The variable

$$x(t) = \log \left[\frac{p(t)}{p_0} \right] \tag{7.2.1}$$

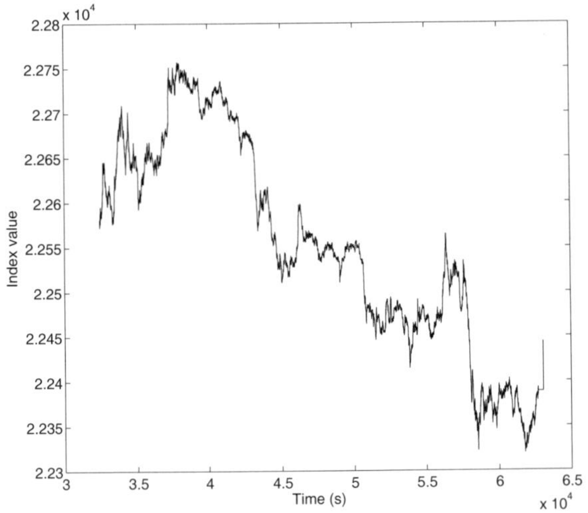

Fig. 7.3 Tick-by-tick price fluctuations represented as a càdlàg step function. FTSE MIB Index: A whole trading day, February 3rd 2011.

is called *logarithmic price* or *log-price* or even *logarithmic return* or *log-return*. With this choice, one has $x(0) = 0$. Now, let t_i be the epoch of the i-th trade and $p(t_i)$ the corresponding price, then the variable

$$\xi_i = \log\left[\frac{p(t_i)}{p(t_{i-1})}\right] \tag{7.2.2}$$

is called *tick-by-tick log-return*. Let $n(t)$ represent the number of trades from the market opening, up to time t, then the relationship between the log-price and the tick-by-tick log returns is

$$x(t) = \sum_{i=1}^{n(t)} \xi_i. \tag{7.2.3}$$

The reason for using these variables in finance is as follows. If price fluctuations are small compared to the price, one can see that the tick-by-tick log-return is very close to the tick-by-tick return $r_i = [p(t_i) - p(t_{i-1})]/p(t_{i-1})$. In symbols, one can write

$$\xi_i = \log\left[\frac{p(t_i)}{p(t_{i-1})}\right] \approx \frac{p(t_i) - p(t_{i-1})}{p(t_{i-1})} = r_i. \tag{7.2.4}$$

However, returns are not additive, one cannot write the return from a time t to a time $t + \Delta t$ as the sum of tick-by-tick returns, whereas this is possible

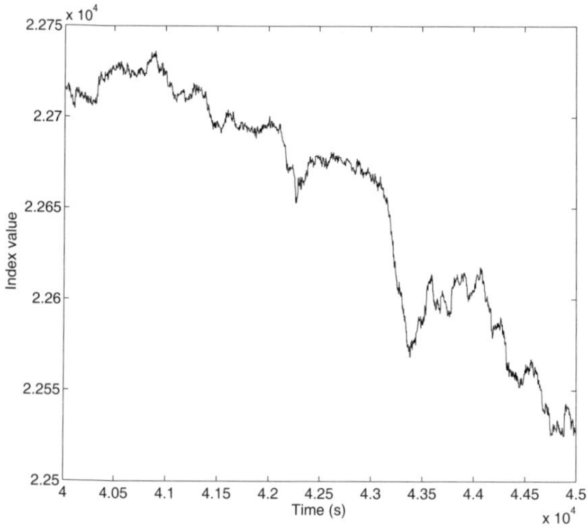

Fig. 7.4 Tick-by-tick price fluctuations represented as a càdlàg step function. A zoom of the data of Figure 7.3.

for log-returns. Equation (7.2.3) can be compared with Equation (6.2.77) in Chapter 6. It becomes natural to interpret $x(t)$ as a realization of a CTRW or of a compound (renewal) process $X(t)$, $n(t)$ as a realization of a counting (renewal) process $N(t)$ and ξ_i as a value of a random variable Y_i. The simplest possible choice for $x(t)$ is the normal compound Poisson process (NCPP) discussed in Chapter 6, with normally distributed tick-by-tick log-returns (jumps) $\xi \sim N(\mu_\xi, \sigma_\xi^2)$ and exponentially distributed durations (sojourn times) $\tau_i = t_i - t_{i-1} \sim \exp(\lambda)$. However, the normal compound Poisson process is falsified by the following empirical findings on high frequency data:

(1) The empirical distribution of tick-by-tick log-returns is leptokurtic, whereas the NCPP assumes a normal distribution which is mesokurtic.
(2) The empirical distribution of durations is non-exponential with excess standard deviation [222, 223, 402, 510, 551], whereas the NCPP assumes an exponential distribution.
(3) The autocorrelation of absolute log-returns decays slowly [510], whereas the NCPP assumes i.i.d. log-returns.

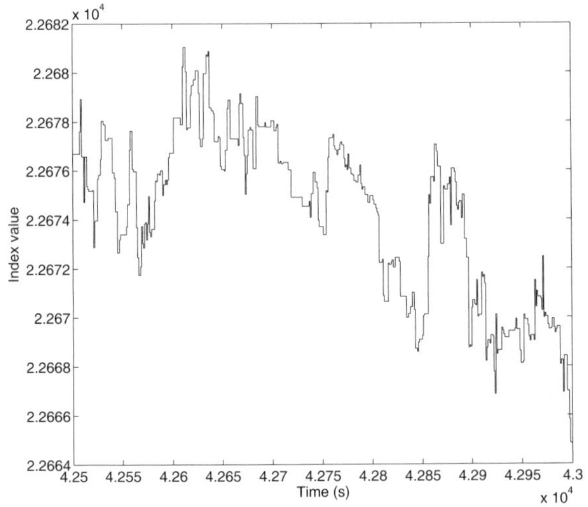

Fig. 7.5 Tick-by-tick price fluctuations represented as a càdlàg step function. A zoom of the data of Figure 7.4.

(4) Log-returns and waiting times are not independent [417, 510], whereas the NCPP assumes their independence.
(5) Volatility and activity vary during the trading day [98], whereas the NCPP assumes they are constant.

Compound renewal processes take into account facts (1) and (2) as well as fact (4) [417], but they are falsified by fact (3), as they assume independent returns and durations, and by point (5), as they assume identically distributed returns and durations.

7.3 Simulation

Simulation of CTRWs is not difficult [249, 256]. A typical algorithm for uncoupled CTRWs uses the following steps:

(1) generate n values for durations from your favorite distribution and store them in a vector;
(2) generate n values for tick-by-tick log-returns from your favorite distribution and store them in a second vector;

(3) generate the epochs by means of a cumulative sum of the duration vector;
(4) generate the log-prices by means of a cumulative sum of the tick-by-tick log-return vector.

If one wishes to simulate $x(t)$, the value of the log-price at time t, it is enough to include a control statement in the above algorithm to ensure that it runs until the sum of durations is less or equal than t. These algorithms generate single realizations of the process either for n jumps or until time t respectively. If many independent runs are performed, one can approximate the distribution of $x(t)$ or any other finite dimensional distribution of the process. An algorithm for the fractional compound process with jumps distributed according to the symmetric α-stable distribution is given in Appendix A.5. In that example, for the variables ξ_i, we have used the standard transformation method by Chambers, Mallows and Stuck [148] for $\alpha \in (0, 2]$

$$\xi_i = \gamma_x \left(\frac{-\log(u)\cos(\phi)}{\cos[(1-\alpha)\phi]} \right)^{1-1/\alpha} \frac{\sin(\alpha\phi)}{\cos\phi}, \tag{7.3.1}$$

where γ_x is a scale factor, u is a uniformly distributed random variable between 0 and 1 and $\phi = \pi(v - 1/2)$, with v uniformly distributed between 0 and 1 and not depending on u. For $\alpha = 2$, Equation (7.3.1) reduces to $\xi_i = 2\gamma\sqrt{-\log(u)}\sin(\phi)$, that is to the Box-Muller algorithm for normally distributed random numbers. The algorithm for the generation of Mittag-Leffler distributed τ_is with $\beta \in (0, 1]$ is (see [249, 318, 350–353, 478])

$$\tau_i = -\gamma_t \log(u) \left(\frac{\sin(\beta\pi)}{\tan(\beta\pi v)} - \cos(\beta\pi) \right)^{1/\beta}, \tag{7.3.2}$$

where u and v are independent uniformly distributed random variables with values between 0 and 1. For $\beta = 1$ Equation (7.3.2) reduces to $\tau_i = -\gamma_t \log(u)$, that is to the standard transformation formula for the exponential distribution. The results of simulations based on this algorithm are represented in Figures 7.6 to 7.11 for the following couples of parameters ($\alpha = 2$, $\beta = 0.5$), ($\alpha = 2$, $\beta = 0.99$), ($\alpha = 1.95$, $\beta = 0.5$), ($\alpha = 1.95$, $\beta = 0.5$), ($\alpha = 1$, $\beta = 0.5$), and ($\alpha = 1$, $\beta = 0.5$). Visual inspection shows that larger jumps in time are more likely for smaller values of β and larger jumps in log-price are expected for smaller values of α.

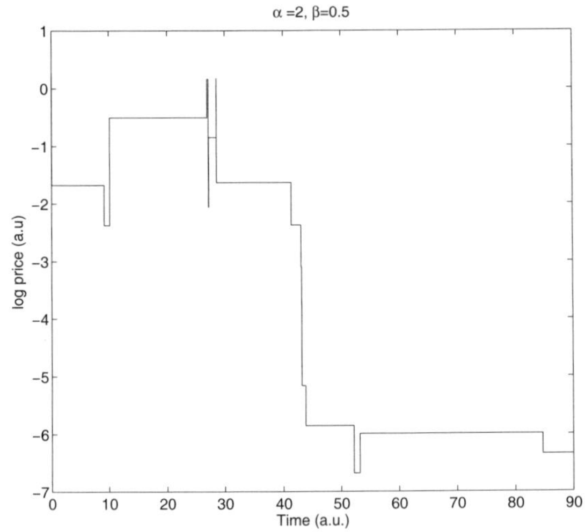

Fig. 7.6 Compound fractional Poisson process simulation for $\alpha = 2$ and $\beta = 0.5$.

7.4 Option Pricing

In Sections 7.1 and 7.2, we gave arguments in favor of using CTRWs as models of tick-by-tick price fluctuations in financial markets. We have also seen the limits of uncoupled CTRWs as market models. Now, let us suppose that we have an underlying asset whose log-price fluctuations are described by Equation (7.2.3). In other words, we assume that log-price fluctuations follow a compound renewal process. Furthermore, we assume that these fluctuations represent the intra-day behavior of an asset, such as a share traded in a stock exchange. For an intra-day time horizon, we can safely assume that the *risk-free* interest rate is $r_F = 0$. This would be the return rate of a zero-coupon bond. Even if such a return rate were $r_Y = 10\%$ on a yearly time horizon, meaning that the State issuing this financial instrument is close to default (so that, it would not be so riskless, after all) or that the inflation rate is quite high, then the interest rate for one day would be $r_d \approx 1/(10 \cdot 200) = 5 \cdot 10^{-4}$ (200 is the typical number of working days in a year) and this number has still to be divided by 8 (number of trading hours) and by 3600 (number of seconds in one hour) in order to get an approximate interest rate for a time horizon of 1 second.

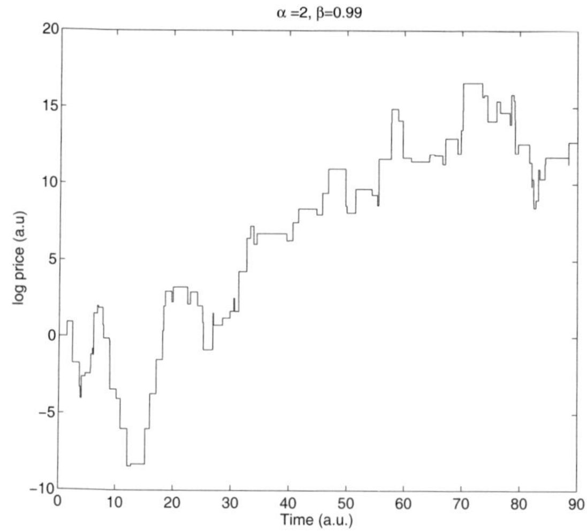

Fig. 7.7 Compound fractional Poisson process simulation for $\alpha = 2$ and $\beta = 0.99$.

This gives $r_s \approx 1.7 \cdot 10^{-8}$. On the other hand, typical tick-by-tick returns in a stock exchange are larger than the tick divided by the price of the share. Even if we assume that the share is worth 100 monetary units, with a 1/100 tick size (the minimum price difference allowed), we will have a return r larger than $1 \cdot 10^{-4}$ and much larger than r_s. Therefore, it is safe to assume a risk free interest rate $r = 0$ for intra-day hedging.

Hedging is performed through special contracts called options whose price is assumed to depend on the price of the underlying contract. A detailed discussion of these contracts is outside the scope of the present book. However, it is possible to present the basic ideas on option pricing, before turning to our high-frequency problem. The interested reader can consult the introductory books by Hull [309] and by Willmott [631]. One of the simplest option contract is the so-called *plain-vanilla European call*. This is the right (not the obligation) to buy an asset at a future time at a given price K called the *strike price* at a future date T called the *maturity*. If, at maturity, the asset price $p(T)$ is larger than K then, in principle, the option holder can exercise the option, pay K to the option writer to get one unit of asset and resell the assets on the market thus realizing a profit

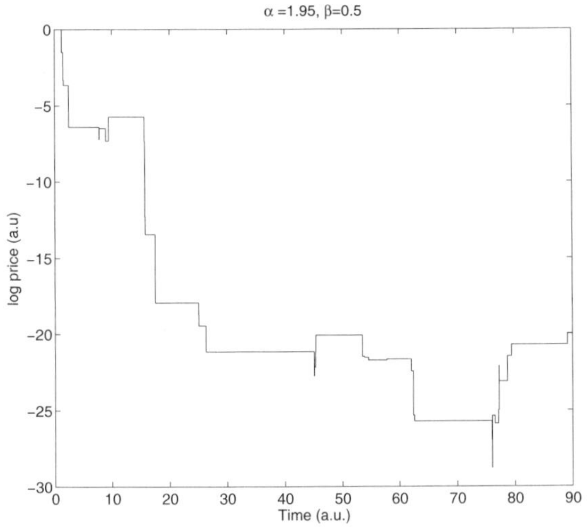

Fig. 7.8 Compound fractional Poisson process simulation for $\alpha = 1.95$ and $\beta = 0.5$.

of $p(T) - K$ for each asset unit. On the other side, if $p(T) < K$, it does not make sense to exercise the option. So, one has that the option payoff at maturity is given by

$$C(T) = \max(p(T) - K, 0). \tag{7.4.1}$$

The problem to be addressed is the following. Suppose you are at time $t < T$ and you want to get a plain vanilla option contract. Which is its fair price? In order to give a feeling on how to solve this problem, we shall first consider a simplified version: the so-called one-period binomial option pricing. The price of an asset is $p_0 = p(0)$ at time $t = 0$ and it can either go up or down at the next time step $t = 1$. Assume that $p_1^+ = p(1) = p_0 u$ with probability q and $p_1^- = p(1) = p_0 d$ with probability $1 - q$, where u is the up factor and d is the down factor. For the sake of simplicity, we shall further assume that the risk free interest rate is $r_F = 0$ during this period. The two factors, u and d cannot assume arbitrary values. Since we want that $0 < p_1^- < p_1^+$, this means that $0 < d < u$. Moreover, we want to avoid arbitrage, a trading strategy according to which one can get money out of nothing. Suppose indeed, that $d \geq 1$ and that we take from a bank p_0 monetary units to buy one share at time $t = 0$, then at

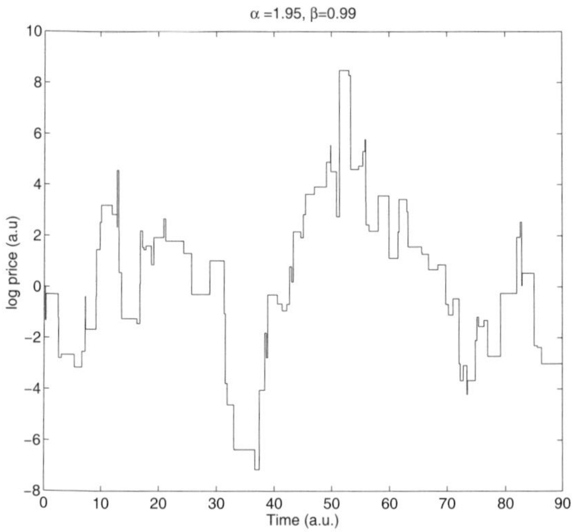

Fig. 7.9 Compound fractional Poisson process simulation for $\alpha = 1.95$ and $\beta = 0.99$.

time $t = 1$, the value of our share will be $p_1 \geq dp_0$, then by selling it and giving back p_0 monetary units to the bank, we will surely get a net profit of $p_1 - p_0 \geq dp_0 - p_0 = (d-1)p_0 \geq 0$ if $d \geq 1$. Therefore, to avoid arbitrage, we must take $d < 1$. Similarly, assume that $u \leq 1$, then one could borrow an asset share at time $t = 0$, then sell it for p_0 units of money and put the money in a bank. Now, at time $t = 1$, we could use this money to pay the share we borrowed at $t = 0$, since $p_1 \leq up_0$, in the end we would realize a certain profit $p_0 - p_1 \geq p_0 - up_0 = (1-u)p_0 \geq 0$ since $u \leq 1$. In this case, to avoid arbitrage, we must have $u > 1$. In the end, we must require that $0 < d < 1 < u$.

Now, if our strike price is $p_1^- < K < p_1^+$, our payoff at time $t = 1$ will be $C_1^+ = C(1) = p_1^+ - K$ with probability q and $C_1^- = C(1) = 0$ with probability $1-q$. It is possible to prove that the option price $C(0)$ at $t = 0$ is given by the following conditional expectation

$$C(0) = \mathbb{E}_{\widetilde{\mathbb{Q}}}[C(1)|I(0)], \tag{7.4.2}$$

where $I(0)$ represents the information available at time $t = 0$ and $\widetilde{\mathbb{Q}}$ represents the equivalent martingale measure. Two probability measures are *equivalent* if each one is absolutely continuous with respect to the other. A

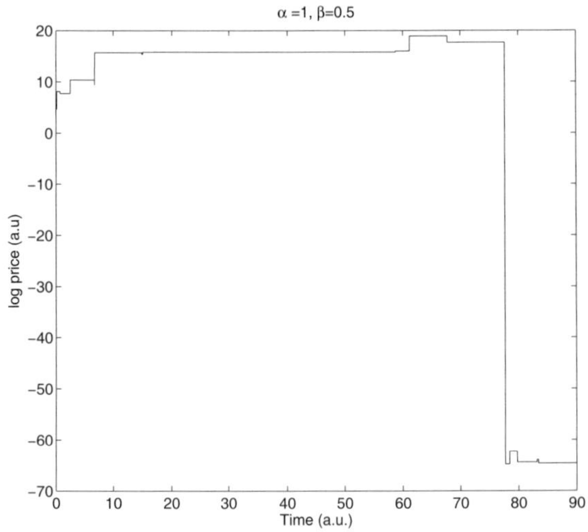

Fig. 7.10 Compound fractional Poisson process simulation for $\alpha = 1$ and $\beta = 0.5$.

probability measure $\mathbb{P}$ is absolutely continuous with respect to measure $\mathbb{Q}$ if its null set is contained in the null set of $\mathbb{Q}$. The null sets of two equivalent measures do coincide. An elementary introduction to these concepts can be found in a book by Mikosch [428]. Among all the equivalent measures, the equivalent martingale measure is the one for which the price process is a martingale, meaning that the price process is integrable and

$$\mathbb{E}_{\tilde{\mathbb{Q}}}[p(1)|I(0)] = p(0). \tag{7.4.3}$$

In our case, the martingale measure is given by $\tilde{q} = (1-d)/(u-d)$ and $1 - \tilde{q} = (u-1)/(u-d)$ and the option price is given by

$$C(0) = \frac{1-d}{u-d} C_1^+. \tag{7.4.4}$$

The martingale measure can be found by simple algebraic manipulations imposing Equation (7.4.3). Indeed, one has that $\mathbb{E}_{\tilde{\mathbb{Q}}}[p(1)|I(0)] = \tilde{q}up_0 + (1-\tilde{q})dp_0$, and imposing (7.4.3) immediately leads to $\tilde{q} = (1-d)/(u-d)$. Note that Equation (7.4.2) means that also C is a martingale under the measure $\tilde{\mathbb{Q}}$.

Technically speaking $I(0)$ in Equation (7.4.2) is the *filtration* at time $t = 0$. A filtration is a non decreasing family of σ-algebras which represents

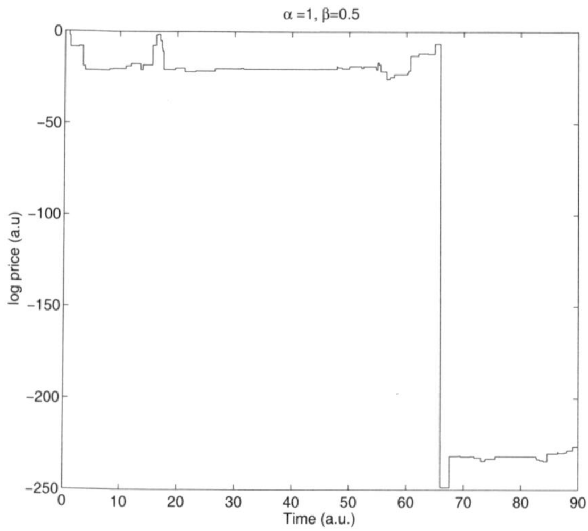

Fig. 7.11 Compound fractional Poisson process simulation for $\alpha = 1$ and $\beta = 0.99$.

the information available at a certain time, see [428] for a rigorous definition of this concept.

Equation (7.4.4) can be derived from the fact that it is possible to replicate the option in terms of a suitable non-financing portfolio coupled with a no-arbitrage argument. This derivation shows that Equations (7.4.2) and (7.4.4) give the *optimal* option price in term of fairness. However, it is not always possible to extend the arguments leading to the martingale option price when more general assumptions on the process followed by the price of the underlying asset are made. However, it is often possible to compute the martingale price in many cases of practical interest and this is done by *quants* in everyday financial practice [443].

In 1976, Merton solved the problem of finding the option martingale price for an underlying whose log-price follows the NCPP [422]. The idea behind Merton's derivation is as follows. Assume that $t = 0$ is a renewal point, that the risk free interest rate is $r_F = 0$ and denote the price at $t = 0$ by $S_0 = S(0)$, the strike price by K and the maturity by T. The NCPP assumption means that the underlying log-price follows the process

$$X(t) = \log(S_0) + \sum_{i=1}^{N(t)} Y_i, \tag{7.4.5}$$

where $Y_i \sim N(\mu, \sigma^2)$ and $N(t)$ is the Poisson process. The price should follow the process $S(t) = S_0 e^{X(t)}$. This is not a martingale, however. In order to find the option martingale price, let us consider the situation in which there are exactly n jumps from 0 and T. In this case, one has to study the processes

$$X_n = \log(S_0) + \sum_{i=1}^{n} Y_i \tag{7.4.6}$$

and

$$S_n = S_0 e^{X_n} = S_0 \prod_{i=1}^{n} e^{Y_i}. \tag{7.4.7}$$

Notice that the random variables e^{Y_i} follow the log-normal distribution. The process defined by X_n is not a martingale, but the equivalent martingale measure can be found by imposing that the process

$$S'_n = S_0 e^{X_n + na} \tag{7.4.8}$$

is a martingale and this leads to

$$a = -\log[\mathbb{E}(e^Y)], \tag{7.4.9}$$

where

$$\mathbb{E}(e^Y) = e^{\sigma^2/2} \tag{7.4.10}$$

so that

$$a = -(\sigma^2/2). \tag{7.4.11}$$

The option price at $t = 0$ with n jumps up to maturity T is thus given by

$$C_n(S_0, K, \sigma^2) = \mathbb{E}_{\mathbb{S}'}[C(T)|I(0)], \tag{7.4.12}$$

where the expected value is computed according to the measure defined by the process (7.4.8). For the plain vanilla European call option with $C(T) = \max(S(T) - K, 0)$, a straightforward calculation leads to

$$C_n(S_0, K, \sigma^2) = N(d_{1,n})S_0 - N(d_{2,n})K, \tag{7.4.13}$$

where

$$N(x) = \frac{1}{\sqrt{2\pi}} \int_{-\infty}^{x} dy \, \mathrm{e}^{-y^2/2} \tag{7.4.14}$$

is the standard normal cumulative distribution function and

$$d_{1,n} = \frac{\log(S_0/K) + n(\sigma^2/2)}{\sqrt{n}\sigma}, \tag{7.4.15}$$

$$d_{2,n} = d_{1,n} - \sigma\sqrt{n}. \tag{7.4.16}$$

Given the independence between jumps and durations, one can now write the option price at $t = 0$ as

$$C(0) = \mathrm{e}^{-\lambda T} \sum_{n=0}^{\infty} \frac{(\lambda T)^n}{n!} C_n(S_0, K, \sigma^2), \tag{7.4.17}$$

where λ is the intensity of the Poisson process. To obtain Equation (7.4.17), it is enough to notice that the probability of having n jumps in the time interval $[0, T]$ is given by the Poisson distribution of parameter λT and that one can go from S_0 to $S(T)$ in any number of steps n.

Equation (7.4.17) can be generalized to general renewal processes simply by replacing the Poisson distribution with a generic counting distribution $P(n, t)$. In the Mittag-Leffler case with $\gamma_t = 1$, one can write (see Equation (6.2.62))

$$C(0) = \sum_{n=0}^{\infty} \frac{t^{\beta n}}{n!} E_{\beta}^{(n)}(-t^{\beta}) \, C_n(S_0, K, \sigma^2). \tag{7.4.18}$$

Merton's result has been recently revisited and it is still the object of active research [151]. Note that this method works when the random variable e^Y has finite moments. This is the case when all the moments of the tick-by-tick log-returns Y are finite as in the normal case discussed above.

It is possible to generalize these results to the case in which the option position is opened at an instant t after the opening of the continuous auction and closed at a an instant T before the continuous auction closes [553]. For a renewal counting process $N(t)$, Equations (7.4.17) and (7.4.18) generalize to

$$C(0) = \sum_{n=0}^{\infty} \mathbb{P}(N(t) = n) \, C_n(S_0, K, \sigma^2). \tag{7.4.19}$$

With all the hypotheses leading to (7.4.19), except for the fact that t is no longer a renewal point, it is natural to write that

$$C(t) = \mathbb{E}_{\mathbb{S}'}(C(T)|I(t)) = \sum_{n=0}^{\infty} \mathbb{P}(N(T) - N(t) = n|I(t))\, C_n(S_0, K, \sigma^2), \tag{7.4.20}$$

where $I(t)$ represents what is known of the process up to time t. However, it is necessary to prove that $\mathbb{S}'$ is an appropriate martingale measure. In order to be more specific, let $\{U_i\}_{i=1}^{\infty}$ be a sequence of i.i.d. positive random variables. Let us define a random product of positive random variables.

Definition 7.1. Let $S(t)$ denote the random multiplicative process

$$S(t) = S_0 \prod_{i=1}^{N(t)} U_i. \tag{7.4.21}$$

Proposition 7.1. *The random multiplicative process $S(t)$ belongs to the class of semi-Markov processes.*

Proof. This is true by construction. However, one can see that

$$\begin{aligned} &\mathbb{P}(S_{n+1} \in A, J_{n+1} \leq t|S_1, \ldots, S_n, T_1, \ldots, T_n) \\ &\quad = \mathbb{P}(S_{n+1} \in A, J_{n+1} \leq t|S_n) = (S_{n+1} \in A|S_n)F_J(t), \end{aligned} \tag{7.4.22}$$

as usual. □

Lemma 7.1. *The expected value of S_n is*

$$\mathbb{E}(S_n) = [\mathbb{E}(U_1)]^n. \tag{7.4.23}$$

Proof. This is an immediate consequence of the fact that the U_is are i.i.d.. □

Definition 7.2. The *natural filtration* $I(t)$ is the σ-field generated by the process $S(t)$, namely

$$I(t) = \sigma(S(u), u \leq t) = \sigma(J_1, \ldots, J_k, U_1, \ldots, U_k, k \leq N(t)) = G(N(t)). \tag{7.4.24}$$

Lemma 7.2. *Consider the ratio $S(t)/S(u)$ with $u < t$, this is given by*

$$\frac{S(t)}{S(u)} = \prod_{i=N(u)+1}^{N(t)} U_i. \tag{7.4.25}$$

The conditional expectation $\mathbb{E}(S(t)/S(u)|I(u))$ *is given by*

$$\begin{aligned}\mathbb{E}(S(t)/S(u)|I(u)) &= \mathbb{E}\left(\prod_{i=N(u)+1}^{N}(t)|I(u)\right)\\ &= \sum_{n=0}^{\infty}\mathbb{P}(N(t)-N(u)=n|I(u))[\mathbb{E}(U_1)]^n\\ &= \sum_{n=0}^{\infty}\mathbb{P}(N(t)-N(u)=n|I(u))\mathbb{E}(S_n) \qquad (7.4.26)\end{aligned}$$

Proof. For $i > N(u)$, the U_is are independent of $I(u)$ and the U_is and J_is are mutually independent. The expected value of the product of n U_is is given by Equation (7.4.23), and between u and t there can be n events with any non-negative n and the expectation must be multiplied by the probability of the event $N(t)-N(u)=n$. This probability depends on $I(u)$ because it depends on the *residual lifetime* (a.k.a. *forward recurrence time*) $T_{N(u)+1}-u$ which, in turn, depends on the previous history as discussed in [491]. □

Corollary 7.1. *If* $\mathbb{E}(U_1)=1$ *then* $\mathbb{E}(S(t)/S(u)|I(u))=1$.

Proof. This is an immediate consequence of Equation (7.4.26). □

With the above results, it is possible to prove the following

Proposition 7.2. *Let* $S(t)$ *be a random multiplicative process defined as in 7.1 and further assume that* $\mathbb{E}(U_1)=1$, *then* $S(t)$ *is a martingale with respect to its natural filtration* $I(t)$.

Proof. Considering the definition, one has that $|S(t)|=S(t)$. Therefore $|S(t)|$ is integrable, that is

$$\mathbb{E}(|S(t)|)=\mathbb{E}(S(t))=\sum_{n=0}^{\infty}\mathbb{P}(N(t)=n)\mathbb{E}(S_n)=\sum_{n=0}^{\infty}\mathbb{P}(N(t)=n)=1<\infty. \qquad (7.4.27)$$

Using the fact that one can take out what is known in conditional expectation, it is possible to write the following chain of equalities for $u < t$.

$$\mathbb{E}(S(t)|I(u)) = \mathbb{E}\left(\prod_{i=1}^{N(t)} U_i|I(u)\right) = S_0 \prod_{i=1}^{N(u)} U_i \,\mathbb{E}\left(\prod_{i=N(u)+1}^{N(t)} U_i|I(u)\right)$$

$$= S(u)\,\mathbb{E}\left(\prod_{i=N(u)+1}^{N(t)} U_i|I(u)\right). \quad (7.4.28)$$

Now, Corollary 7.1 and Equation (7.4.28) lead to

$$\mathbb{E}(S(t)|I(u)) = S(u), \quad (7.4.29)$$

namely our thesis. □

The variables $U_i = \mathrm{e}^{Y_i - \log(\mathbb{E}(\mathrm{e}^{Y_i}))}$ indeed have mean 1 and $\mathbb{S}'$ is an equivalent martingale measure. Finally, one has to remark that Equation (7.4.20) is not the unique option price in this incomplete market [436].

7.5 Other Applications

So far, this chapter focused on the application of uncoupled continuous-time random walks in high-frequency financial data modelling. For the sake of simplicity, we have not discussed the coupled case, but this is covered in reference [417]. Moreover, it is possible to use the program described in Section 7.3 for scenario simulation and speculative option pricing [552]. It is now time to discuss several other applications.

7.5.1 *Application to insurance and economics*

As discussed in Section 6.1, it is not surprising that CTRWs can also be applied otherwise. A standard application is to insurance [163, 389], where the capital $R(t)$ of an insurance company can be written as

$$R(t) = u + ct - \sum_{i=1}^{N(t)} Y_i, \quad (7.5.1)$$

and where u is the initial capital of the company, c is the rate of capital increase, $N(t)$ is the random number of claims Y_i that the company has paid since inception. In this case, ruin is the interesting phenomenon. It takes place the first time that $R(t) = 0$, that is when the capital of the insurance company vanishes. In this framework, one can define the time to ruin as the following hitting time

$$\tau(u) = \inf \left\{ t: \ u + ct - \sum_{i=1}^{N(t)} Y_i < 0 \right\}. \tag{7.5.2}$$

Two interesting quantities are the probability $q(u)$ of ruin in infinite time and the probability of ruin in a finite time T. These two quantities are defined as follows, respectively:

$$q(u) = \mathbb{P}(\tau(u) < \infty), \tag{7.5.3}$$

and

$$q(u, T) = \mathbb{P}(\tau(u) < T). \tag{7.5.4}$$

It is always possible to study these quantities by means of Monte Carlo simulations, using the algorithm of Appendix A.5 or a suitable modification.

Another interpretation of the random variables is in terms of economic growth. Let us be as general as possible and denote by S a suitable "size". This size has the meaning of wealth, firm size, city size, *etc.*, depending on the scientific context. Then, according to Gibrat's approach [257], one can define the log-size $X = \log(S)$ and write it as a sum of exogenous shocks Y_i

$$X_n = X_0 + \sum_{i-1}^{n} Y_i. \tag{7.5.5}$$

For large n, if the shocks have finite first and second moments, X_n approximately follows the normal distribution as a consequence of central limit theorems (see Theorem 6.4 for a simple version). This means that the size S_n approximately follows the log-normal distribution [32]. If the growth shocks arrive at random times, Equation (7.5.5) can be replaced by the familiar equations for CTRWs with non-homogeneous initial position

$$X(t) = X_0 + \sum_{i=1}^{N(t)} Y_i. \tag{7.5.6}$$

This method was used by Italian economists to study firm growth and size distribution [115, 116, 250].

7.5.2 *Applications to physics*

In the previous chapter, constant reference was made to CTRWs as models of (potentially anomalous) tracer diffusion. However, semi-Markov models and equations such as (6.2.84) or (6.2.96) can be applied to other systems, as well. Consider any Markov chain X_n (not necessarily a random walk). Here, the focus will be on Markov chains with discrete state space. Let $q_{i,j}$ denote the one-step transition probability from state i to state j for a Markov chain X_n and let $N(t)$ be a generic counting renewal process. Then, for the process

$$X(t) = X_{N(t)}, \tag{7.5.7}$$

the probabilities $p_{i,j}(t) = \mathbb{P}\{X(t) = j|X(0) = i\}$ are given by

$$p_{i,j}(t) = \bar{F}_J(t)\delta_{ij} + \sum_{n=1}^{\infty} q_{i,j}^{(n)} \mathbb{P}(N(t) = n), \tag{7.5.8}$$

where $q_{i,j}^{(n)}$ is the probability of moving from state i to state j in n steps. A heuristic argument to justify (7.5.8) runs as follows. In the time interval $(0, t)$, $n \geq 0$ events may have occurred. In the case $n = 0$, at time t the process is still in state i and $\mathbb{P}(N(t) = 0) = \mathbb{P}(J > t) = \bar{F}_J(t)$. If $n \geq 1$, the probability of being in state j after n events is given by $q_{i,j}^{(n)}$. Given the independence between the renewal process and the Markov chain, the probability of being in state j at time t and n transitions occurring in the time interval $(0, t)$ is $q_{i,j}^{(n)} \mathbb{P}(N(t) = n)$. Now, all these events are exhaustive and mutually exclusive. Then, total probability and infinite additivity imply that $p_{i,j}(t)$ is given by (7.5.8). These considerations suggest further generalizations taking into account possible dependence within the couple $\{(X_n, T_n)\}_{n \geq 1}$ as well as serial dependence or state dependence of inter-event times.

Example 7.1. In order to illustrate the usefulness of Equation (7.5.8) with an example, one can consider relaxation phenomena [324, 421]. Assume that a physical system exists in two states A and B. Further assume that state A is transient and state B absorbing, so that the deterministic embedded chain X_n is very simple and has the following transition probabilities

$$q_{A,A} = \mathbb{P}(X_1 = A|X_0 = A) = 0,$$

$$q_{A,B} = \mathbb{P}(X_1 = B|X_0 = A) = 1,$$

$$q_{B,A} = \mathbb{P}(X_1 = A|X_0 = B) = 0,$$

and

$$q_{B,B} = \mathbb{P}(X_1 = B|X_0 = B) = 1.$$

If the system is prepared in state A, it will move to B at the first step and it will stay there forever. This is a very boring behavior. However, if the inter-event time J is random and follows an exponential distribution with rate $\lambda = 1$ for the sake of simplicity, as a consequence of Equation (7.5.8), one finds $p_{A,A}(t) = \bar{F}_J(t) = \mathrm{e}^{-t}$. Therefore, the probability of finding the system in the initial state decays exponentially towards zero. This relaxation function is the solution of

$$\frac{d}{dt} p_{A,A}(t) = -p_{A,A}(t), \;\; p_{A,A}(0) = 1. \tag{7.5.9}$$

The response function is defined as $\xi_D(t) = -dp_{A,A}(t)/dt$ and its Laplace transform is $1/(1+s)$. For $s = -i\omega$ this coincides with the Debye model of dielectric relaxation [324]. If, on the contrary, inter-event times follow the Mittag-Leffler distribution, we get $p_{A,A}(t) = \bar{F}_J(t) = E_\beta(-t^\beta)$. As derived in Corollary 6.5, this is the solution of

$$^{\mathrm{C}}D_t^\beta p_{A,A}(t) = -p_{A,A}(t), \;\; p_{A,A}(0) = 1. \tag{7.5.10}$$

In this case, the Laplace transform of the response function $\xi_{CC}(t) = -dp_{A,A}(t)/dt$ is $1/(1+s^\beta)$ and for $s = -i\omega$, one gets the Cole-Cole model of dielectric relaxation [157, 158, 324].

Example 7.2. Another interesting application of the concepts developed in chapter 6 as well as in the present chapter is to random graph dynamics [255, 509]. In such a case, the state space of the Markov chain X_n is given *e. g.* by all the possible $2^{N(N+1)/2}$ undirected graphs with M vertices. The simple random link activation and deletion dynamics analyzed in detail in [255] was introduced in [509]. An edge (link) is selected at random among all the possible $M = 2^{N(N-1)/2}$ edges. If it is present, it is kept with probability $0 < \alpha < 1$, otherwise it is deleted with probability $1 - \alpha$. If it is not there, it is added with probability $1 - \alpha$, otherwise nothing happens with probability α. Given the symmetry of the transition probability matrix of this chain, this irreducible and aperiodic chain has the uniform distribution on the state space of graphs without self loops as invariant and equilibrium distribution. If $\alpha = 0$, one gets a periodic (period 2) version of the chain. If one is interested in the number of occupied links and not on which specific link is occupied (a statistical description according to the definition of [251]), one has the following transition probability for the edge occupation numbers Y_n

$$q_{k,k-1} = \mathbb{P}(Y_{j+1} = k-1|Y_j = k) = \begin{cases} 0, & k = 0, \\ 1-\alpha, & k = M, \\ (1-\alpha)\frac{k}{M}, & \text{otherwise}, \end{cases} \tag{7.5.11}$$

$$q_{k,k} = \alpha, \tag{7.5.12}$$

and

$$q_{k,k+1} = \mathbb{P}(Y_{j+1} = k+1|Y_j = k) = \begin{cases} 1-\alpha, & k = 0, \\ 0, & k = M, \\ (1-\alpha)\frac{M-k}{M}, & \text{otherwise}. \end{cases} \tag{7.5.13}$$

The above transition probabilities are the same as in the case of the α-delayed version of the Ehrenfest urn model. Therefore, the invariant and equilibrium probability distribution is the binomial distribution

$$\pi_k = \lim_{t\to\infty} \mathbb{P}(Y(t) = k|Y(0) = i) = \binom{M}{k} 2^{-M}, \quad k \in \{0, 1, \ldots, M\}. \tag{7.5.14}$$

After defining $Y(t)$ as $Y(t) = Y_{N_\beta(t)}$, it is not difficult to prove that the probabilities $p_{i,j}(t) = \mathbb{P}(Y(t) = j|Y(0) = i)$ obey the following governing equations

$$\begin{aligned} {}^{\mathrm{C}}D_t^\beta p_{i,j}(t) &= -(1-\alpha)\, p_{i,j}(t) \\ &\quad +(1-\alpha)\left(\frac{M-j+1}{M} p_{i,j-1}(t) + \frac{j+1}{M} p_{i,j+1}(t)\right), \end{aligned} \tag{7.5.15}$$

with the following equations for the boundary terms

$${}^{\mathrm{C}}D_t^\beta p_{i,0}(t) = (1-\alpha)\left(-p_{i,0}(t) + \frac{1}{M} p_{i,1}(t)\right), \tag{7.5.16}$$

$${}^{\mathrm{C}}D_t^\beta p_{i,M}(t) = (1-\alpha)\left(-p_{i,M}(t) + \frac{1}{M} p_{i,M-1}(t)\right). \tag{7.5.17}$$

These governing equations can be solved with the appropriate intial conditions. Their solution is

$$p_{i,j}(t) = E_\beta(-t^\beta)\delta_{ij} + \sum_{n=1}^{\infty} q_{i,j}^{(n)} \frac{t^{\beta n}}{n!} E^{(n)}(-t^\beta). \qquad (7.5.18)$$

They are analogous to (6.2.71), (6.3.1) and (7.5.10). In these problems, every time that exponentially distributed durations are replaced by Mittag-Leffler distributed durations, the ordinary derivative can be replaced by the Caputo derivative. In essence, this is the true physical meaning of the Caputo derivative. Incidentally, the advantage of the formulation in terms of Caputo derivatives is now evident, at least in these problems: One can use the same initial conditions as in the Markovian case. A natural question is: what happens when the Mittag-Leffler distribution for durations is replaced by another distribution? Is there another operator that can be used in lieu of the Caputo derivative? Recently, Meerschaert and Toaldo [421] studied these questions. It turns out that Bernstein functions [557] play an important role in finding the answer.

Chapter 8

Selected Applications of Fractional Calculus Models in the Life Sciences

Following the discussion of applications of fractional calculus in finance and economics given in Chapter 7, we now turn our attention towards the use of fractional calculus techniques for modeling three concrete processes arising in a completely different area, namely in the life sciences. Specifically, Section 8.1 will be devoted to the mathematical description of the evolution of an epidemic, Section 8.2 is aimed at the modeling of a biological reaction connected to a fermentation process, and in Section 8.3 we shall present an application from pharmacology.

We remark that, apart from the cases discussed below, many other real life phenomena have also been observed to possess properties indicating that fractional calculus based models might be more appropriate than models based on integer order derivatives; see, *e. g.*, the survey article of Pinto *et al.* [485].

8.1 The Spreading of an Epidemic

In our first example application of fractional calculus techniques in the life sciences, we consider the dynamics of an outbreak of dengue fever, an epidemic disease that has been found to cause problems whose magnitude has increased dramatically over the last two decades. A few years ago, the World Health Organization issued the statement that dengue fever is currently the most important arthropod-borne viral disease of humans, *cf.* page v of [633].

The development of our model is based on the ideas given in [188]. We refer to this publication and the references cited therein for further details on the significance of the disease. Here we merely stress that dengue fever itself is not very lethal; for example, in the outbreak considered below, less than 0.035% of the infected persons eventually did not survive the disease [527]. The importance of fighting the spreading of dengue fever is thus not related to its direct symptoms but to the fact that infected humans are much more likely to develop other very dangerous illnesses like dengue haemorrhagic fever.

8.1.1 *Basic facts about dengue fever and the classical mathematical model*

The virus that causes dengue fever is transmitted from one human to another if first the former and then the latter is bitten by a female individuum of a certain type of mosquitoes, mainly *Aedes aegypti* but also other types like *Aedes albopictus*, see page 14 of [633]. When such a mosquito bites an infected human, it infects itself with a certain probability and then remains in this infected condition for the rest of its life. When a susceptible human being is bitten by an infected mosquito, the infection is transferred to the person once again with a certain probability. The key difference between humans and mosquitoes in this respect is the fact that humans recover from the disease after a certain period of time and become immune against further infections after that. In particular, as pointed out on pages 52 and 56 of [461], this means that humans who have gained such an immunity cannot transfer the virus back to a mosquito.

We stress that the male mosquitoes do not play a role in the process to be modeled here. Hence, for the sake of simplicity, whenever we only speak of "mosquitoes" in the remainder of this section, this is tacitly meant to mean "female mosquitoes".

The classical mathematical model for describing the dynamics of the infection and recovery process is based on first-order differential equations. It decomposes the human population into three groups (susceptible, infected, and recovered and hence immune humans) and the mosquito population into two groups (susceptible and infected; as mentioned above, the mosquitoes do not recover from the infection). To describe these groups, we will use the following notation:

$$\begin{aligned}
N_{\mathrm{h}}(t) &= \text{total number of humans,} \\
R_{\mathrm{h}}(t) &= \text{number of humans having recovered, and hence} \\
&\quad\ \text{obtained resistance (immunity), against the infection,} \\
I_{\mathrm{h}}(t) &= \text{number of infected humans,} \qquad (8.1.1) \\
S_{\mathrm{h}}(t) &= \text{number of susceptible humans,} \\
N_{\mathrm{m}}(t) &= \text{total number of female mosquitoes,} \\
I_{\mathrm{m}}(t) &= \text{number of infected female mosquitoes,} \\
S_{\mathrm{m}}(t) &= \text{number of susceptible female mosquitoes,}
\end{aligned}$$

all of which are to be understood at the time t. Evidently, there holds

$$N_{\mathrm{h}}(t) = R_{\mathrm{h}}(t) + S_{\mathrm{h}}(t) + I_{\mathrm{h}}(t)$$

and

$$N_{\mathrm{m}}(t) = S_{\mathrm{m}}(t) + I_{\mathrm{m}}(t)$$

for all t, so it suffices to take the five functions R_{h}, I_{h}, S_{h}, I_{m} and S_{m} as the independent functions that need to be described, whereas the two others, N_{m} and N_{h}, depend on these five functions.

The classical model from Eq. (1) of [461] then consists of a system of five ordinary differential equations for the five independent functions that takes the form

$$\frac{dS_{\mathrm{h}}(t)}{dt} = \mu_{\mathrm{h}}(N_{\mathrm{h}}(t) - S_{\mathrm{h}}(t)) - \frac{\beta_{\mathrm{h}} b}{N_{\mathrm{h}}(t) + m} S_{\mathrm{h}}(t) I_{\mathrm{m}}(t), \qquad (8.1.2a)$$

$$\frac{dI_{\mathrm{h}}(t)}{dt} = \frac{\beta_{\mathrm{h}} b}{N_{\mathrm{h}}(t) + m} S_{\mathrm{h}}(t) I_{\mathrm{m}}(t) - (\mu_{\mathrm{h}} + \gamma) I_{\mathrm{h}}(t), \qquad (8.1.2b)$$

$$\frac{dR_{\mathrm{h}}(t)}{dt} = \gamma I_{\mathrm{h}}(t) - \mu_{\mathrm{h}} R_{\mathrm{h}}(t), \qquad (8.1.2c)$$

$$\frac{dS_{\mathrm{m}}(t)}{dt} = A - \frac{\beta_{\mathrm{m}} b}{N_{\mathrm{h}}(t) + m} S_{\mathrm{m}}(t) I_{\mathrm{h}}(t) - \mu_{\mathrm{m}} S_{\mathrm{m}}(t), \qquad (8.1.2d)$$

$$\frac{dI_{\mathrm{m}}(t)}{dt} = \frac{\beta_{\mathrm{m}} b}{N_{\mathrm{h}}(t) + m} S_{\mathrm{m}}(t) I_{\mathrm{h}}(t) - \mu_{\mathrm{m}} I_{\mathrm{m}}(t); \qquad (8.1.2e)$$

it can be considered a natural generalization of the original Kermack-McKendrick model (Eq. (29) of [329]) where only one population is considered and not two. Because of the decomposition of the populations of

Table 8.1 Parameters of the equation system (8.1.2).

Parameter	Explanation
μ_{h}	per capita mortality rate (reciprocal of average life time) of humans
μ_{m}	per capita mortality rate of mosquitoes
γ	recovery rate of humans (reciprocal of average time span between being infected and becoming immune)
b	biting rate (average number of bites per mosquito per time unit)
β_{m}	transmission probability from human to mosquito
β_{h}	transmission probability from mosquito to human
m	number of alternative (non-human) blood sources available to mosquitoes
A	(constant) recruitment rate for mosquitoes

humans and mosquitoes into their susceptible, infected and recovered subpopulations, models of this type are known as *SIR models*. The parameters appearing in the equation system (8.1.2) are briefly explained in Table 8.1.

Adding up Eqs. (8.1.2a), (8.1.2b) and (8.1.2c) we find

$$\frac{dN_{\mathrm{h}}(t)}{dt} = \frac{dS_{\mathrm{h}}(t)}{dt} + \frac{dR_{\mathrm{h}}(t)}{dt} + \frac{dI_{\mathrm{h}}(t)}{dt} = \mu_{\mathrm{h}}(N_{\mathrm{h}}(t) - S_{\mathrm{h}}(t) - I_{\mathrm{h}}(t) - R_{\mathrm{h}}(t)) = 0, \tag{8.1.3}$$

i. e. the total number of humans is constant. This is in line with the observation mentioned above that the number of people who die from dengue fever is very small indeed and hence negligible. Similarly, adding Eqs. (8.1.2d) and (8.1.2e) we obtain

$$\frac{dN_{\mathrm{m}}(t)}{dt} = \frac{dS_{\mathrm{m}}(t)}{dt} + \frac{dI_{\mathrm{m}}(t)}{dt} = A - \mu_{\mathrm{m}}(S_{\mathrm{m}}(t) + I_{\mathrm{m}}(t)) = A - \mu_{\mathrm{m}}N_{\mathrm{m}}(t). \tag{8.1.4}$$

Our specific instance of an outbreak of dengue fever that we shall concentrate on occurred in 2009 on the Cape Verde islands. Due to the remote geographic location of the islands, it is safe to assume that outside events did not have a significant effect on the evolution of the epidemic. Moreover, the number of infected people has been documented quite accurately, so a reliable data set is available.

In this environmental system, it is known that the humans were actually the only blood source for the mosquitoes, so we can set

$$m = 0.$$

Moreover, general observations (not only valid for this outbreak) have shown that the total number of mosquitoes also remains constant (see page

57 of [461]), so from Eq. (8.1.4) we conclude

$$A = \mu_{\mathrm{m}} N_{\mathrm{m}}(0) \quad \text{and hence} \quad N_{\mathrm{m}}(t) = N_{\mathrm{m}}(0) \text{ for all } t.$$

As far as the remaining parameters are concerned, it is helpful to note that the mosquitoes on Cape Verde are of a well investigated species; in fact the following values are appropriate [493]:

$$\begin{aligned} \mu_{\mathrm{h}} &= \frac{1}{71 \cdot 365 \cdot 86400 \text{ s}}, & \mu_{\mathrm{m}} &= \frac{1}{10 \cdot 86400 \text{ s}}, \\ \beta_{\mathrm{h}} &= 0.36, & \beta_{\mathrm{m}} &= 0.36, \\ b &= \frac{0.7}{86400 \text{ s}}, & \gamma &= \frac{1}{3 \cdot 86400 \text{ s}}. \end{aligned} \tag{8.1.5}$$

In view of this quite precise knowledge of the relevant data, it was rather surprising to see [493] that this model was unable to reproduce the statistical data collected in the outbreak with the expected sufficient degree of accuracy.

8.1.2 *The fractional version of the model*

Based on this observation, we start our development of a modified system of equations that now includes differential equations of fractional order. The basic idea is very simple; we only replace the first order differential operators in the system (8.1.2) by fractional differential operators. In particular, we do not see any reason why the dynamics of the sub-populations of the mosquitoes should behave in the same way as the sub-populations of the humans; therefore we do not demand the "fractionalized" versions of Eqs. (8.1.2a), (8.1.2b) and (8.1.2c) on the one hand must have the same orders as Eqs. (8.1.2d) and (8.1.2e) on the other hand.

For this particular application scenario, operators of Caputo type are most appropriate because

- they allow us to use the initial conditions in the same form as in the classical case, and
- constant functions vanish under the Caputo differential operation, which means that our generalized versions of Eqs. (8.1.3) and (8.1.4) retain the property that the total populations of humans and mosquitoes, respectively, remain constant, thus matching the measured data.

We thus arrive at the preliminary system

$$^{\mathrm{C}}D^{\alpha_{\mathrm{h}}} S_{\mathrm{h}}(t) = \mu_{\mathrm{h}}(N_{\mathrm{h}}(t) - S_{\mathrm{h}}(t)) - \frac{\beta_{\mathrm{h}} b}{N_{\mathrm{h}}(t)} S_{\mathrm{h}}(t) I_{\mathrm{m}}(t), \tag{8.1.6a}$$

$$^{\mathrm{C}}D^{\alpha_{\mathrm{h}}} I_{\mathrm{h}}(t) = \frac{\beta_{\mathrm{h}} b}{N_{\mathrm{h}}(t)} S_{\mathrm{h}}(t) I_{\mathrm{m}}(t) - (\mu_{\mathrm{h}} + \gamma) I_{\mathrm{h}}(t), \tag{8.1.6b}$$

$$^{\mathrm{C}}D^{\alpha_{\mathrm{h}}} R_{\mathrm{h}}(t) = \gamma I_{\mathrm{h}}(t) - \mu_{\mathrm{h}} R_{\mathrm{h}}(t), \tag{8.1.6c}$$

$$^{\mathrm{C}}D^{\alpha_{\mathrm{m}}} S_{\mathrm{m}}(t) = \mu_{\mathrm{m}} N_{\mathrm{m}}(t) - \frac{\beta_{\mathrm{m}} b}{N_{\mathrm{h}}(t)} S_{\mathrm{m}}(t) I_{\mathrm{h}}(t) - \mu_{\mathrm{m}} S_{\mathrm{m}}(t), \tag{8.1.6d}$$

$$^{\mathrm{C}}D^{\alpha_{\mathrm{m}}} I_{\mathrm{m}}(t) = \frac{\beta_{\mathrm{m}} b}{N_{\mathrm{h}}(t)} S_{\mathrm{m}}(t) I_{\mathrm{h}}(t) - \mu_{\mathrm{m}} I_{\mathrm{m}}(t). \tag{8.1.6e}$$

The fact that exactly one initial condition is available for each of these differential equations then immediately implies that both our parameters α_{h} and α_{m} must lie in the half-open interval $(0, 1]$.

Now, a brief look at the system (8.1.6) reveals a certain technical inconsistency. This problem stems from the fact that we exclusively replaced the first derivatives by fractional derivatives: The biological parameters from Eq. (8.1.5) have been used in an unchanged form for the new fractional system (8.1.6). Because of this and the changes in the orders of the differential operators, a dimensional analysis here tells us that the left hand sides of the equations in the system do not have the same dimensions as their right-hand sides and therefore the equations need to be modified to make the dimensions match. A straightforward way of doing this gives the new system

$$^{\mathrm{C}}D^{\alpha_{\mathrm{h}}} S_{\mathrm{h}}(t) = \mu_{\mathrm{h}}^{\alpha_{\mathrm{h}}}(N_{\mathrm{h}}(t) - S_{\mathrm{h}}(t)) - \frac{\beta_{\mathrm{h}} b^{\alpha_{\mathrm{h}}}}{N_{\mathrm{h}}(t)} S_{\mathrm{h}}(t) I_{\mathrm{m}}(t), \tag{8.1.7a}$$

$$^{\mathrm{C}}D^{\alpha_{\mathrm{h}}} I_{\mathrm{h}}(t) = \frac{\beta_{\mathrm{h}} b^{\alpha_{\mathrm{h}}}}{N_{\mathrm{h}}(t)} S_{\mathrm{h}}(t) I_{\mathrm{m}}(t) - (\mu_{\mathrm{h}}^{\alpha_{\mathrm{h}}} + \gamma^{\alpha_{\mathrm{h}}}) I_{\mathrm{h}}(t), \tag{8.1.7b}$$

$$^{\mathrm{C}}D^{\alpha_{\mathrm{h}}} R_{\mathrm{h}}(t) = \gamma^{\alpha_{\mathrm{h}}} I_{\mathrm{h}}(t) - \mu_{\mathrm{h}}^{\alpha_{\mathrm{h}}} R_{\mathrm{h}}(t), \tag{8.1.7c}$$

$$^{\mathrm{C}}D^{\alpha_{\mathrm{m}}} S_{\mathrm{m}}(t) = \mu_{\mathrm{m}}^{\alpha_{\mathrm{m}}} N_{\mathrm{m}}(t) - \frac{\beta_{\mathrm{m}} b^{\alpha_{\mathrm{m}}}}{N_{\mathrm{h}}(t)} S_{\mathrm{m}}(t) I_{\mathrm{h}}(t) - \mu_{\mathrm{m}}^{\alpha_{\mathrm{m}}} S_{\mathrm{m}}(t), \tag{8.1.7d}$$

$$^{\mathrm{C}}D^{\alpha_{\mathrm{m}}} I_{\mathrm{m}}(t) = \frac{\beta_{\mathrm{m}} b^{\alpha_{\mathrm{m}}}}{N_{\mathrm{h}}(t)} S_{\mathrm{m}}(t) I_{\mathrm{h}}(t) - \mu_{\mathrm{m}}^{\alpha_{\mathrm{m}}} I_{\mathrm{m}}(t). \tag{8.1.7e}$$

This *fractional SIR model* is the system that we will actually use for modeling our problem.

A reduction of the complexity of this system, and hence also a reduction of the cost of numerically solving it, can be obtained by recalling that

$N_{\rm h} = S_{\rm h} + I_{\rm h} + R_{\rm h}$ and $N_{\rm m} = S_{\rm m} + I_{\rm m}$ are constant functions whose values are known in advance from the initial values of $S_{\rm h}, I_{\rm h}, R_{\rm h}, S_{\rm m}$ and $I_{\rm m}$. Thus we can eliminate Eqs. (8.1.7c) and (8.1.7e) from our system, solve the remaining three-dimensional system for the unknowns $S_{\rm h}, I_{\rm h}$ and $S_{\rm m}$ and finally compute $R_{\rm h}$ and $I_{\rm m}$ from these quantities and $N_{\rm h}$ and $N_{\rm m}$.

Remark 8.1. We stress that our approach that lead from the preliminary fractional order system (8.1.6) to the final version of Eq. (8.1.7) is not the only possible way in which the system (8.1.6) can be converted into a physically consistent form. A viable path to construct an alternative solution appears to be, for example, to use Eqs. (8.1.7a), (8.1.7c), (8.1.7d), and (8.1.7e) and replace (8.1.7b) by

$$^{\rm C}D^{\alpha_{\rm h}} I_{\rm h}(t) = \frac{\beta_{\rm h} b^{\alpha_{\rm h}}}{N_{\rm h}(t)} S_{\rm h}(t) I_{\rm m}(t) - (\mu_{\rm h} + \gamma)^{\alpha_{\rm h}} I_{\rm h}(t).$$

However, this modification has a major disadvantage: Adding up the three equations describing the evolution of the sub-populations of the humans would lead to the relation

$$\begin{aligned} ^{\rm C}D^{\alpha_{\rm h}} N_{\rm h}(t) &= {}^{\rm C}D^{\alpha_{\rm h}} S_{\rm h}(t) + {}^{\rm C}D^{\alpha_{\rm h}} I_{\rm h}(t) + {}^{\rm C}D^{\alpha_{\rm h}} R_{\rm h}(t) \\ &= \left[\mu_{\rm h}^{\alpha_{\rm h}} + \gamma^{\alpha_{\rm h}} - (\mu_{\rm h} + \gamma)^{\alpha_{\rm h}}\right] I_{\rm h}(t) \end{aligned}$$

whose right-hand side is nonzero in general (unless we are dealing with the special case $\alpha_{\rm h} = 1$ for which both versions coincide), *i. e.* in contradiction to the recorded data, the total population of the humans predicted by the model is not constant any more.

Another option would be to replace the parameters in (8.1.7) by completely new ones that have the correct dimensions but are generally unrelated to those of the original first-order model (8.1.2) that we listed in Table 8.1. Whether such an approach were more appropriate for modeling our epidemic spreading process remains to be investigated; first steps in this direction have been described in [188].

Because no method for obtaining an exact solution is available, we have used the fractional predictor-corrector method from Subsection 2.3.2 as a numerical method for obtaining an approximate solution. The time step size has been chosen as 864 s (0.01 days); the evolution of the epidemic was simulated for a period of 90 days.

From our original considerations, we know the values for most of the parameters of the system that needs to be solved, but not for the orders α_{h} and α_{m} of the differential operators. Therefore, we performed an exhaustive search of the parameter space $(0,1]^2$ for these parameters, *i. e.* we approximately solved the equations for a large set of such pairs of parameters, and found the best match to the recorded data for $\alpha_{\mathrm{h}} = 1$ and $\alpha_{\mathrm{m}} = 0.77$. In other words, the evolution of the sub-populations of the humans seems to follow a classical law governed by a first order differential operator whereas the dynamics of the mosquito sub-populations appears to be truly fractional. Figure 8.1 shows the final result; in particular it clearly demonstrates that the fractional model provides a much better approximation to the recorded data than the classical model (8.1.2).

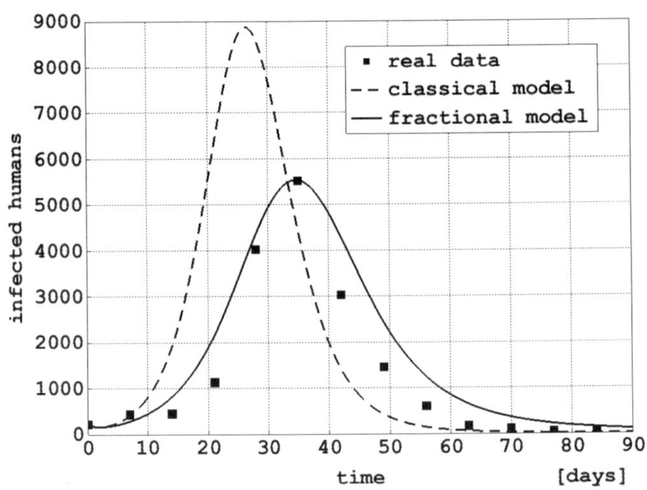

Fig. 8.1 Number of infected humans in the 2009 dengue fever outbreak on the Cape Verde islands: True values as recorded by the government agencies (filled squares) and predictions obtained via the classical model Eq. (8.1.2) (dashed line) and the fractional model (8.1.7) with parameters $\alpha_{\mathrm{h}} = 1$ and $\alpha_{\mathrm{m}} = 0.77$ (continuous line).

8.2 Biological Reactive Systems

A completely different but also very interesting application of fractional calculus based models was recently discussed by Toledo-Hernandez *et al.*

[602, 603]. In these two papers that we shall briefly review here, they demonstrate that fractional order models can be successfully used to describe the behaviour of a system containing biological materials that react with each other.

More precisely, the scenario under consideration deals with fermentation processes. Following the original work [602], we develop the idea by looking at a special case, namely the production of bio-ethanol, a type of alcohol that is made from organically grown materials and that can be used, *e. g.*, as a fuel for combustion engines, but the ideas can also be applied to other similar processes.

8.2.1 *A mathematical model for the bio-ethanol fermentation process*

Neglecting some aspects of minor significance, the fermentation process that leads to the production of bio-ethanol is classically formulated in terms of three components that are responsible for the essentials of the behaviour. These components are the final product of the process (the bio-ethanol), the ingredient from which the ethanol is produced (the so-called substrate; usually some kind of sugar like glucose, fructose or sucrose), and the biomass that is responsible for converting the latter into the former, typically some sort of bacteria like a variant of *Zymomonas mobilis* or *Saccharomyces cerevisiae* [484]. Hence, the mathematical model consists of three ordinary differential equations, one for each of the three key components that are involved.

We denote the concentrations of the substrate, the biomass and the bio-ethanol at time t by $s(t)$, $b(t)$ and $e(t)$, respectively; all of them are given in grammes per litre. The model is then based on the following properties:

- While the process runs, the biomass concentration decreases due to the mortality of the cells that form the biomass; the mortality rate is denoted by m.
- As long as substrate and biomass are present, the biomass consumes the substrate to reproduce itself (*i. e.* to increase its own concentration) and to produce bio-ethanol. The intensity of these reactions is proportional to the amounts of substrate and biomass that are available for

an interaction.

- Bio-ethanol that has been produced will not react any further and hence remains in an unchanged state.

A classical model would thus consist of three first-order differential equations each of which would have a form similar to that of the equations which constitute the classical first-order Volterra-Lotka system of population dynamics. The key observation of Toledo-Hernandez *et al.* now was that the model much better matches experimental data when the first order differential operators are replaced by fractional derivatives of Caputo type. Hence, as proposed in Section 4.1 of [602], the model takes the form

$$^{\mathrm{C}}D^{\beta} b(t) = cb(t)s(t) - mb(t), \tag{8.2.1a}$$

$$^{\mathrm{C}}D^{\sigma} s(t) = -kb(t)s(t), \tag{8.2.1b}$$

$$^{\mathrm{C}}D^{\epsilon} e(t) = pb(t)s(t). \tag{8.2.1c}$$

Remark 8.2. The model describes the evolution of three objects of significantly different characters. It is therefore natural to propose three different orders for the differential operators. This differs from our approach in Section 8.1 where, for example, all equations describing the development of the sub-species of the human beings were postulated to have the same order.

Another difference to the model of Section 8.1 is that the system (8.2.1) has no such property as a conservation of mass. This is natural because during the reaction other products are created as well; however, these products are not relevant for the process and hence the way in which they evolve is not explicitly modeled.

As described above, the parameter $m > 0$ is the mortality rate of the biomass, and the positive coefficients c, k and p denote the growth rate of the biomass, the consumption rate of the substrate (note that no new substrate is generated), and the formation rate of the final product ethanol, respectively. Since the differential operators are not assumed to be of first order, all these rates need to be interpreted in a generalized sense; in particular the units in which they are given involve fractional orders of the time unit as indicated in Table 8.2 because otherwise the units on the left-hand sides of the equations would not match those on the right-hand sides.

Remark 8.3. The model (8.2.1) is somewhat simplistic in that it assumes

Table 8.2 Units of the quantities used in the fermentation model (8.2.1).

Quantity	m	c	k	p
Unit	$1/\mathrm{s}^{\beta}$	$1/(\mathrm{g}\cdot\mathrm{s}^{\beta})$	$1/(\mathrm{g}\cdot\mathrm{s}^{\sigma})$	$1/(\mathrm{g}\cdot\mathrm{s}^{\epsilon})$

a constant mortality rate m for the biomass. In reality, one often observes [484] that the bacteria can only survive if the concentration of the alcohol remains below a certain threshold. Thus, a more realistic model would include a term that depends on $e(t)$ in Eq. (8.2.1a). However, as long as one is interested in the behaviour of the process during the time interval before this threshold is reached, the equation system (8.2.1) appears to be able to reproduce all relevant features.

It is clear that suitable initial conditions must be given in order to assert the uniqueness of the solution. These conditions evidently take the form

$$b(0) = b_0, \tag{8.2.2a}$$

$$s(0) = s_0, \tag{8.2.2b}$$

$$e(0) = 0 \tag{8.2.2c}$$

where the last equation reflects the fact that the product to be manufactured, the bio-ethanol, is not present when the process starts. The quantities b_0 and s_0, *i. e.* the concentrations of the input materials (biomass and substrate) at the beginning of the process, are of course assumed to be known to the user who sets up the process.

8.2.2 *A concrete example and its key properties*

In order to present some realistic results of this simulation, Toledo-Hernandez *et al.* [602] have fitted the parameters of the model (8.2.1) so that they provide a good match to the experimental results of Pinilla *et al.* [484]. We report the values that they have found to be suitable for this specific experiment (*i. e.* the type of reactor used in [484], the special kind of bacteria used, the environmental parameters such as pH value, temperature, *etc.*) in Table 8.3. In particular, they found that the mortality rate m of the bacteria is zero. This reflects the facts that the process is not running for such a long time that a significant number of bacteria will die and, in particular, that the concentration of the bio-ethanol remains below the threshold that the bacteria can tolerate.

Table 8.3 Fitted parameters of Toledo-Hernandez *et al.* for modeling the experiment of Pinilla *et al.* for the production of bio-ethanol from sugarcane using the model (8.2.1).

β	0.5706	c	$24.3 \cdot 10^{-6}$ 1/(g·s$^{\beta}$)	m	0 s$^{-\beta}$
σ	0.2404	k	$83.2 \cdot 10^{-3}$ 1/(g·s$^{\sigma}$)		
ϵ	0.3938	p	$7.75 \cdot 10^{-3}$ 1/(g·s$^{\epsilon}$)		

Using the parameters from Table 8.3, we have computed the solution to the system (8.2.1) subject to the specific choices of the initial values for (8.2.2) given in Table 8.4. The computations were performed using the predictor-corrector method introduced in Subsection 2.3.2. In accordance with the experiments conducted in [484], we have decided to simulate the behaviour of the system for a time period of 60 hours. The time step size was chosen to be one second, so we have taken a total of $60^3 = 216000$ time steps. The results are shown in Figures 8.2 to 8.7.

Table 8.4 Initial values for biomass and substrate concentrations in the example simulations.

b_0	s_0	Results shown in
0.01 g/l	120 g/l	Figure 8.2
0.02 g/l	120 g/l	Figure 8.3
0.01 g/l	160 g/l	Figure 8.4
0.02 g/l	160 g/l	Figure 8.5
0.01 g/l	200 g/l	Figure 8.6
0.02 g/l	200 g/l	Figure 8.7

The graphs shown in Figures 8.2 to 8.7 indicate a monotonic growth of the concentrations of biomass and bio-ethanol and a monotonic decrease of the substrate concentration. This is in accordance with what one would expect from the basic considerations underlying the development of the model (8.2.1).

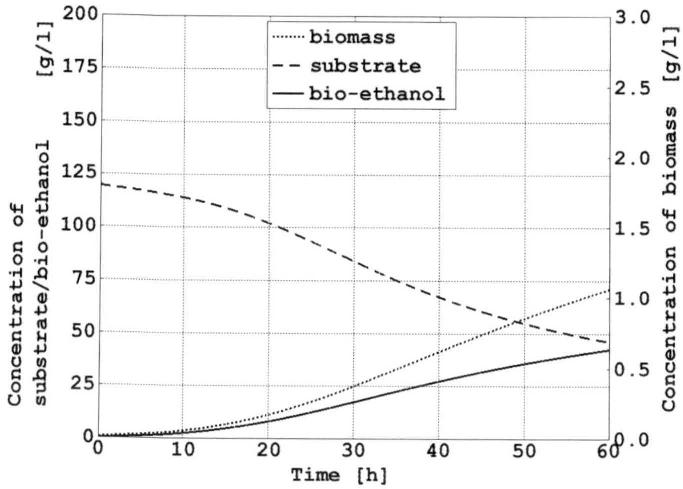

Fig. 8.2 Evolution of concentrations of biomass, substrate and bio-ethanol for model (8.2.1) with initial values chosen as $b_0 = 0.01$ g/l and $s_0 = 120$ g/l.

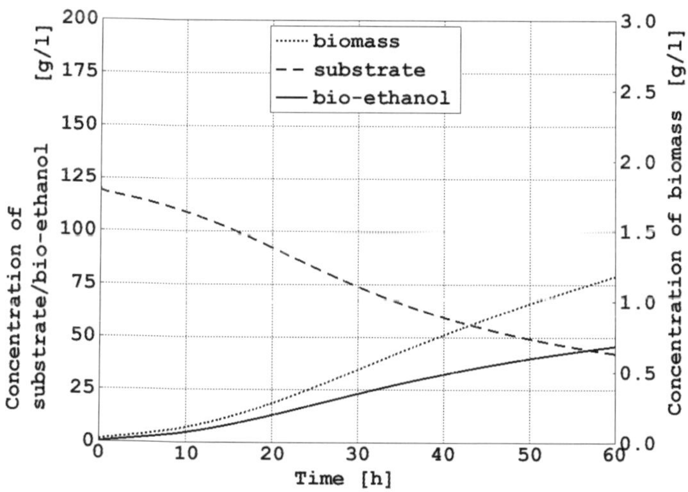

Fig. 8.3 Evolution of concentrations of biomass, substrate and bio-ethanol for model (8.2.1) with initial values chosen as $b_0 = 0.02$ g/l and $s_0 = 120$ g/l.

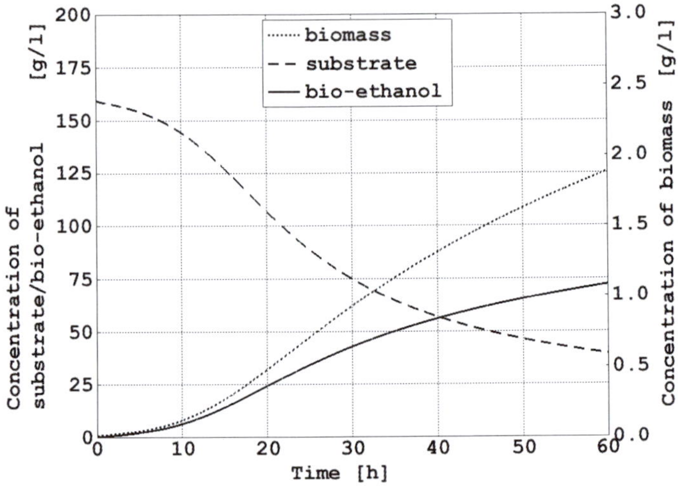

Fig. 8.4 Evolution of concentrations of biomass, substrate and bio-ethanol for model (8.2.1) with initial values chosen as $b_0 = 0.01$ g/l and $s_0 = 160$ g/l.

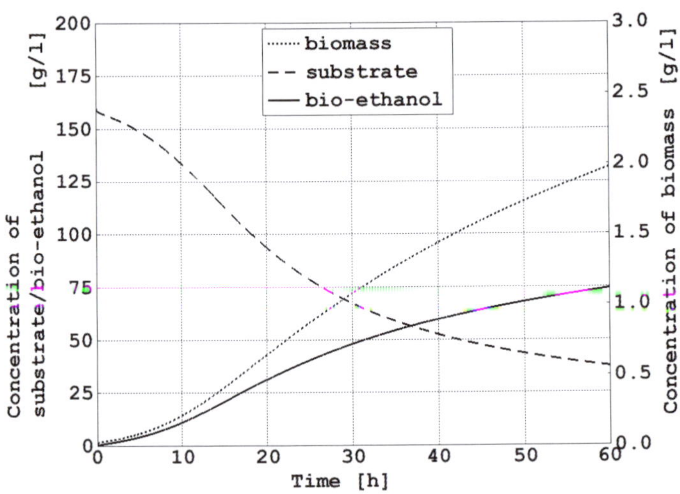

Fig. 8.5 Evolution of concentrations of biomass, substrate and bio-ethanol for model (8.2.1) with initial values chosen as $b_0 = 0.02$ g/l and $s_0 = 160$ g/l.

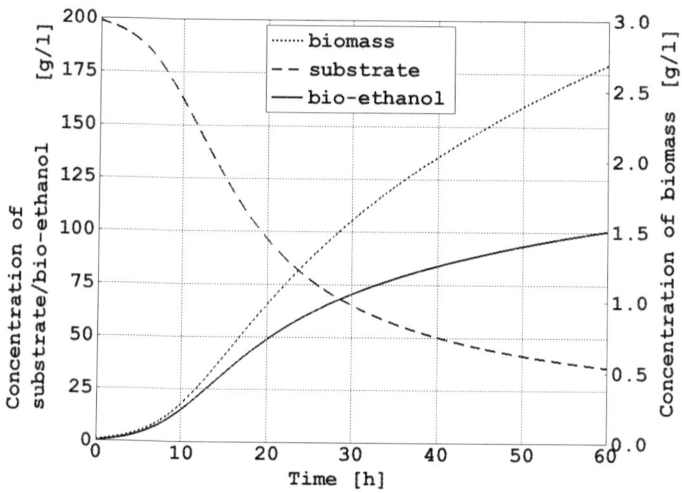

Fig. 8.6 Evolution of concentrations of biomass, substrate and bio-ethanol for model (8.2.1) with initial values chosen as $b_0 = 0.01$ g/l and $s_0 = 200$ g/l.

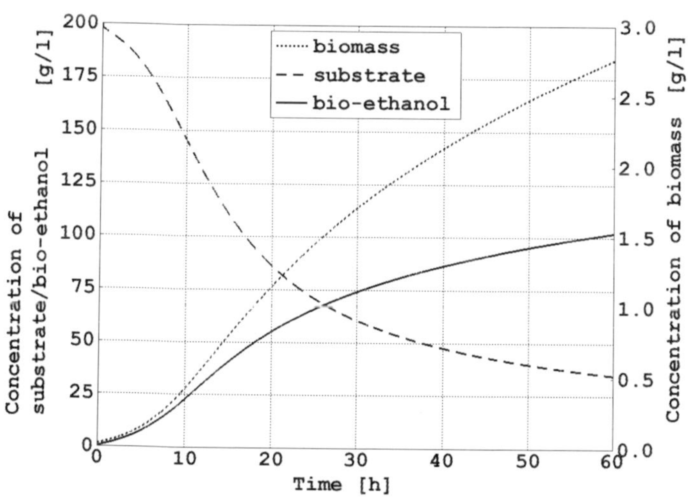

Fig. 8.7 Evolution of concentrations of biomass, substrate and bio-ethanol for model (8.2.1) with initial values chosen as $b_0 = 0.02$ g/l and $s_0 = 200$ g/l.

An alternative perspective and some other forms of insight into the behaviour of the process are given in the representation of Figures 8.8, 8.9 and 8.12. In each of these figures, we have plotted one state variable of the system (biomass concentration, substrate concentration or bio-ethanol concentration) for all six combinations of the initial values that we consider.

From the first of these graphics, *viz.* Figure 8.8, we conclude that an increase in the initial biomass concentration leads to a higher concentration of the biomass throughout the entire process. Moreover, the graph tells us that an increase in the substrate concentration also results in a higher concentration of the biomass over the full time interval. Since the reproduction rate of the bacteria is proportional to both their own concentration and the substrate concentration, this behaviour is completely expected.

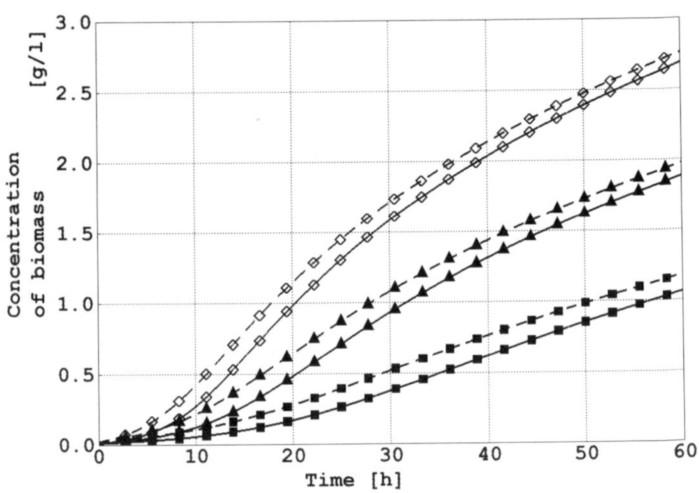

Fig. 8.8 Evolution of biomass concentration for model (8.2.1) with different choices of initial values: $b_0 = 0.01$ g/l (continuous lines) and $b_0 = 0.02$ g/l (dashed lines) in combination with $s_0 = 120$ g/l (filled squares), $s_0 = 160$ g/l (filled triangles) and $s_0 = 200$ g/l (non-filled diamonds).

The second set of graphs of this group, Figure 8.9, demonstrates the evolution of the substrate concentration. Its dependence on the initial concentrations of biomass and substrate is less distinctive. We can observe that a higher biomass concentration at the beginning (and hence, in view of the observation from Figure 8.8, during the complete process) causes a lower substrate concentration over the whole interval of time. This is natural since a large number of bacteria will convert substrate into bio-ethanol more rapidly than a smaller population. However, we can also see that the difference in substrate concentrations for two simulations with identical initial values s_0 and different inital biomass concentrations grows in the early part of the process and then decreases again. Figure 8.10 clarifies this effect by showing the curves for $s_0 = 160$ g/l and the two different choices of b_0 together with the graph of the difference function.

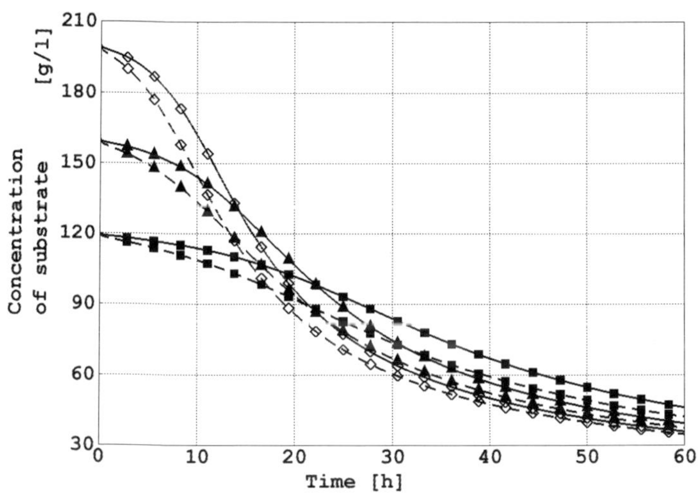

Fig. 8.9 Evolution of substrate concentration for model (8.2.1) with different choices of initial values: $b_0 = 0.01$ g/l (continuous lines) and $b_0 = 0.02$ g/l (dashed lines) in combination with $s_0 = 120$ g/l (filled squares), $s_0 = 160$ g/l (filled triangles) and $s_0 = 200$ g/l (non-filled diamonds).

The influence of the initial substrate concentration s_0 on the development of the substrate concentration is even less evident. Our graphs show that an initially large concentration decreases much more rapidly than a concentration that was small originally. At some point in time, the curves intersect and thus the roles are reversed. Later, the curves once again approach each other; possibly they cross each other again at a time instance outside of the interval where we have simulated the process. This property is better visualized in Figure 8.11 that displays only the curves for an initial biomass concentration of 0.01 g/l and the three choices for s_0.

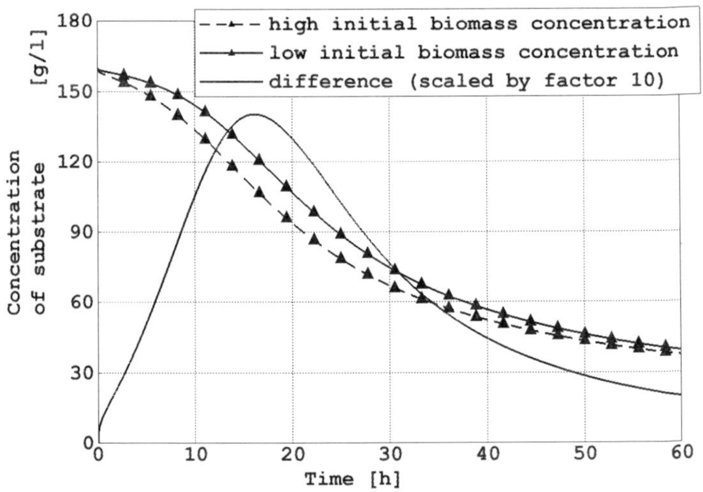

Fig. 8.10 Evolution of substrate concentration for model (8.2.1) with initial values $b_0 = 0.01$ g/l (continuous line, filled triangles) and $b_0 = 0.02$ g/l (dashed line, filled triangles) in combination with $s_0 = 160$ g/l. The third curve (without markers) indicates the difference between the first two curves; for clarity of presentation it has been scaled by a factor of 10.

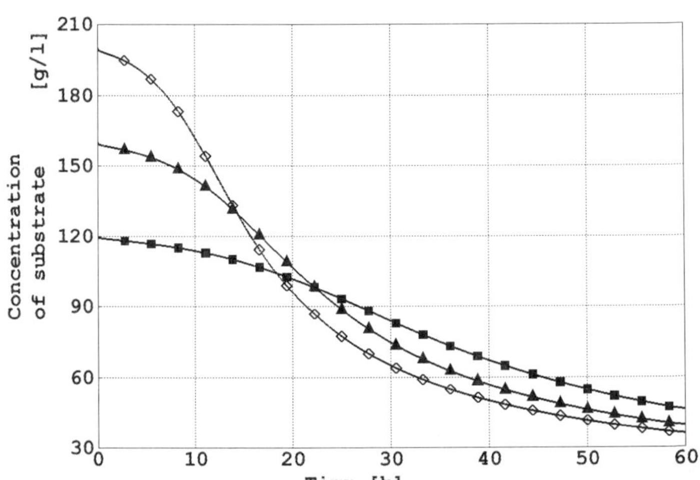

Fig. 8.11 Evolution of substrate concentration for model (8.2.1) with initial values $s_0 = 120$ g/l (filled squares), $s_0 = 160$ g/l (filled triangles) and $s_0 = 200$ g/l (non-filled diamonds) in combination with $b_0 = 0.01$ g/l.

The most important quantity in this process is of course the concentration of the product that is to be manufactured in the process, *i. e.* the bio-ethanol. Figure 8.12 shows its evolution for our six combinations of s_0 and b_0. We observe the following properties:

- A higher initial concentration of the substrate leads to a higher concentration of bio-ethanol throughout the process.
- A higher initial concentration of the biomass also leads to a higher concentration of bio-ethanol throughout the process.
- An increase of the initial substrate concentration influences the final result much more than a raise in the initial biomass concentration.
- In the early and middle part of our time interval, the positive effect on the bio-ethanol production caused by increasing the initial biomass concentration is larger than in the final part (see Figure 8.13).
- In contrast to this, the positive effect on the bio-ethanol production caused by increasing the initial substrate concentration seems to be persistent; in fact, Figure 8.14 indicates that the difference seems to become essentially stationary after a certain amount of time has passed.

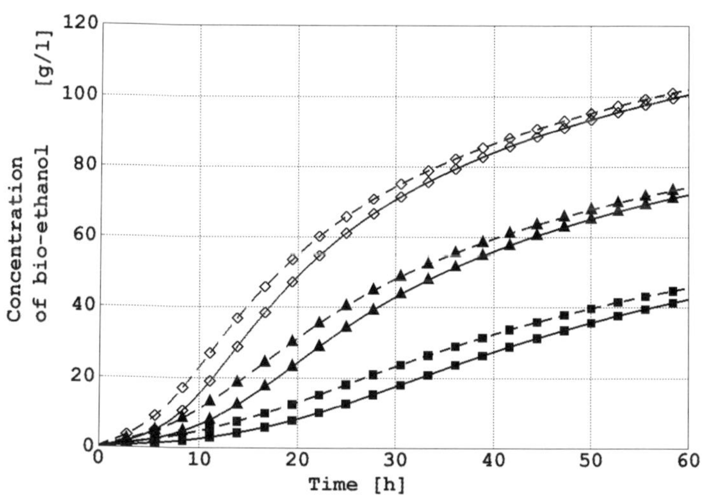

Fig. 8.12 Evolution of bio-ethanol concentration for model (8.2.1) with different choices of initial values: $b_0 = 0.01$ g/l (continuous lines) and $b_0 = 0.02$ g/l (dashed lines) in combination with $s_0 = 120$ g/l (filled squares), $s_0 = 160$ g/l (filled triangles) and $s_0 = 200$ g/l (non-filled diamonds).

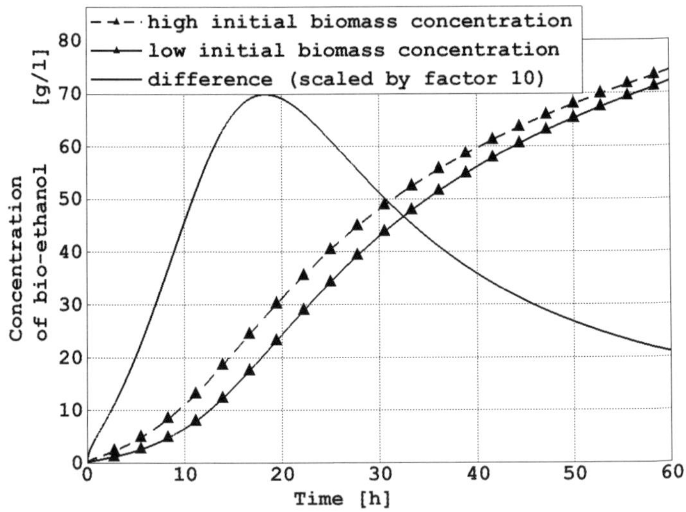

Fig. 8.13 Evolution of bio-ethanol concentration for model (8.2.1) with initial values $b_0 = 0.01$ g/l (continuous line, filled triangles) and $b_0 = 0.02$ g/l (dashed line, filled triangles) combined with $s_0 = 160$ g/l. The curve without markers indicates the difference between the first two curves; for clarity of presentation it has been scaled by a factor of 10.

Remark 8.4. As pointed out by Toledo-Hernandez *et al.* [602], the system (8.2.1) can also be used to model other production processes based on fermentations such as, *e. g.*, the manufacturing of alcoholic beverages like Tequila. The paper [602] provides a comparison of simulations based on the system (8.2.1) with data measured in experiments [39] and concludes that (8.2.1) yields a much better matching data set than the corresponding integer order system.

Remark 8.5. In this section we have recalled the development of a fractional calculus based model of fermentation provided by Toledo-Hernandez *et al.* [602]. Based on this model, the paper [603] proceeds one step further and discusses methods for controlling the fermentation process with the goal of optimizing the concentration of the final product bio-ethanol at the end of the time interval during which the reaction is running. In this extension, it is assumed that additional substrate can be fed into the reactor at a rate that can be determined by the user. We shall not dwell into this issue here and only refer to the original work [603].

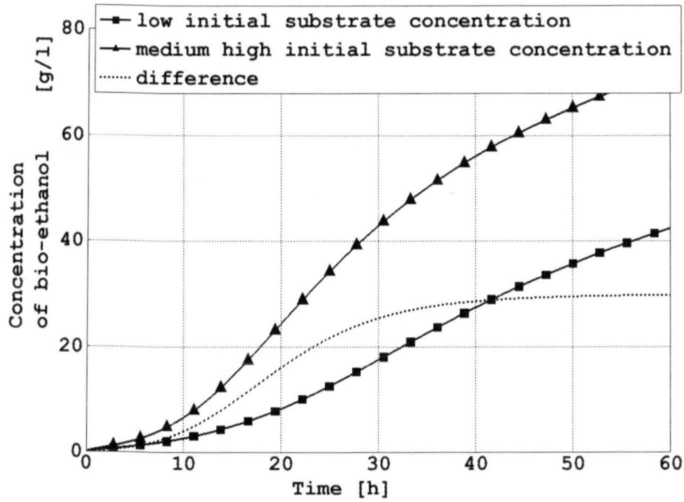

Fig. 8.14 Evolution of bio-ethanol concentration for model (8.2.1) with initial values $s_0 = 120$ g/l (filled squares) and $s_0 = 160$ g/l (filled triangles) in combination with $b_0 = 0.01$ g/l and the associated difference function (dotted line).

8.3 The Evolution of the Concentration of a Drug in the Human Body

Our third and last example application of fractional calculus in the life sciences discusses yet another completely different scenario. Specifically, we look at a human being who takes a certain drug and describe a model for the way in which the active pharmaceutical ingredients of the drug are being absorbed and processed inside the person's body.

8.3.1 *The classical compartmental model of pharmacokinetics*

Our application example in this section is based on the ideas of Dokoumetzidis *et al.* [206]. We start by recalling a classical *compartmental model* of pharmacokinetics; later on we shall extend this model to include an additional feature.

In the model, schematically depicted in Figure 8.15, we assume that a dose of a drug is instantaneously brought into the human organism via an intravenous injection (a so-called bolus). In order to describe what happens to the drug, it has turned out to be useful to divide those parts of the human body into which the drug enters into two categories, usually denoted as *compartments*. The first compartment, commonly known as the central compartment, consists of the general circulation system and the surrounding well perfused tissues. This is also the part into which the drug is directly injected. The second compartment contains the more peripherally located deeper tissues that are less well perfused.

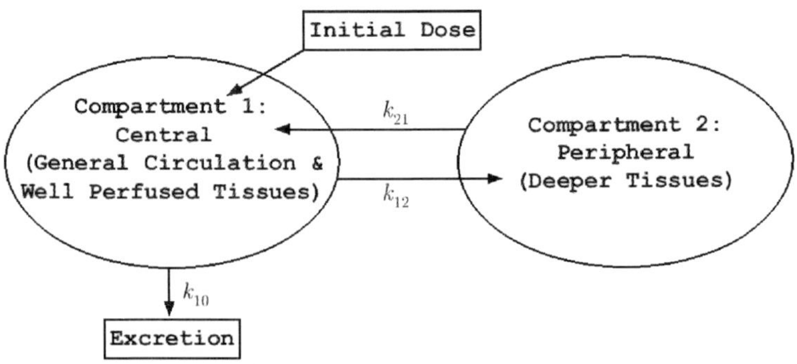

Fig. 8.15 Schematic representation of two-compartmental model of pharmacokinetics, *e. g.* for an intravenously administered drug.

In such a process, it can be observed that a drug transfer takes place from the central to the peripheral compartment and vice versa. Moreover, a part of the drug that is in the central compartment leaves the organism via an excretion process; no such elimination of material happens from the peripheral compartment.

Denoting the amount of material, *i. e.* the mass of the drug's active ingredient, in the compartment j at time t by $A_j(t)$, the classical mathematical model then reads

$$D^1 A_1(t) = -k_{12}A_1(t) + k_{21}A_2(t) - k_{10}A_1(t), \tag{8.3.1a}$$

$$D^1 A_2(t) = \quad k_{12}A_1(t) - k_{21}A_2(t), \tag{8.3.1b}$$

subject to the initial conditions

$$A_1(0) = A^*, \qquad A_2(0) = 0, \tag{8.3.2}$$

where D^1 denotes the classical differential operator of order 1; *cf.* pp. 508–509 of [206]. In this model, the constants k_{ij} describe the mass transfer from compartment i to compartment j (a notation that is interpreted in the reverse way by some authors), in the case $j = 0$ we interpret them as the rate of excretion of the material. The initial condition on A_2 reflects the fact that, at the beginning of the process, no active ingredient is present in the second (peripheral) compartment.

Remark 8.6. A crucial point that needs to be handled very carefully later on is the fact that the coefficients k_{ij} for $i, j = 1, 2$ appear in both equations of the system (8.3.1) and that their signs in Eq. (8.3.1a) are opposite to the signs in Eq. (8.3.1b). This property of the system reflects the fact that the flux coming into compartment j from compartment i must be the same as the flux going out of compartment i towards compartment j.

8.3.2 *Derivation of a fractional order model*

From the standard theory of such differential equation systems of first order, it is well known that the solution can be written in terms of exponential functions. This implies a quite rapid decay of the quantities $A_1(t)$ and $A_2(t)$ as t becomes large. However, such a property can not always be observed in real experiments; rather, one frequently measures an algebraic decay which, from the theory of fractional calculus, is known to be a natural feature of fractional order operators. Following the development of pp. 509–512 of [206], we therefore describe how to rewrite the system in a way that allows to replace the first order differential operators by suitable fractional order generalizations. As pointed out in [205], a naive approach such as the one proposed in [494] can easily lead to a fractional model that violates dimensional consistency (this is an issue that we also had to consider in the fractionalization of the models in Sections 8.1 and 8.2) and the property discussed in Remark 8.6. Thus, one needs to proceed with more care here.

The first step in the generalization of the model (8.3.1) consists in integrating the differential equations and incorporating the initial conditions. This leads to the system

$$A_1(t) = A^* - k_{12} J^1 A_1(t) + k_{21} J^1 A_2(t) - k_{10} J^1 A_1(t), \qquad (8.3.3a)$$

$$A_2(t) = \qquad k_{12} J^1 A_1(t) - k_{21} J^1 A_2(t). \qquad (8.3.3b)$$

where J^α denotes the Riemann-Liouville integral operator of order α, *cf.* Eq.

(1.3.8). In this system, it is now possible to replace the first order integral operators by fractional order operators in such a way that all the consistency conditions mentioned above are observed. Specifically, the orders of the operators that describe the transfer from compartment i to compartment j may be chosen as $\alpha_{ij} \in (0,1]$, but this choice needs to be applied in both equations of the system (8.3.3). This allows us to choose α_{ij} different from α_{ji} if we wish to do so, *i. e.* we may set up the model such that the transfer from compartment i to compartment j does not behave in the same manner as the transfer in the opposite direction — a property that is indeed observed in experiments. We are thus lead to the fractional counterpart of (8.3.3), namely

$$A_1(t) = A^* - k_{12}J^{\alpha_{12}}A_1(t) + k_{21}J^{\alpha_{21}}A_2(t) - k_{10}J^{\alpha_{10}}A_1(t), \quad (8.3.4a)$$

$$A_2(t) = \qquad k_{12}J^{\alpha_{12}}A_1(t) - k_{21}J^{\alpha_{21}}A_2(t). \quad (8.3.4b)$$

Whereas the coefficients k_{ij} in Eq. (8.3.1) had to be interpreted as classical rates, *i. e.* with units s^{-1}, they must be seen in this *fractional compartmental model* (8.3.4) as generalized rates with units $s^{-\alpha_{ij}}$ in order to retain the dimensional consistency between the left-hand sides and the right-hand sides of the equations.

In practice, the experimental data often lead to the observation that the transfer out of the central compartment is governed by a classical first order equation, no matter whether the drugs are transported to the peripheral compartment or to the outside of the organism, whereas the flux from the peripheral compartment back to the central compartment exhibits slower kinetics due to the fact that the ingredients are trapped in the deeper layers of the tissue, and so this part of the process needs to be modeled by a fractional order equation (see page 513 of [206]).

To demonstrate the effects of the fractionalization of the model, we have numerically computed the solutions to the fractional and to the classical versions. The numerical algorithm is simply based on the (fractional or classical) trapezoidal approximation (as in Eqs. (2.1.8) and (2.1.9)) of the integral operators appearing on the right-hand sides of the model equations with a step size of 1/200.

A comparison is provided in Figures 8.16 to 8.19. Specifically, we have solved the classical system (8.3.1) with parameters $A^* = 1$, $k_{10} = 0.5$, $k_{12} = 0.75$ and $k_{21} = 0.6$ and the fractional system (8.3.4) with the same parameters and $\alpha_{10} = \alpha_{12} = 1$ and $\alpha_{21} = 0.7$ which is in line with the above mentioned observations. It should be noted in this context that the quantities k_{21} that appear in both systems are not immediately comparable; they have the same numerical value but different units ($s^{-\alpha_{21}}$ with two different values of α_{21}). Since it is then not clear whether the two systems can be directly compared to each other, we have also included a third system that is based on our fractional system but has $k_{21} = 0.6^{\alpha_{21}} = 0.6^{0.7}$ instead. Our numerical results indicate that this change in the parameter k_{21} for the fractional order system has a very small effect only.

Our first graphs, given in Figures 8.16 and 8.17, show the numerical results for the functions A_1 and A_2, respectively, *i. e.* for the amounts of the drug in each of the two compartments; we look at an initial time interval up to $t_{\max} = 25$ in these two figures. The vertical axis has a logarithmic scale.

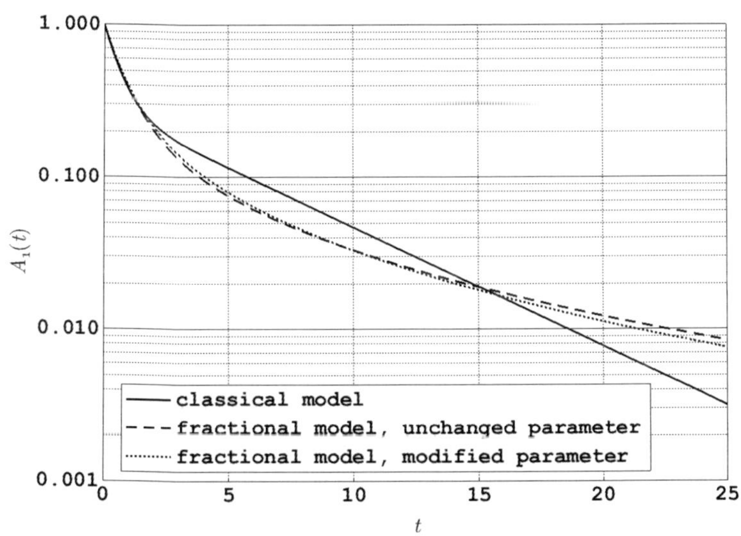

Fig. 8.16 Numerical approximation of the contents of the dose of the drug in the central compartment according to the classical system (8.3.1) with $A^* = 1$, $k_{10} = 0.5$, $k_{12} = 0.75$ and $k_{21} = 0.6$ (continuous line) and corresponding results for the fractional system (8.3.4) with the same parameters and $\alpha_{10} = \alpha_{12} = 1$ and $\alpha_{21} = 0.7$ (dashed line) and with $k_{21} = 0.6^{\alpha_{21}}$ but otherwise as in the previous case (dotted line).

It turns out that in the initial phase of the process the difference between

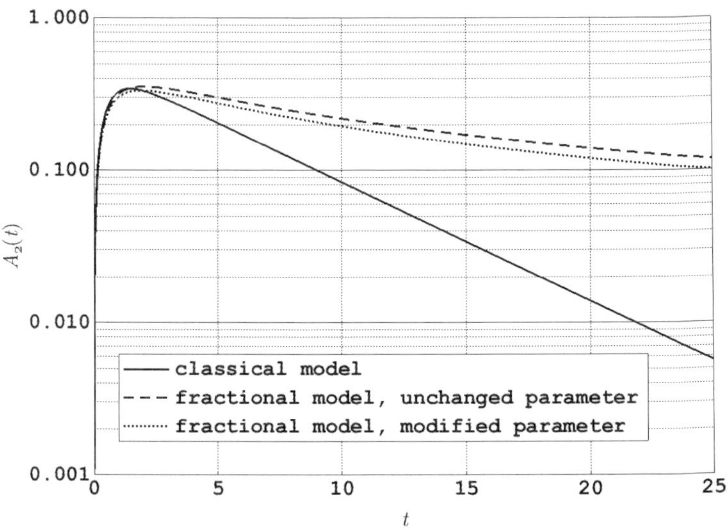

Fig. 8.17 Numerical approximation of the contents of the dose of the drug in the peripheral compartment according to the classical system (8.3.1) with $A^* = 1$, $k_{10} = 0.5$, $k_{12} = 0.75$ and $k_{21} = 0.6$ (continuous line) and corresponding results for the fractional system (8.3.4) with the same parameters and $\alpha_{10} = \alpha_{12} = 1$ and $\alpha_{21} = 0.7$ (dashed line) and with $k_{21} = 0.6^{\alpha_{21}}$ but otherwise as in the previous case (dotted line).

classical and fractional models is rather small. Later on, however, we see in Figure 8.17 that the fractional model predicts a significantly slower decay of the amount of active ingredients of the drug in the peripheral compartment (compartment 2) than the classical model. For the central compartment that is described by the function A_1 we note a similar effect in Figure 8.16 but it sets in later and is less prominent than in the peripheral compartment. Moreover, for the central compartment there is a clearly visible intermediate phase during which the fractional models predicts a slightly larger drug content than the classical model.

Figures 8.18 and 8.19 essentially show the same data as the two previous figures, but for a larger time interval up to $t_{\max} = 100$ and using a logarithmic scale on both axes.

The graphs indicate the long term behaviour of our functions A_1 and A_2. For the function A_2 that describes the peripheral compartment, Figure 8.19 provides a confirmation of the trend that was already evident in Figure 8.17: The drug concentration in the fractional model decays much slower than

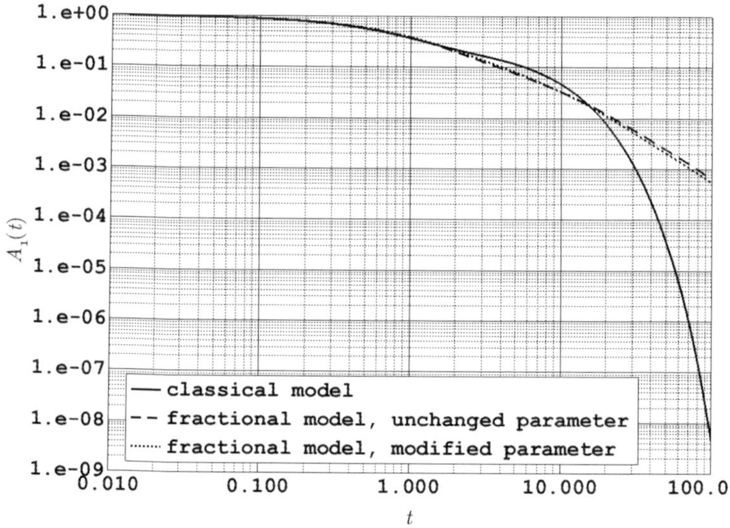

Fig. 8.18 Numerical approximation of the contents of the dose of the drug in the central compartment according to the classical system (8.3.1) with $A^* = 1$, $k_{10} = 0.5$, $k_{12} = 0.75$ and $k_{21} = 0.6$ (continuous line) and corresponding results for the fractional system (8.3.4) with the same parameters and $\alpha_{10} = \alpha_{12} = 1$ and $\alpha_{21} = 0.7$ (dashed line) and with $k_{21} = 0.6^{\alpha_{21}}$ but otherwise as in the previous case (dotted line).

in the classical model. From Figure 8.18, we see that the same property is true for the central compartment. However, this asymptotic behaviour sets in somewhat later than in the peripheral department, and therefore this feature of the models was not so clearly apparent in the smaller time interval displayed in Figure 8.16.

The fact that we have logarithmic scales on both axes in Figures 8.18 and 8.19 actually provides us with some additional insight. We can see that the graphs corresponding to the fractional models show a linear behaviour in the log-log plot for sufficiently large times. Thus, the decay properties are in fact algebraic as one would expect from a fractional order model and hence much slower than the exponential decay that is inherent in a model based on differential equations of first order. Hence our numerical results once again confirm the observation that fractional order models are better suited for describing processes that exhibit algebraic asymptotics than integer order models.

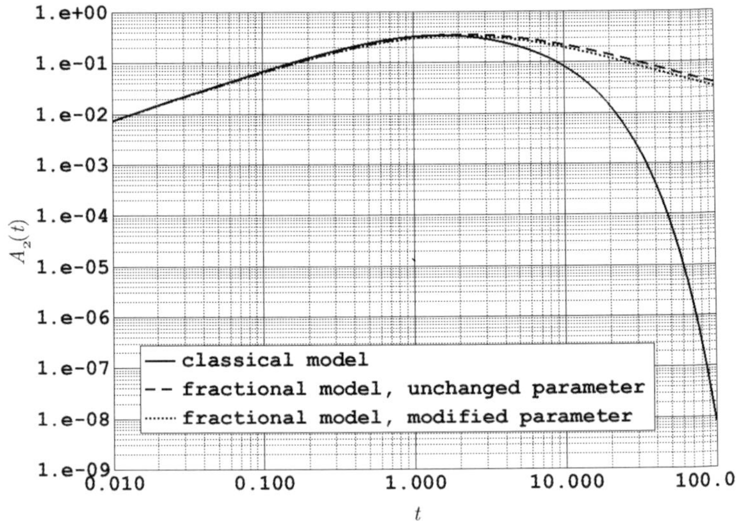

Fig. 8.19 Numerical approximation of the contents of the dose of the drug in the peripheral compartment according to the classical system (8.3.1) with $A^* = 1$, $k_{10} = 0.5$, $k_{12} = 0.75$ and $k_{21} = 0.6$ (continuous line) and corresponding results for the fractional system (8.3.4) with the same parameters and $\alpha_{10} = \alpha_{12} = 1$ and $\alpha_{21} = 0.7$ (dashed line) and with $k_{21} = 0.6^{\alpha_{21}}$ but otherwise as in the previous case (dotted line).

8.3.3 *Further generalizations*

The approach described in Subsections 8.3.1 and 8.3.2 can be extended to cover the case of more than two compartments in an obvious manner. For example, if we want to assume the drug to be given orally, it is necessary to add another compartment. This third compartment represents the gut in which the dose of the drug is initially released. From there, the active ingredients are then dissolved and dispersed into the central compartment; no transfer of the ingredients into the other direction occurs. The interaction between central and peripheral compartments and the excretion of the drug are the same as in the original model that we discussed in detail in the previous subsections. A schematic view is given in Figure 8.20; more details are also stated in [206].

In Figure 8.20 we see that there are some compartments between which the interaction is either only unidirectional (compartments 1 and 2) or completely absent (compartments 1 and 3). As a matter of fact, this is a

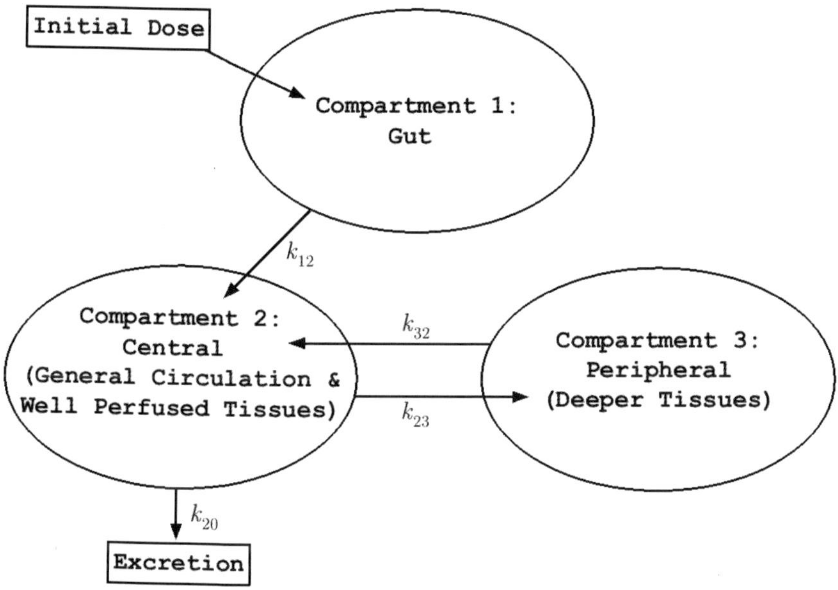

Fig. 8.20 Schematic representation of three-compartmental model of pharmacokinetics, *e. g.* for an orally administered drug.

property that can be observed in many multi-compartmental models; in fact it is quite typical for such models to have many compartments interacting with a small number of other compartments only and to have only a few compartments interacting with many other compartments. Clearly, this feature simplifies the equation system corresponding to the model since many of the coefficients k_{ij} have the value zero. For our specific three compartment model, the approach described in the previous subsections leads us to a fractional order equation system then takes the form

$$A_1(t) = A^* - k_{12}J^{\alpha_{12}}A_1(t), \tag{8.3.5a}$$

$$A_2(t) = k_{12}J^{\alpha_{12}}A_1(t) - k_{20}J^{\alpha_{20}}A_1(t) - k_{23}J^{\alpha_{23}}A_2(t) + k_{32}J^{\alpha_{32}}A_3(t), \tag{8.3.5b}$$

$$A_3(t) = k_{23}J^{\alpha_{23}}A_2(t) - k_{32}J^{\alpha_{32}}A_3(t) \tag{8.3.5c}$$

and that can be handled in a manner that is very much analog to our approach stated above.

Appendix A

Source Codes

In this appendix, we have collected the source codes of a number of the most important numerical methods for fractional differential equations. This appendix is by no means meant to be comprehensive or complete in any sense: our goal is to provide the reader with a selection of generally applicable easy-to-use programs that should be helpful in a large number of potential applications.

Whereas most of the programs are written in FORTRAN (mainly, standard FORTRAN77), *i. e.* in a language for which compilers are generally available on almost every conceivable hardware environment, we have also included some Matlab [599], Mathematica [632] and R [600] codes for users who prefer these platforms.

A.1 The Adams-Bashforth-Moulton Method

We begin with the standard version of the Adams-Bashforth-Moulton method with a uniform mesh. The code is divided into a number of subroutines. The first of these computes the required weights.

Algorithm A.1. Computation of the weights of the method. This routine requires an external routine `GAMMA` (such as, *e. g.*, the one provided in [155]) for the evaluation of Euler's Gamma function.

```
      SUBROUTINE ABMW(ALPHA, APRED, ACORR, BCORR, CCORR, N, IERR)
C
C     CALCULATION OF WEIGHTS FOR FRACTIONAL
C     ADAMS-BASHFORTH-MOULTON METHOD
```

```
C
C     INPUT VARIABLES
C     N     - NUMBER OF GRID POINTS
C     ALPHA - ORDER OF DIFFERENTIAL OPERATOR
C
C     OUTPUT VARIABLES
C     APRED  - PREDICTOR WEIGHTS
C     ACORR  - FUNDAMENTAL CORRECTOR WEIGHTS
C     BCORR  - START-POINT CORRECTOR WEIGHTS
C     CCORR  - END-POINT CORRECTOR WEIGHT
C     IERR   - ERROR CODE
C                0 : SUCCESSFUL EXIT
C               12 : ALPHA < 0
C
      IMPLICIT NONE
C
      INTEGER          N, IERR
      DOUBLE PRECISION ALPHA
      DOUBLE PRECISION APRED(N), ACORR(N), BCORR(N), CCORR
C
      INTEGER          J
      DOUBLE PRECISION GMALP1, GMALP2
C
      DOUBLE PRECISION GAMMA
      EXTERNAL         GAMMA
C
      IERR = 0
      IF (ALPHA .LE. 0.D0) IERR = 12
      IF (IERR .NE. 0) RETURN
C
      GMALP1 = GAMMA(ALPHA + 1.D0)
      GMALP2 = GMALP1 * (ALPHA + 1.D0)
C
      CCORR = 1.D0 / GMALP2
C
      DO 100 J = 1, N
         ACORR(J) = ((J+1)**(ALPHA+1.D0) + (J-1)**(ALPHA+1.D0)
     &              - 2 * J**(ALPHA+1.D0)) * CCORR
         APRED(J) = (J**ALPHA - (J-1)**ALPHA) / GMALP1
         BCORR(J) = ((J-1)**(ALPHA+1.D0) - (J-1-ALPHA) * J**ALPHA)
     &              * CCORR
```

```
  100 CONTINUE
C
      RETURN
      END
```

The next piece of code demonstrates the application of the Adams-Bashforth-Moulton method, using the weights computed above, by means of a concrete example, namely the initial value problem

$$^{C}D^{1.3}y(x) = x^{0.7}E_{1,1.7}(-x) + \exp(-2x) - [y(x)]^2, \quad y(0) = 1,\, y'(0) = -1,$$

where $E_{1,1.7}$ denotes the two-parameter Mittag-Leffler function with parameters 1 and 1.7. The exact solution of this equation is

$$y(x) = \exp(-x).$$

The following program solves this initial value problem with the Adams-Bashforth-Moulton method on the interval $[0, 4]$. The code required here for the evaluation of the Mittag-Leffler function that appears on the right-hand side of the differential equation, *i. e.* the routine ML2P, is given in Appendix A.4 below (see Algorithm A.8).

Algorithm A.2. Standard version of the Adams-Bashforth-Moulton method. The algorithm uses the routine ABMW from Algorithm A.1 to compute the weights of the method.

```
      SUBROUTINE RHS(X, Y, F)
      IMPLICIT NONE
      DOUBLE PRECISION X, Y, F
C     COMPUTE THE RIGHT-HAND SIDE OF THE DIFFERENTIAL EQUATION
C
      DOUBLE PRECISION ML2P
      EXTERNAL         ML2P
C
      F = X**0.7D0 * ML2P(1.D0, 1.7D0, -X) + EXP(-2.D0*X) - Y**2
      RETURN
      END
C
C
C
      DOUBLE PRECISION FUNCTION EXACT(X)
      IMPLICIT NONE
```

```
      DOUBLE PRECISION X
C     COMPUTE THE EXACT SOLUTION
C     OF THE DIFFERENTIAL EQUATION AT THE POINT X
C
      EXACT = EXP(-X)
      RETURN
      END
C
C
      SUBROUTINE ABMS(ALPHA, N, XMIN, XMAX, RHS, CORRIT, APRED,
     &                BPRED, ACORR, BCORR, CCORR, Y, Y0, IERR)
C
C     FRACTIONAL ADAMS METHOD FOR SINGLE CAPUTO FDE OF ORDER ALPHA
C
      IMPLICIT NONE
C
      EXTERNAL         RHS
      INTEGER          N, CORRIT, IERR
      DOUBLE PRECISION ALPHA, XMIN, XMAX, APRED(N), BPRED(N),
     &                 ACORR(N), BCORR(N), CCORR, Y0(*), Y(N)
C
C     INPUT VARIABLES (UNCHANGED ON OUTPUT)
C     ALPHA    -  ORDER OF DE, MUST BE > 0.0
C     N        -  NUMBER OF STEPS, MUST BE > 0
C     XMIN     -  LOWER BOUND OF INTERVAL WHERE SOLUTION IS SOUGHT
C     XMAX     -  UPPER BOUND OF INTERVAL WHERE SOLUTION IS SOUGHT,
C                 MUST BE > XMIN
C     RHS      -  GIVEN SUBROUTINE THAT COMPUTES THE RHS OF THE DE
C                 (SEE BELOW FOR DETAILS)
C     CORRIT   -  NUMBER OF REQUESTED CORRECTOR ITERATIONS
C                 (MUST ALWAYS BE >= 1;
C                  IT IS RECOMMENDED TO USE CORRIT = 1
C                  FOR ALPHA >= 1)
C     APRED,BPRED -  PREDICTOR WEIGHTS AS PROVIDED BY WADAMS
C     ACORR,BCORR,CCORR - CORRECTOR WEIGHTS AS PROVIDED BY ABMW
C     Y0       -  INITIAL CONDITIONS;
C                 Y0(J) MEANS (J-1)ST DERIVATIVE OF Y AT XMIN
C
C     OUTPUT VARIABLES
C     Y        -  Y(J) DENOTES APPROXIMATE SOLUTION
C                 AT POINT XMIN + J*(XMAX-XMIN)/N
```

```
C     IERR    - ERROR FLAG; WILL BE 0 ON SUCCESSFUL TERMINATION
C
C     THE SUBROUTINE   RHS   MUST HAVE THE FOLLOWING STRUCTURE:
C           SUBROUTINE RHS(X, Y, F)
C           DOUBLE PRECISION X, Y, F
C           INPUT VARIABLES: X, Y
C           OUTPUT VARIABLE: F = F(X,Y)
C     IT MUST BE DECLARED "EXTERNAL" IN THE CALLING SUBROUTINE.
C
C
      DOUBLE PRECISION H, HALP, T, KFAC, WRK(3)
      INTEGER          J, K, M, CI
C
      IERR = 0
      IF (      N .LE. 0)    IERR = 11
      IF ( ALPHA .LE. 0.D0) IERR = 12
      IF (CORRIT .LE. 0)    IERR = 14
      IF (  XMAX .LE. XMIN) IERR = 15
      IF (IERR .NE. 0) RETURN
C
      H    = (XMAX - XMIN) / N
      HALP = H ** ALPHA
      M    = INT(ALPHA)
      IF (ALPHA .EQ. DBLE(M)) M = M - 1
C
      DO 1000 J = 1, N
         T = XMIN + J * H
C
C        SET INITIAL CONDITION
         WRK(1) = Y0(1)
         KFAC = 1.D0
         DO 50 K = 1, M
            KFAC = KFAC * K
            WRK(1) = WRK(1) + T**K / KFAC * Y0(K+1)
 50      CONTINUE
C
C        PREDICTOR
         CALL RHS(XMIN, Y0(1), WRK(3))
         WRK(2) = WRK(1) + HALP * BPRED(J) * WRK(3)
         DO 70 K = 1, J-1
            CALL RHS(XMIN + (J-K)*H, Y(J-K), WRK(3))
```

```
            WRK(2) = WRK(2) + HALP * APRED(K) * WRK(3)
 70      CONTINUE
C
C        CORRECTORS
         CALL RHS(XMIN, Y0(1), WRK(3))
         WRK(1) = WRK(1) + HALP * BCORR(J) * WRK(3)
         DO 110 K = 1, J-1
            CALL RHS(XMIN + (J-K)*H, Y(J-K), WRK(3))
            WRK(1) = WRK(1) + HALP * ACORR(K) * WRK(3)
 110     CONTINUE
         DO 120 CI = 1, CORRIT
            CALL RHS(T, WRK(2), WRK(3))
            WRK(2) = WRK(1) + HALP * CCORR * WRK(3)
 120     CONTINUE
C
         Y(J) = WRK(2)
C
 1000 CONTINUE
C
      RETURN
      END
C
C
C
      PROGRAM EXAMPL
      IMPLICIT NONE
C
      DOUBLE PRECISION Y0(2), Y(40), XMIN, XMAX, ALPHA, X,
     &                 APRED(40), ACORR(40), BCORR(40), CCORR,
     &                 EXACT
      INTEGER          J, K, N, CORRIT, IERR
      EXTERNAL         RHS, EXACT
C
C     DEFINE PARAMETERS OF EQUATION
      ALPHA = 1.3D0
      XMIN = 0.D0
      XMAX = 4.D0
C
C     DEFINE INITIAL CONDITIONS
      Y0(1) =  1.D0
      Y0(2) = -1.D0
```

```
C
C     DEFINE PARAMETERS OF NUMERICAL METHOD
C     MAXIMUM NUMBER OF GRID POINTS
      N = 40
C     NUMBER OF CORRECTOR ITERATIONS ACCORDING TO RECOMMENDATION
      CORRIT = 1
C
C     COMPUTE WEIGHTS
      CALL ABMW(ALPHA, APRED, ACORR, BCORR, CCORR, N, IERR)
      IF (IERR .NE. 0) THEN
         WRITE(*,'(A,I4)') 'ERROR IN COMPUTATION OF WEIGHTS: ', IERR
         STOP
      END IF
C
C     SOLVE EQUATION NUMERICALLY WITH 10, 20, 40 GRID POINTS
      DO 100 J = 1, 3
         N = 5 * 2**J
         CALL ABMS(ALPHA, N, XMIN, XMAX, RHS, CORRIT,
     &             APRED, APRED, ACORR, BCORR, CCORR, Y, Y0, IERR)
         IF (IERR .NE. 0) THEN
            WRITE(*, '(A,I3,A,I3)') 'ERROR CODE ', IERR,
     &             ' IN DIFFERENTIAL EQUATION SOLVER FOR N =', N
         ELSE
            WRITE(*, '(A)') '=========================='
            WRITE(*, '(A,I3)') 'NUMBER OF GRID POINTS: ', N
            WRITE(*, '(A)')
     &         'X        EXACT VALUE   APPROX. VALUE       ERROR'
            DO 50 K = 1, N
               X = XMIN + (XMAX-XMIN) / DBLE(N) * DBLE(K)
               WRITE (*, '(F6.4,F13.6,F16.6,F12.6)')
     &            X, EXACT(X), Y(K), EXACT(X) - Y(K)
50          CONTINUE
         END IF
100   CONTINUE
      END
```

Next we provide the parallel version of the method as developed in [187]. It uses the OpenMP programming model [466] for the parallelization. The particular example implemented here solves the initial value problem

$$^{\mathrm{C}}D^{\alpha}y(x) = -y(x), \qquad y(0) = 1, \quad y'(0) = 0,$$

on an interval $[0, X]$. The program should be called with three parameters, namely α (the order of the differential operator which must be in the interval $(0, 2]$; the second of the two initial conditions above is only used if $\alpha > 1$), the number of grid points, and X, the right end point of the interval of interest.

Algorithm A.3. OpenMP-based parallel version of the Adams-Bashforth-Moulton method. The algorithm uses the routine `ABMW` from Algorithm A.1 to compute the weights of the method.

```
      SUBROUTINE ABMSP(ALPHA, N, XMIN, XMAX, RHS, CORRIT,
     &                 APRED, BPRED, ACORR, BCORR, CCORR, Y, Y0,
     &                 WRK, IERR)
C
C     FRACTIONAL ADAMS METHOD FOR SINGLE CAPUTO FDE OF ORDER ALPHA
C     PARALLELIZED WITH OPENMP
C
      IMPLICIT NONE
C
      EXTERNAL         RHS
      INTEGER          N, CORRIT, IERR, FNSHD
      DOUBLE PRECISION ALPHA, XMIN, XMAX, APRED(N), BPRED(N),
     &                 ACORR(N), BCORR(N), CCORR, Y0(*), Y(N),
     &                 WRK(N)
C
C     INPUT VARIABLES (UNCHANGED ON OUTPUT)
C     ALPHA    -  ORDER OF DE, MUST BE > 0.0
C     N        -  NUMBER OF STEPS, MUST BE > 0
C     XMIN     -  LOWER BOUND OF INTERVAL WHERE SOLUTION IS SOUGHT
C     XMAX     -  UPPER BOUND OF INTERVAL WHERE SOLUTION IS SOUGHT,
C                 MUST BE > XMIN
C     RHS      -  GIVEN SUBROUTINE THAT COMPUTES THE RHS OF THE DE
C                 (SEE BELOW FOR DETAILS)
C     CORRIT   -  NUMBER OF REQUESTED CORRECTOR ITERATIONS
C                 (MUST ALWAYS BE >= 1;
C                  RECOMMENDATION: USE CORRIT = 1 FOR ALPHA >= 1)
C     APRED,BPRED -  PREDICTOR WEIGHTS AS PROVIDED BY ABMW
C     ACORR,BCORR,CCORR - CORRECTOR WEIGHTS AS PROVIDED BY ABMW
C     Y0       -  INITIAL CONDITIONS;
C                 Y0(J) MEANS (J-1)ST DERIVATIVE OF Y AT XMIN
C
```

```
C     OUTPUT VARIABLES
C     Y        -  Y(J) DENOTES APPROXIMATE SOLUTION
C                 AT POINT XMIN + J*(XMAX-XMIN)/N
C     WRK      -  WORKSPACE
C     IERR     -  ERROR FLAG; WILL BE 0 ON SUCCESSFUL TERMINATION
C
C     THE SUBROUTINE   RHS   MUST HAVE THE FOLLOWING STRUCTURE:
C           SUBROUTINE RHS(X, Y, F)
C           DOUBLE PRECISION X, Y, F
C           INPUT VARIABLES: X, Y
C           OUTPUT VARIABLE: F = F(X,Y)
C     IT MUST BE DECLARED "EXTERNAL" IN THE CALLING SUBROUTINE.
C
      DOUBLE PRECISION H, HALP, T, KFAC, ICND, PRD, CRR, F, W0
      INTEGER          B, J, K, M, CI, I
      INTEGER          NBLKS, OMPNTH,
     &                 OMP_GET_NUM_THREADS, OMP_GET_THREAD_NUM
      EXTERNAL         OMP_GET_NUM_THREADS, OMP_GET_THREAD_NUM
C
      IERR = 0
      IF (      N .LE. 0)     IERR = 11
      IF ( ALPHA .LE. 0.D0) IERR = 12
      IF (CORRIT .LE. 0)     IERR = 14
      IF (  XMAX .LE. XMIN) IERR = 15
      IF (IERR .NE. 0) RETURN
C
      H    = (XMAX - XMIN) / N
      HALP = H ** ALPHA
      M    = INT(ALPHA)
      IF (ALPHA .EQ. DBLE(M)) M = M - 1
C
      CALL RHS(XMIN, Y0(1), W0)
      W0 = W0 * HALP
      FNSHD = 0
C
C
!$OMP    PARALLEL PRIVATE(I, J, K, ICND, KFAC, NBLKS,
!$OMP&                     T, PRD, CRR, CI, F, B)
!$OMP&             SHARED (OMPNTH, N, H, XMIN, Y0, M, APRED,
!$OMP&                     ACORR, BCORR, CCORR, WRK, Y, CORRIT,
!$OMP&                     HALP, FNSHD, W0)
```

```
C
      OMPNTH = OMP_GET_NUM_THREADS()
      NBLKS = CEILING(DBLE(N)/DBLE(OMPNTH))
C
      DO 1000 B = 0, NBLKS - 1
C
!$OMP   DO
C
         DO 2000 I = 1, OMPNTH
            J = B * OMPNTH + I
            IF (J .LE. N) THEN
               T = XMIN + J * H
C
C              SET INITIAL CONDITION
               ICND = Y0(1)
               KFAC = 1.D0
               DO 50 K = 1, M
                  KFAC = KFAC * K
                  ICND = ICND + T**K / KFAC * Y0(K+1)
 50            CONTINUE
C
C              PREDICTOR AND CORRECTOR: FIRST PART OF SUMMATION
               PRD = ICND + BPRED(J) * W0
               CRR = ICND + BCORR(J) * W0
               DO 70 K = J - B * OMPNTH, J - 1
                  PRD = PRD + APRED(K) * WRK(J-K)
                  CRR = CRR + ACORR(K) * WRK(J-K)
 70            CONTINUE
C
               DO 80 K = I-1, 1, -1
                  DO WHILE (FNSHD .LT. J-K)
!$OMP                FLUSH (Y, WRK, FNSHD)
                  END DO
                  PRD = PRD + APRED(K) * WRK(J-K)
                  CRR = CRR + ACORR(K) * WRK(J-K)
 80            CONTINUE
C
C              CORRECTORS: PROPER CORRECTOR ITERATIONS
               DO 120 CI = 1, CORRIT
                  CALL RHS(T, PRD, F)
                  PRD = CRR + HALP * CCORR * F
```

```
 120          CONTINUE
C
              Y(J) = PRD
              CALL RHS(T, Y(J), WRK(J))
              WRK(J) = WRK(J) * HALP
              FNSHD = J
!$OMP         FLUSH (Y, WRK, FNSHD)
C
          ENDIF
C
 2000     CONTINUE
!$OMP     END DO
C
 1000 CONTINUE
!$OMP END PARALLEL
      RETURN
      END
C
C
      SUBROUTINE RHS (X, Y, F)
      IMPLICIT NONE
      DOUBLE PRECISION X, Y, F
      F = -1.D0 * Y
      RETURN
      END
C
C
C     EXAMPLE MAIN PROGRAM
C
      PROGRAM PARFDE
C
      IMPLICIT NONE
C
      INTEGER MAXPTS
      PARAMETER (MAXPTS=1000000)
C
      INTEGER IARGC
C
      DOUBLE PRECISION Y0(2), Y(MAXPTS), XMIN, XMAX, ALPHA, X,
     &                 APRED(MAXPTS), ACORR(MAXPTS), BCORR(MAXPTS),
     &                 CCORR, WRK(MAXPTS)
```

```
      INTEGER          K, N, CORRIT, IERR, NPTS
      EXTERNAL         RHS, IARGC, GETARG
      CHARACTER *72    ARGV
C
      N = IARGC()
      IF (N .NE. 3) THEN
         WRITE (*, '(A)') 'WRONG NUMBER OF ARGUMENTS.'
         WRITE (*, '(A)') 'USAGE: PARFDE ALPHA NPTS XMAX'
         STOP
      ENDIF
C
      CALL GETARG (1, ARGV)
      READ (ARGV, *) ALPHA
      CALL GETARG (2, ARGV)
      READ (ARGV, *) NPTS
      IF (NPTS .GT. MAXPTS) THEN
         WRITE (*, '(A,I,A,I,A)') 'TOO MANY POINTS: ', NPTS,
     &                      '(MAX: ', MAXPTS, ')'
         STOP
      ENDIF
      CALL GETARG (3, ARGV)
      READ (ARGV, *) XMAX
C
C     DEFINE PARAMETERS OF EQUATION
      XMIN = 0.D0
C
C     DEFINE INITIAL CONDITIONS
      Y0(1) = 1.D0
      Y0(2) = 0.D0
C
C     DEFINE PARAMETERS OF NUMERICAL METHOD
      CORRIT = 1
C
C     COMPUTE WEIGHTS
      CALL ABMW(ALPHA, APRED, ACORR, BCORR, CCORR, NPTS, IERR)
      IF (IERR .NE. 0) THEN
         WRITE (*, '(A,I4)') 'ERROR IN COMPUTATION OF WEIGHTS: ',
     &                        IERR
         STOP
      END IF
C
```

```
C      SOLVE EQUATION NUMERICALLY
       CALL ABMSP(ALPHA, NPTS, XMIN, XMAX, RHS, CORRIT, APRED,
     &            APRED, ACORR, BCORR, CCORR, Y, Y0, WRK, IERR)
       IF (IERR .NE. 0) THEN
          WRITE (*, '(A,I3,A,I3)') 'ERROR CODE ', IERR,
     &              ' IN DIFFERENTIAL EQUATION SOLVER FOR NPTS = ',
     &              NPTS
       ELSE
          WRITE(*, '(A,F14.6)') 'ALPHA = ', ALPHA
          WRITE(*, '(A,I14)')   'NPTS  = ', NPTS
          WRITE(*, '(A,F14.6)') 'XMAX  = ', XMAX
          WRITE(*, '(A)') '=========================='
          WRITE(*, '(A)')
     &       '          X          APPROX. VALUE'
          DO 50 K = 1, NPTS
             X = XMIN + (XMAX-XMIN) / DBLE(NPTS) * DBLE(K)
             WRITE (*, '(F16.8,F19.12)') X, Y(K)
 50       CONTINUE
       END IF
 100   CONTINUE
C
       END
```

A.2 Lubich's Fractional Backward Differentiation Formulas

For the fractional backward differentiation formulas introduced by Lubich, we here list the Matlab implementations due to Weilbeer [624]. This implementation consists of three separate routines. The first of these routines computes the solution at the starting points of the grid.

Algorithm A.4. Matlab implementation of Lubich's fractional BDF: Starting points

```
function[Y] = startpoints(C,S,a,y0,alpha,T)
% STARTPOINTS solves the nonlinear equation system for the first
%   a values of the fractional differential equation
%
%   (1)     D^{\alpha} * y(t) = f(t,y(t))
%
%   using a simple Newton method with a start solution corresponding
```

```
%   to the initial condition of (1).
%
%   Input Data:
%   -----------
%   C     - Convolution weights vector
%   S     - Starting weights matrix
%   a     - number of start weights
%   alpha  - Order of FDE alpha \in (0,1)
%   T     - mesh
%   y0     - Initial conditions
%
%   Output Data:
%   ------------
%   Y     - Solution at a-1 points
%
%   Note:
%   -----
%   This is not a stand-alone function but a subroutine of FBDFP.m

h = T(2)-T(1);
Y = y0(1).*ones(a-1,1);
IC= fliplr(y0); % Initial conditions polynomial

syms y t;

% right-hand side diff w.r.t. y
fj = inline(char(diff(rhs(t,y,alpha),y)),'t','y');
% right-hand side
ff = inline(char(rhs(t,y,alpha)),'t','y');

B = 1/h^alpha*(rot90(hankel(fliplr(C(1:a)))) + S(1:a,1:a));

% Evaluation of the initial conditions polynomial
P = polyval(IC,T(2:a));

B = B(2:a,2:a);
Pp = B*P';

er = 1;

while er > 1e-8
```

```
    for k=1:a-1
        Fj(k) = fj(T(k+1),Y(k));
        Ff(k) = ff(T(k+1),Y(k))+Pp(k);
    end
    X = (B - diag(Fj))\(B*Y - Ff');
    er = abs(max(X));
    Y = Y-X;
end
```

The second routine is devoted to the construction of the matrix that contains the weights of the method. Recall that a linear system of equations needs to be solved in this context; the parameter `c` of the subroutine allows the user to choose between various alternative numerical methods for the solution of this linear system.

Algorithm A.5. Matlab implementation of Lubich's fractional BDF: Computation of weights.

```
function[C, S, a, r] = ltsmp(N,alpha,p,c)
% LTSMP(N,alpha,p,c)
%   LTSMP creates the p-th order lower triangular strip matrix
%   corresponding to the discrete approximation of the fractional
%   differential or integral operator D^alpha at N nodes.
%
%   Input Data:
%   -----------
%   N       - number of nodes of the discretization.
%   alpha   - order of the fractional integration / differentiation.
%             alpha > 0 means integration,
%             alpha < 0 means differentiation.
%             abs(alpha) < 1.
%   p       - order of approximation
%             (between 1 and 6 for stability reasons).
%   c       - choice of numerical method for the
%             calculation of the starting weights.
%       'lu'       - solution of the equation system
%                         using lu decomposition
%       'lui'      - computation of the inverse using
%                         lu decomposition of wcoeff followed by
%                         matrix multiplication A^(-1)*rhs
%       'gmres'    - gmres solution of the equation system
```

```
%       'gmresi'  - inverse using gmres ...
%       'gmresh'  - gmres solution of the equation system using
%                         Householder orthogonalization
%       'qr'      - solution of the equation system using
%                         QR decomposition
%       'svd'     - solution of the equation system using SVD
%       'bpi'     - inverse using Bjoerck-Pereyra Algorithm
%                         (alpha has to be a unit fraction
%                         for this choice)
%
%    Output Data:
%    ------------
%    <C>     - Vector with N convolution weights
%                (omega_0,omega_1,...,omega_N-1)
%    <S1>    - Matrix with N*a starting weights:
%               [ w_0,1  w_0,2  w_0,3 ... w_0,a ]
%               [ w_1,1  w_1,1  w_1,2 ... w_1,a ]
%               [          . . .       ]
%               [ w_N-1,1         . . .    w_N-1,a ]
%    <a>     - Number of starting weights for each n.
%    <r>     - average residual of the start weight computation
%               over all right hand sides.

% Error handling.

if nargin ~= 4
   error('Wrong number of input arguments.')
end

% Calculating length(a) and q (=gamma).
cnt = 1;
q   = 0;
for k=0:p-1
   for l=0:ceil(1/abs(alpha)).*(p-1)
      qt = k+l*abs(alpha);  % potential gamma value.
      de = find(chop(q,10)==chop(qt,10)); % Avoid double entries.
      if qt <= p-1 & isempty(de) == 1
 cnt = cnt+1;
 q(cnt) = qt;
      end
   end
```

```
end
% Sort feasible gamma entries in increasing order.
q = sort(q);
[nil, a] = size(q);          % Counting number of starting weights.

% Convolution weights (via automatic differentiation).
U = zeros(6,N);
U(1:6,1:7) =    [ 1      -1    0      0      0      0     0;
                 3/2     -2   1/2     0      0      0     0;
                11/6     -3   3/2   -1/3     0      0     0;
                25/12    -4    3    -4/3    1/4     0     0;
               137/60    -5    5   -10/3    5/4   -1/5    0;
               147/60    -6  15/2  -20/3   15/4   -6/5   1/6 ];

C(1) = U(p,1)^(-alpha);
for s=1:N-1
   tmp = zeros(1,p+1);
   if s < p+2
      for v=0:s-1
         tmp(v+1) = (alpha*(v-s)-v)*C(v+1)*U(p,s-v+1);
      end
      C(s+1) = 1/(s*U(p,1)) * sum(tmp);
   else
      for v=s-(p+1):s-1
         tmp(v-(s-(p+1))+1) = (alpha*(v-s)-v)*C(v+1)*U(p,s-v+1);
      end
      C(s+1) = 1/(s*U(p,1)) * sum(tmp);
   end
end

% Starting weights
if p ~= 1   % if p=1 no additional starting weights are needed

   % Exact fractional integrals
   s(1,1:a) = gamma(1+q(1,1:a))./gamma(1+q(1,1:a)+alpha);

   % Right-hand sides
   for k=1:a
      T(1:N) = (1:N).^q(1,k);
      L(1:N-1) = (1:N-1).^(q(1,k)+alpha);
      Y = conv(C,T);
```

```
    rhs(2:N,k) = s(1,k)'*L(1:N-1)' - Y(1:N-1)';
end

% Coefficient Matrix
for k=1:a
    Wcoeff(k,:) = (0:a-1).^q(1,k);
end

switch lower(c)

    case 'lu'
    % 1) Solution of the linear equation systems
    %    using the \ operator
    S(1:N,:) = (Wcoeff\rhs(1:N,:)')';

    case 'lui'
    % 2) Solution of the linear equation systems
    %    using the inverse
    Wcoeffinv = inv(Wcoeff);
    S(1:N,:) = (Wcoeffinv*rhs(1:N,:)')';

    case 'gmres'
    % 3) Solution using gmres at each step
    %    on the actual system.
    for k=1:N
        [S(:,k), flag,relres,iter]
            = gmres(Wcoeff,rhs(k,:)',a+1,1e-16,1);
    end
    S = S';

    case 'gmresi'
    % 4) Computing the inverse of Wcoeff using gmres
    for k=1:a
        unitVector = zeros(a,1);
        unitVector(k,1) = 1;
        [Wcoeffinv(:,k), flag]
            = gmres(Wcoeff,unitVector,a+1,1e-16,1);
    end
    for k=1:N
        S(k,:) = (Wcoeffinv*rhs(k,:)')';
    end
```

```
case 'gmresh'
% 5) Solution using gmreshouseholder at each step
%    on the actual system.
for k=1:N
   S(:,k)
      = gmresh(Wcoeff,rhs(k,:)',zeros(a,1),1e-32,a);
end
S = S';

case 'qr'
% 6) Solution using QR decomposition.
[Q R] = qr(Wcoeff);
S(1:N,:) = (R\(Q'*rhs(1:N,:)'))';

case 'svd'
% 7) Solution using SVD.
S = zeros(N,a);
[Uv Sv Vv] = svd(Wcoeff);
for l=1:N
   for k=1:a
      S(l,:) = S(l,:)
                  + (Uv(:,k)'*rhs(l,:)'*Vv(:,k)./Sv(k,k))';
   end
end

case 'bpi'
% 8) Calculating the inverse using the Bjoerck-Pereyra
%    algorithm (alpha has to be a unit fraction)
[num, num2] = rat(alpha);
if abs(num) ~= 1
   fprintf(1,
      '\n alpha is not a unit fraction. Cannot use bpi\n');
else
   X = (0:a-1).^(abs(alpha));
   % Calculating the inverse
   Wcoeffinv = bpif(X,'p');
   % Solving for the N right-hand sides
   S(1:N,:) = (Wcoeffinv*rhs(1:N,:)')';
end

otherwise
```

```
        fprintf(1,
            '\n Wrong choice for parameter c.\n');
    end
    % residual tester
    Residual(1:N,1) = norm(Wcoeff*S(1:N,:)'- rhs(1:N,:)');
    r = sum(abs(Residual))/length(Residual);

else
    a = 1;
    S = zeros(N,1);
    r = 0;
end

fprintf(1,'\n The average residual is %g.\n',r);
```

Finally we give the main part of the algorithm that actually computes the required approximate solutions, using the two auxiliary routines listed above.

Algorithm A.6. Matlab implementation of Lubich's fractional BDF: Main part.

```
function[Y] = fbdfp(N,alpha,y0,b,p,in,ie,c)

% FBDFP(N,alpha,y0,b,p,in,ie,c)
%   Fractional Backward Difference Formula of order p.
%   FBDFP solves the fractional order differential equation (FDE)
%
%   (1)     D^alpha * y(t) = f(t,y(t))   ,alpha \in (0,1)
%
%   at N equispaced mesh points on the interval [0,b].
%   The method has a theoretical convergence order of p. The
%   parameters 'in' (iteration number) and 'ie' (iteration error)
%   determine the maximal number of iterations or the minimal
%   change of two consecutive iterations at each mesh points.
%   The parameter c determines the numerical method used to generate
%   the discretized fractional differential operator D^alpha.
%   The right-hand side of (1) has to be stored in an external
%   matlab function in the file rhs.m.
%
%   Input Data:
```

```
%   -----------
%   N       - Number of mesh points.
%   alpha   - Order of FDE alpha \in (0,1).
%   b       - Right interval end (0 assumed to be left start).
%   y0      - Starting values y^(k)(0) = c_k as vector
$             y0 = [c_0, c_1, ... , c_k].
%   p       - desired convergence order (p = 1-6).
%   in      - Maximal number of iteration at each mesh point.
%   ie      - Minimal change of two consecutive iteration steps.
%   c       - choice of numerical method for the calculation
%             of the starting weights
%
%   rhs.m  - externally stored right-hand side f(t,y(t)) of FDE.
%
%   Output Data:
%   ------------
%   Y       - Solution at the N mesh points.
%
%   See also FBDFPS, LTSMP, LTSMPS, RHS, STARTPOINTS.

if nargin ~= 8
   error('Wrong number of input arguments.')
end

Y      = zeros(N,1);                  % Solution.
Yt     = zeros(N,1);                  % Auxiliary variables.
Z      = zeros(N,1);

Y(1,1) = y0(1);                       % Initial Conditions.
IC     = fliplr(y0);

T = linspace(0,b,N);                  % Equidistant mesh points.
h = T(2)-T(1);                        % Step size.
[C S a] = ltsmp(N,-alpha,p,c);        % Calculating lower
                                      % triangular strip matrix.

if p ~= 1
   Sp = startpoints(C,S,a,y0,alpha,T); % Startpoints calculation.
   Y(2:a) = Sp(1:a-1);                 % Setting start values.
end
```

```
for k=1+a:N

   K  = zeros(1,k);
   Ym = Y(1:k);
   ICm(1:k) = polyval(IC,T(1:k));

   K(1:k) = C(k:-1:1);
   K(1:a) = K(1:a) + S(k,1:a);

   St = 1/C(1).*K*Ym;
   A  = 1/C(1).*K*ICm';

   % First guess using previous solution point

   Z(k) = (h^(alpha))/C(1)*rhs(T(k),Y(k-1),alpha) - St + A;

   er = 5;                        % auxiliary variables for iteration.
   cnt = 1;

   while er >= ie & cnt < in      % Fixed point iteration.
      Yt(k) = (h^(alpha))/C(1)*rhs(T(k),Z(k),alpha) - St + A;
      er    = abs(Yt(k)-Z(k));
      Z(k)  = Yt(k);
      cnt   = cnt+1;
   end

   Y(k) = Z(k);

   if imag(Y(k)) ~= 0             % termination criterion.
      Y(k:N) = 0;
      break;
   end

end
```

A.3 Time-fractional Diffusion Equations

Our last differential equation solver is a finite difference method for a time-fractional diffusion equation

$$^{\mathrm{C}}D_t^\alpha y(x,t) + \Phi(x,t)\frac{\partial^2}{\partial x^2}y(x,t) = f(x,t), \qquad x \in [a,b], \quad t \in [0,T],$$

with $0 < \alpha < 1$, the initial condition

$$y(x,0) = y_0(x)$$

and mixed Dirichlet-Neumann boundary conditions

$$g_1(t)y(a,t) + h_1(t)\frac{\partial}{\partial x}y(a,t) = r_1(t),$$
$$g_2(t)y(b,t) + h_2(t)\frac{\partial}{\partial x}y(b,t) = r_2(t)$$

as discussed in Subsection 3.2.2. This algorithm is given in a form suitable for direct use within the Mathematica software package [632]. It is based on the approximation developed in Eqs. (2.1.10) and (2.1.11) for the fractional derivative with respect to t, using a stepsize of $\Delta t = T/n_t$, and on a standard centered difference approximation with step size $\Delta x = (b-a)/n_x$ for the second-order derivative with respect to x, where n_t and n_x are preassigned positive integers.

The algorithm requires that $\Phi(a,t) \neq 0$ and $\Phi(b,t) \neq 0$ for all $t \in [0,T]$. The algorithm computes (n_x+1)-dimensional vectors u[0], u[1], ..., u[n_t] (in Mathematica notation) with the interpretation that the component u[k][[j]] is the approximation of the exact solution $y((j-1)\cdot\Delta_x, k\cdot\Delta_t)$ where $k \in \{0,1,\ldots n_t\}$ and $j \in \{1,2,\ldots,n_x+1\}$.

Algorithm A.7. Mathematica implementation of a backward differentiation method for the solution of the time-fractional diffusion equation in one space dimension.

```
(* Load package for handling of tridiagonal linear systems *)
<< LinearAlgebra`Tridiagonal`


(* Initialize parameters of the equation (given data) *)


(* boundaries of intervals *)
(* values are to be understood as examples *)
```

```
a = 0;
b = 1;
T = 1;

(* functions in boundary conditions *)
(* values are to be understood as examples *)
r1[t_] := 0;
r2[t_] := 1;
g1[t_] := 1;
g2[t_] := 1;
h1[t_] := 0;
h2[t_] := 0;

(* function in initial condition *)
(* value is to be understood as example *)
u0[x_] := x;

(* functions in the differential equation itself *)
(* values are to be understood as examples *)

phi[x_, t_] := -1;
f[x_, t_] := 0;

(* order of fractional differential operator with respect to time *)

alpha = 0.5;
Clear[weight];

(* initialize parameters of the numerical scheme:
   number of steps in coordinate directions *)
(* values for nx and nt are to be understood as examples *)
nx = 67;
nt = 80;
deltax = (b - a)/nx;
deltat = T/nt;

(* auxiliaries *)
ga = Gamma[2 - alpha];

(* define weights of fractional BDF *)
weight[j_, k_] := weight[j, k] = Which[j == 0, 1,
```

```
    j == k, k^(-alpha) (1 - alpha - k) + (k - 1)^(1 - alpha),
    True, (j - 1)^(1 - alpha) -
     2 j^(1 - alpha) + (j + 1)^(1 - alpha)]/ga

(* main part of algorithm *)

(* assign initial values *)

u[0] = Table[u0[a + (j - 1) deltax], {j, 1, nx + 1}];

(* time step loop *)
Do[
 tk = k deltat;

 (* construction of tridiagonal matrix of linear system
    for each time step *)
 lower =
  Append[Table[phi[a + j deltax, tk], {j, 1, nx - 1}],
         -h2[tk]] / deltax^2;
 main = Append[
   Prepend[Table[-2 phi[a + j deltax, tk] / deltax^2 +
      weight[0, k] / deltat^alpha, {j, 1, nx - 1}],
    g1[tk] / deltax +
     h1[tk] (-1/deltax^2 + 1/(2 ga deltat^alpha phi[a, tk] ))],
   g2[tk] / deltax +
    h2[tk] (1/deltax^2 - 1/(2 ga deltat^alpha phi[b, tk] ))];
 upper = Prepend[Table[phi[a + j deltax, tk], {j, 1, nx - 1}],
    h1[tk]]/deltax^2;

 (* construction of vector on right-hand side of system *)
 rhs = Prepend[
   Append[
    Table[f[a + j deltax, tk]
          + u[0][[j]] (1 - alpha) /ga/tk^alpha
          - Sum[weight[mu, k] u[k - mu][[j]], {mu, 1, k}] /
                deltat^alpha, {j, 2, nx}],
    r2[tk]/deltax -
     0.5 h2[tk] / phi[b, tk] (
       f[b, tk] + u0[b] tk^(-alpha) (1 - alpha)/ga
        - Sum[weight[mu, k] u[k - mu][[nx + 1]], {mu, 1, k}]/
         deltat^alpha)],
```

```
    r1[tk]/deltax +
    0.5 h1[tk] / phi[a, tk] (
      f[a, tk] + u0[a] tk ^(-alpha) (1 - alpha)/ga
       - Sum[weight[mu, k] u[k - mu][[1]], {mu, 1, k}]
           / deltat^alpha)];

 (* approximate solution at present time step is obtained
    by solving this tridiagonal system *)

 u[k] = TridiagonalSolve[lower, main, upper, rhs]
, {k, 1, nt}]
```

A.4 Computation of the Mittag-Leffler Function

In this section, we give a fast routine for the computation of the two-parameter Mittag-Leffler function $E_{\alpha,\beta}(x)$ for $\alpha \geq 0$, $\beta > 0$ and real values of x. The routine has been developed to run fast; its accuracy is very high only for $x \leq 1$. If highly accurate results are required for $x > 1$ or for complex arguments x then we recommend to use the routine of Gorenflo *et al.* [270, 271]. A Matlab version of such a code is available [489]. Alternatively, one can also design a numerical method on the basis of the recent approaches of [253] and [252].

Algorithm A.8. FORTRAN77 implementation of a routine for the fast evaluation of two-parameter Mittag-Leffler functions. This routine requires an external routine `GAMMA` (such as, *e. g.*, the one provided in [155]) for the evaluation of Euler's Gamma function.

```
      DOUBLE PRECISION FUNCTION ML2P(ALPHA, BETA, X)
C
      IMPLICIT NONE
      DOUBLE PRECISION ALPHA, BETA, X
C
C     FAST COMPUTATION OF TWO-PARAMETER MITTAG-LEFFLER FUNCTION
C
C     PARAMETERS
C     ALPHA - INPUT - FIRST PARAMETER OF MITTAG-LEFFLER FUNCTION
C     BETA  - INPUT - SECOND PARAMETER OF MITTAG-LEFFLER FUNCTION
C     X     - INPUT - ARGUMENT OF THE MITTAG-LEFFLER FUNCTION
```

```
C
      INTEGER          K, IERR
      DOUBLE PRECISION TMP, NEW, GMARG, GAMMA
      EXTERNAL         GAMMA
C
      TMP = 1.D0 / GAMMA(BETA)
      K = 1
C
 100  GMARG = ALPHA * K + BETA
      NEW =  X**K / GAMMA(GMARG)
      TMP = TMP + NEW
      K = K + 1
      IF (GMARG .LT. 5.D0) GOTO 100
      IF (K .LT. 100  .AND.  ABS(NEW) * 1.D16 .GT. TMP) GOTO 100
C
      ML2P = TMP
C
      RETURN
      END
```

A.5 Monte Carlo Simulation of CTRW

We conclude this appendix with two implementations (in R and in Matlab) of a Monte Carlo program to simulate CTRWs according to the algorithm described in Section 7.3. The program below, generates and plots a single realization of a CTRW with a given number of jumps and durations. Even if this is a very simple algorithm, it consists of three parts. The first part is the generator of independent and indentically distributed Mittag-Leffler deviates according to equation (7.3.2). Then, Lévy α-stable deviates are generated following equation (7.3.1). Finally, cumulative sums give the position coordinates and the epochs and positions are plotted as a function of the epochs. This routine can be easily modified with suitable external cycles to generate many realizations up to a given time t and estimate the probability density $f_{X(t)}(x, t)$ from the histogram of realized positions $X(t)$. This was explicitly done in reference [249].

Algorithm A.9. Matlab implementation for the Monte Carlo simulation of CTRW.

```
%Plot of a single CTRW realization

%Generation of Mittag-Leffler deviates
%See Fulger, Scalas, Germano 2008 and references therein

n=100; %number of points
gammat=1; %scale parameter
beta=0.99; %ML parameter

u1=rand(n,1); %uniform deviates
v1=rand(n,1); %uniform deviates

%Generation of symmetric alpha stable deviates

tau=-gammat*log(u1).*(sin(beta*pi).
/tan(beta*pi*v1)-cos(beta*pi)).^(1/beta);

gammax=1; %scale parameter
alpha=1.95; %Levy parameter

u2=rand(n,1); %uniform deviates
v2=rand(n,1); %uniform deviates
phi=pi*(v2-0.5);

xi=gammax*(sin(alpha*phi)./cos(phi)).
*(-log(u2).*cos(phi)./cos((1-alpha)*phi)).^(1-1/alpha);

%Random walk

x=cumsum(xi');
x=[0 x];

%Epochs

t=cumsum(tau');
t=[0 t];

stairs(t,x) %plots ctrw
```

Algorithm A.10. R implementation for the Monte Carlo simulation of CTRW.

```
# Plot of a single CTRW realization

# Generation of Mittag-Leffler deviates
# See Fulger, Scalas, Germano 2008 and references therein

n <- 10000 #number of points
gammat <- 1 #scale parameter
beta <- 0.95 #ML parameter

u1 <- runif(n) #uniform deviates
v1 <- runif(n) #uniform deviates

tau <- -gammat * log(u1) * (sin(beta * pi)/tan(beta * pi * v1) -
cos(beta * pi))^(1/beta)

# Generation of symmetric alpha-stable deviates
# See Fulger, Scalas, Germano 2008 and references therein

gammax <- 1 #scale parameter
alpha <- 1.95 #Levy parameter

u2 <- runif(n)
v2 <- runif(n)
phi <- pi*(v2 - 0.5)

xi <- gammax *(sin(alpha*phi)/cos(phi))*
(-log(u2)*cos(phi)/cos((1-alpha)*phi))^(1-1/alpha)

# histogram of xi
# hist(xi)

# Random walk

x <- cumsum(xi)

# Epochs

t <- cumsum(tau)
```

```
# Stairplot

plot(t,x,type="s")
```

Bibliography

[1] Abdeljawad, T. and Baleanu, D. (2011a). Caputo q-fractional initial value problems and a q-analogue Mittag-Leffler function, *Commun. Nonlin. Sci. Numer. Sim.* **16**, pp. 4682–4688.

[2] Abdeljawad, T. and Baleanu, D. (2011b). Fractional differences and integration by parts, *J. Comput. Anal. Appl.* **13**, pp. 574–582.

[3] Abel, N. H. (1881). Solution de quelques problèmes à l'aide d'intégrales définies, in L. Sylow and S. Lie (eds.), *Œuvres Complètes de Niels Henrik Abel, Nouvelle Edition*, Vol. 1 (Grøndahl, Christiania), pp. 11–27.

[4] Abramowitz, M. and Stegun, I. A. (1965). *Handbook of Mathematical Functions* (Dover Publ., New York).

[5] Adamchick, V. S. (1997). On Stirling numbers and Euler sums, *J. Comput. Appl. Math.* **79**, pp. 119–130.

[6] Adolfsson, K., Enelund, M. and Larsson, S. (2004). Adaptive discretization of fractional order viscoelasticity using sparse time history, *Comput. Meth. Appl. Mech. Engin.* **193**, pp. 4567–4590.

[7] Adomian, G. (1989). *Nonlinear Stochastic Systems Theory and Applications to Physics* (Kluwer, Dordrecht).

[8] Adomian, G. (1994). *Solving Frontier Problems of Physics: The Decomposition Method* (Kluwer, Dordrecht).

[9] Agarwal, R. P., Benchohra, M. and Hamani, S. (2010a). A survey on existence results for boundary value problems of nonlinear fractional differential equations and inclusions, *Acta Appl. Math.* **109**, pp. 973–1033.

[10] Agarwal, R. P., Bohner, M., O'Regan, D. and Peterson, A. (2002). Dynamic equations on time scales: A survey, *J. Comput. Appl. Math.* **141**, pp. 1–26.

[11] Agarwal, R. P., Lakshmikantham, V. and Nieto, J. J. (2010b). On the concept of solution for fractional differential equations with uncertainty, *Nonlinear Anal.: Theory, Meth. Appl.* **72**, pp. 2859–2862.

[12] Agarwal, R. P., O'Regan, D. and Stanek, S. (2010c). Positive solutions for Dirichlet problems of singular nonlinear fractional differential equations, *J.*

Math. Anal. Appl. **371**, pp. 57–68.

[13] Agarwal, R. P., Zhou, Y. and He, Y. (2010d). Existence of fractional neutral functional differential equations, *Comput. Math. Appl.* **59**, pp. 1095–1100.

[14] Agila, A., Baleanu, D., Eid, R. and Irfanoglu, B. (2016). Applicatons of the extended fractional Euler-Lagrange equations model to freely oscillating dynamical systems, *Rom. J. Phys.* **61**, pp. 350–359.

[15] Agrawal, O. P. (2001). A new Lagrangian equation of motion for fractionally damped systems, *J. Appl. Mech.* **68**, pp. 339–341.

[16] Agrawal, O. P. (2002a). Formulation of Euler-Lagrange equations for fractional variational problems, *J. Math. Anal. Appl.* **52**, pp. 1247–1253.

[17] Agrawal, O. P. (2002b). Solution for a fractional diffusion-wave equation defined in a bounded domain, *Nonlinear Dyn.* **29**, pp. 145–155.

[18] Agrawal, O. P. (2004). A general formulation and solution scheme for fractional optimal control problems, *Nonlinear Dyn.* **38**, pp. 323–337.

[19] Agrawal, O. P. (2006). Fractional variational calculus and the transversality conditions, *J. Phys. A.: Math. Gen.* **39**, pp. 10375–10384.

[20] Agrawal, O. P. (2007a). Fractional variational calculus in terms of Riesz fractional derivatives, *J. Phys. A: Math. Theor.* **40**, pp. 6287–6303.

[21] Agrawal, O. P. (2007b). Generalized Euler-Lagrange equations and transversality conditions for FVPs in terms of the Caputo derivative, *J. Vibration Contr.* **13**, pp. 1217–1237.

[22] Agrawal, O. P. (2008). A formulation and numerical scheme for fractional optimal control problems, *J. Vibr. Contr.* **14**, pp. 1291–1299.

[23] Agrawal, O. P. (2010). Generalized variational problems and Euler-Lagrange equations, *Comput. Math. Appl.* **59**, pp. 1852–1864.

[24] Agrawal, O. P. and Baleanu, D. (2007). A Hamiltonian formulation and a direct numerical scheme for fractional optimal control problems, *J. Vibration Contr.* **13**, pp. 1269–1281.

[25] Agrawal, O. P., Defterli, O. and Baleanu, D. (2010). Fractional optimal control problems with several state and control variables, *J. Vibration Contr.* **16**, pp. 1967–1976.

[26] Agrawal, O. P. and Gregory, J. (2003). O(h^2) global pointwise algorithms for delay quadratic problems in the calculus of variations, *SIAM J. Numer. Anal.* **41**, pp. 1773–1784.

[27] Agrawal, O. P., Gregory, J. and Spector, K. P. (1997). A Bliss-type multiplier rule for constrained variational problems with time delay, *J. Math. Anal. Appl.* **20**, pp. 702–711.

[28] Agrawal, O. P., Muslih, S. I. and Baleanu, D. (2011). Generalized variational calculus in terms of multi-parameters fractional derivatives, *Commun. Nonlin. Sci. Numer. Sim.* **16**, pp. 4756–4767.

[29] Ahmad, B. and Nieto, J. J. (2009). Existence results for a coupled system of

nonlinear fractional differential equations with three-point boundary conditions, *Comput. Math. Appl.* **58**, pp. 1838–1843.

[30] Ahmed, E. and Elgazzar, A. S. (2007). On fractional order differential equations model for nonlocal epidemics, *Physica A* **379**, pp. 607–614.

[31] Aigner, M. (1975). *Kombinatorik I* (Springer, Berlin).

[32] Aitchison, J. and Brown, J. A. (1957). *The Lognormal Distribution* (Cambridge Univ. Press, Cambridge).

[33] Alaimia, M. R. and Tatar, N. E. (2005). Blow up for the wave equation with a fractional damping, *J. Appl. Anal.* **11**, pp. 133–144.

[34] Almeida, R. and Torres, D. F. (2009). Calculus of variations with fractional derivatives and fractional integrals, *Appl. Math. Lett.* **22**, pp. 1816–1820.

[35] Ames, W. F. (1992). *Numerical Methods for Partial Differential Equations* (Acad. Press, Boston).

[36] Anderson, D. R. and Zafer, A. (2009). Nonlinear oscillation of second-order dynamic equations on time scales, *Appl. Math. Lett.* **22**, pp. 1591–1597.

[37] Apostol, T. M. (1957). *Mathematical Analysis: A Modern Approach to Advanced Calculus* (Addison-Wesley, London).

[38] Applebaum, D. (2009). *Lévy Processes and Stochastic Calculus* (Cambridge Univ. Press, Cambridge).

[39] Arellano-Plaza, M., Herrera-López, E. J., Díaz-Montaño, D. M., Moran, A. and Ramírez-Córdova, J. J. (2007). Unstructured kinetic model for Tequila batch fermentation, *Internat. J. Math. Computers Simul.* **1**, pp. 1–6.

[40] Arena, P., Caponetto, R., Fortuna, L. and Porto, D. (2000). *Nonlinear Noninteger Order Circuits and Systems: Theory and Applications* (World Scientific Publ. Comp., Singapore).

[41] Artin, E. (1964). *The Gamma Function* (Holt, Rinehart and Winston, New York).

[42] Atanacković, T. M., Budinčević, M. and Pilipović, S. (2005). On a fractional distributed-order oscillator, *J. Phys. A: Math. Gen.* **38**, pp. 6703–6713.

[43] Atanacković, T. M., Konjik, S. and Pilipović, S. (2008). Variational problems with fractional derivatives: Euler-Lagrange equations, *J. Phys. A: Math. Theor.* **41**, pp. 095201–095212.

[44] Atici, F. M. and Eloe, P. W. (2007). A transform method in discrete fractional calculus, *Int. J. Differ. Equ.* **2**, pp. 165–176.

[45] Atici, F. M. and Eloe, P. W. (2009). Initial value problem in discrete fractional calculus, *Proc. Amer. Math. Soc.* **137**, pp. 981–989.

[46] Audounet, J., Giovangigli, V. and Roquejoffre, J. M. (1998). A threshold phenomenon in the propagation of a point source initiated flame, *Physica D* **121**, pp. 295–316.

[47] Bachelier, L. (1900). Théorie de la spéculation, *Annales Scientifiques de l'École Normale Supérieure* **3**, pp. 21–86.

[48] Bagley, R. L. and Torvik, P. J. (2000a). On the existence of the order domain and the solution of distributed order equations – Part I, *Internat. J. Appl. Math.* **2**, pp. 865–882.

[49] Bagley, R. L. and Torvik, P. J. (2000b). On the existence of the order domain and the solution of distributed order equations – Part II, *Internat. J. Appl. Math.* **2**, pp. 965–987.

[50] Baillie, R. T. and King, M. L. (1996). Fractional differencing and long memory processes, *J. Econometrics* **73**, pp. 1–3.

[51] Balachandran, K. and Trujillo, J. J. (2010). The nonlocal Cauchy problem for nonlinear fractional integrodifferential equations in Banach spaces, *Nonlinear Anal.: Theory, Meth. Appl.* **72**, pp. 4587–4593.

[52] Balakrishnan, V. (1985). Anomalous diffusion in one dimension, *Physica A* **132**, pp. 569–580.

[53] Baleanu, D. (1999). Killing-Yano tensors and Nambu tensors, *Nuovo Cimento B* **114**, pp. 1065–1072.

[54] Baleanu, D. (2004). Constrained systems and Riemann-Liouville fractional derivative, in A. Le Méhauté, J. A. Tenreiro Machado, J. C. Trigeassou and J. Sabatier (eds.), *Proceedings of the 1st IFAC Workshop on Fractional Differentiation and its Applications* (ENSEIRB, Bordeaux), pp. 597–602.

[55] Baleanu, D. (2005a). About fractional calculus of singular Lagrangians, *J. Adv. Comput. Intell., Intel. Inform.* **9**, pp. 395–398.

[56] Baleanu, D. (2005b). Constrained systems and Riemann-Liouville fractional derivatives, in A. Le Méhauté, J. A. Tenreiro Machado, J. C. Trigeassou and J. Sabatier (eds.), *Fractional Differentiation and its Applications* (Ubooks, Neusäß), pp. 69–80.

[57] Baleanu, D. (2006). Fractional Hamiltonian analysis of irregular systems, *Signal Process.* **86**, pp. 2632–2636.

[58] Baleanu, D. (2008a). Fractional constrained systems and Caputo derivatives, *J. Comput. Nonlin. Dyn.* **3**, Article No. 021102.

[59] Baleanu, D. (2008b). New applications of fractional variational principles, *Rep. Math. Phys.* **61**, pp. 199–206.

[60] Baleanu, D. (2009a). About fractional quantization and fractional variational principles, *Commun. Nonlin. Sci. Numer. Sim.* **14**, pp. 2520–2523.

[61] Baleanu, D. (2009b). Fractional variational principles in action, *Phys. Scr.* **T136**, Article No. 014006.

[62] Baleanu, D. and Agrawal, O. P. (2006). Fractional Hamilton formalism within Caputo's derivative, *Czech. J. Phys.* **56**, pp. 1087–1092.

[63] Baleanu, D. and Avkar, T. (2004). Lagrangians with linear velocities within Riemann-Liouville fractional derivatives, *Nuovo Cimento B* **119**, pp. 73–79.

[64] Baleanu, D., Defterli, O. and Agrawal, O. P. (2009a). A central difference numerical scheme for fractional optimal control problems, *J. Vibration*

Contr. **15**, pp. 583–597.

[65] Baleanu, D., Garra, R. and Petras, I. (2013). A fractional variational approach to the fractional Basset-type equation, *Comm. Theor. Phys.* **72**, pp. 57–64.

[66] Baleanu, D., Golmankhaneh, A. K. and Golmankhaneh, A. K. (2009b). The dual action of fractional multi time Hamilton equations, *Int. J. Theor. Phys.* **48**, pp. 2558–2569.

[67] Baleanu, D., Golmankhaneh, A. K. and Golmankhaneh, A. K. (2009c). Fractional Nambu mechanics, *Int. J. Theor. Phys.* **48**, pp. 1044–1052.

[68] Baleanu, D., Guvenc, Z. B. and Tenreiro Machado, J. A. (2010a). *New Trends in Nanotechnology and Fractional Calculus Applications* (Springer, Heidelberg).

[69] Baleanu, D., Jarad, F. and Muslih, S. (2006). Difference discrete and fractional variational principles, *Facta Univ.: Phys. Chem. Tech.* **4**, pp. 175–183.

[70] Baleanu, D., Jihad, A. and Petras, I. (2012a). Fractional order two-electric pendulum, *Rom. Rep. Phys.* **64**, pp. 907–914.

[71] Baleanu, D., Jihad, H. and Petras, I. (2014). Fractional Bateman-Feshbach Tikochinsky oscillator, *Comm. Theor. Phys.* **61**, pp. 221–225.

[72] Baleanu, D., Jihad, H. and Petras, I. (2015). Numerical solution of the fractional Euler-Lagrange's equations of a thin elastica model, *Nonlinear Dyn.* **81**, pp. 97–102.

[73] Baleanu, D., Maraaba, T. and Jarad, F. (2008a). Fractional variational principles with delay, *J. Phys. A: Math. Theor.* **41**, Article No. 315403.

[74] Baleanu, D. and Muslih, S. I. (2005a). About fractional supersymmetric quantum mechanics, *Czech. J. Phys.* **55**, pp. 1063–1066.

[75] Baleanu, D. and Muslih, S. I. (2005b). Lagrangian formulation of classical fields within Riemann-Liouville fractional derivatives, *Phys. Scr.* **72**, pp. 119–121.

[76] Baleanu, D. and Muslih, S. I. (2007). Fractional Euler-Lagrange and fractional Hamilton equations for super symmetric classical model, *Fractals* **15**, pp. 379–383.

[77] Baleanu, D., Muslih, S. I. and Rabei, E. M. (2008b). On fractional Euler-Lagrange and Hamilton equations and the fractional generalization of total time derivative, *Nonlinear Dyn.* **53**, pp. 67–74.

[78] Baleanu, D., Muslih, S. I., Rabei, E. M., Golmankhaneh, A. K. and Golmankhaneh, A. K. (2010b). On fractional dynamics on the extended phase space, *J. Comput. Nonlin. Dyn.* **5**, Article No. 041011.

[79] Baleanu, D., Muslih, S. I., Rabei, E. M., Golmankhaneh, A. K. and Golmankhaneh, A. K. (2011). On fractional Hamiltonian systems possessing first-class constraints within Caputo derivatives, *Rom. Rep. Phys.* **63**, pp.

3–8.

[80] Baleanu, D., Mustafa, O. G. and Agarwal, R. P. (2010c). Asymptotically linear solutions for some linear fractional differential equations, *Abstr. Appl. Anal.* Article No. 865139.

[81] Baleanu, D., Petras, I., Jihad, A., Elagan, S. and Bilgen, A. (2012b). Fractional Euler-Lagrange equation of Caldirola-Kanai oscillator, *Rom. Rep. Phys.* **64**, pp. 1171–1177.

[82] Baleanu, D., Petras, I., Jihad, A. and Velasco, V. M. (2012c). Fractional Pais-Uhlenbeck oscillator, *Int. J. Theor. Phys.* **51**, pp. 1253–1258.

[83] Baleanu, D., Sadati, S. J., Ranjbar, A., Ghaderi, R., Abdeljawad, T. and Jarad, F. (2010d). Razumikhin stability theorem for fractional systems with delay, *Abstr. Appl. Anal.* Article No. 124812.

[84] Baleanu, D. and Trujillo, J. J. (2008). On exact solutions of a class of fractional Euler-Lagrange equations, *Nonlinear Dyn.* **52**, pp. 331–335.

[85] Baleanu, D. and Trujillo, J. J. (2010). A new method of finding the fractional Euler-Lagrange and Hamilton equations within Caputo fractional derivatives, *Commun. Nonlin. Sci. Numer. Sim.* **15**, pp. 1111–1115.

[86] Baleanu, D. and Vacaru, S. (2011a). Constant curvature coefficients and exact solutions in fractional gravity and geometric mechanics, *Centr. Eur. J. Phys.* **9**, pp. 1267–1279.

[87] Baleanu, D. and Vacaru, S. (2011b). Fedosov quantization of fractional Lagrange spaces, *Int. J. Theor. Phys.* **50**, pp. 233–243.

[88] Baleanu, D. and Vacaru, S. (2011c). Fractional almost Kähler-Lagrange geometry, *Nonlinear Dyn.* **64**, pp. 365–373.

[89] Baleanu, D. and Vacaru, S. (2011d). Fractional curve flows and solitonic hierarchies in gravity and geometric mechanics, *J. Math. Phys.* **52**, Article No. 053514.

[90] Bastos, N. R., Ferreira, R. A. and Torres, D. F. (2011). Necessary optimality conditions for fractional difference problems of the calculus of variations, *Discr. Contin. Dyn. Syst., Ser. A* **29**, pp. 417–437.

[91] Beghin, L. and Orsingher, E. (2009). Fractional Poisson process and related random motions, *Electron. J. Probab.* **14**, pp. 1790–1826.

[92] Ben-Avraham, D. and Havlin, S. (2005). *Diffusion and Reactions in Fractals and Disordered Systems* (Cambridge Univ. Press, Cambridge).

[93] Benson, D. A. (1998). *The Fractional Advection-Dispersion Equation: Development and Application*, Ph.D. thesis, University of Nevada at Reno.

[94] Benson, D. A., Wheatcraft, S. W. and Meerschaert, M. M. (2000). Application of a fractional advection-dispersion equation, *Water Resour. Res.* **36**, pp. 1403–1412.

[95] Bering, K. (2000). A note on non-locality and Ostrogradski's construction, Tech. rep., arXiv:hep-th/0007192.

[96] Bernstein, S. (1928). Sur les fonctions absolument monotones, *Acta Math.* **52**, pp. 1–66.

[97] Bertoin, J. (1998). *Lévy Processes* (Cambridge Univ. Press, Cambridge).

[98] Bertram, W. (2004). An empirical investigation of Australian Stock Exchange data, *Physica A* **341**, pp. 533–546.

[99] Bhalekar, S. and Daftardar-Gejji, V. (2010). Fractional ordered Liu system with time-delay, *Commun. Nonlin. Sci. Numer. Sim.* **15**, pp. 2178–2191.

[100] Bhalekar, S., Daftardar-Gejji, V., Baleanu, D. and Magin, R. (2011). Fractional Bloch equation with delay, *Comput. Math. Appl.* **61**, pp. 1355–1365.

[101] Bhrawy, A., Doha, E. H., Tenreiro Machado, J. A. and Ezz-Eldien, S. S. (2015). An efficient numerical scheme for solving multi-dimensional fractional optimal control problems with a quadratic performance index, *Asian J. Contr.* **17**, pp. 2389–2402.

[102] Bingham, N. H., Goldie, C. M. and Teugels, J. L. (1987). *Regular Variation* (Cambridge Univ. Press, Cambridge).

[103] Birk, C. and Song, C. (2009). A continued-fraction approach for transient diffusion in unbounded medium, *Comput. Methods Appl. Mech. Eng.* **198**, pp. 2576–2590.

[104] Birk, C. and Song, C. (2010). A local high-order doubly asymptotic open boundary for diffusion in a semi-infinite layer, *J. Comput. Phys.* **229**, pp. 6156–6179.

[105] Bisquert, J. (2005). Interpretation of a fractional diffusion equation with nonconserved probability density in terms of experimental systems with trapping or recombination, *Phys. Rev. E* **72**, Article No. 011109.

[106] Biswas, R. J. and Sen, S. (2011). Fractional optimal control problems with specified final time, *J. Comput. Nonlin. Dyn.* **6**, Article No. 021009.

[107] Björck, Å. and Pereyra, V. (1970). Solution of Vandermonde systems of equations, *Math. Comput.* **24**, pp. 893–903.

[108] Blank, L. (1997). Numerical treatment of differential equations of fractional order, *Nonlinear World* **4**, pp. 473–490.

[109] Bochner, S. (1932). *Vorlesungen über Fouriersche Integrale* (Akademische Verlagsgesellschaft, Leipzig).

[110] Bohner, M. and Guseinov, G. S. (2004). Partial differentiation on time scales, *Dyn. Syst. Appl.* **13**, pp. 351–379.

[111] Bohner, M. and Peterson, A. (2001). *Dynamic equations on time scales* (Birkhäuser, Basel).

[112] Bolster, D., Benson, D. A., Le Borgne, T. and Dentz, M. (2010). Anomalous mixing and reaction induced by superdiffusive nonlocal transport, *Phys. Rev. E* **82**, Article No. 021119.

[113] Bonilla, B., Kilbas, A. A. and Trujillo, J. J. (2003). *Cálculo Fraccionario y Ecuaciones Diferenciales Fraccionarias* (UNED, Madrid).

[114] Bonilla, B., Rivero, M., Rodríguez-Germá, L. and Trujillo, J. J. (2007). Fractional differential equations as alternative models to the nonlinear fractional differential equations, *Appl. Math. Comput.* **187**, pp. 79–88.

[115] Bottazzi, G. and Secchi, A. (2003). Why are distributions of firm growth rates tent shaped? *Economics Letters* **80**, pp. 415–420.

[116] Bottazzi, G. and Secchi, A. (2006). Explaining the distribution of firm growth rates, *The RAND Journal of Economics* **37**, pp. 235–256.

[117] Bounaim, A., Holm, S., Chen, W. and Odegard, A. (2004). Sensitivity of the ultrasonic CARI technique for breast tumor detection using a FETD scheme, *Ultrasonics* **42**, pp. 919–25.

[118] Braß, H. (1995). On the principle of avoiding the singularity in quadrature, *Z. Angew. Math. Mech.* **75**, pp. S617–S618.

[119] Brunner, H. and van der Houwen, P. J. (1986). *The Numerical Solution of Volterra Equations* (North-Holland, Amsterdam).

[120] Butzer, P. L. and Hauss, M. (1991a). On Stirling functions of the second kind, *Stud. Appl. Math.* **84**, pp. 71–91.

[121] Butzer, P. L. and Hauss, M. (1991b). Stirling functions of first and second kind: Some new applications, in S. Baron and D. Leviatan (eds.), *Approximation, Interpolation and Summability* (AMS, Providence), pp. 89–108.

[122] Butzer, P. L. and Hauss, M. (1999). Application of sampling theory to combinatorical analysis, Stirling numbers, special functions and the Riemann zeta functions, in J. R. Higgins and R. L. Stens (eds.), *Sampling Theory in Fourier Analysis: Advanced Topics* (Oxford Univ. Press, Oxford), pp. 1–37.

[123] Butzer, P. L., Hauss, M. and Schmidt, M. (1989). Factorial functions and Stirling numbers of fractional orders, *Results Math.* **16**, pp. 16–48.

[124] Butzer, P. L. and Jansche, S. (1997). A direct approach to the Mellin transform, *J. Fourier Anal. Appl.* **3**, pp. 325–375.

[125] Butzer, P. L., Kilbas, A. A. and Trujillo, J. J. (2002a). Compositions of Hadamard-type fractional integration operators and the semigroup property, *J. Math. Anal. Appl.* **269**, pp. 387–400.

[126] Butzer, P. L., Kilbas, A. A. and Trujillo, J. J. (2002b). Fractional calculus in the Mellin setting and Hadamard-type fractional integrals, *J. Math. Anal. Appl.* **269**, pp. 1–27.

[127] Butzer, P. L., Kilbas, A. A. and Trujillo, J. J. (2002c). Mellin transform analysis and integration by parts for Hadamard-type fractional integrals, *J. Math. Anal. Appl.* **270**, pp. 1–15.

[128] Butzer, P. L., Kilbas, A. A. and Trujillo, J. J. (2003a). Generalized Stirling functions of second kind and representations of fractional order differences via derivatives, *J. Difference Equ. Appl.* **9**, pp. 503–533.

[129] Butzer, P. L., Kilbas, A. A. and Trujillo, J. J. (2003b). Stirling functions of

the second kind in the setting of difference and fractional calculus, *Numer. Funct. Anal. Optimization* **24**, pp. 673–711.

[130] Butzer, P. L., Markett, C. and Schmidt, M. (1991). Stirling numbers, central factorial numbers, and representations of the Riemann zeta function, *Results Math.* **19**, pp. 257–275.

[131] Butzer, P. L., Rodríguez-Germá, L., Kilbas, A. A. and Trujillo, J. J. (2007). Stirling functions of the first kind in the setting of fractional calculus, *J. Difference Equ. Appl.* **13**, pp. 683–721.

[132] Butzer, P. L., Schmeisser, G. and Stens, R. L. (2001). An introduction to sampling analysis, in F. Marvasti (ed.), *Nonuniform Sampling: Theory and Practice* (Kluwer Acad. Publ., New York), pp. 17–121.

[133] Butzer, P. L., Splettstoesser, W. and Stens, R. L. (1988). The sampling theorem and linear prediction in signal analysis, *Jahresber. Dtsch. Math.-Ver.* **90**, pp. 1–70.

[134] Cabré, X. and Roquejoffre, J. M. (2009). Front propagation in Fisher-KPP equations with fractional diffusion, *C. R., Math., Acad. Sci. Paris* **347**, pp. 1361–1366.

[135] Caffarelli, L. and Silvestre, L. (2007). An extension problem related to the fractional Laplacian, *Commun. Partial Differ. Equations* **32**, pp. 1245–1260.

[136] Caffarelli, L. and Vázquez, J. L. (2011). Asymptotic behaviour of a porous medium equation with fractional diffusion, *Discret. Contin. Dyn. Syst.* **29**, pp. 1393–1404.

[137] Caniego, F. J., Espejo, R., Martín, M. A. and San José Martínez, F. (2005). Multifractal scaling of soil spatial variability, *Ecological Model.* **182**, pp. 291–303.

[138] Cao, X., Burrage, K. and Abdullah, F. (2006). A variable stepsize implementation for fractional differential equations, Tech. rep., University of Queensland, Brisbane.

[139] Caponetto, R., Dongola, G. and Fortuna, L. (2010). *Fractional Order Systems: Modeling and Control Applications* (World Scientific Publ. Comp., Singapore).

[140] Caputo, M. (1966). Linear models of dissipation whose q is almost frequency independent, *Annals of Geophysics* **19**, pp. 383–393.

[141] Caputo, M. (1969). *Elasticità e Dissipazione* (Zanichelli, Bologna).

[142] Caputo, M. (1995). Mean fractional-order-derivatives: Differential equations and filters, *Ann. Univ. Ferrara Sez. VII (N. S.)* **41**, pp. 73–84.

[143] Caputo, M. (2001). Distributed order differential equations modelling dielectric induction and diffusion, *Fract. Calc. Appl. Anal.* **4**, pp. 421–442.

[144] Caputo, M. (2003). Diffusion with space memory modelled with distributed order space fractional differential equations, *Annals of Geophysics* **46**, pp.

223–234.

[145] Caputo, M. and Mainardi, F. (1971). A new dissipation model based on memory mechanism, *Pure Appl. Geophys.* **91**, pp. 134–147.

[146] Castro, A., Córdoba, D., Gancedo, F. and Orive, R. (2009). Incompressible flow in porous media with fractional diffusion, *C. R., Math., Acad. Sci. Paris* **347**, pp. 1361–1366.

[147] Çinlar, E. (1975). *Introduction to Stochastic Processes* (Prentice-Hall, Englewood Cliffs).

[148] Chambers, J. M., Mallows, C. L. and Stuck, B. W. (1976). A method for simulating stable random variables, *J. Amer. Statist. Ass.* **71**, pp. 340–344.

[149] Champeney, D. C. (1987). *A Handbook of Fourier Theorems* (Cambridge Univ. Press, Cambridge).

[150] Chatterjee, A. (2005). Statistical origins of fractional derivatives in viscoelasticity, *J. Sound Vibration* **284**, pp. 1239–1245.

[151] Cheang, G. H. and Chiarella, C. (2011). A modern view on Merton's jump-diffusion model, University of Technology, Sydney, Quantitative Finance Research Centre, Research paper 287.

[152] Chelluri, R., Richmond, L. B. and Temme, N. M. (2000). Asymptotic estimates for generalized Stirling numbers, *Analysis* **20**, pp. 1–13.

[153] Chen, Y. Q. and Moore, K. L. (2002). Analytical stability bound for a class of delayed fractional-order dynamic systems, *Nonlinear Dyn.* **29**, pp. 191–200.

[154] Chen, W., and Baleanu, D. and Tenreiro Machado, J. A. (eds.) (2010). Fractional differentiation and its applications, *Comput. Math. Appl.* **59**, pp. 1585–1896.

[155] Cody, W. J. and Stoltz, L. (1989). GAMMA, in *SPECFUN 2.5*, `http://www.netlib.org/specfun/gamma`.

[156] Coimbra, C. F. M. (2003). Mechanics with variable-order differential operators, *Annalen der Physik* **12**, pp. 692–703.

[157] Cole, K. S. and Cole, R. H. (1941). Dispersion and absorption in dielectrics I. Alternating current characteristics, *J. Chem. Phys.* **9**, pp. 341–352.

[158] Cole, K. S. and Cole, R. H. (1942). Dispersion and absorption in dielectrics II. Direct current characteristics, *J. Chem. Phys.* **10**, pp. 98–105.

[159] Compte, A. (1996). Stochastic foundations of fractional dynamics, *Phys. Rev. E* **53**, pp. 4191–4193.

[160] Comtet, L. (1974). *Advanced Combinatorics: The Art of Finite and Infinite Expansion* (Reidel, Dordrecht).

[161] Copson, E. T. (1943). Some applications of Marcel Riesz's integral of fractional order, *Proc. Royal Soc. Edinburgh A* **61**, pp. 260–272.

[162] Corcino, R. B. and Hsu, L. C. (2001). Combinatorial and statistical applications of generalized Stirling numbers, *J. Math. Res. Exposition* **21**, pp.

337–343.

[163] Cramér, H. (1930). *On the Mathematical Theory of Risk* (Skandia Jubilee Volume, Stockholm).

[164] Cresson, J. (2007). Fractional embedding of differential operators and Lagrangian systems, *J. Math. Phys.* **48**, Article No. 033504.

[165] Cresson, J. (2010). Inverse problem of fractional calculus of variations for partial differential equations, *Commun. Nonlin. Sci. Numer. Sim.* **15**, pp. 987–996.

[166] Cresson, J. and Inizan, P. (2009). About fractional Hamiltonian systems, *Phys. Scr.* **T136**, Article No. 014007.

[167] Cuesta, E. (2010). A variable step size numerical method based on fractional type quadratures for linear integro-differential equations, *Adv. Engineer. Softw.* **41**, pp. 64–69.

[168] Cuesta, E., Lubich, C. and Palencia, C. (2006). Convolution quadrature time discretization of fractional diffusion-wave equations, *Math. Comput.* **75**, pp. 673–696.

[169] Cuesta, E. and Palencia, C. (2003). A numerical method for an integro-differential equation with memory in Banach spaces: Qualitative properties, *SIAM J. Numer. Anal.* **41**, pp. 1232–1241.

[170] Dahlquist, G. (1956). Convergence and stability in the numerical integration of ordinary differential equations, *Math. Scand.* **4**, pp. 33–53.

[171] Dahlquist, G. (1959). Stability and error bounds in the numerical integration of ordinary differential equations, Tekn. Högskol. Handl. 130, Royal Inst. of Technology, Stockholm.

[172] Dahlquist, G. (1978). On accuracy and unconditional stability of linear multistep methods for second order differential equations, *BIT* **18**, pp. 133–136.

[173] Das, S. (2008). *Functional Fractional Calculus for System Identification and Controls* (Springer, Berlin).

[174] Davis, P. J. and Rabinowitz, P. (1984). *Methods of Numerical Integration* (Acad. Press, Orlando).

[175] de Boor, C. (1978). *A Practical Guide to Splines* (Springer, New York).

[176] de Donder, T. (1935). *Théorie invariantive de calculus de variations* (Gauthier-Villars, Paris).

[177] de Finetti, B. (1975). *Theory of Probability*, Vol. II (J. Wiley & Sons, New York).

[178] Deng, W. H. (2007a). Numerical algorithm for the time fractional Fokker-Planck equation, *J. Comput. Phys.* **227**, pp. 1510–1522.

[179] Deng, W. H. (2007b). Short memory principle and a predictor-corrector approach for fractional differential equations, *J. Comput. Appl. Math.* **206**, pp. 174–188.

[180] Diaz, G. and Coimbra, C. F. M. (2009). Nonlinear dynamics and control of a variable order oscillator with application to the van der Pol equation, *Nonlinear Dyn.* **56**, pp. 145–157.

[181] Diethelm, K. (1997a). An algorithm for the numerical solution of differential equations of fractional order, *Electron. Trans. Numer. Anal.* **5**, pp. 1–6.

[182] Diethelm, K. (1997b). Generalized compound quadrature formulae for finite-part integrals, *IMA J. Numer. Anal.* **17**, pp. 479–493.

[183] Diethelm, K. (2003). Efficient solution of multi-term fractional differential equations using P(EC)^{m}E methods, *Computing* **71**, pp. 305–319.

[184] Diethelm, K. (2008a). An analysis of some nonclassical methods for the numerical approximation of Caputo-type fractional derivatives, *Numer. Algorithms* **47**, pp. 361–390.

[185] Diethelm, K. (2008b). Multi-term fractional differential equations, multi-order fractional differential systems and their numerical solution, *J. Europ. Syst. Autom.* **42**, pp. 665–676.

[186] Diethelm, K. (2010). *The Analysis of Fractional Differential Equations* (Springer, Heidelberg).

[187] Diethelm, K. (2011). An efficient parallel algorithm for the numerical solution of fractional differential equations, *Fract. Calc. Appl. Anal.* **14**, pp. 475–490.

[188] Diethelm, K. (2013). A fractional calculus based model for the simulation of an outbreak of dengue fever, *Nonlinear Dyn.* **71**, pp. 613–619.

[189] Diethelm, K., Ford, J. M., Ford, N. J. and Weilbeer, M. (2005a). A comparison of backward differentiation approaches for ordinary and partial differential equations of fractional order, in A. Le Méhauté, J. A. Tenreiro Machado, J. C. Trigeassou and J. Sabatier (eds.), *Fractional Differentiation and its Applications* (Ubooks, Neusäß), pp. 557–569.

[190] Diethelm, K., Ford, J. M., Ford, N. J. and Weilbeer, M. (2006). Pitfalls in fast numerical solvers for fractional differential equations, *J. Comput. Appl. Math.* **186**, pp. 482–503.

[191] Diethelm, K. and Ford, N. J. (2001). Numerical solution methods for distributed order differential equations, *Fract. Calc. Appl. Anal.* **4**, pp. 531–542.

[192] Diethelm, K. and Ford, N. J. (2002a). Analysis of fractional differential equations, *J. Math. Anal. Appl.* **265**, pp. 229–248.

[193] Diethelm, K. and Ford, N. J. (2002b). Numerical solution of the Bagley-Torvik equation, *BIT* **42**, pp. 490–507.

[194] Diethelm, K. and Ford, N. J. (2004). Multi-order fractional differential equations and their numerical solution, *Appl. Math. Comput.* **154**, pp. 621–640.

[195] Diethelm, K. and Ford, N. J. (2009). Numerical analysis for distributed-

order differential equations, *J. Comput. Appl. Math.* **225**, pp. 96–104.

[196] Diethelm, K. and Ford, N. J. (2012). Volterra integral equations and fractional calculus: Do neighbouring solutions intersect? *J. Integral Equations Appl.* **24**, pp. 25–37.

[197] Diethelm, K., Ford, N. J. and Freed, A. D. (2002). A predictor-corrector approach for the numerical solution of fractional differential equations, *Nonlinear Dyn.* **29**, pp. 3–22.

[198] Diethelm, K., Ford, N. J. and Freed, A. D. (2004). Detailed error analysis for a fractional Adams method, *Numer. Algorithms* **36**, pp. 31–52.

[199] Diethelm, K., Ford, N. J., Freed, A. D. and Luchko, Y. (2005b). Algorithms for the fractional calculus: A selection of numerical methods, *Comput. Methods Appl. Mech. Eng.* **194**, pp. 743–773.

[200] Diethelm, K. and Freed, A. D. (1999a). The FracPECE subroutine for the numerical solution of differential equations of fractional order, in S. Heinzel and T. Plesser (eds.), *Forschung und wissenschaftliches Rechnen 1998*, no. 52 in GWDG-Berichte (Gesellschaft für wissenschaftliche Datenverarbeitung, Göttingen), pp. 57–71.

[201] Diethelm, K. and Freed, A. D. (1999b). On the solution of nonlinear fractional differential equations used in the modeling of viscoplasticity, in F. Keil, W. Mackens, H. Voß and J. Werther (eds.), *Scientific Computing in Chemical Engineering II — Computational Fluid Dynamics, Reaction Engineering, and Molecular Properties* (Springer, Heidelberg), pp. 217–224.

[202] Diethelm, K. and Weilbeer, M. (2004a). Initial-boundary value problems for time-fractional diffusion-wave equations and their numerical solution, in A. Le Méhauté, J. A. Tenreiro Machado, J. C. Trigeassou and J. Sabatier (eds.), *Proceedings of the 1st IFAC Workshop on Fractional Differentiation and its Applications* (ENSEIRB, Bordeaux), pp. 551–557.

[203] Diethelm, K. and Weilbeer, M. (2004b). A numerical approach for Joulin's model of a point source initiated flame, *Fract. Calc. Appl. Anal.* **7**, pp. 191–212.

[204] Doetsch, G. (1937). *Theorie und Anwendung der Laplace-Transformation* (Springer, Berlin).

[205] Dokoumetzidis, A., Magin, R. and Macheras, P. (2010a). A commentary on fractionalization of multi-compartmental models, *J. Pharmacokinet. Pharmacodyn.* **37**, pp. 203–207.

[206] Dokoumetzidis, A., Magin, R. and Macheras, P. (2010b). Fractional kinetics in multi-compartmental systems, *J. Pharmacokinet. Pharmacodyn.* **37**, pp. 507–524.

[207] Dong, J. P. and Xu, M. Y. (2007). Some solutions to the space fractional Schrödinger equation using momentum representation method, *J. Math. Phys.* **48**, p. 072105.

[208] Driver, R. D. (1977). *Ordinary and Delay Differential Equations* (Springer, Berlin).

[209] Duarte Ortigueira, M. (2011). *Fractional Calculus for Scientists and Engineers* (Springer, Berlin).

[210] Duarte Ortigueira, M. and Tenreiro Machado, J. A. (eds.) (2003b). Fractional signal processing and applications, *Signal Process.* **83**, pp. 2285–2480.

[211] Duarte Ortigueira, M. and Tenreiro Machado, J. A. (eds.) (2006a). Fractional calculus applications in signals and systems, *Signal Process.* **86**, pp. 2503–2819.

[212] Duarte Ortigueira, M., Vinagre, B., Tenreiro Machado, J. A. and Trujillo, J. J. (eds.) (2011). Advances in fractional signals and systems, *Signal Process.* **91**, pp. 349–606.

[213] Dzherbashyan, M. M. (1966). *Integral Transforms and Representations of Functions in the Complex Plane* (Nauka, Moscow).

[214] Dzherbashyan, M. M. (1970). A boundary value problem for a Sturm-Liouville type differential operator of fractional order, *Izv. Akad. Nauk Arm. SSR, Ser. Mat.* **5**, pp. 71–96.

[215] Edwards, J. T., Ford, N. J. and Simpson, A. C. (2002). The numerical solution of linear multi-term fractional differential equations: Systems of equations, *J. Comput. Appl. Math.* **148**, pp. 401–418.

[216] Edwards, J. T., Roberts, J. A. and Ford, N. J. (1997). A comparison of Adomian's decomposition method and Runge-Kutta methods for approximate solution of some predator prey model equations, Numerical Analysis Report 309, Manchester Centre for Computational Mathematics.

[217] Einstein, A. (1905). Über die von der molekularkinetischen Theorie der Wärme geforderte Bewegung von in ruhenden Flüssigkeiten suspendierten Teilchen, *Annalen der Physik* **17**, pp. 549–560.

[218] El-Nabulsi, R. A. and Torres, D. F. M. (2008). Fractional action-like variational problems, *J. Math. Phys.* **49**, pp. 053521–053527.

[219] Elliott, D. (1993). An asymptotic analysis of two algorithms for certain Hadamard finite-part integrals, *IMA J. Numer. Anal.* **13**, pp. 445–462.

[220] Engheta, N. (1996). On fractional calculus and fractional multipoles in electromagnetism, *IEEE Trans. Antennas and Propagation* **44**, pp. 554–566.

[221] Engheta, N. (1998). Fractional curl operator in electromagnetics, *Microwave and Optical Technology Letters* **17**, pp. 86–91.

[222] Engle, R. and Russel, J. (1997). Forecasting the frequency of changes in quoted foreign exchange prices with the autoregressive conditional duration model, *J. of Empirical Finance* **4**, pp. 187–212.

[223] Engle, R. and Russel, J. (1998). Autoregressive conditional duration: A new model for irregularly spaced transaction data, *Econometrica* **66**, pp.

1127–1162.

[224] Erdélyi, A. (1940). Transformation of hypergeometric integrals by means of fractional integration by parts, *Quart. J. Math.* **11**, pp. 293–303.

[225] Erdélyi, A. (1965). An application of fractional integrals, *J. d'Analyse Math.* **14**, pp. 113–126.

[226] Erdélyi, A., Magnus, W., Oberhettinger, F. and Tricomi, F. G. (1953). *Higher Transcendental Functions*, Vol. I (McGraw-Hill, New York).

[227] Erdélyi, A., Magnus, W., Oberhettinger, F. and Tricomi, F. G. (1954). *Tables of Integral Transforms*, Vol. II (McGraw-Hill, New York).

[228] Euler, L. (1738). De progressionibus transcentibus, sev quarum termini algebraice dari nequeunt, *Comment. Acad. Sci. Imperiales Petropolitanae* **1**, pp. 38–57.

[229] Ezz-Eldien, S. S., Hafez, R. M., Bhrawy, A., Baleanu, D. and El-Kalaawy, A. A. (2016). New numerical approach for fractional variational problems using shifted Legendre orthonormal polynomials, *J. Optim. Theor. Appl.* In press, DOI:10.1007/s10957-016-0886-1.

[230] Faddeev, L. and Jackiw, R. (1988). Hamiltonian reduction of unconstrained and constrained systems, *Phys. Rev. Lett.* **60**, pp. 1692–1694.

[231] Feller, W. (1943). On a general class of "contagious" distributions, *Ann. Math. Stat.* **14**, pp. 389–400.

[232] Feller, W. (1971). *Introduction to Probability Theory and its Applications*, Vol. II (J. Wiley & Sons, New York).

[233] Fischer, H. (2010). *A History of the Central Limit Theorem: From Classical to Modern Probability Theory* (Springer, New York).

[234] Fitzgerald, W. J. and Leung, C. M. (1996). A critical review of the Adomian method, Tech. Rep. CUED/F-INFENG/TR. 277, University of Cambridge.

[235] Flajolet, P. and Prodinger, H. (1999). On Stirling numbers for complex arguments and Hankel contours, *SIAM J. Discrete Math.* **12**, pp. 155–159.

[236] Flomenbom, O. and Klafter, J. (2005). Closed-form solutions for continuous time random walks on finite chains, *Phys. Rev. Lett.* **95**, Article No. 098105.

[237] Flomenbom, O. and Silbey, R. J. (2007). Path PDFs for semi-Markovian random walks, *Phys. Rev. E* **76**, Article No. 041101.

[238] Fogedby, H. C. (1994). Langevin equations for continuous time Lévy flights, *Phys. Rev. E* **50**, pp. 1657–1660.

[239] Ford, N. J. and Connolly, J. A. (2006). Comparison of numerical methods for fractional differential equations, *Commun. Pure Appl. Anal.* **5**, pp. 289–307.

[240] Ford, N. J. and Connolly, J. A. (2009). Systems-based decomposition schemes for the approximate solution of multi-term fractional differential equations, *J. Comput. Appl. Math.* **229**, pp. 382–391.

[241] Ford, N. J. and Simpson, A. C. (2000). The approximate solution of frac-

tional differential equations of order greater than 1, in M. Deville and R. Owens (eds.), *Proceedings of the 16th IMACS World Congress on Scientific Computation, Applied Mathematics, and Simulation* (Lausanne), Article No. 213-1.

[242] Ford, N. J. and Simpson, A. C. (2001). The numerical solution of fractional differential equations: Speed versus accuracy, *Numer. Algorithms* **26**, pp. 333–346.

[243] Ford, N. J., Xiao, J. and Yan, Y. (2011). A finite element method for time fractional partial differential equations, *Fract. Calc. Appl. Anal.* **14**, pp. 454–474.

[244] Fourier, J. (1888). *Théorie Analytique de la Chaleur* (Gauthier-Villars, Paris).

[245] Frederico, G. S. and Torres, D. F. (2010). Fractional Noether's theorem in the Riesz-Caputo sense, *Appl. Math. Comput.* **217**, pp. 1023–1033.

[246] Fujita, Y. (1990a). Cauchy problems of fractional order and stable processes, *Japan J. Appl. Math.* **7**, pp. 459–476.

[247] Fujita, Y. (1990b). Integro-differential equation which interpolates the heat equation and the wave equation. I, *Osaka J. Math.* **27**, pp. 309–321.

[248] Fujita, Y. (1990c). Integro-differential equation which interpolates the heat equation and the wave equation. II, *Osaka J. Math.* **27**, pp. 797–804.

[249] Fulger, D., Scalas, E. and Germano, G. (2008). Monte Carlo simulation of uncoupled continuous-time random walks yielding a stochastic solution of the space-time fractional diffusion equation, *Phys. Rev. E* **77**, Article No. 021122.

[250] Gaffeo, E., Delli Gatti, D., Di Guilmi, C., Giulioni, G., Gallegati, M. and Palestrini, A. (2004). Business cycle fluctuations and firms' size distribution dynamics, *Adv. Complex Syst.* **7**, pp. 223–240.

[251] Garibaldi, U. and Scalas, E. (2010). *Finitary Probabilistic Methods in Econophysics* (Cambridge University Press, Cambridge, UK).

[252] Garrappa, R. (2015). Numerical evaluation of two and three parameter Mittag-Leffler functions, *SIAM J. Numer. Anal.* **53**, pp. 1350–1369.

[253] Garrappa, R. and Popolizio, M. (2013). Evaluation of generalized Mittag-Leffler functions on the real line, *Adv. Comput. Math.* **39**, pp. 205–225.

[254] Gaume, L. A. and Barbon, J. L. F. (2001). Nonlinear vacuum phenomena in noncommutative QED, *Int. J. Mod. Phys. A* **16**, pp. 1123–1146.

[255] Georgiou, N., Kiss, I. Z. and Scalas, E. (2015). Solvable non-Markovian dynamic network, *Phys. Rev. E* **92**, p. 042801.

[256] Germano, G., Politi, M., Scalas, E. and Schilling, R. L. (2009). Stochastic calculus for uncoupled continuous-time random walks, *Phys. Rev. E* **79**, Article No. 066102.

[257] Gibrat, R. (1931). *Les Inégalités Economiques, Application: Aux Inégalités*

des Richesses, à la Concentrations des Entreprises, aux Populations des Villes, aux Statistiques des Familles, etc., d'une Loi Nouvelle, al Loi de l'Effet Proportionnel (Librairie du Recueil Sirey, Paris).

[258] Gitman, D. M. and Tytin, I. V. (1990). *Quantization of Fields with Constraints* (Springer, Berlin).

[259] Gnedenko, B. V. and Kolmogorov, A. N. (1954). *Limit Distributions for Sums of Independent Random Variables* (Addison-Wesley, Cambridge, MA).

[260] Goldfain, E. (2010). Fractional dynamics and the TeV regime of field theory, *Commun. Nonlin. Sci. Numer. Sim.* **13**, pp. 666–676.

[261] Golmankhaneh, A. K. (2009). Relativistic scalar fields for non-conservative systems, *Phys. Scr.* **T136**, Article No. 014008.

[262] Golmankhaneh, A. K., Golmankhaneh, A. K., Baleanu, D. and Baleanu, M. C. (2010). Hamiltonian structure of fractional first order Lagrangian, *Int. J. Theor. Phys.* **49**, 2, pp. 365–375.

[263] Golmankhaneh, A. K., Golmankhaneh, A. K., Baleanu, D. and Baleanu, M. C. (2011). Fractional odd-dimensional mechanics, *Adv. Difference Equ.* Article No. 526472.

[264] Gomis, J., Kamimura, K. and Llosa, J. (2001). Hamiltonian formalism for space-time noncommutative theories, *Phys. Rev. D* **63**, Article No. 045003.

[265] Gomis, J., Kamimura, K. and Ramirez, T. (2004). Physical degrees of freedom of non-local theories, *Nucl. Phys. B* **696**, pp. 263–291.

[266] Gomis, J. and Mehen, T. (2000). Space-time noncommutative field theories and unitarity, *Nucl. Phys. B* **591**, pp. 265–276.

[267] Gorenflo, R. (1965). Computation of an integral transform of Abel's type in the presence of noise by quadratic programming, in *Proceedings of IFIP Congress 1965, Vol. II* (New York), pp. 575–576.

[268] Gorenflo, R. and Abdel-Rehim, E. A. A. (2004). From power laws to fractional diffusion: The direct way, *Vietnam J. Math.* **32 (SI)**, pp. 65–75.

[269] Gorenflo, R., Kilbas, A. A., Mainardi, F. and Rogosin, S. (2014). *Mittag-Leffler Functions, Related Topics and Applications* (Springer, Berlin).

[270] Gorenflo, R., Loutchko, J. and Luchko, Y. (2002). Computation of the Mittag-Leffler function $E_{\alpha,\beta}(z)$ and its derivative, *Fract. Calc. Appl. Anal.* **5**, pp. 491–518.

[271] Gorenflo, R., Loutchko, J. and Luchko, Y. (2003). Corrections to the paper "Computation of the Mittag-Leffler function $E_{\alpha,\beta}(z)$ and its derivative", *Fract. Calc. Appl. Anal.* **6**, pp. 111–112.

[272] Gorenflo, R. and Vessella, S. (1991). *Abel Integral Equations* (Springer, Berlin).

[273] Graham, I. G. (1982). Galerkin methods for second kind integral equations with singularities, *Math. Comput.* **39**, pp. 519–533.

[274] Graham, R. L., Knuth, D. E. and Patashnik, O. (1994). *Concrete Mathematics: A Foundation for Computer Science* (Addison-Wesley, Boston).

[275] Grünwald, A. K. (1867). Über "begrenzte" Derivationen und deren Anwendung, *Z. Math. Phys.* **12**, pp. 441–480.

[276] Hackbusch, W. (1995). *Integral Equations: Theory and Numerical Treatment* (Birkhäuser, Basel).

[277] Hadamard, J. (1892). Essai sur l'étude des fonctions donnés par leur développement de Taylor, *J. Math. Pures Appl.* **8**, pp. 101–186.

[278] Hadamard, J. (1952). *Lectures on Cauchy's Problem in Linear Partial Differential Equations* (Dover Publ., New York), reprint.

[279] Hairer, E., Lubich, C. and Schlichte, M. (1988). Fast numerical solution of weakly singular Volterra equations, *J. Comput. Appl. Math.* **23**, pp. 87–98.

[280] Hairer, E., Nørsett, S. P. and Wanner, G. (1993). *Solving Ordinary Differential Equations I: Nonstiff Problems* (Springer, Berlin).

[281] Hairer, E. and Wanner, G. (1996). *Solving Ordinary Differential Equations II: Stiff and Differential-algebraic Problems* (Springer, Berlin).

[282] Hardy, G. H. (1917). On some properties of integrals of fractional order, *Messenger Math.* **47**, pp. 145–150.

[283] Hardy, G. H. and Littlewood, J. E. (1917). Some properties of fractional integrals, *Proc. London Math. Soc.* **24**, pp. 37–43.

[284] Hardy, G. H. and Littlewood, J. E. (1928). Some properties of fractional integrals I, *Math. Z.* **27**, pp. 565–570.

[285] Hardy, G. H. and Littlewood, J. E. (1931). Some properties of fractional integrals II, *Proc. London Math. Soc.* **34**, pp. 403–410.

[286] Hartley, T. T. and Lorenzo, C. F. (1999). Fractional system identification: An approach using continuous order-distributions, NASA Technical Memorandum 1999-209640, NASA Glenn Research Center, Cleveland.

[287] Hauss, M. (1989). *Über die Theorie der fraktionierten Stirlingzahlen und deren Anwendung*, Diploma thesis, Lehrstuhl A für Mathematik, RWTH Aachen.

[288] Hauss, M. (1995). *Verallgemeinerte Stirling-, Bernoulli- und Euler-Zahlen, deren Anwendungen und schnell konvergente Reihen für Zeta-Funktionen*, Ph.D. thesis, RWTH Aachen.

[289] He, J. H. (1998). Approximate solution of non linear differential equations with convolution product nonlinearities, *Computer Meth. Appl. Mech. Eng.* **167**, pp. 69–73.

[290] He, J. H. (2007). Variational iteration method. Some recent results and new interpretations, *J. Comput. Appl. Math.* **207**, pp. 3–17.

[291] Heaviside, O. (1892). *Electrical Papers* (Macmillan, London).

[292] Heaviside, O. (1893). On operators in physical mathematics, *Proc. Royal Soc.* **52**, pp. 504–510.

[293] Henneaux, M. and Teitelboim, C. (1993). *Quantization of Gauge Systems* (Princeton Univ. Press, Princeton).

[294] Hernández, E. and Balachandran, K. (2010). Existence results for abstract degenerate neutral functional differential equations, *Bull. Aust. Math. Soc.* **81**, pp. 329–342.

[295] Herrmann, R. (2010). Common aspects of q-deformed Lie algebras and fractional calculus, *Physica A* **389**, pp. 4613–4622.

[296] Herrmann, R. (2011). *Fractional Calculus: An Introduction for Physicists* (World Scientific Publ. Comp., New Jersey).

[297] Herzallah, M. A. and Baleanu, D. (2009). Fractional-order Euler-Lagrange equations and formulation of Hamiltonian equations, *Nonlinear Dyn.* **69**, pp. 385–391.

[298] Herzallah, M. A. and Baleanu, D. (2011). Fractional-order variational calculus with generalized boundary conditions, *Adv. Differ. Equ.* Article No. 357580.

[299] Herzallah, M. A. and Baleanu, D. (2012). Fractional Euler-Lagrange equations revisited, *Nonlinear Dyn.* **58**, pp. 977–982.

[300] Herzallah, M. A., Muslih, S. I., Baleanu, D. and Rabei, E. M. (2011). Hamilton-Jacobi and fractional like action with time scaling, *Nonlinear Dyn.* **66**, pp. 549–555.

[301] Heymans, N. and Podlubny, I. (2005). Physical interpretation of initial conditions for fractional differential equations with Riemann-Liouville fractional derivatives, *Rheol. Acta* **45**, pp. 765–771.

[302] Higgins, J. R. (1996). *Sampling Theory in Fourier and Signal Analysis: Foundations* (Clarendon Press, Oxford).

[303] Higham, N. J. (1987). Error analysis of the Björck-Pereyra algorithms for solving Vandermonde systems, *Numer. Math.* **50**, pp. 613–632.

[304] Hilfer, R. (2000). *Applications of Fractional Calculus in Physics* (World Scientific Publ. Comp., Singapore).

[305] Hilfer, R. (2008). Threefold introduction to fractional derivatives, in R. Klages, G. Radons and I. M. Sokolov (eds.), *Anomalous Transport: Foundations and Applications* (Wiley-VCH, Weinheim), pp. 17–74.

[306] Hilfer, R. and Anton, L. (1995). Fractional master equations and fractal time random walks, *Phys. Rev. E* **51**, pp. R848–R851.

[307] Hilger, S. (1988). *Ein Maßkettenkalkül mit Anwendung auf Zentrumsmannigfaltigkeiten*, Ph.D. thesis, Universität Würzburg.

[308] Hsu, L. C. and Shiue, P. J. (1998). A unified approach to generalized Stirling numbers, *Adv. Appl. Math.* **20**, pp. 366–384.

[309] Hull, J. C. (2005). *Options, Futures and Other Derivatives* (Prentice Hall, Upper Saddle River).

[310] Iserles, A. (1996). *A First Course in the Numerical Analysis of Differential*

Equations (Cambridge Univ. Press, Cambridge).

[311] Jacob, N. (2005). *Pseudo-differential Operators and Markov Processes*, Vol. III (Imperial College Press, London).

[312] Jacod, J. and Shiryaev, A. (2002). *Limit Theorems for Stochastic Processes* (Springer, New York).

[313] Janicki, A. and Weron, A. (1994). *Simulation and Chaotic Behavior of α-Stable Stochastic Processes* (Marcel Dekker, New York).

[314] Janssen, J. and Manca, R. (2007). *Semi-Markov Risk Models for Finance, Insurance and Reliability* (Springer, New York).

[315] Jarad, F., Abdeljawad, T. and Baleanu, D. (2010a). Fractional variational optimal control problems with delayed arguments, *Nonlinear Dyn.* **62**, pp. 609–614.

[316] Jarad, F., Abdeljawad, T. and Baleanu, D. (2010b). Fractional variational principles with delay within Caputo derivatives, *Rep. Math. Phys.* **65**, pp. 17–28.

[317] Jarad, F., Abdeljawad, T. and Baleanu, D. (2012). Higher order fractional variational optimal control problems with delayed arguments, *Appl. Math. Comput.* **218**, pp. 9234–9240.

[318] Jayakumar, K. (2003). Mittag-Leffler process, *Math. Comput. Modelling* **37**, pp. 1427–1434.

[319] Jelicic, Z. D. and Petrovacki, N. (2009). Optimality conditions and a solution scheme for fractional optimal control problems, *Struct. Multidisc. Optim.* **38**, pp. 571–581.

[320] Jordan, C. (1950). *Calculus of Finite Differences* (Chelsea Publ. Comp., New York).

[321] Joulin, G. (1985). Point-source initiation of lean spherical flames of light reactants: An asymptotic theory, *Combustion Science and Technology* **43**, pp. 99–113.

[322] Jumarie, G. (2007). Lagrangian mechanics of fractional order, Hamilton-Jacobi fractional PDE and Taylor's series of nondifferentiable functions, *Chaos Soliton Frac.* **32**, pp. 969–987.

[323] Jumarie, G. (2009). From Lagrangian mechanics fractal in space to space fractal Schrödinger's equation via fractional Taylor's series, *Chaos Soliton Frac.* **41**, pp. 1590–1604.

[324] Jurlewicz, A. and Weron, K. (2000). Infinitely divisible waiting-time distributions underlying the empirical relaxation responses, *Acta Physica Polonica* **31**, pp. 1077–1084.

[325] Kalia, R. N. (ed.) (1993). *Recent Advances in Fractional Calculus* (Global Publ. Co., Sauk Rapids).

[326] Kanatchikov, I. V. (1999). De Donder-Weyl theory and a hypercomplex extension of quantum mechanics to field theory, *Rep. Math. Phys.* **43**, pp.

157–170.

[327] Kaneko, H., Noren, R. and Xu, Y. (1992). Numerical solutions for weakly singular Hammerstein equations and their superconvergence, *J. Integral Equations Appl.* **4**, pp. 391–406.

[328] Kaymakcalan, B. (1993). Existence and comparison results for dynamic systems on time scales, *J. Math. Anal. Appl.* **172**, pp. 243–255.

[329] Kermack, W. O. and McKendrick, A. G. (1927). A contribution to the mathematical theory of epidemics, *Proc. Royal Soc. London Ser. A* **115**, pp. 700–721.

[330] Kern, J. (1974). Lagrange geometry, *Arch. Math.* **25**, pp. 438–443.

[331] Kilbas, A. A. (2005). Fractional calculus of the generalized Wright function, *Fract. Calc. Appl. Anal.* **8**, pp. 113–126.

[332] Kilbas, A. A., Pierantozzi, T., Trujillo, J. J. and Vázquez, L. (2004). On the solution of fractional evolution equations, *J. Phys. A: Math. Gen.* **37**, pp. 3271–3282.

[333] Kilbas, A. A. and Saigo, M. (2004). *H-Transforms: Theory and Applications* (Chapman & Hall, Boca Raton).

[334] Kilbas, A. A., Saigo, M. and Trujillo, J. J. (2002). On the generalized Wright function, *Fract. Calc. Appl. Anal.* **5**, pp. 437–460.

[335] Kilbas, A. A., Srivastava, H. M. and Trujillo, J. J. (2006). *Theory and Applications of Fractional Differential Equations* (Elsevier, Amsterdam).

[336] Kilbas, A. A. and Trujillo, J. J. (2001). Differential equations of fractional order: Methods, results and problems. I, *Appl. Anal.* **78**, pp. 153–192.

[337] Kilbas, A. A. and Trujillo, J. J. (2002). Differential equations of fractional order: Methods, results and problems. II, *Appl. Anal.* **81**, pp. 435–493.

[338] Kirane, M. and Laskri, Y. (2005). Nonexistence of global solutions to a hyperbolic equation with a space-time fractional damping, *Appl. Math. Comput.* **167**, pp. 1304–1310.

[339] Kirane, M., Medve, M. and Tatar, N. (2011). Semilinear Volterra integrodifferential problems with fractional derivatives in the nonlinearities, *Abstr. Appl. Anal.* **2011**, pp. 1248–1251.

[340] Kiryakova, V. S. (1994). *Generalized Fractional Calculus and Applications* (Longman, Harlow).

[341] Klafter, J., Lim, S. C. and Metzler, R. (eds.) (2011). *Fractional Dynamics in Physics: Recent Advances* (World Scientific Publ. Comp., Singapore).

[342] Klafter, J., Shlesinger, M. F. and Zumofen, G. (1996). Beyond Brownian motion, *Phys. Today* **49**, pp. 33–39.

[343] Klimek, M. (2001). Fractional sequential mechanics: models with symmetric fractional derivative, *Czech. J. Phys.* **51**, pp. 1348–1354.

[344] Klimek, M. (2002). Lagrangean and Hamiltonian fractional sequential mechanics, *Czech. J. Phys.* **52**, pp. 1247–1253.

[345] Knabner, P. and Angermann, L. (2003). *Numerical Methods for Elliptic and Parabolic Partial Differential Equations* (Springer, Berlin).

[346] Kober, H. (1940). On fractional integrals and derivatives, *Quart. J. Math.* **11**, pp. 193–211.

[347] Koeller, R. C. (1986). Polynomial operators, Stieltjes convolution, and fractional calculus in hereditary mechanics, *Acta Mech.* **58**, pp. 251–264.

[348] Kolokoltsov, V. (2011). *Markov Processes, Semigroups and Generators* (De Gruyter, Berlin).

[349] Konotop, V. V. and Vázquez, L. (1994). *Nonlinear Random Waves* (World Scientific Publ. Comp., Singapore).

[350] Kozubowski, T. J. (1998). Mixture representation of Linnik distribution revisited, *Stat. Probab. Lett.* **38**, pp. 157–160.

[351] Kozubowski, T. J. (2000a). Computer simulation of geometric stable distributions, *J. Comput. Appl. Math.* **116**, pp. 221–229.

[352] Kozubowski, T. J. (2000b). Exponential mixture representation of geometric stable distributions, *Annals of the Institute of Statistical Mathematics* **52**, pp. 231–238.

[353] Kozubowski, T. J. and Rachev, S. T. (1999). Univariate geometric stable distributions, *J. Comput. Anal. Appl.* **1**, pp. 177–217.

[354] Lakshmikantham, V., Leela, S. and Devi, J. V. (2009). *Theory of Fractional Dynamic Systems* (Cambridge Sci. Publ., Cambridge).

[355] Lang, W. (2000). On generalizations of the Stirling number triangles, *J. Integer Seq.* **3**, pp. 1–19.

[356] Laplace, P. S. (1812). *Théorie Analytique des Probabilités* (Courcier, Paris).

[357] Larsson, S. and Thomée, V. (2003). *Partial Differential Equations with Numerical Methods* (Springer, Berlin).

[358] Laskin, N. (2000). Fractional quantum mechanics, *Phys. Rev. E* **62**, pp. 3135–3145.

[359] Laskin, N. (2003). Fractional Poisson process, *Commun. Nonlin. Sci. Numer. Sim.* **8**, pp. 201–213.

[360] Laurent, H. (1884). Sur le calcul des dérivées a indices quelconques, *Nouv. Ann. Math.* **3**, pp. 240–252.

[361] Le Méhauté, A., Tenreiro Machado, J. A., Trigeassou, J. C. and Sabatier, J. (eds.) (2005). *Fractional Differentiation and its Applications* (Ubooks, Neusäß).

[362] Lederman, C., Roquejoffre, J. M. and Wolanski, N. (2002). Mathematical justification of a nonlinear integro-differential equation for the propagation of spherical flames, *C. R., Math., Acad. Sci. Paris* **334**, pp. 569–574.

[363] LePage, W. R. (1980). *Complex Variables and the Laplace Transform for Engineers* (Dover Publ., New York).

[364] Letnikov, A. V. (1868a). On historical development of differentiation with

an arbitrary index, *Mat. Sb.* **3**, pp. 85–112, in Russian.

[365] Letnikov, A. V. (1868b). Theory of differentiation with an arbitrary index, *Mat. Sb.* **3**, pp. 1–66, in Russian.

[366] Letnikov, A. V. (1872). An explanation of the main propositions of differentiation theory with an arbitrary index, *Mat. Sb.* **6**, pp. 413–445, in Russian.

[367] Lévy, P. P. (1954). *Théorie de l'addition des variables aléatoires* (Gauthier-Villars, Paris).

[368] Li, C. and Chen, G. (2004). Chaos in the fractional order Chen system and its control, *Chaos, Solitons and Fractals* **22**, pp. 549–554.

[369] Li, J. R. (2010). A fast time stepping method for evaluating fractional integrals, *SIAM J. Sci. Comput.* **31**, pp. 4696–4714.

[370] Li, X. and Xu, C. (2009). A space-time spectral method for the time fractional diffusion equation, *SIAM J. Numer. Anal.* **47**, pp. 2108–2131.

[371] Li, Y. M. and Xu, C. J. (2007). Finite difference/spectral approximations for the time-fractional diffusion equation, *J. Comput. Phys.* **225**, pp. 1533–1552.

[372] Liang, S. and Jeffrey, D. J. (2009). Comparison of homotopy analysis method and homotopy perturbation method through an evolution equation, *Commun. Nonlin. Sci. Numer. Sim.* **14**, pp. 4057–4064.

[373] Liao, S. J. (2003). *Beyond Perturbation: Introduction to the Homotopy Analysis Method* (Chapman & Hall, Boca Raton).

[374] Lim, S. C. and Muniandy, S. V. (2004). Stochastic quantization of nonlocal fields, *Phys. Lett. A* **324**, pp. 396–405.

[375] Liouville, J. (1832). Mémoire sur quelques questions de géométrie et de mécanique, et sur un nouveau genre de calcul pour résoudre ces questions, *J. de l'Ecole Roy. Polytéchn.* **13**, pp. 1–69.

[376] Liouville, J. (1835). Sur le changement de la variable indépendante dans calcul des différentielles à indices quelconques, *J. Ecole Polytech.* **15**, pp. 17–54.

[377] Llosa, J. and Vives, J. (1994). Hamiltonian-formalism for nonlocal Lagrangians, *J. Math. Phys.* **35**, pp. 2856–2877.

[378] Loeb, D. E. (1992). A generalization of the Stirling numbers, *Discrete Math.* **103**, pp. 325–376.

[379] López-Fernández, M., Lubich, C. and Schädle, A. (2008). Adaptive, fast, and oblivious convolution in evolution equations, *SIAM J. Sci. Comput.* **30**, pp. 1015–1037.

[380] Love, E. R. (1971). Fractional derivatives of imaginary order, *J. London Math. Soc.* **2**, pp. 241–259.

[381] Lu, J. F. and Hanyga, A. (2005). Wave field simulation for heterogeneous porous media with singular memory drag force, *J. Comput. Phys.* **208**, pp.

651–674.

[382] Lubich, C. (1983a). On the stability of linear multistep methods for Volterra convolution equations, *IMA J. Numer. Anal.* **3**, pp. 439–465.

[383] Lubich, C. (1983b). Runge-Kutta theory for Volterra and Abel integral equations of the second kind, *Math. Comput.* **41**, pp. 87–102.

[384] Lubich, C. (1985). Fractional linear multistep methods for Abel-Volterra integral equations of the second kind, *Math. Comput.* **45**, pp. 463–469.

[385] Lubich, C. (1986). Discretized fractional calculus, *SIAM J. Math. Anal.* **17**, pp. 704–719.

[386] Lubich, C. (1988a). Convolution quadrature and discretized operational calculus. I, *Numer. Math.* **52**, pp. 129–145.

[387] Lubich, C. (1988b). Convolution quadrature and discretized operational calculus. II, *Numer. Math.* **52**, pp. 413–425.

[388] Luchko, Y. F., Rivero, M., Trujillo, J. J. and Velasco, M. P. (2010). Fractional models, non-locality, and complex systems, *Comput. Math. Appl.* **59**, pp. 1048–1056.

[389] Lundberg, P. (1903). *Approximerad Framställning av Sannolikehetsfunktionen. Återförsäkering av Kollektivrisker* (Almqvist & Wicksell, Uppsala).

[390] Luo, A. C. and Afraimovich, V. S. (eds.) (2011). *Long-range Interactions, Stochasticity and Fractional Dynamics* (Springer, Berlin).

[391] Magdziarz, M. and Weron, K. (2006). Anomalous diffusion schemes underlying the Cole-Cole relaxation: The role of the inverse α-stable subordinator, *Physica A* **367**, pp. 1–6.

[392] Magin, R. (2006). *Fractional Calculus in Bioengineering* (Begell House, Redding).

[393] Magin, R., Abdullah, O., Baleanu, D. and Zhou, X. H. (2008). Anomalous diffusion expressed through fractional order differential operators in the Bloch-Torrey equation, *J. Magn. Res.* **190**, pp. 255–270.

[394] Magin, R., Duarte Ortigueira, M., Podlubny, I. and Trujillo, J. J. (2011). On the fractional signals and systems, *Signal Process.* **91**, pp. 350–371.

[395] Magin, R., Feng, X. and Baleanu, D. (2009). Solving the fractional order Bloch equation, *Concep. Magn. Res.* **34A**, pp. 16–23.

[396] Mainardi, F. (1997a). Fractional calculus: Integral and differential equations of fractional order, in A. Carpinteri and F. Mainardi (eds.), *Fractals and Fractional Calculus in Continuum Mechanics* (Springer, Wien), pp. 291–348.

[397] Mainardi, F. (1997b). Fractional calculus: Some basic problems in continuum and statistical mechanics, in A. Carpinteri and F. Mainardi (eds.), *Fractals and Fractional Calculus in Continuum Mechanics* (Springer, Wien), pp. 291–348.

[398] Mainardi, F. (2010). *Fractional Calculus and Waves in Linear Viscoelas-*

ticity: An Introduction to Mathematical Models (Imperial College Press, London).

[399] Mainardi, F. (2012). An historical perspective on fractional calculus in linear viscoelasticity, *Fract. Calc. Appl. Anal.* **15**, pp. 712–717.

[400] Mainardi, F., Gorenflo, R. and Scalas, E. (2004). A fractional generalization of the Poisson process, *Vietnam J. Math.* **32 (SI)**, pp. 53–64.

[401] Mainardi, F., Luchko, Y. and Pagnini, G. (2001). The fundamental solution of the space-time fractional diffusion equation, *Fract. Calc. Appl. Anal.* **4**, pp. 153–192.

[402] Mainardi, F., Raberto, M., Gorenflo, R. and Scalas, E. (2000). Fractional calculus and continuous-time finance: The waiting-time distribution, *Physica A* **287**, pp. 468–481.

[403] Malinowska, A. B. and Torres, D. F. M. (2010). Generalized natural boundary conditions for fractional variational problems in terms of the Caputo derivative, *Comput. Math. Appl.* **59**, pp. 3110–3116.

[404] Maravall, D. (1956). Nuevos tipos de ecuaciones integrales e integrodiferenciales. Nuevos fenómenos de oscilaciones, *Rev. R. Acad. Cienc. Exactas, Fis. Nat.* **1**, pp. 1–151.

[405] Maravall, D. (1971). Ecuaciones diferenciales lineales no enteras y oscilaciones fraccionarias, *Rev. R. Acad. Cienc. Exactas, Fis. Nat.* **65**, pp. 245–258.

[406] Markett, C. (1994). Triple sums and the Riemann zeta function, *J. Number Theory* **48**, pp. 113–132.

[407] Martinez, C. and Sanz, M. (2001). *The Theory of Fractional Powers of Operators* (Elsevier, Amsterdam).

[408] Masoliver, J., Montero, M., Perelló, J. and Weiss, G. H. (2006). The continuos time random walk formalism in financial markets, *J. of Economic Behavior and Organization* **61**, pp. 577–591.

[409] Mathai, A. M. (1993). *A Handbook of Generalized Special Functions for Statistical and Physical Sciences* (Clarendon Press, Oxford).

[410] Mathai, A. M. and Haubold, H. J. (2008). *Special Functions for Applied Scientists* (Springer, Berlin).

[411] Mathai, A. M. and Saxena, R. K. (1978). *The H-Function with Applications in Statistics and other Disciplines* (J. Wiley & Sons, New York).

[412] Matignon, D. and Montseny, G. (eds.) (1998). *Fractional Differential Systems: Models, Methods, and Applications* (SMAI, Paris).

[413] McBride, A. C. (1979). *Fractional Calculus and Integral Transforms of Generalized Functions* (Pitman, London).

[414] McLean, W. and Mustapha, K. (2007). A second-order accurate numerical method for a fractional wave equation, *Numer. Math.* **105**, pp. 481–510.

[415] Medicis, A. and Leroux, R. (1995). Generalized Stirling numbers, convolu-

tion formulae and p, q-analogues, *Canad. J. Math.* **47**, pp. 474–499.

[416] Meerschaert, M. M., Nane, E. and Vellaisamy, P. (2011). The fractional Poisson process and the inverse stable subordinator, *Electron. J. Probab.* **16**, pp. 1600–1620.

[417] Meerschaert, M. M. and Scalas, E. (2006). Coupled continuous time random walks in finance, *Physica A* **370**, pp. 114–118.

[418] Meerschaert, M. M. and Scheffler, H. P. (2001). *Limit Distributions for Sums of Independent Random Vectors: Heavy Tails in Theory and Practice* (J. Wiley & Sons, New York).

[419] Meerschaert, M. M., Scheffler, H.-P. and Tadjeran, C. (2006). Finite difference methods for two-dimensional fractional dispersion equation, *J. Comput. Phys.* **211**, pp. 249–261.

[420] Meerschaert, M. M. and Tadjeran, C. (2004). Finite difference approximations for fractional advection-dispersion flow equations, *J. Comput. Appl. Math.* **272**, pp. 65–77.

[421] Meerschaert, M. M. and Toaldo, B. (2015). Relaxation patterns and semi-Markov dynamics, arXiv:1506.02951 [math.PR].

[422] Merton, R. C. (1976). Option pricing when underlying stock returns are discontinuous, *J. of Financial Economics* **3**, pp. 125–144.

[423] Metzler, R., Barkai, E. and Klafter, J. (1999). Anomalous diffusion and relaxation close to thermal equilibrium: A fractional Fokker-Planck equation approach, *Phys. Rev. Lett.* **82**, pp. 3563–3567.

[424] Metzler, R., Glöckle, W. G. and Nonnenmacher, T. F. (1994). Fractional model equation for anomalous diffusion, *Physica A* **211**, pp. 13–24.

[425] Metzler, R. and Klafter, J. (2000). The random walk guide to anomalous diffusion: A fractional dynamics approach, *Phys. Rep.* **339**, pp. 1–77.

[426] Metzler, R. and Klafter, J. (2004). The restaurant at the end of the random walk: Recent developments in the description of anomalous transport by fractional dynamics, *J. Phys. A: Math. General* **37**, pp. R161–R208.

[427] Metzler, R. and Nonnenmacher, T. F. (1997). Fractional diffusion: exact representations of spectral functions, *J. Phys. A: Math. Gen.* **30**, pp. 1089–1093.

[428] Mikosch, T. (1999). *Elementary Stochastic Calculus With Finance in View* (World Scientific Publ. Comp., Singapore).

[429] Miller, K. S. and Ross, B. (1989). Fractional difference calculus, in H. M. Srivastava and S. Owa (eds.), *Univalent Functions, Fractional Calculus, and Their Applications* (Ellis Horwood, Chichester), pp. 139–152.

[430] Miller, K. S. and Ross, B. (1993). *An Introduction to the Fractional Calculus and Fractional Differential Equations* (J. Wiley & Sons, New York).

[431] Miller, R. K. and Feldstein, A. (1971). Smoothness of solutions of Volterra integral equations with weakly singular kernels, *SIAM J. Math. Anal.* **2**,

pp. 242–258.

[432] Miron, R. and Anastasiei, M. (1994). *The Geometry of Lagrange Spaces: Theory and Applications* (Kluwer, Dordrecht).

[433] Mittag-Leffler, G. M. (1903). Sur la nouvelle fonction E_α, *C. R., Math., Acad. Sci. Paris* **137**, pp. 554–558.

[434] Mittag-Leffler, G. M. (1904). Sopra la funzione $E_\alpha(x)$, *Rend. Accad. Lincei* **13**, pp. 3–5.

[435] Mittag-Leffler, G. M. (1905). Sur la représentation analytique d'une branche uniforme d'une fonction monogene. Cinquième note, *Acta Math.* **29**, pp. 101–181.

[436] Miyahara, Y. (2011). *Option Pricing in Incomplete Markets* (World Scientific Publ. Comp., Singapore).

[437] Momani, S. and Odibat, Z. (2007). Homotopy perturbation method for nonlinear partial differential equations of fractional order, *Phys. Lett. A* **365**, pp. 345–350.

[438] Monje, C. A., Chen, Y., Vinagre, B., Xue, D. and Feliu, V. (2010). *Fractional-order Systems and Controls: Fundamentals and Applications* (Springer, New York).

[439] Montroll, E. W. and Scher, H. (1973). Random walks on lattices IV: Continuous-time random walks and influence of absorbing boundary conditions, *J. Stat. Phys.* **9**, pp. 101–135.

[440] Montroll, E. W. and Weiss, G. H. (1965). Random walks on lattices II, *J. Math. Phys.* **6**, pp. 167–181.

[441] Montseny, G., Audounet, J. and Matignon, D. (2000). Diffusive representation for pseudo-differentially damped nonlinear systems, in A. Isidori, F. Lamnabhi-Lagarrigue and W. Respondek (eds.), *Nonlinear Control, Vol. 2* (Springer, Berlin), pp. 163–182.

[442] Mozyrska, D. and Torres, D. F. M. (2010). Minimal modified energy control for fractional linear control systems with the Caputo derivative, *Carpat. J. Math.* **26**, pp. 210–221.

[443] Musiela, M. and Rutkowski, M. (2002). *Martingale Methods in Financial Modelling* (Springer, New York).

[444] Muslih, S. I., Agrawal, O. P. and Baleanu, D. (2010a). Fractional Schrödinger equation and its solution, *Int. J. Theor. Phys.* **49**, pp. 1746–1752.

[445] Muslih, S. I. and Baleanu, D. (2005). Hamiltonian formulation of systems with linear velocities within Riemann-Liouville fractional derivatives, *J. Math. Anal. Appl.* **304**, pp. 599–606.

[446] Muslih, S. I. and Baleanu, D. (2007). Fractional Euler-Lagrange equations of motion in fractional space, *J. Vibr. Contr.* **13**, pp. 1209–1216.

[447] Muslih, S. I., Baleanu, D. and Rabei, E. M. (2007). Fractional Hamilton's

equations of motion in fractional time, *Cent. Eur. J. Phys.* **5**, pp. 549–557.

[448] Muslih, S. I., Saddallah, M., Baleanu, D. and Rabei, E. (2010b). Lagrangian formulation of Maxwell's field in fractional d-dimensional space-time, *Rom. J. Phys.* **55**, pp. 659–663.

[449] Naber, M. (2004). Time fractional Schrödinger equation, *J. Math. Phys.* **45**, pp. 3339–3352.

[450] Nakhushev, A. M. (2003). *Fractional Calculus and its Applications* (Fizmatlit, Moscow).

[451] Nambu, Y. (1973). Generalized Hamiltonian dynamics, *Phys. Rev. D* **7**, pp. 2405–2412.

[452] Nasiri, S. Y., Sobouti, Y. and Taati, F. (2006). Phase space quantum mechanics-direct, *J. Math. Phys.* **47**, Article No. 092106.

[453] Nesterenko, V. V. (1989). Singular Lagrangians with higher derivatives, *J. Phys. A: Math. Gen.* **22**, pp. 1673–1687.

[454] Nielsen, N. (1965). *Handbuch der Theorie der Gammafunktionen* (Chelsea Publ. Comp., New York).

[455] Nieto, J. J. (2010). Maximum principles for fractional differential equations derived from Mittag-Leffler functions, *Appl. Math. Letters* **23**, pp. 1248–1251.

[456] Nigmatulin, R. I. (1979). Spatial averaging in the mechanics of heterogeneous and dispersed systems, *Int. J. of Multiphase Flow* **5**, pp. 353–385.

[457] Nigmatullin, R. R. (2005). Theory of dielectric relaxation in non-crystalline solids: From a set of micromotions to the averaged collective motion in the mesoscale region, *Physica B* **358**, pp. 201–215.

[458] Nigmatullin, R. R., Arbuzov, A. A., Salehli, F., Gis, A., Bayrak, I. and Catalgil-Giz, H. (2007). The first experimental confirmation of the fractional kinetics containing the complex power-law exponents: Dielectric measurements of polymerization reaction, *Physica B* **388**, pp. 418–434.

[459] Nigmatullin, R. R. and Baleanu, D. (2010). Is it possible to derive Newtonian equation of motion with memory? *J. Theor. Phys.* **49**, pp. 701–708.

[460] Nigmatullin, R. R. and Le Méhauté, A. (2005). Is there geometrical/physical meaning of the fractional integral with complex exponent? *J. Non-Cryst. Sol.* **351**, pp. 2888–2899.

[461] Nishiura, H. (2006). Mathematical and statistical analyses of the spread of dengue, *Dengue Bulletin* **30**, pp. 51–67.

[462] Nolan, J. P. (2012). *Stable Distributions: Models for Heavy Tailed Data* (Birkhäuser, Boston).

[463] Odibat, Z., Momani, S. and Erturk, V. S. (2008). Generalized differential transform method: Application to differential equations of fractional order, *Appl. Math. Comput.* **197**, pp. 467–477.

[464] Oldham, K. B. (1969). A new approach to the solution electrochemical

problems involving diffusion, *Anal. Chem.* **41**, pp. 1904–1910.

[465] Oldham, K. B. and Spanier, J. (1974). *The Fractional Calculus* (Acad. Press, New York).

[466] OpenMP Architecture Review Board (2010). The OpenMP API specification for parallel programming, `http://www.openmp.org`.

[467] O'Shaughnessy, L. (1918). Problem ♯ 433, *Amer. Math. Monthly* **25**, pp. 172–173.

[468] Osler, T. J. (1970). Leibniz rule for fractional derivatives generalized and an application to infinite series, *SIAM J. Appl. Math.* **18**, pp. 658–674.

[469] Oustaloup, A. (1981). Fractional order sinusoidal oscillators: Optimization and their use in highly linear FM modulation, *IEEE Trans. Circuits Systems* **28**, pp. 1007–1009.

[470] Oustaloup, A. (1983). *Systèmes asservis linéaires d'ordre fractionnaire: Théorie et pratique* (Masson, Paris).

[471] Oustaloup, A. (1991). *La Commande CRONE: Commande Robuste d'Ordre Non Entier* (Hermès, Paris).

[472] Oustaloup, A. (1995). *La Dérivation Non Entiére. Théorie, Synthése et Applications* (Hermès Sci., Paris).

[473] Oustaloup, A., Cois, O., Lanusse, P., Melchior, P., Moreau, X. and Sabatier, J. (2006). The CRONE approach: Theoretical developments and major applications, *IFAC Proceedings Volumes* **39 (11)**, pp. 324–354.

[474] Oustaloup, A. and Mathieu, B. (1999). *La commande CRONE: du scalaire au multivariable* (Hermés, Paris).

[475] Oustaloup, A., Sabatier, J. and Lanusse, P. (1999). From fractal robustness to CRONE control, *Fract. Calc. Appl. Anal.* **2**, pp. 1–30.

[476] Ozdemir, N., Agrawal, O. P., Karadeniz, D. and Iskender, B. B. (2009). Fractional optimal control problem of an axis-symmetric diffusion-wave propagation, *Phys. Scr.* **T136**, Article No. 014024.

[477] Pachepsky, Y., Benson, D. A. and Rawls, W. (2000). Simulating scale-dependent solute transport in soils with the fractional advective-dispersive equation, *Soil Sci. Soc. Am. J.* **64**, pp. 1234–1243.

[478] Pakes, A. G. (1998). Mixture representations for symmetric generalized Linnik laws, *Stat. Probab. Lett.* **37**, pp. 213–221.

[479] Perron, O. (1919a). Über Integration von gewöhnlichen Differentialgleichungen durch Reihen, *Sitz.-Ber. Heidelberger Akademie der Wissenschaften A: Math.-Nat. Klasse* Abhandlung 2.

[480] Perron, O. (1919b). Über Integration von gewöhnlichen Differentialgleichungen durch Reihen, II, *Sitz.-Ber. Heidelberger Akademie der Wissenschaften A: Math.-Nat. Klasse* Abhandlung 8.

[481] Perron, O. (1919c). Über Integration von gewöhnlichen Differentialgleichungen durch Reihen, III, *Sitz.-Ber. Heidelberger Akademie der Wis-*

senschaften A: Math.-Nat. Klasse Abhandlung 12.

[482] Perron, O. (1920). Über Integration partieller Differentialgleichungen durch Reihen, *Sitz.-Ber. Heidelberger Akademie der Wissenschaften A: Math.-Nat. Klasse* Abhandlung 9.

[483] Picozzi, S. and West, B. J. (2002). Fractional Langevin model of memory in financial time series, *Phys. Rev. E* **66**, pp. 046118–046129.

[484] Pinilla, L., Torres, R. and Ortiz, C. (2011). Bioethanol production in batch mode by a native strain of Zymomonas mobilis, *World J. Microbiol. Biotechnol.* **27**, pp. 2521–2528.

[485] Pinto, C. M. A., Mendes Lopes, A. and Tenreiro Machado, J. A. (2012). A review of power laws in real life phenomena, *Commun. Nonlin. Sci. Numer. Sim.* **17**, pp. 3558–3578.

[486] Platonov, M. L. (1997). The relations between generalized Stirling numbers constructed on different basis, *Combinatorial and Asymptotic Analysis. Krasnoyarsk. Gos. Univ.* **2**, pp. 142–152.

[487] Podlubny, I. (1999). *Fractional Differential Equations* (Acad. Press, San Diego).

[488] Podlubny, I. (2000). Matrix approach to discrete fractional calculus, *Fract. Calc. Appl. Anal.* **3**, pp. 359–386.

[489] Podlubny, I. (2009). Mittag-Leffler function, `http://www.mathworks.com/matlabcentral/fileexchange/8738-mittag-leffler-function`.

[490] Podlubny, I., Petras, I., Vinagre, B., Chen, Y. Q., O'Leary, P. and Dorcak, L. (2003). Realization of fractional order controllers, *Acta Montan. Slovaca* **8**, pp. 233–235.

[491] Politi, M., Kaizoji, T. and Scalas, E. (2011). Full characterization of the fractional Poisson process, *Europhysics Letters* **96**, p. 20004.

[492] Pollard, H. (1948). The completely monotonic character of the Mittag-Leffler function $e_a(-x)$, *Bull. Amer. Math. Soc.* **54**, pp. 1115–1116.

[493] Pooseh, S., Rodrigues, H. S. and Torres, D. F. M. (2011). Fractional derivatives in dengue epidemics, in T. E. Simos, G. Psihoyios, C. Tsitouras and Z. Anastassi (eds.), *Numerical Analysis and Applied Mathematics ICNAAM 2011* (Amer. Inst. Phys., Melville), pp. 739–742.

[494] Popović, J. K., Atanacković, M. T., Pilipović, A. S., Rapaić, M. R., Pilipović, S. and Atanacković, T. M. (2010). A new approach to the compartmental analysis in pharmacokinetics: Fractional time evolution of Diclofenac, *J. Pharmacokinet. Pharmacodyn.* **37**, pp. 119–134.

[495] Post, E. L. (1919). Discussion of the solution of $(d/dx)^{1/2}y = y/x$ (Problem ♯ 433), *Amer. Math. Monthly* **26**, pp. 37–39.

[496] Prudnikov, A. P., Brychkov, Y. A. and Marichev, O. I. (1986). *Integrals and Series, Vols. I, II and III* (Gordon and Breach, New York).

[497] Pskhu, A. V. (2005). *Partial Differential Equations of Fractional Order*

(Nauka, Moscow).
[498] Quirós, F., Rossi, J. D. and Vázquez, J. L. (2002). Complete blow-up and thermal avalanche for heat equations with nonlinear boundary conditions, *Commun. Partial. Differ. Equations* **27**, pp. 395–424.
[499] Rabei, E. M. and Ababenh, B. S. (2008). Hamilton-Jacobi fractional mechanics, *J. Math. Anal. Appl.* **344**, pp. 799–805.
[500] Rabei, E. M., Ajlounid, A. W. and Gassib, H. B. (2006). Quantization of Brownian motion, *Int. J. Theor. Phys.* **45**, pp. 1619–1629.
[501] Rabei, E. M., Alhalholy, T. S. and Rousan, A. A. (2004a). Potentials of arbitrary forces with fractional derivatives, *J. Mod. Phys.* **19**, pp. 3083–3092.
[502] Rabei, E. M., Alhalholy, T. S. and Taani, A. A. (2004b). On Hamiltonian formulation of non-conservative systems, *Turkish J. Phys.* **28**, pp. 213–221.
[503] Rabei, E. M., Almayteh, I., Muslih, S. I. and Baleanu, D. (2008). Hamilton-Jacobi formulation of systems within Caputo's fractional derivative, *Phys. Scr.* **77**, Article No. 015101.
[504] Rabei, E. M., Altarazi, I. A., Muslih, S. I. and Baleanu, D. (2009a). Fractional WKB approximation, *Nonlinear Dyn.* **57**, pp. 171–175.
[505] Rabei, E. M., Muslih, S. I. and Baleanu, D. (2010). Quantization of fractional systems using WKB approximation, *Commun. Nonlin. Sci. Numer. Sim.* **15**, pp. 807–811.
[506] Rabei, E. M., Nawafleh, K. I., Hijjawi, R. S., Muslih, S. I. and Baleanu, D. (2007). The Hamilton formalism with fractional derivatives, *J. Math. Anal. Appl.* **327**, pp. 891–897.
[507] Rabei, E. M., Rawashdeh, I. M., Muslih, S. I. and Baleanu, D. (2011). Hamilton-Jacobi formulation for systems in terms of Riesz's fractional derivatives, *Int. J. Theor. Phys.* **50**, pp. 1569–1576.
[508] Rabei, E. M., Tarawneh, D. M., Muslih, S. I. and Baleanu, D. (2009b). Heisenberg's equations of motion with fractional derivatives, *J. Vibr. Contr.* **13**, pp. 1239–1247.
[509] Raberto, M., Rapallo, F. and Scalas, E. (2011). Semi-Markov graph dynamics, *PLOS ONE* **6**, Article No. e23370.
[510] Raberto, M., Scalas, E. and Mainardi, F. (2002). Waiting-times and returns in high-frequency financial data: An empirical study, *Physica A* **314**, pp. 749–755.
[511] Rabinowitz, P. (1979). On avoiding the singularity in the numerical integration of proper integrals, *BIT* **19**, pp. 104–110.
[512] Rabotnov, Y. N. (1969). *Creep Problems in Structural Members* (North-Holland, Amsterdam).
[513] Rabotnov, Y. N. (1973). *On the Use of Singular Operators in the Theory of Viscoelasticity* (Lecture Notes for the CISM course on Rheology, Bologna).

[514] Razminia, A., Baleanu, D., Majd, V. and Johari, V. (2013). Conditional optimization problems: fractional order case, *J. Optim.Theor. Appl.* **156**, pp. 45–55.

[515] Répaci, A. (1990). Nonlinear dynamical systems: On the accuracy of Adomian's decomposition method, *Appl. Math. Lett.* **3**, 4, pp. 35–39.

[516] Repin, O. N. and Saichev, A. I. (2000). Fractional Poisson law, *Radiophys. Quantum Electron.* **43**, pp. 738–741.

[517] Richmond, R. and Merlini, D. (1997). Stirling numbers for complex arguments, *SIAM J. Discrete Math.* **10**, pp. 73–82.

[518] Riemann, B. (1876). Versuch einer allgemeinen Auffasung der Integration und Differentiation, in *Gesammelte Werke* (Teubner, Leipzig), pp. 331–344.

[519] Riesz, M. (1936). L'intégrale de Riemann-Liouville et solution invariantive du probléme de cauchy pour l'equation des ondes, *C. R. Congrès Intern. Math.* **2**, pp. 44–45.

[520] Riesz, M. (1949). L'intégrale de Riemann-Liouville et le problème de Cauchy, *Acta Math.* **81**, pp. 1–223.

[521] Riewe, F. (1996). Nonconservative Lagrangian and Hamiltonian mechanics, *Phys. Rev. E* **53**, pp. 1890–1899.

[522] Riewe, F. (1997). Mechanics with fractional derivatives, *Phys. Rev. E* **55**, pp. 3581–3592.

[523] Riordan, J. (1979). *Combinatorial Identities* (Krieger Publ., Huntington).

[524] Rivero, M., Rodríguez-Germá, L., Trujillo, J. J. and Velasco, M. P. (2010). Fractional operators and some special functions, *Comput. Math. Appl.* **59**, pp. 1822–1834.

[525] Rivero, M., Trujillo, J. J., Vázquez, L. and Velasco, M. P. (2011). Fractional dynamics of populations, *Appl. Math. Comput.* **218**, pp. 1089–1095.

[526] Rivero, M., Trujillo, J. J. and Velasco, M. P. (2009). On deterministic fractional models, in D. Baleanu, Z. B. Güvenç and J. A. Tenreiro Machado (eds.), *New Trends in Nanotechnology and Fractional Calculus Applications* (Springer-Verlag, Berlin), pp. 123–150.

[527] Rodrigues, H. S., Monteiro, M. T. T., Torres, D. F. M. and Zinober, A. (2012). Dengue disease, basic reproduction number and control, *Internat. J. Comput. Math.* **89**, pp. 334–346.

[528] Rodríguez-Germá, L., Trujillo, J. J., Vázquez, L. and Velasco, M. P. (2008). Fractional calculus framework to avoid singularities of differential equations, *Fract. Calc. Appl. Anal.* **11**, pp. 431–441.

[529] Roman, H. E. and Alemany, P. A. (1994). Continuous-time random walks and the fractional diffusion equation, *J. Phys. A* **27**, pp. 3407–3410.

[530] Ross, B. (ed.) (1974). *Fractional Calculus and its Applications: Proceedings of the International Conference, New Haven, June 1974* (Springer, New York).

[531] Ross, B. (1977). The development of fractional calculus 1695–1900, *Hist. Math.* **4**, pp. 75–89.

[532] Rossikhin, Y. A. and Shitikova, M. V. (2007). Comparative analysis of viscoelastic models involving fractional derivatives of different orders, *Fract. Calc. Appl. Anal.* **10**, pp. 111–121.

[533] Rucinski, A. and Voigt, B. (1990). A local limit theorem for generalized Stirling numbers, *Rev. Roumaine Math. Pures Appl.* **35**, pp. 161–172.

[534] Saad, Y. (2003). *Iterative Methods for Sparse Linear Systems* (SIAM, Philadelphia).

[535] Saad, Y. and Schultz, M. H. (1986). GMRES: A generalized minimal residual algorithm for solving nonsymmetric linear systems, *SIAM J. Sci. Statist. Comput.* **7**, pp. 856–869.

[536] Sabatier, J., Agrawal, O. P. and Tenreiro Machado, J. A. (eds.) (2007). *Advances in Fractional Calculus: Theoretical Developments and Applications in Physics and Engineering* (Springer, Berlin).

[537] Sabatier, J., Melchior, P., Tenreiro Machado, J. A. and Vinagre, B. (eds.) (2008). Fractional order systems: Applications in modelling, identification and control, *J. Europ. Syst. Autom.* **42**, pp. 622–1051.

[538] Sadati, S. J., Baleanu, D., Ranjbar, A., Ghaderi, R. and Abdeljawad, T. (2010). Mittag-Leffler stability theorem for fractional nonlinear systems with delay, *Abstr. Appl. Anal.* Article No. 108651.

[539] Saichev, A. I. and Zaslavsky, G. M. (1997). Fractional kinetic equations: solutions and applications, *Chaos* **7**, pp. 753–764.

[540] Samko, S. G. (1967). Solution of generalized Abel equation by means of an equation with Cauchy kernel, *Soviet Math. Dokl.* **6**, pp. 1259–1262, in Russian.

[541] Samko, S. G. (2002). *Hypersingular Integrals and Their Applications* (Taylor and Francis, New York).

[542] Samko, S. G., Kilbas, A. A. and Marichev, O. I. (1993). *Fractional Integrals and Derivatives. Theory and Applications* (Gordon and Breach, Yverdon).

[543] Samorodnitsky, G. and Taqqu, M. S. (1994). *Stable non-Gaussian Random Processes: Stochastic Models with Infinite Variance* (Chapman & Hall, London).

[544] San José Martínez, F., Pachepsky, Y. A. and Rawls, W. J. (2007). Fractional advective-dispersive equation as a model of solute transport in porous media, in J. Sabatier, O. P. Agrawal and J. A. Tenreiro Machado (eds.), *Advances in Fractional Calculus - Theoretical Developments and Applications in Physics and Engineering* (Springer, Berlin), pp. 199–212.

[545] San José Martínez, F., Pachepsky, Y. A. and Rawls, W. J. (2010). Modelling solute transport in soil columns using advective-dispersive equations with fractional spatial derivatives, *Adv. Engineer. Soft.* **41**, pp. 4–8.

[546] Sanz-Serna, J. M. (1988). A numerical method for a partial integro-differential equation, *SIAM J. Numer. Anal.* **25**, pp. 319–327.

[547] Sato, K.-I. (1999). *Lévy Processes and Infinitely Divisible Distributions* (Cambridge Univ. Press, Cambridge).

[548] Scalas, E. (2006). The application of continuous-time random walks in finance and economics, *Physica A* **362**, pp. 225–239.

[549] Scalas, E. (2011). A class of CTRWs: Compound fractional Poisson processes, in J. C. Klafter, S. C. Lim and R. Metzler (eds.), *Fractional Dynamics: Recent Advances* (World Scientific Publ. Comp., Singapore), pp. 351–372.

[550] Scalas, E., Gorenflo, R. and Mainardi, F. (2000). Fractional calculus and continuous-time finance, *Physica A* **284**, pp. 376–384.

[551] Scalas, E., Gorenflo, R. and Mainardi, F. (2004a). Uncoupled continuous-time random walks: Solutions and limiting behavior of the master equation, *Phys. Rev. E* **69**, Article No. 011107.

[552] Scalas, E., Gorenflo, R., Mainardi, F. and Meerschaert, M. M. (2004b). Speculative option valuation and the fractional diffusion equation, in A. Le Méhauté, J. A. Tenreiro Machado, J. C. Trigeassou and J. Sabatier (eds.), *Proceedings of the 1st IFAC Workshop on Fractional Differentiation and its Applications* (ENSEIRB, Bordeaux).

[553] Scalas, E. and Politi, M. (2013). A note on intraday option pricing, *International Journal of Applied Nonlinear Science* **1**, pp. 76–86.

[554] Scher, H. and Lax, M. (1973a). Stochastic transport in a disordered solid I. Theory, *Phys. Rev. B* **7**, pp. 4491–4502.

[555] Scher, H. and Lax, M. (1973b). Stochastic trasport in a disordered solid II. Impurity conduction, *Phys. Rev. B* **7**, pp. 4506–4519.

[556] Schiesser, W. E. (1991). *The Numerical Method of Lines* (Acad. Press, San Diego).

[557] Schilling, R., Song, R. and Vondraček, Z. (2010). *Bernstein Functions: Theory and Applications* (De Gruyter, Berlin).

[558] Schmidt, A. and Gaul, L. (2006). On a critique of a numerical scheme for the calculation of fractionally damped dynamical systems, *Mech. Res. Commun.* **33**, pp. 99–107.

[559] Schoutens, W. (2003). *Lévy Processes in Finance* (J. Wiley & Sons, New York).

[560] Schumer, R., Meerschaert, M. M. and Baeumer, B. (2009). Fractional advection-dispersion equations for modeling transport at the earth surface, *J. Geophysical Research* **114**, pp. F00A07–F00A21.

[561] Schwartz, L. M. (1950). *Théorie des Distributions* (Hermann, Paris).

[562] Seiberg, N., Susskind, L. and Toumbas, T. (2000). Space/time non-commutativity and causality, *J. High Energy Phys.* **6**, p. 021.

[563] Shen, S., Liu, F., Chen, J., Turner, I. and Anh, V. (2012). Numerical techniques for the variable order time fractional diffusion equation, *Appl. Math. Comput.* **218**, pp. 10861–10870.

[564] Shlesinger, M. F. (1996). Random processes, in G. L. Trigg (ed.), *Encyclopedia of Applied Physics*, Vol. 16 (VCH, New York), pp. 45–70.

[565] Shlesinger, M. F., Zaslavsky, G. M. and Klafter, J. (1993). Strange kinetics, *Nature* **363**, pp. 31–37.

[566] Singh, S. J. and Chatterjee, A. (2006). Galerkin projections and finite elements for fractional order derivatives, *Nonlinear Dyn.* **45**, pp. 183–206.

[567] Singh-Chandel, R. S. (1977). Generalized Stirling numbers and polynomials, *Publ. Inst. Math., Beograd (N. S.)* **22**, pp. 43–48.

[568] Sinha, V. P. and Dhawan, G. K. (1969). On generalized Stirling numbers and polynomials, *Riv. Math. Univ. Parma* **10**, pp. 95–100.

[569] Sixdeniers, J. M., Penson, K. A. and Solomon, A. L. (2001). Extended Bell and Stirling numbers from hypergeometric exponentiation, *J. Integer Seq.* **4**, pp. 1–11.

[570] Sneddon, I. N. (1957). *Elements of Partial Differential Equations* (McGraw-Hill, New York).

[571] Sneddon, I. N. (1966). *Mixed boundary value problems in potential theory* (North-Holland, Amsterdam).

[572] Sneddon, I. N. (1979). *The use of operators of fractional integration in applied mathematics* (Polish Acad. Sci., Warszawa-Poznan).

[573] Sokolov, I. M., Chechkin, A. V. and Klafter, J. (2004). Distributed order fractional kinetics, *Acta Physica Polonica B* **35**, pp. 1323–1341.

[574] Srivastava, H. M. (1968). On an extension of the Mittag-Leffler function, *Yokohama Math. J.* **16**, pp. 77–88.

[575] Srivastava, H. M. and Buschman, R. G. (1977). *Convolution Integral Equations with Special Function Kernels* (J. Wiley & Sons, New York).

[576] Srivastava, H. M., Gupta, K. C. and Goyal, S. P. (1982). *The H-Functions of One and Two Variables with Applications* (South Asian Publ., New Delhi and Madras).

[577] Srivastava, H. M. and Kashyap, B. R. (1982). *Special Functions in Queuing Theory and Related Stochastic Processes* (Acad. Press, New York).

[578] Stanislavsky, A. A. (2004). Probability interpretation of the integral of fractional order, *Theor. Math. Phys.* **138**, pp. 418–431.

[579] Stanislavsky, A. A. (2006). Hamiltonian formalism of fractional systems, *Eur. Phys. J. B* **49**, pp. 93–101.

[580] Stenger, F. (1976). Approximation via Whittaker's cardinal function, *J. Approx. Theory* **17**, pp. 222–240.

[581] Stirling, J. (1730). *Methodus Differentialis, sive Tractatus de Summatione et Interpolatione Serium Infinitarum* (Gul. Bowyer, London).

[582] Su, H., Chen, W., Li, C. and Chen, Y. (2010). Fractional differential models for anomalous diffusion, *Phys. A: Stat. Mech. Appl.* **389**, pp. 2719–2724.

[583] Sun, H. G., Chen, W. and Chen, Y. Q. (2009). Variable-order fractional differential operators in anomalous diffusion modeling, *Physica A* **388**, pp. 4586–4592.

[584] Szegő, G. (1975). *Orthogonal Polynomials*, 4th edn. (Amer. Math. Soc., Providence).

[585] Tarasov, V. E. (2005). Fractional hydrodynamics equations for fractal media, *Anal. Phys.* **318**, pp. 286–307.

[586] Tarasov, V. E. (2006). Fractional statistical mechanics, *Chaos* **16**, Article No. 033108.

[587] Tarasov, V. E. (2010). Fractional dynamics of relativistic particle, *Int. J. Theor. Phys.* **49**, pp. 293–303.

[588] Tarasov, V. E. (2011). *Fractional Dynamics: Application of Fractional Calculus to Dynamics of Particles, Fields and Media* (Springer, Berlin).

[589] Tarasov, V. E. and Zaslavsky, G. M. (2006). Nonholonomic constraints with fractional derivatives, *J. Phys. A: Math. Gen.* **39**, pp. 9797–9815.

[590] Tarasov, V. E. and Zaslavsky, G. M. (2008). Conservation laws and Hamilton's equations for systems with long-range interaction and memory, *Commun. Nonlin. Sci. Numer. Sim.* **13**, pp. 1860–1878.

[591] Tatar, N. E. (2010). Existence results for an evolution problem with fractional nonlocal conditions, *Comput. Math. Appl.* **60**, pp. 2971–2982.

[592] Tatar, N. E. (2011). Mild solutions for a problem involving fractional derivatives in the nonlinearity and in the non-local conditions, *Adv. Difference Equ.* Article No. 18.

[593] Tatari, M. and Dehghan, M. (2007). On the convergence of He's variational iteration method, *J. Comput. Appl. Math.* **207**, pp. 121–128.

[594] Tenreiro Machado, J. A. (2001). Discrete-time fractional-order controllers, *Fract. Calc. Appl. Anal.* **4**, pp. 47–66.

[595] Tenreiro Machado, J. A. (ed.) (2002). *Fractional order systems*, Nonlinear Dyn. 29, pp. 1–385.

[596] Tenreiro Machado, J. A. (2003). A probabilistic interpretation of the fractional-order differentiation, *Fract. Calc. Appl. Anal.* **6**, pp. 73–80.

[597] Tenreiro Machado, J. A. (2009). Fractional derivatives: probability interpretation and frequency response of rational approximations, *Commun. Nonlin. Sci. Numer. Sim.* **14**, pp. 3492–3497.

[598] Tenreiro Machado, J. A., Kiryakova, V. and Mainardi, F. (2011). Recent history of fractional calculus, *Commun. Nonlin. Sci. Numer. Sim.* **16**, pp. 1140–1153.

[599] The MathWorks, Inc. (2011). Matlab, `http://www.mathworks.com/products/matlab/`.

[600] The R Project for Statistical Computing (2011). R, `http://www.r-project.org/`.

[601] Toader, S. and Toader, G. (1998). Generalized Stirling numbers, *Studia Univ. Babes-Bolyai Math.* **33**, pp. 50–53.

[602] Toledo-Hernandez, R., Rico-Ramirez, V., Iglesias-Silva, G. A. and Divekar, U. M. (2014a). A fractional calculus approach to the dynamic optimization of biological reactive systems. Part I: Fractional models for biological reactions, *Chemical Engineering Science* **117**, pp. 217–228.

[603] Toledo-Hernandez, R., Rico-Ramirez, V., Rico-Martinez, R., Hernandez-Castro, S. and Divekar, U. M. (2014b). A fractional calculus approach to the dynamic optimization of biological reactive systems. Part II: Numerical solution of fractional optimal control problems, *Chemical Engineering Science* **117**, pp. 239–247.

[604] Tomowski, Z., Hilfer, R. and Srivastava, H. M. (2010). Fractional and operational calculus with generalized fractional derivative operators and Mittag-Leffler type functions, *Int. Trans. Spec. Func.* **21**, pp. 797–814.

[605] Torvik, P. J. and Bagley, R. L. (1984). On the appearance of the fractional derivative in the behavior of real materials, *J. Appl. Mech.* **51**, pp. 294–298.

[606] Tricaud, C. and Chen, Y. Q. (2009). Time-optimal control of fractional dynamic systems, *Proceedings of the IEEE Conference on Decision and Control* , pp. 5027–5032.

[607] Trujillo, J. J. and Baleanu, D. (eds.) (2011). Fractional models and their applications, *Adv. Difference Equ.* .

[608] Uchaikin, V. V. (2008). *Method of Fractional Derivatives* (Artishok Press, Ulyanovsk).

[609] Umarov, S. and Gorenflo, R. (2005). Cauchy and non-local multi-point problems for distributed order pseudo-differential equations, Part I, *J. Anal. Appl.* **24**, pp. 449–466.

[610] Usero, D. and Vázquez, L. (2003). Fractional derivative. A new formulation for damped systems, in L. Vázquez, R. S. MacKay and M. P. Zorzano (eds.), *Localization and Energy Transfer in Nonlinear Systems* (World Scientific Publ. Comp., Singapore), pp. 296–303.

[611] Vacaru, S. (2008). Finsler and Lagrange geometries in Einstein and string gravity, *Int. J. Geom. Methods. Mod. Phys.* **5**, pp. 473–511.

[612] Vacaru, S. (2012a). Fractional dynamics from Einstein gravity, general solutions, and black holes, *Int. J. Theor. Phys.* **51**, pp. 1338–1359.

[613] Vacaru, S. (2012b). Fractional nonholonomic Ricci flows, *Chaos Solitons Fract.* **45**, pp. 1266–1276.

[614] Vacaru, S., Stavrinos, P., Gaburov, E. and Gonţa, D. (2006). *Clifford and Riemann-Finsler Structures in Geometric Mechanics and Gravity* (Geometry Balkan Press, Bucharest), `http://www.mathem.pub.ro/dgds/mono/`

D-07-Vac.pdf.

[615] Vainikko, G. and Uba, P. (1981). A piecewise polynomial approximation to the solution of an integral equation with weakly singular kernel, *J. Austral. Math. Soc., Ser. B* **22**, pp. 431–438.

[616] Valério, D. and Sá da Costa, J. (2011). Variable-order fractional derivatives and their numerical approximations, *Signal Process.* **91**, pp. 470–483.

[617] Vasilév, V. V. and Simak, L. A. (2008). *Fractional Calculus and Approximation Methods for Modeling Dynamic Systems* (Ukr. Nat. Acad. Sci., Kiev), in Russian.

[618] Vázquez, L. (2003). Fractional diffusion equations with internal degrees of freedom, *J. Comput. Math.* **21**, pp. 491–494.

[619] Vázquez, L. and Vilela Mendes, R. (2003). Fractionally coupled solutions of the diffusion equation, *Appl. Math. Comput.* **141**, pp. 125–130.

[620] Wagner, C. G. (1996). Generalized Stirling and Lah numbers, *Discrete Math.* **160**, pp. 199–218.

[621] Walker, H. F. (1988). Implementation of the GMRES method using Householder transformations, *SIAM J. Sci. Statist. Comput.* **9**, pp. 152–163.

[622] Wang, J. and Zhou, Y. (2011). A class of fractional evolution equations and optimal controls, *Nonlinear Anal.: Real World Appl.* **12**, pp. 262–272.

[623] Wang, S. W. and Xu, M. Y. (2007). Generalized fractional Schrödinger equation with space-time fractional derivatives, *J. Math. Phys.* **48**, Article No. 043502.

[624] Weilbeer, M. (2006). *Efficient Numerical Methods for Fractional Differential Equations and their Analytical Background*, Ph.D. thesis, Technische Universität Braunschweig.

[625] Weyl, H. (1917). Bemerkungen zum Begriff des Differentialquotienten gebrochener Ordnung, *Zürich Naturf. Ges.* **62**, pp. 296–302.

[626] Weyl, H. (1935). Geodesic fields in the calculus of variation for multiple integrals, *Ann. Math.* **36**, pp. 607–629.

[627] Whittaker, E. T. (1915). On the functions which are represented by the expansions of the interpolation-theory, *Proc. Royal Soc. Edinburgh* **35**, pp. 181–194.

[628] Whittaker, J. M. (1935). *Interpolatory Function Theory* (Cambridge Univ. Press, Cambridge).

[629] Widder, D. V. (1941). *The Laplace Transform* (Princeton Univ. Press, Princeton).

[630] Williams, D. (1991). *Probability with Martingales* (Cambridge Univ. Press, Cambridge).

[631] Wilmott, P. (1998). *Derivatives: The Theory and Practice of Financial Engineering* (J. Wiley & Sons, New York).

[632] Wolfram Research, Inc. (2011). Mathematica, http://www.wolfram.com/

mathematica/.

[633] World Health Organization (2009). *Dengue — Guidelines for Diagnosis, Treatment, Prevention and Control*, new edn. (WHO Press, Geneva), http://whqlibdoc.who.int/publications/2009/9789241547871_eng.pdf.

[634] Wu, J. (2006). Lower bounds for an integral involving fractional Laplacians and the generalized Navier-Stokes equations in Besov spaces, *Commun. Math. Phys.* **263**, pp. 803–831.

[635] Wyss, W. (1986). The fractional diffusion equation, *J. Math. Phys* **27**, pp. 2782–2795.

[636] Wyss, W. and Schneider, W. R. (1989). Fractional diffusion and wave equations, *J. Math. Phys* **30**, pp. 569–580.

[637] Xue, D., Chen, Y. and Atherton, D. P. (2008). *Linear Feedback Control: Analysis and Design with MATLAB* (SIAM, Philadelphia).

[638] Ye, H. and Ding, Y. (2009a). A fractional-order differential equation model of HIV infection of CD4+ T-cells, *Math. Comput. Model.* **50**, pp. 386–392.

[639] Ye, H. and Ding, Y. (2009b). Nonlinear dynamics and chaos in a fractional-order HIV model, *Math. Probl. Eng.* Article No. 378614.

[640] Yuan, L. and Agrawal, O. P. (2002). A numerical scheme for dynamic systems containing fractional derivatives, *J. Vibration Acoustics* **124**, pp. 321–324.

[641] Yuste, S. B. and Acedo, L. (2005). An explicit finite difference method and a new von Neumann type stability analysis for fractional diffusion equations, *SIAM J. Numer. Anal.* **42**, pp. 1862–1874.

[642] Zaslavsky, G. M. (1994). Fractional kinetic equation for Hamiltonian chaos, *Physica D* **76**, pp. 110–122.

[643] Zaslavsky, G. M. (2002). Chaos, fractional kinetics, and anomalous transport, *Phys. Rep.* **371**, pp. 461–580.

[644] Zayernouri, M. and Karniadakis, G. E. (2013). Fractional Sturm-Liouville eigen-problems: Theory and numerical approximations, *J. Comput. Phys.* **47**, pp. 2108–2131.

[645] Zayernouri, M. and Karniadakis, G. E. (2014). Exponetially accurate spectral and spectral element methods for fractional ODEs, *J. Comput. Phys.* **257**, pp. 460–480.

[646] Zemanian, A. H. (1968). *Generalized Integral Transformations* (J. Wiley & Sons, New York).

[647] Zhang, C. (1994). *Quelques études en théorie des équations fonctionnelles et en analyse combinatoire*, Ph.D. thesis, Université Louis Pasteur Strasbourg.

[648] Zhang, X., Lv, M., Crawford, J. and Young, I. M. (2007). The impact of boundary on the fractional advection–dispersion equation for solute transport in soil: Defining the fractional dispersive flux with the Caputo derivatives, *Adv. Water Resour.* **30**, pp. 1205–1217.

[649] Zhang, Y. (2009). Finite difference approximations for space-time fractional partial differential equation, *J. Numer. Math.* **17**, pp. 319–326.

[650] Zoia, A., Rosso, A. and Kardar, M. (2007). Fractional Laplacian in bounded domains, *Phys. Rev. E* **76**, Article No. 021116.

[651] Zolotarev, V. M. (1986). *One-dimensional Stable Distributions* (Amer. Math. Soc., Providence).

[652] Zygmund, A. (1934). Some points in the theory of trigonometric and power series, *Trans. Amer. Math. Soc.* **36**, pp. 586–617.

[653] Zygmund, A. (1945). A theorem on fractional derivatives, *Duke Math J.* **3**, pp. 455–464.

Index